Engineering Fundamentals

Engineering Fundamentals

PRINCIPLES, PROBLEMS, & SOLUTIONS

DONALD G. NEWNAN, Ph.D. *San Jose State College*

BRUCE E. LAROCK, Ph.D. *University of California, Davis*

Wiley-Interscience, A DIVISION OF JOHN WILEY & SONS

New York · London · Sydney · Toronto

Copyright © 1970, by John Wiley & Sons, Inc.

All rights reserved. No part of this book may be reproduced by any means, nor transmitted, nor translated into a machine language without the written permission of the publisher.

10 9 8 7 6 5

Library of Congress Catalogue Card Number: 78-96960

SBN 471 63450 6

Printed in the United States of America

Acknowledgments

Many individuals and groups have aided us in one way or another during the preparation of this book. We appreciate the response of the National Council of Engineering Examiners and the State Boards of Registration for Engineers of the fifty United States to our letters and inquiries, and are especially grateful to the many boards which supplied us with sets of engineering examination problems. Siamak Parsanejad and Steve Lionberger materially assisted in the preparation of problem solutions. We are also grateful to our colleagues Harry Dwyer, Hugh Edgar, Lincoln Jones, Dean Newnan, and Ted Zsutty, each of whom reviewed a chapter of the book. Thanks are also due Mrs. Kay King, who so ably typed a large portion of the manuscript. The final responsibility for the correctness of the text resides with the authors, however.

Donald G. Newnan
Bruce E. Larock

San Jose, California
Davis, California

Contents

Chapter 1	Introductory Comments	1
Chapter 2	Mathematics	3
Chapter 3	Statics	79
Chapter 4	Dynamics	151
Chapter 5	Mechanics of Materials	217
Chapter 6	Fluid Mechanics	285
Chapter 7	Thermodynamics	340
Chapter 8	Chemistry	386
Chapter 9	Electricity	442
Chapter 10	Engineering Economy	494
Chapter 11	Other Problems	526

Appendix A Centroidal Coordinates and Moments of Inertia for Common Shapes	579
Appendix B Compound Interest Factors	581
Index	585

Engineering Fundamentals

Chapter 1

Introductory Comments

This book has been prepared for the express purpose of encouraging and aiding the engineer in the systematic review of the fundamental principles of engineering science. An orderly program of continuing education and professional registration are currently desirable and, in many cases, necessary elements in the professional development of the engineer. Ready access to a compact collection of basic engineering principles, to a large number of problems illustrating the use of these principles, and to a complete solution to every one of these many problems should materially assist the engineer in the review process.

A program of continuing education is becoming increasingly important to an engineer's steady professional advancement. It is a fact that many engineers do not ultimately practice in the branch of engineering for which they originally trained; it is also known that the pace of technical progress is actually accelerating. The engineer with a firm grounding in fundamental engineering principles will be better able to cope with these realities *if* his fund of basic knowledge is kept current or is periodically reviewed and updated as the need arises. This foundation of basic knowledge will enable the engineer to grasp the new technologies more easily by preparing him to begin private study or formal course work of a relatively advanced technical nature.

Registration is uniformly recognized in the United States as a landmark along any engineer's path of professional development. In each of the fifty states laws regulate the practice of engineering and the right of an individual to designate himself as an "engineer." These laws were enacted to protect the public from dealing with people who might call themselves engineers but who in reality are not qualified to perform competent engineering work. Thus a certification of professional competency by a State Board of Registration for Engineers is a desirable goal.

One normally follows several steps in becoming a registered professional engineer. A usual first step is graduation from an accredited engineering school. Practical engineering experience usually is acceptable in lieu of a college degree, however. He then passes a state-administered examination on engineering fundamentals. He next engages in the professional practice of engineering under responsible supervision for several years. Finally, he passes a one- or two-day examination on the professional practice of some specified branch of engineering.

The initial examination, normally called the Engineer-In-Training (EIT) examination, which takes 8 hours, tests one's grasp of basic engineering concepts. This book is devoted to a review of these fundamentals; it was developed around representative problems which have appeared on EIT examinations. The orderly review of engineering fundamentals is thus a logical step toward either continuing education or professional registration.

Each chapter begins with a short review of those fundamental principles which are conceptually important and practically useful in the field. The goal is to present basic ideas compactly and directly; lengthy, detailed derivations are avoided. The result is a selective overview of the discipline rather than an encyclopedic coverage.

The remainder of each chapter presents numerous questions and engineering problems upon which the reader can test his knowledge. Each problem has been assigned a grading weight (Wt.) between 1 and 6 to give an indication of the length or difficulty of the problem; the weight appears immediately after the problem number. A unique feature of this book is the presentation of a complete solution to every problem. If the reader has difficulty in solving a problem, the answer alone may be of little assistance because it often provides no hint on how to approach the problem systematically. And if he fails to obtain the given answer, the reason for this difference is frequently unclear. By providing a complete solution to each problem these difficulties are avoided. One should not forget, however, that in many instances there is more than one approach to a problem and that all correct approaches should result in the same final answers.

Chapter 2

Mathematics

According to some, mathematics is *the* most fundamental branch of all science. Indeed, the goal of much scientific and technical work is to express in precise mathematical terms the behavior of our universe and its smaller component parts. In varying degrees mathematics is certainly used in all areas of the disciplines which together make up engineering fundamentals. For this reason it is highly desirable to have a familiarity with and a working knowledge of some of the basic relations of algebra, trigonometry, geometry, and calculus. We present here a *brief* review of fundamental principles; a thorough review, including proofs, could properly be the subject of an entire book in itself and for this reason is considered to be outside the scope of this volume.

ALGEBRA

The basic rules of algebra apply equally well to real and complex numbers, that is, numbers expressible in the form $a_1 + ia_2$, where a_1 and a_2 are real numbers, zero or nonzero, and $i^2 = -1$. The basic rules, given in additive and multiplicative form, are three:

Commutative:	$a + b = b + a$	$ab = ba$
Distributive:	$a(b + c) = ab + ac$	
Associative:	$a + (b + c) = (a + b) + c$	$a(bc) = (ab)c$

The laws of exponents and logarithms are intimately related. For positive numbers a and b and any positive or negative exponents x and y, the rules for exponents are as follows:

$$b^{-x} = \frac{1}{b^x} \qquad b^x b^y = b^{x+y}$$

$$(ab)^x = a^x b^x \qquad b^{xy} = (b^x)^y$$

4 / Mathematics

If $b^y = x$ for positive b and x, then $y = \log_b x$ is the definition of the logarithm of x to the base b. The logarithm is therefore a kind of exponent. The most commonly used base numbers are $b = 10$ for common logarithms and $b = e = 2.718\cdots$ for natural logarithms. (When $b = 10$ it is often not written down; when $b = e$ often $\log_e = \ln$ is written.) Regardless of the value of b these laws hold for logarithms:

$$b^{\log_b x} = x \qquad \log_b b^x = x$$

$$\log_b (xy^n) = \log_b x + n \log_b y \qquad \text{for any value of } n$$

To change, for example, the base of a logarithm from any base b to the base e,

$$\log_b x = \frac{\log_e x}{\log_e b} = \frac{\ln x}{\ln b} = \log_b e \times \ln x$$

since $(\log_b e)(\log_e b) = \log_e (b^{\log_b e}) = \log_e e = 1$.

An entire branch of mathematics, linear algebra, has grown out of an interest in solving sets of linear, simultaneous algebraic equations. The field is a generalization of solving the equation $ax = b$, which is linear in the one unknown, x, and has the obvious solution $x = b/a$. For two simultaneous equations in the unknowns x and y, the equations are often solved by eliminating y and solving for x:

$$a_{11}x + a_{12}y = b_1$$
$$a_{21}x + a_{22}y = b_2$$

From the first equation $y = (1/a_{12})(b_1 - a_{11}x)$. Insertion of this expression for y into the second equation yields an equation of the form of the single linear equation, which is easily solved.

The foregoing problem and also larger sets of simultaneous linear equations can be solved by using determinants. The determinant of the coefficients in the problem is

$$D = \begin{vmatrix} a_{11} & a_{12} \\ a_{21} & a_{22} \end{vmatrix} = a_{11}a_{22} - a_{12}a_{21}$$

If D is nonzero, then Cramer's rule gives the solution for x and y as

$$x = \frac{D_1}{D} \qquad y = \frac{D_2}{D}$$

D_1 is formed from D by replacing a_{11} and a_{21} by b_1 and b_2 respectively. To find D_2, replace the second column of a's by the b's. The same procedure is followed for three or more unknown variables. A 3×3 determinant can be reduced to a 2 × 2 determinant by expanding it in terms of minors along one column or row:

$$\begin{vmatrix} a_{11} & a_{12} & a_{13} \\ a_{21} & a_{22} & a_{23} \\ a_{31} & a_{32} & a_{33} \end{vmatrix} = a_{11} \begin{vmatrix} a_{22} & a_{23} \\ a_{32} & a_{33} \end{vmatrix} - a_{12} \begin{vmatrix} a_{21} & a_{23} \\ a_{31} & a_{33} \end{vmatrix} + a_{13} \begin{vmatrix} a_{21} & a_{22} \\ a_{31} & a_{32} \end{vmatrix}$$

Systems of three linear equations in three unknowns can still be solved by successive elimination, but the use of Cramer's rule and determinants is often more efficient. For four or more unknowns the required bookkeeping becomes formidable by either method, but it is then preferable to use Cramer's rule because it is more systematic.

Quadratic equations are always solvable by algebra. If $ax^2 + bx + c = 0$, then the two solutions are

$$x = \frac{1}{2a}[-b \pm (b^2 - 4ac)^{1/2}]$$

If $b^2 < 4ac$, the roots of the equation are complex numbers. Formulas also exist which give the solutions to third- and fourth-order equations, but it is usually easier to try solving the equation by (*a*) attempting to factor the equation algebraically (often not successful), (*b*) graphing the equation and noting the points of intersection with the x-axis, or (*c*) by numerical trial-and-error substitution.

Another useful formula in algebra is the binomial theorem, which is a special form of the Taylor's series of calculus:

$$(a + b)^n = a^n + \frac{n}{1!}a^{n-1}b + \frac{n(n-1)}{2!}a^{n-2}b^2 + \cdots$$
$$+ \frac{n(n-1)\cdots(n-r+1)}{r!}a^{n-r}b^r + \cdots + b^n$$

For a positive integer n this expansion has $(n + 1)$ terms. In the formula the convenient "factorial" notation $n! = n(n-1)(n-2)\cdots(3)(2)(1)$ has been used.

TRIGONOMETRY

Trigonometry deals with the relations between the angles and the sides of triangles. The periodic functions defined by these relations, however, have vastly wider applications. In the use of these functions we often deal with angle measurement, of which there are two kinds. One system divides one revolution into 360° (degrees). Each degree is further divisible into 60' (minutes), and each minute into 60" (seconds), although often a fraction of a

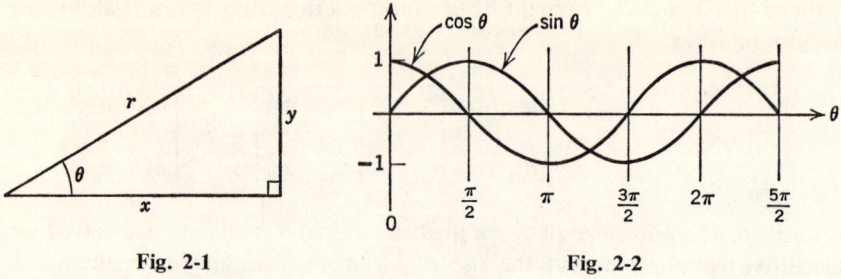

Fig. 2-1   Fig. 2-2

degree is written as a decimal (e.g., $30' = 0.5°$). The unit of measurement in the second system is the radian (rad); 2π rad equal one revolution, or $180° = \pi$ rad $= 3.14159 \cdots$ rad.

The two most basic trigonometric functions are the sine and cosine (Fig. 2-1), which are defined as follows:

$$\sin \theta = \frac{y}{r} \qquad \cos \theta = \frac{x}{r}$$

Here x and y may assume any value, but r is always positive. The other four basic functions are

$$\tan \theta = \frac{\sin \theta}{\cos \theta} = \frac{y}{x} \qquad \cot \theta = \frac{1}{\tan \theta} = \frac{x}{y}$$

$$\sec \theta = \frac{1}{\cos \theta} = \frac{r}{x} \qquad \csc \theta = \frac{1}{\sin \theta} = \frac{r}{y}$$

The sine and cosine are odd and even periodic functions, respectively, with periods of 2π (Fig. 2-2). By learning the variations of these two functions, one can easily deduce the variation of the other functions.

Much can be done in trigonometry by remembering a small number of fundamental identities. Among them are these:

$$\sin^2 \theta + \cos^2 \theta = 1 \qquad 1 + \tan^2 \theta = \sec^2 \theta \qquad 1 + \cot^2 \theta = \csc^2 \theta$$

$$\sin (\theta \pm \phi) = \sin \theta \cos \phi \pm \cos \theta \sin \phi$$

$$\cos (\theta \pm \phi) = \cos \theta \cos \phi \mp \sin \theta \sin \phi$$

From these last two identities the double angle ($\sin 2\theta$, $\cos 2\theta$) formulas can be derived by letting $\theta = \phi$. The half-angle ($\sin \theta/2$, $\cos \theta/2$) formulas can also be derived by replacing θ and ϕ by $\theta/2$ and rearranging the resulting expressions.

In solving for the unknown parts of a plane triangle (Fig. 2-3), two or three basic formulas are often useful. These are

Sum of angles: $\alpha + \beta + \gamma = 180°$

Law of sines: $\dfrac{a}{\sin \alpha} = \dfrac{b}{\sin \beta} = \dfrac{c}{\sin \gamma}$

Law of cosines: $a^2 = b^2 + c^2 - 2bc \cos \alpha$

Note that if $\alpha = 90°$ the triangle is a right triangle, and the law of cosines then becomes a statement of Pythagorean formula.

GEOMETRY

Here we group together some elements of elementary plane geometry, which describes some spatial properties of various shaped objects, and analytic geometry, which employs algebraic notation in its more detailed description of some of these same objects.

The triangle and rectangle are basic geometric figures; they may also be considered as special cases of the trapezoid. From Fig. 2-4 we can consider the trapezoid as the sum of two triangles. The area A of each shape is as follows:

Trapezoid: $A = \dfrac{h}{2}(a + b)$

Rectangle: $A = hb \quad (a = b)$

Triangle: $A = \dfrac{hb}{2} \quad (a = 0)$

Another important shape is the regular polygon having n sides. The central angle subtended by one side is the vertex angle; its value is $2\pi/n$. The included angle between two successive sides of the polygon is $(n - 2)\pi/n$.

The most important nonpolygonal geometric shape is the circle. For a circle of radius r and diameter $d = 2r$, the circumference is $c = \pi d$ and the

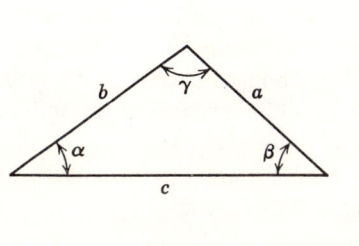

Fig. 2-3

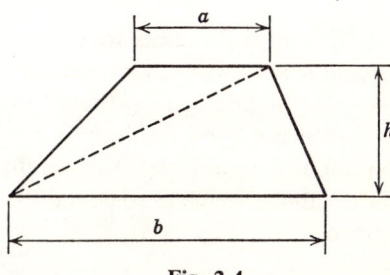

Fig. 2-4

8 / Mathematics

enclosed area $A = \pi r^2$. In three dimensions its counterpart is the sphere, which has a surface area $S = 4\pi r^2$ and an enclosed volume $V = \frac{4}{3}\pi r^3$.

Plane analytic geometry describes algebraically the properties of one- and two-dimensional geometric forms in an (x, y) plane.

The general equation of a straight line is $Ax + By + C = 0$. This equation is often more usefully written in one of the three following forms:

Point-slope: $\quad y - y_1 = m(x - x_1)$

Slope-intercept: $\quad y = mx + b$

Two-intercept: $\quad \dfrac{x}{a} + \dfrac{y}{b} = 1$

For a straight line passing through the points $P_1 = (x_1, y_1)$ and $P_2 = (x_2, y_2)$, the slope $m = (y_2 - y_1)/(x_2 - x_1)$; the intercepts a and b are the coordinate values occurring where the line intersects the x- and y-axes, respectively. The distance between points P_1 and P_2 is $D = [(x_2 - x_1)^2 + (y_2 - y_1)^2]^{1/2}$. Also, parallel lines have equal slopes, whereas perpendicular lines have negative reciprocal slopes.

Included in the general equation of second degree $Ax^2 + Bxy + Cy^2 + Dx + Ey + F = 0$ are a set of geometric shapes called the conic sections. The different conic sections can be recognized by investigating $B^2 - 4AC$:

If $B^2 - 4AC > 0$, the section is a hyperbola

If $B^2 - 4AC = 0$, the section is a parabola

If $B^2 - 4AC < 0$, the section is an ellipse

In the last case if $A = C$ and they are not zero, the section is a circle. If A, B, and C are all zero, the straight line again results.

The basic equation for the hyperbola can be found from the general second-degree equation. For a hyperbola centered at the coordinate origin with limbs opening left and right (Fig. 2-5a), the equation is

$$\frac{x^2}{a^2} - \frac{y^2}{b^2} = 1$$

The difference of distances from the two focuses $(\pm c, 0)$ to a point on the hyperbola is always constant, where $c^2 = a^2 + b^2$. The limbs are asymptotic to the straight lines $y = \pm(b/a)x$. For a hyperbola centered at the point (h, k), replace x by $(x - h)$ and y by $(y - k)$. (This procedure for shifting the location of figures applies generally for all conic sections.)

The parabola, which geometrically is the locus of points equidistant from a point and a line (Fig. 2-5b), may be written in type form as

$$y^2 = 4px$$

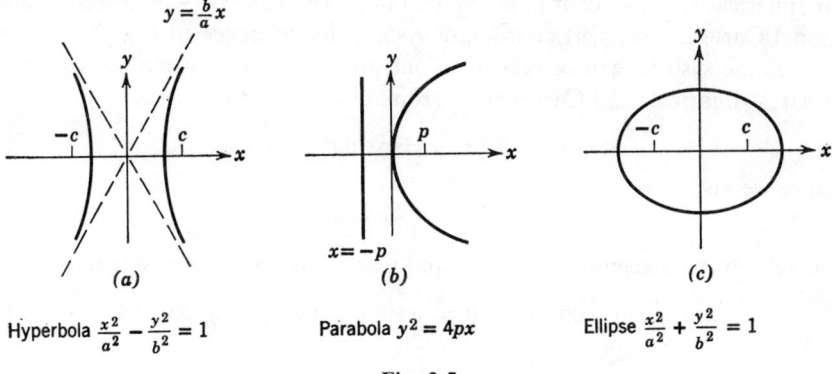

(a) Hyperbola $\frac{x^2}{a^2} - \frac{y^2}{b^2} = 1$ Parabola $y^2 = 4px$ Ellipse $\frac{x^2}{a^2} + \frac{y^2}{b^2} = 1$

Fig. 2-5

when the vertex of the parabola is at the coordinate origin and the parabola opens to the right. Here the parabola is equidistant from the focus point $(p, 0)$ and the directrix line $x = -p$. Change the sign of p to obtain a parabola opening to the right; interchange the roles of y and x to obtain parabolas opening upward or downward.

The type equation for an ellipse, centered on the coordinate origin (Fig. 2-5c), is

$$\frac{x^2}{a^2} + \frac{y^2}{b^2} = 1$$

The semimajor and semiminor axes are a and b. The focuses of the ellipse are at $(\pm c, 0)$, where $c^2 = a^2 - b^2$. Any point on the ellipse is such that the sum of the distances to that point from the two focuses is a constant. If $a = b = r$, the ellipse then becomes a circle of radius r.

Sometimes it is more convenient to use the polar (r, θ), cylindrical (r, θ, z),

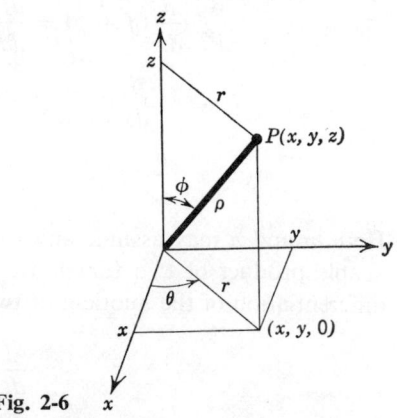

Fig. 2-6

10 / Mathematics

or spherical (ρ, θ, ϕ) coordinate system in place of the two- or three-dimensional Cartesian (x, y, z) coordinate system. By reference to Fig. 2-6, these coordinate systems can be related to one another. The relations between the polar, cylindrical, and Cartesian coordinate systems are

$$x = r \cos \theta \qquad y = r \sin \theta \qquad z = z$$

Since we also have

$$z = \rho \cos \phi \qquad r = \rho \sin \phi$$

the relations between the spherical and Cartesian coordinate systems are

$$x = \rho \sin \phi \cos \theta \qquad y = \rho \sin \phi \sin \theta \qquad z = \rho \cos \phi$$

CALCULUS

At a point on the curve $y = f(x)$ the slope of the curve is the ratio of the change in $f(x)$ to the change in x when the change in x approaches zero in the limit; mathematically,

$$\text{Slope} = \frac{dy}{dx} = \frac{df(x)}{dx} = f'(x)$$

This is also called the rate of change of y with respect to x or the first derivative of $f(x)$. Second and higher derivatives are in turn defined as the rate of change of the next lower ordered derivative.

Derivatives of basic functions of x are given now. Let f and g be functions of x; c, m, and n are constants.

$$\frac{d}{dx}(c) = 0$$

$$\frac{d}{dx}(cx^n) = cnx^{n-1} \qquad \text{for} \qquad n \neq 0$$

$$\frac{d}{dx}(f \pm g) = \frac{df}{dx} \pm \frac{dg}{dx}$$

$$\frac{d}{dx}(f^m) = mf^{m-1}\frac{df}{dx}$$

$$\frac{d}{dx}(f^m g^n) = f^m \frac{d}{dx}(g^n) + g^n \frac{d}{dx}(f^m)$$

Here m and n may assume any value. If $m = 1$, $n = 1$ this rule treats the simple product of two functions; if $m = 1$, $n = -1$ the rule governs the differentiation of the quotient of two functions. If $x = x(t)$,

$$\frac{df}{dt} = \frac{df}{dx}\frac{dx}{dt}$$

This is the chain rule of differentiation.

$$\frac{d}{dx}(\sin f) = \cos f \frac{df}{dx}$$

$$\frac{d}{dx}(\cos f) = -\sin f \frac{df}{dx}$$

At a maximum or a minimum of the function $f(x)$, the rate of change of f is zero, that is, $f'(x) = 0$. At that point f is a maximum if $f'' < 0$; it is a minimum if $f'' > 0$. Often, however, other physical considerations will indicate whether the function has a maximum or minimum, and the second derivative test will not really be needed then. If $f'' = 0$, the point will usually (but not always) be a point of inflection, a point where the curvature of the function changes from concave upward to concave downward.

If the function $f(x)$ approaches the value c as x approaches the value x_0, then we say c is the limiting value of $f(x)$ at the point x_0 and express this mathematically as

$$\lim_{x \to x_0} f(x) = c$$

The algebra of limits is no different from ordinary algebra:

$$\lim (f + g) = \lim f + \lim g$$

$$\lim (fg) = (\lim f)(\lim g)$$

$$\lim \frac{f}{g} = \frac{\lim f}{\lim g} \quad \text{if} \quad \lim g \neq 0$$

In the use of this last equation, however, the indeterminate forms $0/0$ or ∞/∞ may be encountered. In this case L'Hospital's rule is useful: Let f and g be functions having continuous derivatives with $g'(x_0) \neq 0$. If upon approaching the limit point $x = x_0$ we have

$$\lim_{x \to x_0} f(x) = \lim_{x \to x_0} g(x) = 0$$

or

$$\lim_{x \to x_0} f(x) = \lim_{x \to x_0} g(x) = \pm \infty$$

then

$$\lim_{x \to x_0} \left[\frac{f(x)}{g(x)} \right] = \lim_{x \to x_0} \left[\frac{f'(x)}{g'(x)} \right]$$

An example is

$$\lim_{x \to 0} \frac{\sin x}{x} = \lim_{x \to 0} \frac{\cos x}{1} = 1$$

12 / Mathematics

Continuous functions may be expressed as power series expansions around a point $x = c$ by use of the Taylor's series

$$f(x) = f(c) + f'(c)\frac{(x-c)}{1!} + f''(c)\frac{(x-c)^2}{2!} + \cdots + f^{(n)}(c)\frac{(x-c)^n}{n!} + \cdots$$

This series is particularly useful when a polynomial representation for a function is desired for x near c. The series can often then be truncated after only a few terms with little loss in accuracy.

Integration is the inverse of the process of differentiation. It may also be defined as the limit of a sequence; by this process the integral may be used to evaluate the exact area under a curve (Fig. 2-7). The area under this curve between a and b is $A = \int_a^b f(x)\,dx$.

Some basic integration formulas follow, using the same notation as was used for derivatives.

$$\int \frac{df(x)}{dx}\,dx = f(x) + C \quad (C = \text{constant of integration})$$

$$\int 0 \times dx = C$$

$$\int cf(x)\,dx = c\int f(x)\,dx$$

$$\int (f \pm g)\,dx = \int f\,dx \pm \int g\,dx$$

$$\int x^n\,dx = \frac{x^{n+1}}{n+1} + C \quad (n \neq -1)$$

$$\int \frac{dx}{x} = \ln x + C$$

$$\int e^x\,dx = e^x + C$$

$$\int \sin x\,dx = -\cos x + C$$

$$\int \cos x\,dx = \sin x + C$$

$$\int f\,dg = fg - \int g\,df$$

This last formula, involving the two functions $f(x)$ and $g(x)$, is called integration by parts and is a powerful tool of integration. Evaluating any integral

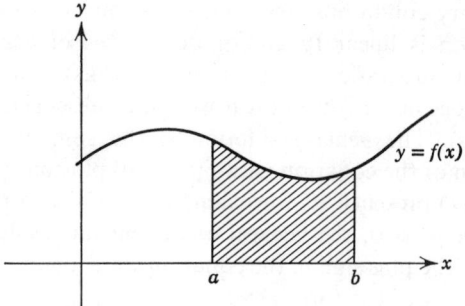

Fig. 2-7

between definite limits will eliminate the constant of integration. We should also note that, for $a \leq b \leq c$,

$$\int_a^c f(x)\,dx = \int_a^b f(x)\,dx + \int_b^c f(x)\,dx$$

and

$$\int_a^b f(x)\,dx = -\int_b^a f(x)\,dx$$

DIFFERENTIAL EQUATIONS

Only the barest introduction to ordinary differential equations can be given here. In some of the problems in this chapter examples of solving differential equations are given, but for any comprehensive review of the subject it is suggested that the reader consult an appropriate mathematics text.

First-order ordinary differential equations are separable if they can be put in the form

$$M(x)\,dx + N(y)\,dy = 0$$

The equation is solved by direct integration; as in all differential equations a boundary condition must be specified before the constant of integration can be evaluated. If $M = M(x, y)$ and $N = N(x, y)$, a solution $F(x, y)$ can be found if the differential equation is exact, that is, if $\partial M/\partial y = \partial N/\partial x$. Then the solution $F(x, y)$ must satisfy the requirements $M = \partial F/\partial x$, $N = \partial F/\partial y$.

Many additional terms and types of basic ordinary differential equations are important to engineering but cannot be reviewed here. Just one representative important differential equation will be mentioned, however, to illustrate some concepts. The equation is

$$\frac{d^2y}{dx^2} + p^2 y = f(x) \qquad (p = \text{constant})$$

plus two boundary conditions since this equation involves a second derivative. The equation is linear (y and/or derivatives of y are not multiplied together to form quadratic or higher order terms) and nonhomogeneous (the right term does not involve y; if it were not present, the equation would be homogeneous). The general solution is the sum of a complementary solution [solution of the equation with $f(x) = 0$] plus one particular solution [solution with $f(x)$ present]. If, for example, $f(x) = x$, a particular solution is $y = x/p^2$. Since $p^2 > 0$, the general complementary solution is $y = A \sin px + B \cos px$. If the plus sign in the equation were a minus sign, the solution would involve terms of the form $e^{\pm px}$.

STATISTICS AND PROBABILITY

Statistics is used as an aid in drawing conclusions from masses of data. The computation of certain properties of the data is useful in answering the questions "How big is it?" and "How much variation in size is there?"

The arithmetic mean, the median, and the mode are valuable tools in ascertaining the answer to the first question. The arithmetic mean $\bar{x}$ is the average value of N individual values x_i:

$$\bar{x} = \frac{1}{N} \sum_{i=1}^{N} x_i$$

The median value is the middle value when all data are arranged in order by magnitude; half the values are larger than the median value, half are smaller. The mode, on the other hand, is the value that occurs most frequently. The standard deviation σ is a statistic which helps to answer the second question; it gives the rms deviation from the mean value $\bar{x}$:

$$\sigma = \left[\frac{1}{N} \sum_{i=1}^{N} (x_i - \bar{x})^2 \right]^{1/2}$$

For small values of N it is preferable to replace the divisor N by $(N - 1)$. The square of the standard deviation σ^2 is called the variance.

Much of basic probability theory deals with mutually exclusive events or independent events. If only one of a set of possible events can occur, the events are mutually exclusive. Events are independent if the occurrence or nonoccurrence of one event does not affect the probability of occurrence of the other events. In this connection it is sometimes necessary to compute the number of permutations or set arrangements of n things taken r at a time, which is $n!/(n-r)!$, and the number of combinations (no set arrangement) of n things taken r at a time, which can be expressed in the equivalent forms

$$\binom{n}{r} = \frac{n!}{r!(n-r)!}.$$

The probability of success of an event plus the probability of failure of an event is obviously unity, a fact of great simplicity and use. The probability of success itself is the ratio of the number of ways of achieving success over the total number of possible events (successes and failures). For independent events the following two rules are also useful:

1. The probability of *A or B* occurring equals the *sum* of the probability of occurrence of *A* and the probability of occurrence of *B*.
2. The probability of *A and B* occurring equals the *product* of the two individual probabilities.

2-1 Wt. 1

The number below that has four significant figures is

(a) 1414.0
(b) 1.4140
(c) 0.141
(d) 0.01414
(e) 0.0014

Solution

A significant figure is any of the digits 1 through 9 as well as 0 except when 0 is used to fix the decimal point or to fill the places of unknown or discarded digits.

Part	No. of Significant Figures
a	5
b	5
c	3
d	4 ▶
e	2

Answer is (*d*)

2-2 Wt. 2

Expand $(1 - 2x^{-1/2})^{-2}$ to five terms.

Solution

Using the binomial expansion

$$(1 - A)^{-n} = 1 + nA + \frac{n(n+1)}{2!} A^2 + \frac{n(n+1)(n+2)}{3!} A^3$$
$$+ \frac{n(n+1)(n+2)(n+3)}{4!} A^4 + \cdots$$

16 / Mathematics

in which we have $A = 2x^{-1/2}$ and $n = 2$ here, we obtain

$$(1 - 2x^{-1/2})^{-2} = 1 + 2(2x^{-1/2}) + \frac{(2)(3)}{2}(2x^{-1/2})^2$$

$$+ \frac{(2)(3)(4)}{(3)(2)}(2x^{-1/2})^3 + \frac{(2)(3)(4)(5)}{(4)(3)(2)}(2x^{-1/2})^4 + \cdots$$

$$= 1 + 4x^{-1/2} + 12x^{-1} + 32x^{-3/2} + 80x^{-2} + \cdots \blacktriangleright$$

2-3 Wt. 1

A common mathematical constant which is equal to

$$1 + 1 + \frac{1}{2!} + \frac{1}{3!} + \frac{1}{4!} + \cdots$$

is identified by the symbol_____.

Solution

From Taylor's expansion

$$f(x) = f(0) + \frac{f'(0)}{1!}x + \frac{f''(0)}{2!}x^2 + \cdots$$

so if $f(x) = e^x$

$$e^x = 1 + \frac{x}{1!} + \frac{x^2}{2!} + \frac{x^3}{3!} + \cdots$$

If $x = 1$

$$e = 1 + \frac{1}{1!} + \frac{1}{2!} + \frac{1}{3!} + \cdots = 2.71828$$

Answer is symbol e ▶

2-4 Wt. 1

If $x = +6$ and -4, the equation satisfying both these values would be

(a) $2x^2 + 3x - 24 = 0$
(b) $x^2 + 10x - 24 = 0$
(c) $x^2 - 2x - 24 = 0$
(d) $x^2 - 4x - 32 = 0$
(e) $x^2 - 3x + 18 = 0$

Solution

If a quadratic equation has solutions of $+6, -4$, it must have the factored form

$$(x - 6)(x + 4) = 0$$

Expanding this, we obtain

$$x^2 - 6x + 4x - 24 = 0$$
$$x^2 - 2x - 24 = 0 \blacktriangleright$$

Answer is (c)

2-5 Wt. 1

If $X^{3/4} = 8$, X equals

(a) 6
(b) 9
(c) −9
(d) 16
(e) 20

Solution

We want to determine X^1 so we raise the entire equation to the $\frac{4}{3}$ power

$$X^1 = (X^{3/4})^{4/3} = 8^{4/3} = (8^{1/3})^4 = 2^4 = 16 \blacktriangleright$$

Answer is (d)

2-6 Wt. 1

If $i = \sqrt{-1}$, the quantity i^{27} is equal to

(a) 0
(b) i
(c) $-i$
(d) 1
(e) -1

Solution

$(i)^n$ is a periodic function, that is,

$$i^1 = i \quad i^2 = -1 \quad i^3 = -i \quad i^4 = +1 \quad i^5 = i \quad \ldots$$

More generally,

$$i^{4n+1} = i \quad i^{4n+2} = -1 \quad i^{4n+3} = -i \quad i^{4n+4} = +1$$

for any integer n.

Here $27 = 4n + 3$ and $i^{27} = -i \blacktriangleright$

Answer is (c)

18 / Mathematics

2-7 Wt. 1

If $x = \log_a N$, then

(a) $x = N^a$
(b) $a = x^N$
(c) $x = a^N$
(d) $N = a^x$
(e) $N = x^a$

Solution

If x is the logarithm of N to the base a ($x = \log_a N$), then by definition $N = a^x$ ▶

$$\text{Answer is } (d)$$

2-8 Wt. 1

The logarithm of the number, -0.028, is

(a) positive
(b) negative
(c) zero
(d) not a real number
(e) a complex number

Solution

$$\text{Answer is } (d) \text{ and } (e) \blacktriangleright$$

2-9 Wt. 1

$\text{Log}_{10}\ 1{,}000{,}000$ is equal to

(a) 1.600
(b) 3.000
(c) 3.303
(d) 5.303
(e) 6.000

Solution

$$\log_{10} 10^6 = 6 \log_{10} 10 = 6 \times 1.000 = 6.000 \blacktriangleright$$

$$\text{Answer is } (e)$$

2-10 Wt. 1

$\text{Log}_{10}\ (100)^2 - \ln_e 2.718$ is equal to

(a) 2
(b) 3
(c) 4
(d) 5
(e) 6

Solution

$$\log_{10}(100)^2 = \log_{10} 10^4 = 4 \log_{10} 10 = 4 \times 1 = 4$$

e is approximately 2.718; therefore $\ln_e 2.718 = \ln_e e = 1$,

$$\log_{10}(100)^2 - \ln_e 2.718 = 4 - 1 = 3 \blacktriangleright$$

Answer is (*b*)

2-11 Wt. 4

$\ln_e (2.718)^{xy}$ is equal to

(*a*) xy
(*b*) exy
(*c*) $2.718xy$
(*d*) $x + y$
(*e*) $0.434xy$

Solution

$$e = 2.718 \qquad \ln_e e^{xy} = xy \ln_e e = xy(1) = xy \blacktriangleright$$

Answer is (*a*)

2-12 Wt. 1

The $\log_{10}$ of 2 is 0.30103. The log of $\tfrac{1}{2}$ is

(*a*) $9.30103 - 10$
(*b*) $0.30103 \div 2$
(*c*) $1 - 0.30103$
(*d*) $9.69897 - 10$
(*e*) $1 \div 0.30103$

Solution

$$\log \tfrac{1}{2} = \log 1 - \log 2 = 0.00000 - 0.30103$$

which may be written as

$$= 10. - 0.30103 - 10$$
$$= 9.69897 - 10$$

Answer is (*d*)

2-13 Wt. 1

If $\log_a 10 = 0.250$, $\log_{10} a$ equals

(*a*) 4
(*b*) 0.50
(*c*) 2
(*d*) 0.25
(*e*) 1000

Solution

Log$_a$ 10 = 0.250 can be written as $10 = a^{0.250}$. Taking log$_{10}$,

$$\log_{10} 10 = \log_{10} a^{0.250}$$
$$1 = 0.250 \log_{10} a$$

Since $1 = 0.250 \log_{10} a$,

$$\log_{10} a = \frac{1}{0.250} = 4 \blacktriangleright$$

Answer is (a)

2-14 Wt. 1

Log$_{10}$ $(0.001)^3$ is equal to

(a) −9.000
(b) −3.000
(c) −1.000
(d) +0.100
(e) +0.300

Solution

$$\log_{10} (0.001)^3 = 3 \log_{10} 0.001 = 3 \times -3.000 = -9.000 \blacktriangleright$$

Answer is (a)

2-15 Wt. 1

The value of $(0.20)^{½}$ is

(a) 0.0016
(b) 0.004
(c) 0.04
(d) 0.4
(e) 4.0

Solution

$$(0.20)^{½} = 0.20^2 = 0.04 \blacktriangleright$$

Answer is (c)

2-16 Wt. 1

The value of $(0.01)^{3/2}$ is

(a) 0.10
(b) 0.15
(c) 0.015
(d) 0.005
(e) 0.001

Solution
$$(0.01)^{3/2} = [(0.01)^{1/2}]^3 = [0.1]^3 = 0.001 \blacktriangleright$$

Answer is (e)

2-17 Wt. 1

The value of $(0.001)^{2/3}$ is

(a) 10.0
(b) 1.00
(c) 0.10
(d) 0.01
(e) 0.001

Solution
$$(0.001)^{2/3} = [(0.001)^{1/3}]^2 = [0.1]^2 = 0.01 \blacktriangleright$$

Answer is (d)

2-18 Wt. 1

The value of $(0.085)^{3/2}$ is nearest to

(a) 0.290
(b) 0.092
(c) 0.035
(d) 0.025
(e) 0.008

Solution
$$1.5 \log 0.085 = 1.5(-2 + 0.929) = -2 + 0.394$$

(Remember the mantissa is always positive.) The antilog of -2.394 is $0.025 \blacktriangleright$

Slide rule solution

Set the left index of the slide opposite 0.085 on the LLOO scale. Opposite 1.5 on the B scale read the answer (0.025) on the LLOO scale $\blacktriangleright$

Answer is (d)

2-19 Wt. 1

If $x = \frac{1}{2} \ln \dfrac{1+u}{1-u}$ (ln = natural logarithm), solve for u.

22 / Mathematics

Solution

$$2x = \ln \frac{1+u}{1-u}$$

$$e^{2x} = \exp\left(\ln \frac{1+u}{1-u}\right) = \frac{1+u}{1-u}$$

$$(1-u)e^{2x} = 1+u$$

$$u(e^{2x}+1) = e^{2x}-1$$

$$u = \frac{e^{2x}-1}{e^{2x}+1} = \frac{e^x - e^{-x}}{e^x + e^{-x}} = \tanh x \blacktriangleright$$

2-20 Wt. 1

$\text{Log}_e (1000)$ is approximately

(a) 2.303
(b) 3.000
(c) 4.343
(d) 6.908
(e) 12.718

Slide rule solution

Set the left index of the slide opposite e on the LL3 scale (note that e is at the left end of the LL3 scale). Opposite 1000 on the LL3 scale read the answer (6.91) on the C scale ▶

Alternative solution

$$\ln_e 1000 = \frac{\log_{10} 1000}{\log e} = \frac{3}{\log e}$$

where $\log e$ is the modulus of common logs with respect to natural logs and is equal to 0.4343

$$\ln_e 1000 = \frac{3}{0.4343} = 6.91 \blacktriangleright$$

Answer is (d)

2-21 Wt. 1

$\text{Log}_{10} (\sin 30°)$ is approximately

(a) 9.70 − 10
(b) 3.30 − 10
(c) 0.50
(d) 0.30
(e) 0.03

Solution

Sin $30° = \frac{1}{2}$, so the problem becomes $\log_{10} 0.50 = ?$ The characteristic is -1 or $9 - 10$. The mantissa is 0.70. Hence

$$\log_{10} 0.5 = 9.70 - 10 \blacktriangleright$$

Answer is (*a*)

2-22 Wt. 2

Solve the following equation mathematically for X:

$$\log_{10} (X - 1) + \log_{10} X = 1$$

Solution

Since the sum of logarithms of numbers is their product and $\log 10 = 1$, then

$$(X - 1)(X) = 10 \quad \text{or} \quad X^2 - X = 10$$
$$X^2 - X - 10 = 0$$
$$X = \frac{1 \pm \sqrt{1 + 40}}{2} = \frac{1 \pm 6.4}{2} = 3.7 \blacktriangleright$$

Note that there is only one real root. The other answer from the quadratic equation (-2.7) is not real, for the log of a negative number has no meaning.

2-23 Wt. 1

The sum of the interior angles of a polygon with seven sides is

(*a*) 540°
(*b*) 630°
(*c*) 720°
(*d*) 810°
(*e*) 900°

Solution

The sum of the interior angles of a regular polygon of n sides is $(n - 2) \times 180°$. Therefore, $(7 - 2) \times 180 = 900°$ ▶

Answer is (*e*)

2-24 Wt. 1

Each interior angle of a regular polygon with eight sides is nearest to

(*a*) 100°
(*b*) 80°
(*c*) 150°
(*d*) 125°
(*e*) 135°

24 / Mathematics

Solution

The sum of the interior angles of a polygon is equal to $(n - 2) \times 180°$. A regular polygon means all sides are equal, hence all angles are equal.

$$\frac{(n - 2) \times 180°}{8} = \frac{6}{8}(180) = 135° \blacktriangleright$$

Answer is (*e*)

2-25 Wt. 1

The number of degrees in 1 rad of angular measurement is approximately

(*a*) 50
(*b*) 55
(*c*) 57
(*d*) 59
(*e*) 60

Solution

There are 2π rad in 360°. 1 rad = $360/2\pi = 57.3°$ ▶

Answer is (*c*)

2-26 Wt. 3

The trigonometric functions of a given angle θ can be represented on a diagram by a line. Assume that $R = 1$. For example:

$$\sin \theta = \frac{a}{c} = \frac{a}{R} = \frac{a}{1} = a$$

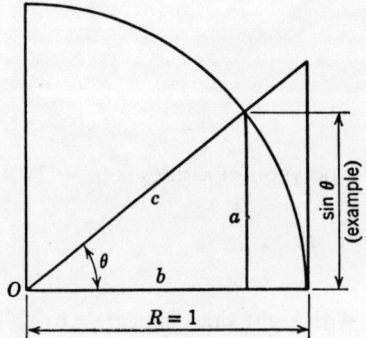

Fig. 2-8

Using the diagram, add any construction you think necessary, and indicate the linear representations of each of the following:

(*a*) $\cos \theta$
(*b*) $\tan \theta$

(c) sec θ
(d) vers θ
(e) csc θ
(f) cot θ

Solution

The problem here is to make the denominator of the function equal to 1. The numerator is then equal to the function

$$\cos\theta = \frac{b}{c} = \frac{b}{R} = \frac{b}{1} = b$$

$$\tan\theta = \frac{a_1}{R} = \frac{a_1}{1} = a_1$$

$$\sec\theta = \frac{c_1}{R} = \frac{c_1}{1} = c_1$$

$$\text{vers }\theta = 1 - \cos\theta = R - b = d$$

$$\csc\theta = \frac{c_2}{R} = \frac{c_2}{1} = c_2$$

$$\cot\theta = \frac{b_1}{R} = \frac{b_1}{1} = b_1$$

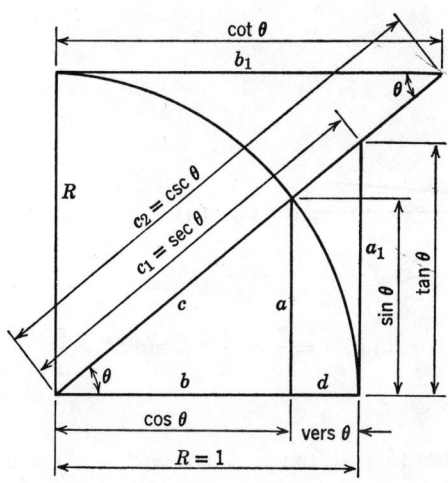

Fig. 2-9

26 / Mathematics

2-27 Wt. 3

What is the tangent of an angle whose cosine is X?

Solution

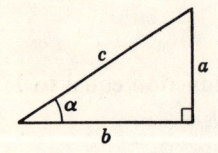

Fig. 2-10

$$\cos \alpha = \frac{b}{c} = X$$

Let $c = 1$ and $b = X$. Then

$$a = ((1)^2 - X^2)^{1/2} = (1 - X^2)^{1/2}$$

$$\tan \alpha = \frac{(1 - X^2)^{1/2}}{X} \blacktriangleright$$

2-28 Wt. 1

The sine of 5° is most nearly equal to the

(*a*) cotangent of 5°
(*b*) cotangent of 85°
(*c*) tangent of 85°
(*d*) cosine of 5°
(*e*) sine of 85°

Solution

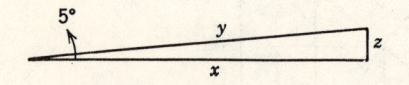

Fig. 2-11

For small angles $x \sim y$.

$$\sin (5°) = \frac{z}{y} \sim \frac{z}{x} \qquad \tan 5° = \frac{z}{x}$$

Therefore

$$\sin (5°) \sim \tan (5°) \qquad \tan 5° = \cot (90° - 5°) = \cot (85°) \blacktriangleright$$

Answer is (*b*)

2-29 Wt. 1

The cosine of 120° is equal to

(a) $\sqrt{3}/2$
(b) $1/\sqrt{3}$
(c) $\frac{1}{2}$
(d) $-\frac{1}{2}$
(e) $-\sqrt{3}/2$

Solution

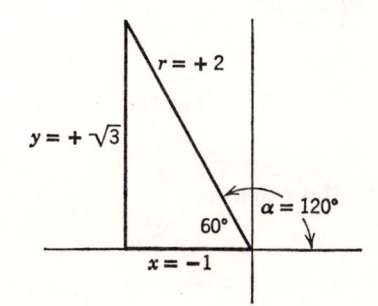

Fig. 2-12

$$\cos \alpha = -\cos 60° = \frac{x}{r} = -\frac{1}{2}$$

Hence

$$\cos 120° = -\tfrac{1}{2} \blacktriangleright$$

Answer is (d)

2-30 Wt. 1

The tangent of an angle of 210° is

(a) 1
(b) $\frac{1}{2}$
(c) $1/\sqrt{3}$
(d) $-\frac{1}{2}$
(e) $-\sqrt{3}$

Solution

From Fig. 2-13, $\tan(180° + \theta) = \tan \theta$

$$\tan 210° = +\tan 30° = \frac{1}{\sqrt{3}} \blacktriangleright$$

Answer is (c)

28 / Mathematics

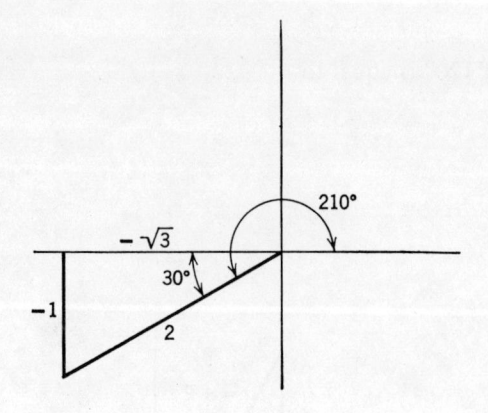

Fig. 2-13

2-31 Wt. 1

The cosecant of 960° is equal to

(a) $-(2\sqrt{3})/3$
(b) 1
(c) $\frac{1}{2}$
(d) -2
(e) $-\sqrt{3}/2$

Solution

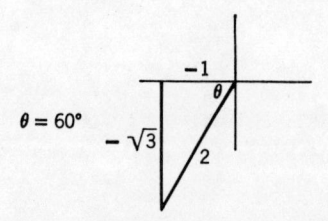

Fig. 2-14

$$\csc 960° = \csc (5\pi + 60)$$
$$= \csc \theta = \frac{r}{y} = \frac{2}{-\sqrt{3}} = -\frac{2\sqrt{3}}{3} \blacktriangleright$$

Answer is (a)

2-32 Wt. 1

The cosine of an angle of 300° is

(a) -1.000
(b) -0.866
(c) -0.500

(d) 0.500
(e) 0.866

Solution

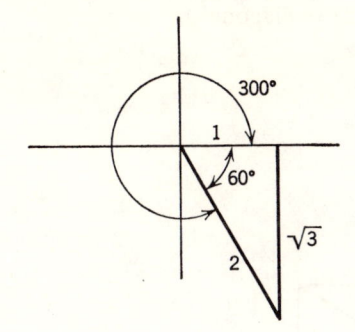

Fig. 2-15

$\cos 300° = \cos 60° = 0.500$ ▶

Answer is (d)

2-33 Wt. 1

If the sine of angle A is given as K, the tangent of angle A would be equal to

(a) $1 - K$
(b) $1/K$
(c) $\sqrt{1 - K^2}$
(d) $\dfrac{1}{\sqrt{1 - K^2}}$
(e) $\dfrac{K}{\sqrt{1 - K^2}}$

Solution

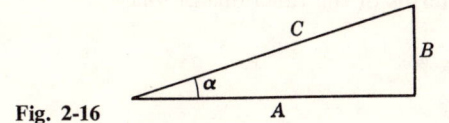

Fig. 2-16

$\sin \alpha = B/C = K$. If we let $C = 1$, then $B = K$. Since $A^2 + B^2 = C^2$, $A^2 + K^2 = 1^2$ or $A = \sqrt{1 - K^2}$. Then the triangle appears as:

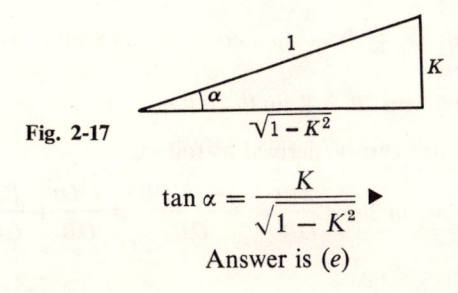

Fig. 2-17

$$\tan \alpha = \frac{K}{\sqrt{1 - K^2}}$$ ▶

Answer is (e)

2-34 Wt. 1

The diagonal of a cube whose side is a in. long can be found by which of the following equations? ($D =$ diagonal.)

(a) $D = \sqrt{a^3}$
(b) $D = a^3/2$
(c) $D = \sqrt{a^2/3}$
(d) $D = a^{3/2}$
(e) $D = \sqrt{3a^2}$

Solution

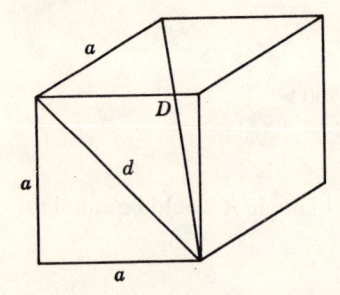

Fig. 2-18

The diagonal of a side of the cube $d = \sqrt{a^2 + a^2} = \sqrt{2a^2}$. From this we can find the length of the diagonal of the cube as

$$D = \sqrt{a^2 + d^2} = \sqrt{a^2 + 2a^2} = \sqrt{3a^2} \blacktriangleright$$

Answer is (e)

2-35 Wt. 1

Sin 2θ expressed in terms of the functions of θ is

(a) $2 \sin \theta \cos \theta$
(b) $1 - 2 \sin^2 \theta$
(c) $\cos^2 \theta - \sin^2 \theta$
(d) $\sin \theta \cos \theta$
(e) $1 - \cos^2 \theta$

Solution

Using the identity $\sin (\alpha + \beta) = \sin \alpha \cos \beta + \cos \alpha \sin \beta$ with $\alpha = \beta = \theta$, we find

$$\sin 2\theta = 2 \sin \theta \cos \theta \blacktriangleright$$

Alternatively, this result can be derived as follows:

$$\sin (\theta + \theta) = \sin 2\theta = \frac{AB}{OB} = \frac{AE + EB}{OB} = \frac{CD}{OB} + \frac{EB}{OB}$$

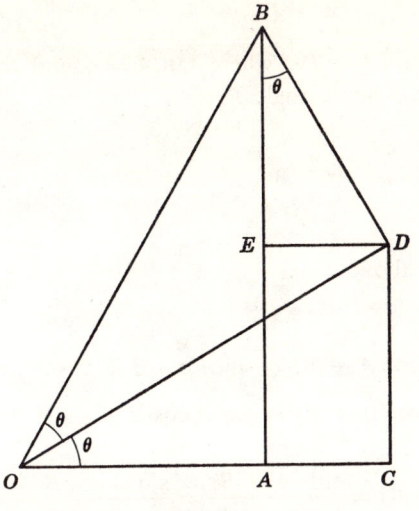

Fig. 2-19

But $CD = OD \sin \theta$ and $EB = BD \cos \theta$. $\sin 2\theta = \dfrac{OD}{OB} \sin \theta + \dfrac{BD}{OB} \cos \theta$. Also $\dfrac{OD}{OB} = \cos \theta$ and $\dfrac{BD}{OB} = \sin \theta$. Therefore

$$\sin 2\theta = \cos \theta \sin \theta + \sin \theta \cos \theta = 2 \sin \theta \cos \theta \blacktriangleright$$

Answer is (a)

2-36 Wt. 1

Which of the following is incorrect?

(a) $\cos^2 A = \tan A \cot A$
(b) $\sin A = \cos A \tan A$
(c) $\cos 2A = \cos^2 A - \sin^2 A$
(d) $2 \sin^2 A = 1 - \cos 2A$
(e) $\sin (A + B) = \sin A \cos B + \cos A \sin B$

Solution

An identity is an equality which is valid for all values of the variable(s) in the equation. Considering the first expression,

$$\tan A \cot A = 1 \quad \text{or} \quad \cos^2 A = 1$$

which is not true for all values of A ▶

The four other expressions are correct.

Answer is (a)

2-37 Wt. 1

The value of $\tan(A + B)$, where $\tan A = \frac{1}{3}$ and $\tan B = \frac{1}{4}$, is (Note: Assume A and B are acute angles)

(a) $\frac{7}{12}$
(b) $\frac{1}{11}$
(c) $\frac{7}{11}$
(d) $\frac{7}{13}$
(e) none of the above

Solution

$$\sin(A + B) = \sin A \cos B + \cos A \sin B$$
$$\cos(A + B) = \cos A \cos B - \sin A \sin B$$

So

$$\tan(A + B) = \frac{\sin(A + B)}{\cos(A + B)} = \frac{\sin A \cos B + \cos A \sin B}{\cos A \cos B - \sin A \sin B}$$

Dividing by $\cos A$ and $\cos B$,

$$\tan(A + B) = \frac{\dfrac{\sin A \cos B}{\cos A \cos B} + \dfrac{\cos A \sin B}{\cos A \cos B}}{\dfrac{\cos A \cos B}{\cos A \cos B} - \dfrac{\sin A \sin B}{\cos A \cos B}}$$

$$= \frac{\tan A + \tan B}{1 - \tan A \tan B}$$

$$= \frac{\frac{1}{3} + \frac{1}{4}}{1 - \frac{1}{3} \times \frac{1}{4}} = \frac{\frac{4}{12} + \frac{3}{12}}{1 - \frac{1}{12}} = \frac{\frac{7}{12}}{\frac{11}{12}} = \frac{7}{11} \blacktriangleright$$

Note that the problem could also be solved by determining angle A (whose tangent is 1/3) and angle B (whose tangent is 1/4). Then we could find the tangent of $(A + B)$.

$$\tan^{-1} \tfrac{1}{3} = 18°26' \qquad \tan^{-1} \tfrac{1}{4} = 14°2'$$

$$\tan(18°26' + 14°2') = \tan(32°28') = 0.636 = \tfrac{7}{11} \blacktriangleright$$

Answer is (c)

2-38 Wt. 3

A circle can be circumscribed around a triangle ABC, as shown. What is the area of the triangle? (AC is a diameter of the circle.)

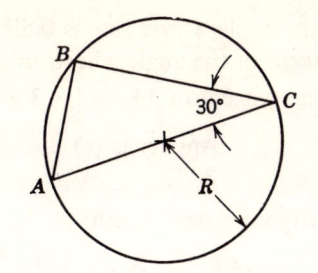

Fig. 2-20

Solution

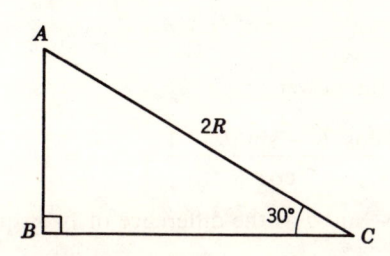

Fig. 2-21

A geometric theorem states that a triangle inscribed in a semicircle is a right triangle, so angle $ABC = 90°$.

$\sin 30° = \dfrac{AB}{2R} = \dfrac{1}{2}$. Therefore $AB = R$. $\cos 30° = \dfrac{BC}{2R} = \dfrac{\sqrt{3}}{2}$. Therefore $BC = \sqrt{3}R$.

$$\text{Area} = \tfrac{1}{2}(AB)(BC) = \tfrac{1}{2}R\sqrt{3}R$$
$$= 0.866R^2 \blacktriangleright$$

2-39 Wt. 1

Tan (arc sin 0.5) is equal to

(a) $\tfrac{1}{4}$
(b) $\tfrac{1}{2}$
(c) $1/\sqrt{3}$
(d) 1
(e) $\sqrt{3}$

Solution

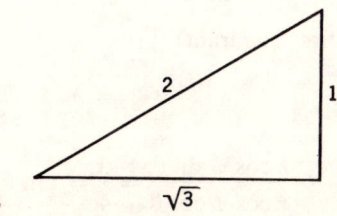

Fig. 2-22

34 / Mathematics

Arc sin 0.5 means "the angle whose sine is 0.5" so the problem could be stated: What is the tangent of the angle whose sine is 0.5? The angle whose sine is 0.5 is 30°, and the tangent of 30° = $1/\sqrt{3}$ ▶

Answer is (c)

2-40 Wt. 3

Prove the following trigonometric identity:

$$\left(\frac{\cos^4\theta - \sin^4\theta + 1}{2\cos^2\theta}\right)^{1/2} = 1$$

Solution

Raise both sides to the power $\frac{2}{1}$:

$$\frac{\cos^4\theta - \sin^4\theta + 1}{2\cos^2\theta} = 1^{2/1} = 1$$

We can factor $\cos^4\theta - \sin^4\theta$ as the difference of two squares.

$$\cos^4\theta - \sin^4\theta = (\cos^2\theta + \sin^2\theta)(\cos^2\theta - \sin^2\theta)$$

But $\sin^2\theta + \cos^2\theta = 1$, so the term reduces to $(1)(\cos^2\theta - \sin^2\theta)$. Substituting back:

$$\frac{\cos^2\theta - \sin^2\theta + 1}{2\cos^2\theta} = 1$$

Since

$$\sin^2\theta + \cos^2\theta = 1$$
$$(-\sin^2\theta + 1) = \cos^2\theta$$

and we have

$$\frac{\cos^2\theta + (\cos^2\theta)}{2\cos^2\theta} = 1 \quad \text{Q.E.D.} \blacktriangleright$$

2-41 Wt. 3

Given the following relations:

$$r \cos\phi \cos\theta = 4 \quad (1)$$
$$r \cos\phi \sin\theta = 3 \quad (2)$$
$$r \sin\phi = 5 \quad (3)$$

Find r and θ (in the first quadrant).

Solution

(2)/(1):

$$\frac{r \cos\phi \sin\theta}{r \cos\phi \cos\theta} = \frac{3}{4} = \tan\theta$$

Therefore
$$\theta = \tan^{-1}(\tfrac{3}{4}) = 36.9° \blacktriangleright$$

(3)/(1):
$$\frac{r \sin \phi}{r \cos \phi \cos \theta} = \frac{5}{4} \qquad \tan \phi \frac{1}{\cos \theta} = \frac{5}{4}$$

but $\dfrac{1}{\cos \theta} = \dfrac{1}{\cos 36.9°} = \dfrac{1}{0.8} = 1.25$, $1.25 \tan \phi = \dfrac{5}{4}$, $\tan \phi = 1$, $\phi = 45°$, $r \sin \phi = 5$, and $r \sin 45° = 5$. Therefore

$$r = \frac{5}{0.707} = 7.07 \blacktriangleright$$

Alternative solution

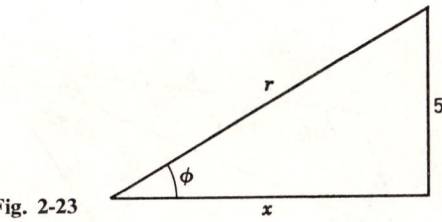

Fig. 2-23

$$r \sin \phi = 5$$
$$x^2 + 5^2 = r^2 \qquad x = (r^2 - 5^2)^{1/2}$$

Therefore
$$r \cos \phi = (r^2 - 5^2)^{1/2}$$

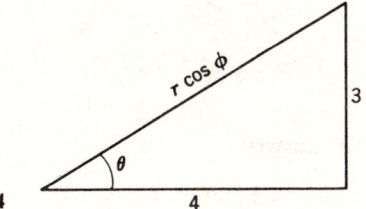

Fig. 2-24

$$3^2 + 4^2 = [(r^2 - 5^2)^{1/2}]^2 = r^2 - 5^2$$
$$r^2 = 9 + 16 + 25 = 50$$
$$r = \sqrt{50} = 5\sqrt{2} = 7.07 \blacktriangleright$$
$$\tan \theta = \tfrac{3}{4}$$

Therefore
$$\theta = \tan^{-1}(\tfrac{3}{4}) = 36.9° \blacktriangleright$$

36 / Mathematics

2-42 Wt. 5

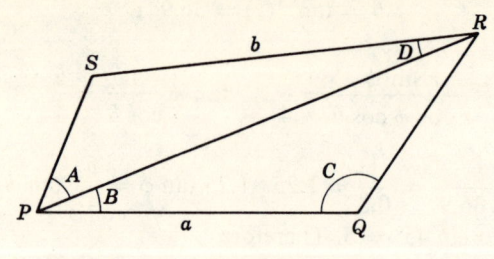

Fig. 2-25

Express side b in the figure in terms of side a and angles A, B, C, and D.

Solution

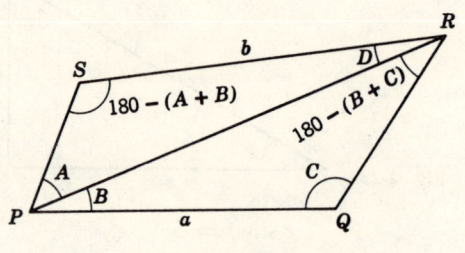

Fig. 2-26

Using the law of sines in triangle PQR, we have

$$\frac{\sin C}{PR} = \frac{\sin [180 - (B + C)]}{a}$$

$$PR = \frac{a \sin C}{\sin (B + C)}$$

In triangle PSR we can write

$$\frac{\sin A}{b} = \frac{\sin [180 - (A + D)]}{PR} = \frac{\sin (A + D)}{\frac{a \sin C}{\sin (B + C)}}$$

$$b = \frac{a \sin A \sin C}{\sin (A + D) \sin (B + C)} \blacktriangleright$$

2-43 Wt. 5

Two points lie on a horizontal line directly south of a tower 100 ft high. The angles of depression to the points are 28°10′ and 42°50′. Find the distance between the points.

Solution

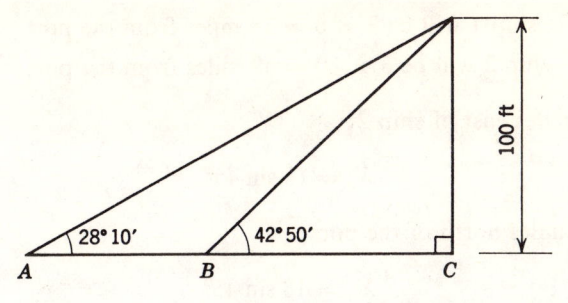

Fig. 2-27

The distance $\overline{AB} = \overline{AC} - \overline{BC}$

$$\tan(28°10') = \frac{100 \text{ ft}}{\overline{AC}} \qquad \tan(42°50') = \frac{100 \text{ ft}}{\overline{BC}}$$

$$\overline{AB} = \frac{100 \text{ ft}}{\tan(28°10')} - \frac{100 \text{ ft}}{\tan(42°50')}$$

$$= 100 \text{ ft}[\cot(28°10') - \cot(42°50')]$$

$$= 100 \text{ ft}(1.8676 - 1.0786)$$

$$= 100 \text{ ft}(0.7890) = 78.90 \text{ ft} \blacktriangleright$$

2-44 Wt. 3

Two ships leave the same port at the same time, one sailing due northeast at a rate of 6 mph and the other sailing due north at the rate of 10 mph. Find the distance between the two ships after 3 hr of sailing.

Solution

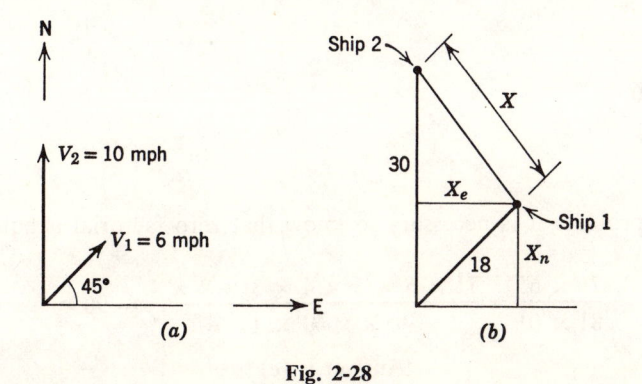

Fig. 2-28

38 / Mathematics

After 3 hr

$$\text{ship 1 will be } 3 \times 6 = 18 \text{ miles from the port}$$
$$\text{ship 2 will be } 3 \times 10 = 30 \text{ miles from the port}$$

Ship 1 is X_e miles east of ship 2:

$$X_e = 18 \sin 45°$$

Ship 1 is X_n miles north of the port:

$$X_n = 18 \sin 45°$$

Solving for X,

$$X^2 = X_e^2 + (30 - X_n)^2$$
$$X^2 = (18 \sin 45°)^2 + (30 - 18 \sin 45°)^2$$
$$X^2 = \left(18 \frac{\sqrt{2}}{2}\right)^2 + \left(30 - 18 \frac{\sqrt{2}}{2}\right)^2$$
$$X^2 = 162 + (30 - 9\sqrt{2})^2$$
$$X^2 = 162 + 900 + 162 - 2(30)(9\sqrt{2})$$
$$X^2 = 1224 - 764 = 460$$

Since X must be positive, $X = 21.5$ miles ▶

2-45 Wt. 1

$\dfrac{7! \times 6!}{8! \times 0!}$ is equal to

(a) ∞
(b) 0
(c) 720
(d) 5760
(e) 90

Solution

In this problem it is necessary to know that zero factorial is equal to 1.

$$\frac{7! \times 6!}{8! \times 0!} = \frac{\cancel{7!} \times 6 \times 5 \times \cancel{4} \times 3 \times \cancel{2} \times 1}{\cancel{8} \times \cancel{7!} \times 1} = 90 \blacktriangleright$$

Answer is (*e*)

2-46 Wt. 3

Reduce $\dfrac{X^{-1} + Y^{-1}}{X^{-2} + Y^{-2}}$ to its simplest form.

Solution

$$\frac{X^{-1} + Y^{-1}}{X^{-2} + Y^{-2}} = \frac{\dfrac{1}{X} + \dfrac{1}{Y}}{\dfrac{1}{X^2} + \dfrac{1}{Y^2}} = \frac{\dfrac{X+Y}{XY}}{\dfrac{Y^2 + X^2}{X^2 Y^2}} = \frac{(X+Y)(X^2 Y^2)}{XY(X^2 + Y^2)} = XY\frac{X+Y}{X^2 + Y^2} \blacktriangleright$$

2-47 Wt. 3

What is the value of the determinant D when

$$D = \begin{vmatrix} 1 & 1 & 1 \\ 2 & -1 & 1 \\ 1 & 2 & -1 \end{vmatrix}$$

Solution

Expanding by minors along the top row,

$$D = \begin{vmatrix} 1 & 1 & 1 \\ 2 & -1 & 1 \\ 1 & 2 & -1 \end{vmatrix} = 1\begin{vmatrix} -1 & 1 \\ 2 & -1 \end{vmatrix} - 1\begin{vmatrix} 2 & 1 \\ 1 & -1 \end{vmatrix} + 1\begin{vmatrix} 2 & -1 \\ 1 & 2 \end{vmatrix}$$

$$= 1(1-2) - 1(-2-1) + 1(4+1)$$
$$= 1(-1) - 1(-3) + 1(+5)$$
$$= -1 + 3 + 5$$
$$= +7 \blacktriangleright$$

2-48 Wt. 6

What is the value of each of the unknowns in the following three equations?

$$\frac{L}{2} + \frac{m}{3} + \frac{n}{4} = 62$$

$$\frac{L}{4} + \frac{m}{5} + \frac{n}{6} = 38$$

$$\frac{L}{3} + \frac{m}{4} + \frac{n}{5} = 47$$

40 / Mathematics

Solution

The first step is to clear the denominator:

$$6L + 4m + 3n = 62(12) = 744 \quad \text{(A)}$$
$$15L + 12m + 10n = 38(60) = 2280 \quad \text{(B)}$$
$$20L + 15m + 12n = 47(60) = 2820 \quad \text{(C)}$$

Multiplying A by (-3) and adding B:

$$\begin{array}{r} -18L - 12m - 9n = -2232 \\ 15L + 12m + 10n = 2280 \\ \hline -3L \quad\quad + n = 48 \end{array} \quad \text{or } n = 48 + 3L$$

Multiplying A by (-4) and adding C:

$$\begin{array}{r} -24L - 16m - 12n = -2976 \\ 20L + 15m + 12n = 2820 \\ \hline -4L \quad -m \quad\quad = -156 \end{array}$$

or $\quad 4L + m = 156 \quad$ or $m = 156 - 4L$

Substituting these two equations into A we get:

$$6L + 4(156 - 4L) + 3(48 + 3L) = 744$$
$$6L + 624 - 16L + 144 + 9L = 744$$
$$-L + 768 = 744 \quad \text{or } L = 24 \blacktriangleright$$
$$m = 156 - 4(24) \quad m = 60 \blacktriangleright$$
$$n = 48 + 3(24) \quad n = 120 \blacktriangleright$$

Alternative solution

Use Cramer's rule and determinants (one must check that D is nonzero):

$$D = \begin{vmatrix} \frac{1}{2} & \frac{1}{3} & \frac{1}{4} \\ \frac{1}{4} & \frac{1}{5} & \frac{1}{6} \\ \frac{1}{3} & \frac{1}{4} & \frac{1}{5} \end{vmatrix} = \frac{1}{2}\begin{vmatrix} \frac{1}{5} & \frac{1}{6} \\ \frac{1}{4} & \frac{1}{5} \end{vmatrix} - \frac{1}{3}\begin{vmatrix} \frac{1}{4} & \frac{1}{6} \\ \frac{1}{3} & \frac{1}{5} \end{vmatrix} + \frac{1}{4}\begin{vmatrix} \frac{1}{4} & \frac{1}{5} \\ \frac{1}{3} & \frac{1}{4} \end{vmatrix}$$

$$D = -\frac{1}{1200} + \frac{1}{540} - \frac{1}{960} = -\frac{1}{43,200} \quad \frac{1}{D} = -43{,}200 \neq 0$$

$$L = D_L \frac{1}{D} \quad m = D_m \frac{1}{D}$$

$$D_L = \begin{vmatrix} 62 & \frac{1}{3} & \frac{1}{4} \\ 38 & \frac{1}{5} & \frac{1}{6} \\ 47 & \frac{1}{4} & \frac{1}{5} \end{vmatrix} = -\frac{62}{600} - \frac{38}{240} + \frac{47}{180} = -\frac{1}{1800} \quad L = 24 \blacktriangleright$$

$$D_m = \begin{vmatrix} \frac{1}{2} & 62 & \frac{1}{4} \\ \frac{1}{4} & 38 & \frac{1}{6} \\ \frac{1}{3} & 47 & \frac{1}{5} \end{vmatrix} = \frac{62}{600} + \frac{38}{60} - \frac{47}{48} = -\frac{1}{720} \quad m = 60 \blacktriangleright$$

Now substitute in the first equation of the problem

$$\frac{n}{4} = 62 - 20 - 12 = 30 \qquad n = 120 \blacktriangleright$$

Check: Use the second equation of the problem: $6 + 12 + 20 = 38$.

2-49 Wt. 3

Solve for the correct value of Y in the following system of simultaneous equations:

$$5X + 2Y + 4Z = 4 \qquad (1)$$
$$3X - Y + 2Z = -11 \qquad (2)$$
$$7X - 3Y - 3Z = 8 \qquad (3)$$

Solution

$$
\begin{array}{rrrr}
(1) & 5X + 2Y + 4Z = & 4 \\
-2 \times (2) & -6X + 2Y - 4Z = & 22 \\
\hline
(4) & -X + 4Y = & 26 \\
\end{array}
$$

$$
\begin{array}{rrrr}
3 \times (2) & 9X - 3Y + 6Z = & -33 \\
2 \times (3) & 14X - 6Y - 6Z = & 16 \\
\hline
(5) & 23X - 9Y = & -17 \\
\end{array}
$$

$$
\begin{array}{rrrr}
23 \times (4) & -23X + 92Y = & 598 \\
(5) & 23X - 9Y = & -17 \\
\hline
& 83Y = & 581 \qquad Y = 7 \blacktriangleright \\
\end{array}
$$

(Additional calculations show that $X = 2$ and $Z = -5$.)

Alternative solution

The problem can be solved by determinants:

$$Y = \frac{\begin{vmatrix} 5 & 4 & 4 \\ 3 & -11 & 2 \\ 7 & 8 & -3 \end{vmatrix}}{\begin{vmatrix} 5 & 2 & 4 \\ 3 & -1 & 2 \\ 7 & -3 & -3 \end{vmatrix}} = \frac{5\begin{vmatrix} -11 & 2 \\ 8 & -3 \end{vmatrix} - 3\begin{vmatrix} 4 & 4 \\ 8 & -3 \end{vmatrix} + 7\begin{vmatrix} 4 & 4 \\ -11 & 2 \end{vmatrix}}{5\begin{vmatrix} -1 & 2 \\ -3 & -3 \end{vmatrix} - 3\begin{vmatrix} 2 & 4 \\ -3 & -3 \end{vmatrix} + 7\begin{vmatrix} 2 & 4 \\ -1 & 2 \end{vmatrix}}$$

$$Y = \frac{5[33 - 16] - 3[-12 - 32] + 7[8 - (-44)]}{5[3 - (-6)] - 3[-6 - (-12)] + 7[4 - (-4)]}$$

$$Y = \frac{5(17) - 3(-44) + 7(52)}{5(9) - 3(6) + 7(8)} = \frac{85 + 132 + 364}{45 - 18 + 56} = \frac{581}{83} = 7 \blacktriangleright$$

42 / Mathematics

2-50 Wt. 1

The value of the determinant $\begin{vmatrix} 1 & 1 & 0 & 0 \\ 1 & 1 & 1 & 0 \\ 0 & 1 & 1 & 1 \\ 0 & 0 & 1 & 1 \end{vmatrix}$ is

(a) +1
(b) 0
(c) −2
(d) −1
(e) none of the above values

Solution

Expanding by minors,

$$\begin{vmatrix} 1 & 1 & 0 & 0 \\ 1 & 1 & 1 & 0 \\ 0 & 1 & 1 & 1 \\ 0 & 0 & 1 & 1 \end{vmatrix} = 1\begin{vmatrix} 1 & 1 & 0 \\ 1 & 1 & 1 \\ 0 & 1 & 1 \end{vmatrix} - 1\begin{vmatrix} 1 & 1 & 0 \\ 0 & 1 & 1 \\ 0 & 1 & 1 \end{vmatrix}$$

$$= 1\begin{vmatrix} 1 & 1 \\ 1 & 1 \end{vmatrix} - 1\begin{vmatrix} 1 & 1 \\ 0 & 1 \end{vmatrix} - 1\begin{vmatrix} 1 & 1 \\ 1 & 1 \end{vmatrix} + 1\begin{vmatrix} 0 & 1 \\ 0 & 1 \end{vmatrix}$$

$$= -1(1) + 1(0) = -1 \blacktriangleright$$

Answer is (d)

2-51 Wt. 2

If 7 men can build 5 boats in 3 days, how many men would be required to build 15 boats in 4 days?

Solution

$$\frac{\text{man-days}}{\text{boat}} = \frac{7(3)}{5} = 4.2$$

For 15 boats in 4 days:

$$\frac{\text{men}(4)}{15} = 4.2$$

$$\text{men} = \frac{4.2(15)}{4} = 15.75$$

Therefore 16 men are required ▶

2-52 Wt. 1

A certain job can be performed by Group X in 100 hr. Group Y can perform the same job in 25 hr, and Group Z requires 20 hr. If the three groups, X, Y, and Z, work together, the number of hours required to complete the job is nearest

(a) 8
(b) 10
(c) 12
(d) 14
(e) 16

Solution

Let N = number of hours to complete the job. The hourly progress is:

$$\text{Group } X = \frac{N}{100}$$

$$\text{Group } Y = \frac{N}{25}$$

$$\text{Group } Z = \frac{N}{20}$$

The combined effort is:

$$\frac{N}{100} + \frac{N}{25} + \frac{N}{20} = 1 \qquad (0.01 + 0.04 + 0.05)N = 1$$

$$N = \frac{1}{0.10} = 10 \text{ hr} \blacktriangleright$$

Answer is (b)

2-53 Wt. 3

The area of a right triangle is 210 ft². If the hypotenuse is 29 ft, what are the lengths of the other two sides?

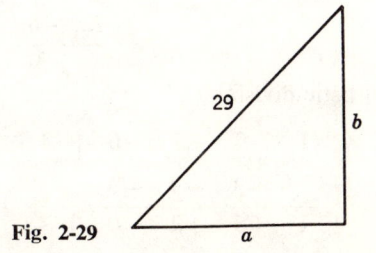

Fig. 2-29

Solution

$$\tfrac{1}{2}ab = 210 \quad (1)$$
$$a^2 + b^2 = 29^2 \quad (2)$$

From equation (1) $a = 420/b$. Inserting this in (2), we have

$$\left(\frac{420}{b}\right)^2 + b^2 = 29^2$$

Let $B = b^2$. Then

$$B^2 - 29^2 B + (420)^2 = 0$$
$$B = \tfrac{1}{2}[29^2 \pm \sqrt{(29^2)^2 - 4(420)^2}]$$
$$= \tfrac{1}{2}[841 \pm \sqrt{841^2 - 840^2}]$$
$$= \tfrac{1}{2}[841 \pm \sqrt{(841 - 840)(841 + 840)}]$$
$$= \tfrac{1}{2}[841 \pm 41] = 441, 400$$

and

$$b = \sqrt{B} = 21, 20$$

Hence

$$a = 420/b = 20, 21$$

The lengths of the other two sides are 20 and 21 ft ▶

2-54 Wt. 3

If 3 is a root of the equation $x^3 - 7x - 6 = 0$, find the remaining roots.

Solution

If $x = 3$ is a root, then $(x - 3)$ is a root of the equation. By long division,

$$\begin{array}{r}
x^2 + 3x + 2 \\
x - 3 \overline{\smash{)}x^3 - 7x - 6} \\
\underline{x^3 - 3x^2} \\
3x^2 \\
\underline{3x^2 - 9x} \\
2x \\
\underline{2x - 6} \\
0
\end{array}$$

Alternatively, by synthetic division,

$$\begin{array}{rrrr|r}
1 & 0 & -7 & -6 & \underline{-3} \\
  & -3 & -9 & -6 & \\ \hline
1 & +3 & +2 & 0 &
\end{array}$$

or $x^2 + 3x + 2$ (+0 remainder). Factoring $x^2 + 3x + 2 = 0$, we obtain

$$(x + 2)(x + 1) = 0$$

Therefore the remaining roots are $x = -2$ and $x = -1$ ▶

2-55 Wt. 1

The surface area of a tetrahedron is described by

(a) 4 equilateral triangles
(b) 6 squares
(c) 12 pentagons
(d) 3 trapeziums
(e) 8 pentagons

Solution

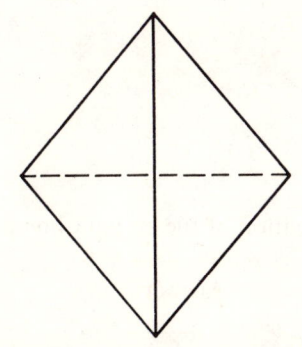

Fig. 2-30

A tetrahedron (triangular pyramid) is bounded by four equilateral triangles ▶

Answer is (a)

2-56 Wt. 1

In construction terminology, a hyperbolic paraboloid refers to a

(a) structure curved in one direction
(b) waterway of uniform channel cross section
(c) special type of concrete tower
(d) structure curved in two directions
(e) hydraulic structure with no curved surfaces

Solution

A hyperbolic paraboloid is a warped surface generated by a line touching two skew lines and remaining parallel to a plane not parallel to either line.

46 / Mathematics

Parabola — Hyperbola
Hyperbolic Paraboloid

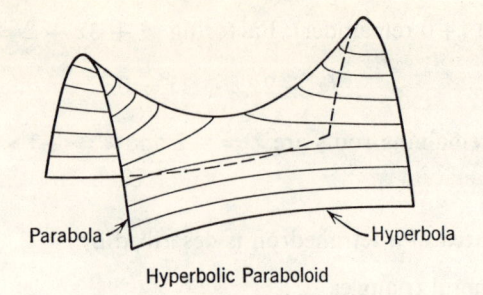

Fig. 2-31

Answer is (d)

2-57 Wt. 1

The equation $y = mx + b$ is an algebraic expression for a

(a) parabola
(b) straight line
(c) circular arc
(d) hyperbola
(e) sine curve

Solution

$y = mx + b$ is the equation of the straight line passing through the point $(0, b)$ with slope m ▶

Answer is (b)

2-58 Wt. 1

The equation of a straight line which has a slope of $+2$ and passes through a point with x and y coordinates of 4 and 5, respectively, is

(a) $x + 2y = 14$
(b) $xy = 20$
(c) $2x + y = 13$
(d) $4y = 5x$
(e) $y = 2x - 3$

Solution

The point-slope equation for a straight line is $y - y_1 = m(x - x_1)$, where m is the slope and x_1 and y_1 are the coordinates of a point on the line.

$$y - 5 = 2(x - 4) \qquad y - 5 = 2x - 8 \qquad y = 2x - 3 \blacktriangleright$$

Answer is (e)

2-59 Wt. 5

Find the equation (in rectangular coordinates) of the plane passing through these three points:

$$(1, 3, 5) \quad (2, 4, 4) \quad \text{and} \quad (3, 4, 2)$$

Solution

The general equation of a plane may be written

$$Ax + By + Cz + D = 0$$

Substituting the three points:

$$A + 3B + 5C + D = 0 \tag{1}$$
$$2A + 4B + 4C + D = 0 \tag{2}$$
$$3A + 4B + 2C + D = 0 \tag{3}$$

Solving for A, B, and D in terms of C, we have:

Equation (3) − (2),

$$A - 2C = 0 \quad A = 2C \tag{4}$$

Equation (3) − (1),

$$2A + B - 3C = 0 \tag{5}$$

Substituting (4) into (5),

$$2(2C) + B - 3C = 0 \quad B = -C \tag{6}$$

Substituting (4) and (6) back into (1) gives

$$2C + 3(-C) + 5C + D = 0 \quad D = -4C$$

From the general equation

$$2Cx - Cy + Cz - 4C = 0$$
$$C(2x - y + z - 4) = 0$$

The equation of the plane is $2x - y + z - 4 = 0$ ▶

2-60 Wt. 1

The curve which is represented by the equation $\dfrac{x^2}{a^2} - \dfrac{y^2}{b^2} = 1$ is a

(a) straight line
(b) circle
(c) ellipse
(d) parabola
(e) hyperbola

48 / Mathematics

Solution

The simple equations for the various curves are:

$$\text{straight line} \quad y = mx + b$$
$$\text{circle} \quad x^2 + y^2 = a^2$$
$$\text{ellipse} \quad \frac{x^2}{a^2} + \frac{y^2}{b^2} = 1$$
$$\text{parabola} \quad y^2 = ax$$
$$\blacktriangleright \text{hyperbola} \quad \frac{x^2}{a^2} - \frac{y^2}{b^2} = 1$$

Answer is (*e*)

2-61 Wt. 1

$b^2x^2 - a^2y^2 - a^2b^2 = 0$ is the equation of a

(*a*) parabola
(*b*) ellipse
(*c*) circle
(*d*) hyperbola
(*e*) straight line

Solution

Divide $b^2x^2 - a^2y^2 - a^2b^2 = 0$ through by (a^2b^2) to obtain

$$\frac{x^2}{a^2} - \frac{y^2}{b^2} - 1 = 0$$

This is the general equation of a hyperbola ▶

Answer is (*d*)

2-62 Wt. 6

(*a*) The figure shown is a parabola. Show on the figure all of the following:

(1) focus
(2) latus rectum
(3) vertex
(4) principal axis
(5) directrix

(*b*) What relationship, if any, exists between the focal length and the latus rectum?

(*c*) Locate the horizontal offsets to points *a*, *b*, and *c* on the curve, if point *d* is located as shown.

Fig. 2-32

Solution

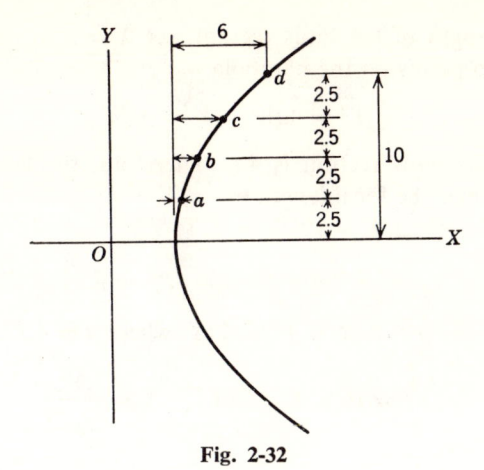

(a)

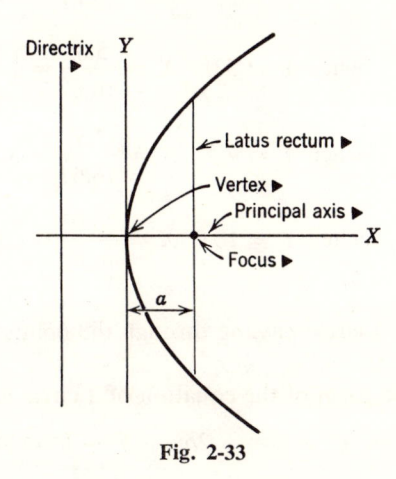

Fig. 2-33

(Note that the Y-axis has been shifted in this illustration.)

A parabola is generated when a point moves so that its distance from a fixed line and a fixed point is equal. The fixed line is called the *directrix* and the fixed point the *focus*. The *principal axis* is the axis of symmetry. The point midway between the focus and directrix is called the *vertex*. The chord drawn through the focus parallel to the directrix is the *latus rectum*.

(b) As drawn, the equation for a parabola with the X-axis as its principal axis and its vertex at the origin is

$$Y^2 = 4aX$$

where a, the focal length, is the distance from the vertex to the focus. To

50 / Mathematics

determine the length of the latus rectum, set $X = a$ in the equation, and solve for the two points on the parabola.

$$Y^2 = 4a^2 \qquad Y = \pm 2a$$

The length of the latus rectum is $4a$. Therefore, the length of the latus rectum is four times the focal length ▶

(c) In our case, when $Y = 10$, $X = 6$, so

$$10^2 = 4a(6) \qquad a = \tfrac{100}{24} = 4.17$$

The equation of the parabola is $Y^2 = 4aX$, where $a = 4.17$, so

$$Y^2 = 16.67X \qquad \text{or} \qquad X = \frac{Y^2}{16.67}$$

Horizontal offsets:

$$\text{point } a \qquad \text{when } Y = 2.5, \quad X = \frac{2.5^2}{16.67} = 0.375 \; \blacktriangleright$$

$$\text{point } b \qquad \text{when } Y = 5.0, \quad X = \frac{5.0^2}{16.67} = 1.5 \; \blacktriangleright$$

$$\text{point } c \qquad \text{when } Y = 7.5, \quad X = \frac{7.5^2}{16.67} = 3.37 \; \blacktriangleright$$

$$\text{when } Y = 10.0, \quad X = \frac{10.0^2}{16.67} = 6.0; \text{ check.}$$

2-63 Wt. 5

Find the center of a circle passing through the points $(3, 1)$, $(-1, 2)$, and $(-2, -2)$.

Given: The general form of the equation of a circle is

$$x^2 + y^2 - 2ax - 2by + a^2 + b^2 = R^2$$

where (a, b) is the center and R is the radius.

Solution

Substitute for x and y the values $(x, y) = (3, 1)$, $(-1, 2)$, $(-2, -2)$ into $x^2 + y^2 - 2ax - 2by + a^2 + b^2 = R^2$ and solve the three equations for a and b. Note that the general equation given is the same as $(x - a)^2 + (y - b)^2 = R^2$.

Substituting the values of x and y:

point $(3, 1)$	$9 + 1 - 6a - 2b + a^2 + b^2 = R^2$	(1)
point $(-1, 2)$	$1 + 4 + 2a - 4b + a^2 + b^2 = R^2$	(2)
point $(-2, -2)$	$4 + 4 + 4a + 4b + a^2 + b^2 = R^2$	(3)

Equation (1) − (2),
$$8a - 2b = 5 \quad (4)$$
Equation (1) − (3),
$$10a + 6b = 2 \quad (5)$$
Multiplying (4) by 3 and adding it to (5) gives
$$34a = 17 \quad a = \tfrac{1}{2} \blacktriangleright$$
Substituting $a = \tfrac{1}{2}$ back into (4) gives
$$4 - 2b = 5$$
$$-2b = 1 \quad b = -\tfrac{1}{2} \blacktriangleright$$
Thus the center of the circle is at $(\tfrac{1}{2}, -\tfrac{1}{2})$ ▶

2-64 Wt. 3

Derive the equation of the largest circle that is tangent to both coordinate axes and has its center on the line: $2X + Y - 6 = 0$.

Solution

From the problem statement we see that the center of the circle is at the intersection of two straight lines: $2X + Y - 6 = 0$ and $X - Y = 0$ or $X + Y = 0$. Solving the two cases we get:

$$\begin{array}{r} 2X + Y - 6 = 0 \\ X - Y = 0 \\ \hline 3X - 6 = 0 \\ X = 2 \\ Y = 2 \end{array}$$

The circle has its center at (2, 2) with radius = 2.

$$\begin{array}{r} 2X + Y - 6 = 0 \\ -X - Y = 0 \\ \hline X - 6 = 0 \\ X = 6 \\ Y = -6 \end{array}$$

The circle has its center at (6, −6) with radius = 6.

The problem calls for the largest circle; hence the second case is the desired solution. See Fig. 2-34 on the next page.

The general equation for a circle is $(X - a)^2 + (Y - b)^2 = R^2$. Substituting, we get $(X - 6)^2 + (Y + 6)^2 = 6^2$. Therefore the equation of the circle is
$$(X - 6)^2 + (Y + 6)^2 = 36 \blacktriangleright$$

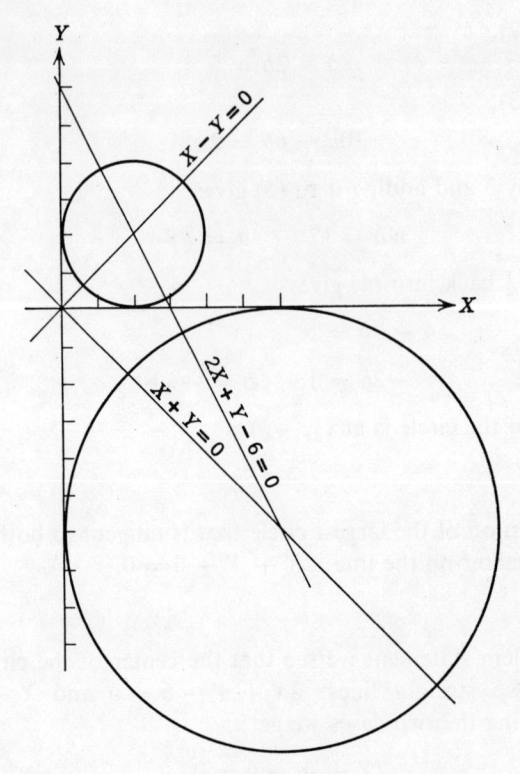

Fig. 2-34

2-65 Wt. 5

The cable of a suspension bridge hangs in the shape of an arc of a parabola (AB). The supporting towers are 70 ft high and 200 ft apart and the lowest point on the cable is 20 ft above the roadway. Find the length of the supporting rod (L) 50 ft from the middle of the bridge.

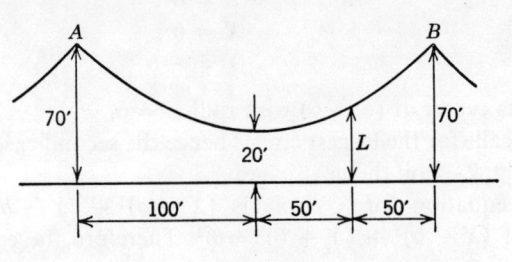

Fig. 2-35

Solution

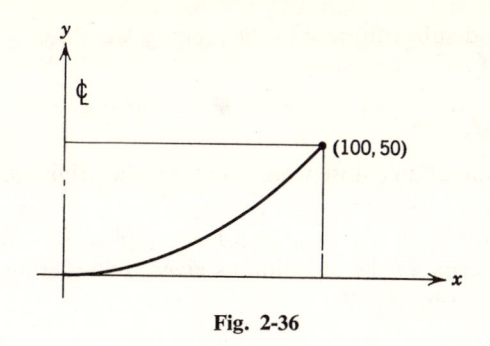

Fig. 2-36

When $x = \pm 100$, $y = 70 - 20 = 50$

$$y = kx^2$$
$$50 = k(100)^2$$

Therefore $k = 0.005$ and $y = 0.005x^2$.
When $x = 50$ ft,

$$y = 0.005(50)^2 = 0.005(2500) = 12.5 \text{ ft}$$

The length of the supporting rod is

$$L = 12.5 + 20 = 32.5 \text{ ft} \blacktriangleright$$

2-66 Wt. 3

The equations of an ellipse in parametric form are:

$$x = a \cos \theta \qquad y = b \sin \theta$$

Find the equation of the tangent line to the curve at the point ($x = 1, y = 2$) in terms of the constants a and b.

Solution

If x and y are rectangular coordinates and each is expressed in terms of a variable or parameter, a parametric equation results. To obtain the rectangular equation from the parametric equations, the parameter (θ) must be eliminated.

Remembering that $\cos^2 \theta + \sin^2 \theta = 1$, we solve the first equation for $\cos \theta$ and the second for $\sin \theta$:

$$\cos \theta = \frac{x}{a} \qquad \sin \theta = \frac{y}{b}$$

54 / Mathematics

Squaring both and substituting into the identity we obtain

$$\frac{x^2}{a^2} + \frac{y^2}{b^2} = 1 \quad \text{or} \quad b^2x^2 + a^2y^2 = a^2b^2$$

This is the equation of an ellipse whose center is the origin and whose focuses are on the x-axis.

In the equation $b^2x^2 + a^2y^2 = a^2b^2$, y is an implicit function of x. We will differentiate the terms of the equation as given, considering y as a function of x, and then solve for dy/dx.

$$\frac{d}{dx}(b^2x^2) + \frac{d}{dx}(a^2y^2) - \frac{d}{dx}(a^2b^2) = 0 \quad 2b^2x + 2a^2y\frac{dy}{dx} = 0$$

$$\frac{dy}{dx} = \frac{-2b^2x}{2a^2y} = -\frac{b^2x}{a^2y}$$

Since dy/dx is the slope of the line tangent to the curve at a point, the slope $m = (-b^2x)/(a^2y)$.

The equation of the tangent line is $y - y_1 = m(x - x_1)$.

$$y - y_1 = \frac{-b^2x}{a^2y}(x - x_1) \quad a^2y^2 - a^2y_1y = -b^2x^2 + b^2x_1x$$

$$b^2x_1x + a^2y_1y = a^2y^2 + b^2x^2$$

But $a^2y^2 + b^2x^2 = a^2b^2$ from the original equation, so $b^2x_1x + a^2y_1y = a^2b^2$. When $x_1 = 1$ and $y_1 = 2$, the equation of the tangent line is

$$b^2x + 2a^2y = a^2b^2 \blacktriangleright$$

2-67 Wt. 1

In the equation $x = \dfrac{t^2 + t}{2t^2 + 1}$, the limit of x as t approaches infinity is

(a) ∞
(b) 2
(c) 1
(d) $\frac{1}{2}$
(e) 0

Solution

$$\lim_{t \to \infty} \frac{t^2 + t}{2t^2 + 1} = ?$$

The usual rule is to divide both the numerator and denominator by the highest power of the variable occurring in either. In this case, divide by t^2.

This step is valid since $t \neq 0$.

$$\lim_{t \to \infty} \frac{t^2 + t}{2t^2 + 1} = \lim_{t \to \infty} \frac{1 + (1/t)}{2 + (1/t^2)} = \frac{1}{2} \blacktriangleright$$

The limit of each term in the numerator and denominator containing t is zero. $(1/\infty = 0.)$

Answer is (d)

2-68 Wt. 2

In the equation $y = \dfrac{-x^3 + 3x + 2}{x^2 + 2x + 1}$, the limit of y as x approaches a value of -1 is

(a) 0
(b) 1
(c) 2
(d) 3
(e) ∞

Solution

$$\lim_{x \to -1} \frac{-x^3 + 3x + 2}{x^2 + 2x + 1} = \frac{-(-1)^3 + 3(-1) + 2}{(-1)^2 + 2(-1) + 1} = \frac{+1 - 3 + 2}{+1 - 2 + 1} = \frac{0}{0}$$

which is indeterminate.

Not obtaining a solution above, we try factoring the numerator and denominator:

$$\lim_{x \to -1} \frac{-x^3 + 3x + 2}{x^2 + 2x + 1} = \lim_{x \to -1} \frac{(x + 1)(x + 1)(-x + 2)}{(x + 1)(x + 1)}$$
$$= \lim_{x \to -1} (-x + 2) = +3 \blacktriangleright$$

We find that y is not continuous at $x = -1$ but does approach a limit of $+3$ as x approaches -1.

Answer is (d)

2-69 Wt. 1

In the equation of $y = \dfrac{\ln(1 - z)}{z}$, the limit of y as z approaches a value of zero is

(a) ∞
(b) 3
(c) 1
(d) 0
(e) -1

56 / Mathematics

Solution

$$\lim_{z \to 0} y = \frac{\ln(1-z)}{z} = \frac{\ln(1-0)}{0} = \frac{0}{0}$$

The form is indeterminate. Applying L'Hospital's rule,

$$\lim_{z \to 0} \frac{\ln(1-z)}{z} = \lim_{z \to 0} \frac{\frac{d}{dz}[\ln(1-z)]}{\frac{d}{dz}(z)} = \lim_{z \to 0} \frac{\frac{-1}{1-z}}{1}$$

$$= \lim_{z \to 0} \frac{-1}{1-z} = -1 \blacktriangleright$$

Answer is (e)

2-70 Wt. 1

The slope of the curve $y = x^3 - 4x$ as it passes through the origin ($x = 0; y = 0$) is equal to

(a) +4
(b) +2
(c) 0
(d) −2
(e) −4

Solution

$$\left.\frac{dy}{dx}\right|_{x=0} = [3x^2 - 4]_{x=0} = -4$$

Therefore the slope of the curve at $x = 0$ is -4 ▶

Answer is (e)

2-71 Wt. 1

The vertex of the parabola $2y^2 - 9x - 12y = 0$ is at the point

(a) (−2, 3)
(b) (−7/8, 3)
(c) (0, 0)
(d) (0, 6)
(e) (∞, ∞)

Solution

Since y, not x, is squared, the principal axis of the parabola is horizontal. At the vertex the slope of the parabola is then vertical, that is, $dx/dy = 0$

at that point.
$$2y^2 - 9x - 12y = 0$$

Regarding x as a function of y and differentiating implicitly,

$$4y - 9\frac{dx}{dy} - 12 = 0$$

$$\frac{dx}{dy} = \frac{4y - 12}{9} = 0$$

$$y = 3$$

Substituting $y = 3$ into the original equation,

$$2(3)^2 - 9x - 12(3) = 0$$

$$x = -2$$

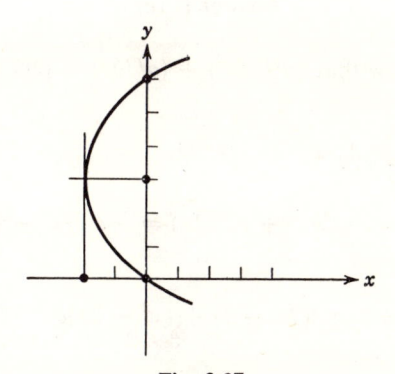

Fig. 2-37

The vertex is thus at $x = -2, y = 3$ ▶

Answer is (a)

2-72 Wt. 1

If the first derivative of the equation of a curve is a constant, the curve is a

(a) circle
(b) hyperbola
(c) parabola
(d) straight line
(e) sine wave

Solution

If $dy/dx = m$, $y = \int m \, dx = m \int dx = mx + b$, so $y = mx + b$, a straight line ▶

Answer is (d)

58 / Mathematics

2-73 Wt. 1

For the position-time function $x = 3t^2 + 2t$, the velocity in the x direction at $t = 1$ is

(a) 4
(b) 5
(c) 6
(d) 7
(e) 8

Solution

$$\text{Velocity} = \frac{dx}{dt} = [6t + 2]_{t=1} = 6 + 2 = 8 \blacktriangleright$$

Answer is (e)

2-74 Wt. 3

If $f(x) = x^2 e^{-x}$, show that $(d^2f/dx^2) + (df/dx) - [2(1 - x)/x^2]f = 0$.

Solution

$$f(x) = x^2 e^{-x}$$

$$\frac{df}{dx} = -x^2 e^{-x} + e^{-x}(2x) = -e^{-x}(x^2 - 2x)$$

$$\frac{d^2f}{dx^2} = -e^{-x}(2x - 2) + (x^2 - 2x)e^{-x}$$

Therefore

$$\frac{d^2f}{dx^2} + \frac{df}{dx} = -e^{-x}(x^2 - 2x + 2x - 2 - x^2 + 2x) = -e^{-x}(2x - 2)$$

If

$$\frac{d^2f}{dx^2} + \frac{df}{dx} - \frac{2(1-x)}{x^2}f = 0$$

then we must have

$$-e^{-x}(2x - 2) = \frac{2(1-x)}{x^2}(x^2 e^{-x})$$

Finally,

$$2(1 - x)e^{-x} = 2(1 - x)e^{-x} \quad \text{Q.E.D.} \blacktriangleright$$

2-75 Wt. 3

The area of a circle is 400 ft². What is the area of an inscribed square?

Solution

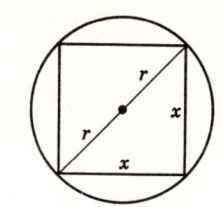

Fig. 2-38

$$x^2 + x^2 = (2r)^2 \qquad \sqrt{2}\, x = 2r \qquad x = \frac{2}{\sqrt{2}} r$$

$$A_\odot = \pi r^2$$

Therefore

$$r = \sqrt{\frac{A}{\pi}} = \left(\frac{400}{\pi}\right)^{1/2} = 11.28 \text{ ft}$$

$$x = \frac{2}{\sqrt{2}} (11.28) = 15.95 \text{ ft}$$

$$A_\square = x^2 = (15.95)^2 = 254.5 \text{ ft}^2 \blacktriangleright$$

2-76 Wt. 5

The stiffness of a rectangular timber is proportional to the width and the cube of the depth. Find the dimensions of the cross section of the stiffest beam that can be made out of a circular log whose diameter is 20 in.

Solution

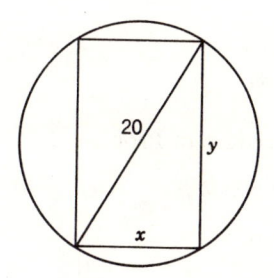

Fig. 2-39

Here x = beam width, y = beam depth, and log diameter = 20 in.

To maximize stiffness, the function xy^3 must be maximized. From the figure, $y = (20^2 - x^2)^{1/2}$. Writing xy^3 as a function of one variable,

$$xy^3 = x(20^2 - x^2)^{3/2}$$

$$f(x) = x(20^2 - x^2)^{3/2} \qquad \text{for } 0 \leq x \leq 20$$

Since $f(0) = f(20) = 0$ and $f(x) > 0$ for intermediate values of x, it is clear that $f(x)$ must attain a maximum value. To find the maximum, we

60 / Mathematics

set $f'(x) = 0$ and solve for x. [Alternatively, we could check that $f(x)$ is indeed maximized by verifying that $f''(x) < 0$ at that point.] Using the chain rule for differentiation of a product,

$$f'(x) = x\tfrac{3}{2}(20^2 - x^2)^{\frac{1}{2}} \times (-2x) + (20^2 - x^2)^{\frac{3}{2}} \times (1)$$
$$= -3x^2(20^2 - x^2)^{\frac{1}{2}} + (20^2 - x^2)^{\frac{3}{2}} = 0$$

Dividing by $(20^2 - x^2)^{\frac{1}{2}}$, we get $-3x^2 + 20^2 - x^2 = 0$, $-4x^2 + 20^2 = 0$, and $x^2 = 400/4 = 100$. Then

$$x = 10 \text{ in.} \blacktriangleright$$

Also, $y = (20^2 - 10^2)^{\frac{1}{2}} = (300)^{\frac{1}{2}}$.

$$y = 17.32 \text{ in.} \blacktriangleright$$

Answer: Cut beam 10 in. × 17.32 in.

2-77 Wt. 3

A bin with a square base, straight sides, and no top is to be constructed from 432 ft² of lumber. Find the dimensions of the bin such that the capacity of the bin is a maximum. What is the maximum volume?

Solution

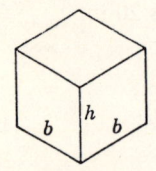

Fig. 2-40

We follow the procedure of Problem 2-76.

$$\text{Volume} = b^2 h \qquad \text{Area} = b^2 + 4bh = 432 \text{ ft}^2$$

From the area relation,

$$h = \frac{108}{b} - \frac{b}{4}$$

then

$$V = b^2 \left(\frac{108}{b} - \frac{b}{4} \right) = 108b - \frac{b^3}{4}$$

Taking the first derivative and equating it to zero,

$$\frac{dV}{db} = 108 - \tfrac{3}{4}b^2 = 0$$

For maximum volume,

$$b = \sqrt{\tfrac{4}{3}(108)} = 12 \text{ ft}$$
$$h = \tfrac{108}{12} - \tfrac{12}{4} = 9 - 3 = 6 \text{ ft}$$
$$V = b^2 h = (12)^2(6) = 864 \text{ ft}^3 \blacktriangleright$$

Checking the second derivative, $d^2V/db^2 = -\tfrac{3}{2}b < 0$ for all values of b; hence V is a maximum.

2-78 Wt. 4

Circular cylindrical cans of volume V_0 are to be manufactured with both ends closed. Determine the ratio between the diameter and height that will require the minimum amount of metal to make each can.

Solution

$$\text{Total surface area} = 2 \text{ end areas} + \text{side surface area}$$
$$= 2\left(\frac{\pi}{4}d^2\right) + \pi\,dh$$

The surface area is to be a minimum. For min/max values we must find the first derivative of the function. But the function contains two variables (d and h), so one must be defined in terms of the other to eliminate one variable.

$$V_0 = \frac{\pi}{4}d^2 h$$

Therefore

$$h = \frac{4V_0}{\pi d^2}$$

Hence

$$\text{Surface area} = 2\left(\frac{\pi}{4}d^2\right) + \pi d\left(\frac{4V_0}{\pi d^2}\right) = \frac{\pi}{2}d^2 + \frac{4V_0}{d}$$

Taking the first derivative and equating it to zero,

$$f'(d) = \frac{\pi}{2}(2d) + 4V_0\left(\frac{-1}{d^2}\right) = 0$$

$$\pi d = \frac{4V_0}{d^2} \quad \text{and} \quad d^3 = \frac{4V_0}{\pi}$$

[Check: $f''(d) = \pi + 8V_0/d^3 > 0$, therefore minimum.]

62 / Mathematics

Since $V_0 = (\pi/4) d^2 h$, we also have $d^2 h = 4V_0/\pi$. Therefore

$$d^2 h = d^3 \quad \text{or} \quad h = d$$

$$\text{Ratio } d/h = 1 \blacktriangleright$$

2-79 Wt. 2

Compute the area under the curve $y = x^2$ between the values $x = +1$ ft and $x = +7$ ft.

Solution

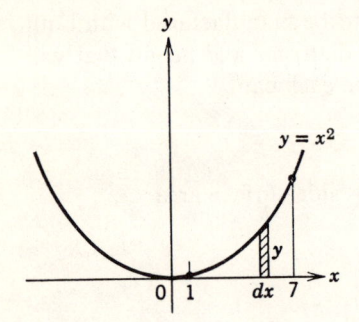

Fig. 2-41

$$dA = y\, dx$$

$$A = \int dA = \int y\, dx$$

Substituting for y in terms of x,

$$A = \int_1^7 x^2\, dx = \frac{x^3}{3}\bigg|_1^7$$

$$= \tfrac{343}{3} - \tfrac{1}{3} = \tfrac{342}{3}$$

$$= 114 \text{ ft}^2 \blacktriangleright$$

2-80 Wt. 4

Find the area bounded by the parabola $y^2 = 2x$ and the line $x = 8$.

Solution

From figure 2-42 $dA = (8 - x)\, dy$. Expressing x as a function of y,

$$\text{Area} = 2\int_0^{+4} (8 - \tfrac{1}{2}y^2)\, dy = 2[8y - \tfrac{1}{6}y^3]_0^4$$

$$= 2(32 - \tfrac{64}{6}) = 2(21\tfrac{1}{3}) = 42\tfrac{2}{3} \blacktriangleright$$

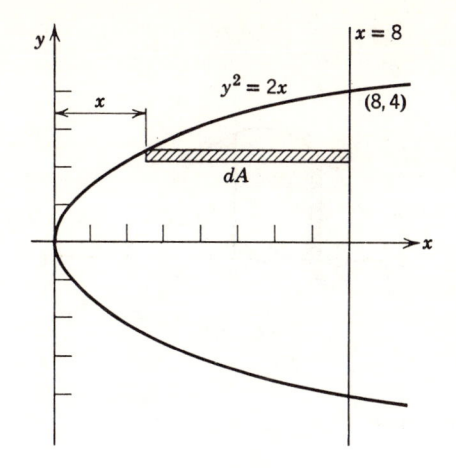

Fig. 2-42

2-81 Wt. 3

Find the area formed by the boundaries $y = 1$, $x = 1$, and $y = e^{-x}$.

Solution

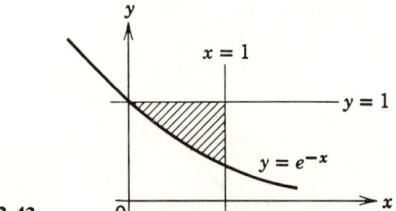

Fig. 2-43

$$\text{Area} = 1^2 - \int_0^1 e^{-x}\,dx = 1 - [-e^{-x}]\Big|_0^1 = 1 - [-e^{-1} - (-e^{-0})]$$

$$= 1 + e^{-1} - 1 = e^{-1}$$

$$= \frac{1}{2.718} = 0.368 \blacktriangleright$$

2-82 Wt. 6

The equation of a certain curve is $y = 10x - x^2$. Calculate

(a) the area between the curve and the x-axis from $x = 0$ to $x = 6$
(b) the x coordinate of the centroid of the area described in (a)

64 / Mathematics

Solution

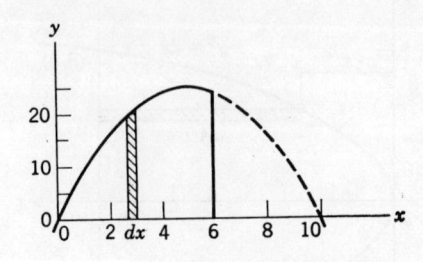

Fig. 2-44

The curve is a parabola.

$$\text{Area} = \int_0^6 y \, dx = \int_0^6 (10x - x^2) \, dx = \left[5x^2 - \frac{x^3}{3}\right]_0^6 = 180 - 72 = 108 \blacktriangleright$$

$$A\bar{x} = \int_0^6 xy \, dx = \int_0^6 x(10x - x^2) \, dx = \int_0^6 (10x^2 - x^3) \, dx$$

$$= \left[\frac{10x^3}{3} - \frac{x^4}{3}\right]_0^6 = 720 - 324 = 396$$

$$\bar{x} = \frac{A\bar{x}}{A}$$

$$= \tfrac{396}{108} = 3.67 \blacktriangleright$$

2-83 Wt. 3

Prove mathematically that the average value of the amplitude of a half sine-wave loop is $2/\pi$ times the maximum value.

Solution

$$\bar{Y} = \frac{\int y \, d\theta}{\int d\theta} = \frac{\int_0^\pi A \sin \theta \, d\theta}{\int_0^\pi d\theta} = \frac{-A[\cos \theta]_0^\pi}{\pi} = \frac{-A(-1 - 1)}{\pi}$$

$$\bar{Y} = \frac{2}{\pi} A \blacktriangleright$$

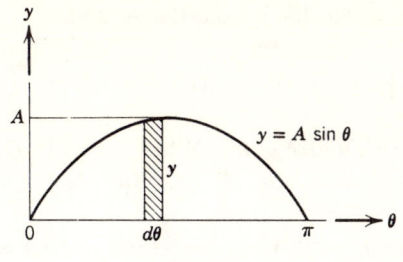

Fig. 2-45

2-84 Wt. 4

The expansion for cos x is

$$\cos x = 1 - \frac{x^2}{2!} + \frac{x^4}{4!} - \frac{x^6}{6!} + \cdots$$

Determine the expansion for sin x to the seventh power of x.

Solution

If we integrate the series we get

$$\int \cos x \, dx = \int 1 \, dx - \int \frac{x^2}{2!} \, dx + \int \frac{x^4}{4!} \, dx - \int \frac{x^6}{6!} \, dx + \cdots$$

$$\sin x = x - \frac{x^3}{3!} + \frac{x^5}{5!} - \frac{x^7}{7!} + \cdots \blacktriangleright$$

2-85 Wt. 6

An open-topped hemispherical bowl, with an inside radius of 10 ft, is filled with water. The water flows out of a hole in the bottom. The rate of flow is 5 cfm at the instant the water level has dropped 4 ft. What is the rate of change of the height of the water surface at this instant in feet per minute?

Solution

We want to find dh/dt when $h = 6$ ft. Consider the volume V of water in the bowl as a solid of revolution. Revolving the area under the curve

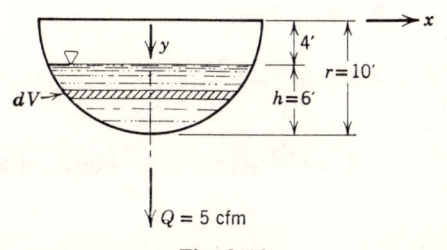

Fig. 2-46

66 / Mathematics

$x = f(y) = (r^2 - y^2)^{1/2}$ about the y-axis, the volume is

$$V = \int dV = \int \pi x^2 \, dy = \pi \int_{r-h}^{r} (r^2 - y^2) \, dy = \frac{\pi h^2}{3}(3r - h)$$

Since $r = 10$ ft, $V = (\pi/3)(30h^2 - h^3)$. When $h = 6$ ft, $Q = dV/dt = 5$ cfm. By the chain rule of differentiation, $dV/dt = (dV/dh)(dh/dt)$.

$$\left.\frac{dV}{dh}\right|_{h=6 \text{ ft}} = \frac{\pi}{3}(60h - 3h^2)\bigg|_{h=6 \text{ ft}} = \frac{\pi}{3}(360 - 108) = 84\pi \text{ ft}^3/\text{ft}$$

Now

$$84\pi \frac{dh}{dt} = 5$$

$$\frac{dh}{dt} = \frac{5}{84\pi} = 0.019 \text{ fpm} \blacktriangleright$$

Alternative solution

Circle

$$x^2 + y^2 = r^2$$

$$x^2 + y^2 = 100$$

When water has dropped 4 ft, $y = 4$, so

$$x^2 = 100 - 16 = 84$$

The rate of flow $Q = \text{Area}(A) \times \text{Velocity}(v)$. Area $= \pi x^2$. Thus the rate of change of height (v) is

$$\frac{Q}{A} = \frac{5}{\pi x^2} = \frac{5}{\pi(84)} = 0.019 \text{ fpm} \blacktriangleright$$

2-86 Wt. 4

If $u = e^{3y} \cos 2x$, what is du/dt if both x and y are functions of t?

Solution

$$\frac{du}{dt} = \frac{\partial u}{\partial x}\frac{dx}{dt} + \frac{\partial u}{\partial y}\frac{dy}{dt}$$

$$\frac{du}{dt} = e^{3y}(-\sin 2x)2\frac{dx}{dt} + \cos 2x(e^{3y})3\frac{dy}{dt}$$

$$\frac{du}{dt} = e^{3y}\left(-2\frac{dx}{dt}\sin 2x + 3\frac{dy}{dt}\cos 2x\right) \blacktriangleright$$

2-87 Wt. 4

Integrate $I = \int e^x \cos x \, dx$.

Solution

This requires integration by parts. The general equation is

$$\int u \, dv = uv - \int v \, du$$

Here choose

$$u = e^x \qquad dv = \cos x \, dx \qquad du = e^x \, dx \qquad v = \int \cos x \, dx = \sin x$$

Substituting in the general equation,

$$\int e^x \cos x \, dx = e^x \sin x - \int \sin x \, e^x \, dx \tag{1}$$

Integrate the new integral by parts, where now we choose

$$u = e^x \qquad dv = \sin x \, dx \qquad du = e^x \, dx \qquad v = \int \sin x \, dx = -\cos x$$

$$\int e^x \sin x \, dx = -e^x \cos x - \int -\cos x \, e^x \, dx$$

$$= -e^x \cos x + \int \cos x \, e^x \, dx \tag{2}$$

Substituting into (1),

$$\int e^x \cos x \, dx = e^x \sin x + e^x \cos x - \int \cos x \, e^x \, dx$$

$$2 \int e^x \cos x \, dx = e^x(\sin x + \cos x)$$

$$\int e^x \cos x \, dx = \tfrac{1}{2} e^x (\sin x + \cos x) + C \blacktriangleright$$

2-88 Wt. 5

Given the two differential equations:

$$\frac{dy}{dt} - \lambda z = 0 \tag{1}$$

$$\frac{dz}{dt} + \lambda y = 0 \tag{2}$$

Solve for y and z if, at $t = 0$, $y = 0$ and $z = 1$ (λ is a constant).

68 / Mathematics

Solution

Differentiation of (1) with respect to t gives

$$\frac{d^2y}{dt^2} - \lambda \frac{dz}{dt} = 0$$

Substituting into (2),

$$\frac{\frac{d^2y}{dt^2}}{\lambda} + \lambda y = 0 \qquad \frac{d^2y}{dt^2} + \lambda^2 y = 0$$

Set $y = e^{mt}$; then $e^{mt}(m^2 + \lambda^2) = 0$, so $m = \pm i\lambda$ where $i = \sqrt{-1}$; therefore the general solution is

$$y = C_1 e^{i\lambda t} + C_2 e^{-i\lambda t}$$

Using $e^{\pm i\lambda t} = \cos \lambda t \pm i \sin \lambda t$, we can write

$$y = C_1[\cos \lambda t + i \sin \lambda t] + C_2[\cos \lambda t - i \sin \lambda t]$$
$$= A \cos \lambda t + B \sin \lambda t$$

where $A = C_1 + C_2$ and $B = i(C_1 - C_2)$. When $t = 0$, $y = 0$, so $A = 0$ and hence $y = B \sin \lambda t$. From the first equation of the problem we see that $z = (dy/dt)/\lambda$, so $z = B \cos \lambda t$. Now when $t = 0$, $z = 1$, hence $1 = B$, and the final answer is

$$z = \cos \lambda t \blacktriangleright$$
$$y = \sin \lambda t \blacktriangleright$$

2-89 Wt. 6

The rate of decay of radioactive elements is usually assumed to be proportional to the number of atoms that have not decayed, where λ is the proportionality constant. If at time $t = 0$ there are X_0 atoms of a given element, derive an expression for the number of atoms, X, that have not decayed as a function of time t, λ, and X_0.

Solution

X = number of atoms that have not decayed. The rate of decay of X is proportional to X, or $dX/dt = -\lambda X$. Rearranging this equation,

$$\frac{dX}{X} = -\lambda \, dt$$

Integrating,

$$\ln X = -\lambda t + A$$
$$X = e^{-\lambda t} e^A$$

When $t = 0$, $X = X_0 = e^A$. Therefore

$$X = X_0 e^{-\lambda t}$$

2-90 Wt. 6

A 1000-ft³ storage tank is filled with natural gas at 80°F and 1 atm pressure. The tank is flushed out with nitrogen gas at 80°F and 1 atm pressure, at a constant rate of 300 cfm. The flushing process is carried out at constant temperature and pressure, under conditions of perfect mixing in the tank at all times. Find the time required to reach a gas composition of 95 vol. % nitrogen in the tank.

Solution

Let g = quantity of pure natural gas in the tank at any time
  x = quantity of nitrogen added to the tank = quantity of mixture removed from the tank
  volume of tank = 1000 ft³
  5% of volume of tank = 0.05 × 1000 = 50 ft³

Suppose a volume Δx of the mixture is removed from the tank. The amount of natural gas thus removed will be $(g/1000) \Delta x$. Hence the change in the amount of natural gas in the tank is given by $\Delta g = -(g/1000) \Delta x$. Then the ratio of the quantity of natural gas removed to the volume of nitrogen added is

$$\frac{\Delta g}{\Delta x} = -\frac{g}{1000}$$

When $\Delta x \to 0$ we obtain the instantaneous rate of change of g with respect to x:

$$\frac{dg}{dx} = \frac{-g}{1000}$$

Separating the variables of the differential equation,

$$\frac{dg}{g} + \frac{dx}{1000} = 0$$

Integrating,

$$\int \frac{dg}{g} + \int \frac{dx}{1000} = C \qquad \ln g + \frac{x}{1000} = C$$

When $g = 1000$, $x = 0$. Therefore $C = \ln 1000$.

$$\frac{x}{1000} = \ln 1000 - \ln g = \ln \frac{1000}{g}$$

We want to find x when g is 5% by volume or 50 ft.³ Hence

$$\frac{x}{1000} = \ln \frac{1000}{50} = \ln 20$$

(Set 20 on LL3. Read the value of $x/1000$ opposite it on the D scale.)

$$\frac{x}{1000} = 3.0 \qquad x = 3000 \text{ ft}^3 \text{ nitrogen}$$

Since nitrogen flows in at 300 cfm, the time required to reach 5% by volume natural gas = 3000/300 = 10 min ▶

2-91 Wt. 3

A factory has measured the diameter of 100 random samples of its product. The results, arranged in ascending order, were:

(45 results between 0.859 and 0.900, inclusive) · · ·
0.901   0.902   0.902   0.902   0.903   0.903
0.904   0.904   0.904   0.904 · · ·
(45 more different results from 0.905 to 0.958, inclusive)

The sum of all 100 observations is 91.170. No observed value among those not numerically shown above occurred more than twice. The smallest value observed was 0.859; the largest, 0.958. From the information given, find the mean, the median, and the mode for these data.

Solution

Mean: The arithmetic mean is what is commonly called "average," that is, the sum of the values divided by the number of values.

$$\bar{X} = \frac{1}{N}\sum_{i=1}^{N} X_i = \frac{91.170}{100} = 0.9117 \quad \text{or} \quad 0.912 \blacktriangleright$$

Median: The median of a set of data is the middle value in order of size if N is odd or the value midway between the two middle items if N is even. Here the median is halfway between 0.903 and 0.903 or 0.903 ▶

Mode: This is the most frequent value. In this case it is 0.904 which occurred four times ▶

2-92 Wt. 1

The geometric mean of the numbers 4 and 49 is

(a) 14.0
(b) 22.0
(c) 26.5
(d) 33.0
(e) 49.3

Solution

The geometric mean (G) may be defined as the Nth root of the product of N values:

$$G = \sqrt[N]{x_1 x_2 \cdots x_n}$$

In this case

$$G = \sqrt[2]{(4)(49)} = \sqrt{4} \times \sqrt{49} = 2(7) = 14 \blacktriangleright$$

Alternative solution

To compute the geometric mean (G), the logarithms of the numbers are averaged to find the logarithm of the geometric mean:

$$\log G = \frac{\Sigma \log x}{N} = \frac{\log 4 + \log 49}{2} = \frac{0.6021 + 1.6902}{2} = 1.1461$$

$$G = 14 \blacktriangleright$$

Answer is (a)

2-93 Wt. 3

What is the probability of rolling either a 7 or 11 with one roll of a pair of dice?

Solution

There are 36 possibilities in one roll of a pair of dice. The possibilities of making *either* 7 *or* 11 are:

```
1 - 6     6 - 1
2 - 5     5 - 2
3 - 4     4 - 3
6 - 5     5 - 6
```

They are eight in number. Therefore, the probability of making either a 7 or an 11 on one roll of the dice is $\frac{8}{36} = 0.222$ ▶

2-94 Wt. 3

What is the probability of obtaining at least one 6 in three throws of a die?

Solution

Let P_1 = probability of getting one or more 6's
P_2 = probability of getting no 6's
For any situation the sum of probabilities for all possibilities is 1:

$$P_1 + P_2 = 1$$

The probability of *not* getting a 6 on any single roll is $\frac{5}{6}$; thus the probability that a 6 will not turn up in three rolls is:

$$P_2 = \left(\tfrac{5}{6}\right)\left(\tfrac{5}{6}\right)\left(\tfrac{5}{6}\right) = \tfrac{125}{216} = 0.58$$

$$P_1 = 1 - P_2 = 1 - 0.58 = 0.42$$

Therefore, the probability of obtaining at least one 6 in three throws of a die is 0.42 ▶

2-95 Wt. 3

A submarine fires four torpedoes simultaneously at a target. If the probability for each torpedo to hit the target is $\frac{1}{4}$, what is the probability that the target is hit?

Solution

We know that the probability that a given torpedo hits the target is $\frac{1}{4}$. Since the sum of all possibilities is 1, and the torpedo must either hit or miss the target, the probability that any given torpedo misses the target is $1 - \frac{1}{4} = \frac{3}{4}$.

The probability that the target is hit = 1 − probability that the target is missed, so let us determine the probability that the target is missed. To fail to score a hit we know that the first, second, third, and fourth torpedoes must *all* miss the target.

Also, the probability of a series of independent events = product of probabilities of individual events. Thus:

$$P_{\text{target is missed}} = (P_{\text{1st torpedo misses}})(P_{\text{2d torpedo misses}})(P_{\text{3rd torpedo misses}})$$

$$(P_{\text{4th torpedo misses}})$$

$$= \left(\tfrac{3}{4}\right)\left(\tfrac{3}{4}\right)\left(\tfrac{3}{4}\right)\left(\tfrac{3}{4}\right) = \left(\tfrac{3}{4}\right)^4 = \tfrac{81}{256} = 0.316$$

$$P_{\text{target is hit}} = 1 - P_{\text{target is missed}} = 1 - 0.316 = 0.684 \blacktriangleright$$

2-96 Wt. 1

An elementary game is played by rolling a die and drawing a ball from a bag containing 3 white and 7 black balls. The player wins whenever he rolls a number less than 4 and draws a black ball. What is the probability of winning in the first attempt?

(a) 7/20
(b) 12/10
(c) 1/2
(d) 7/10
(e) 13/20

Solution

The probability of rolling a number less than 4 with a die $= \frac{3}{6} = \frac{1}{2}$. The probability of drawing a black ball $= 7/(7 + 3) = 7/10$. The probability of a series of independent events = product of probabilities of the individual events:

$$\tfrac{1}{2} \times \tfrac{7}{10} = \tfrac{7}{20} \blacktriangleright$$

Answer is (a)

2-97 Wt. 5

A certain number of balls are in a box. The balls are identical in all respects except color. Some of the balls are red and the remainder are black. Two balls are picked at random, one after the other, and it is found that they are both red. If the probability of this happening is exactly 1/2, and there are no more than 10 balls in the box, what is the exact number of black balls in the box?

Solution

Probability of series of independent events = product of probabilities of individual events; thus $P = (p_1)(p_2)$. Let $R =$ number of red balls and $B =$ number of black balls.

The probability p_1 of selecting a red ball on the first try is $R/(R + B)$, and the probability p_2 of selecting another red ball on the second try is $(R - 1)/((R + B) - 1)$. Thus

$$P = \left(\frac{R}{R + B}\right)\left(\frac{R - 1}{R + B - 1}\right) = \frac{1}{2}$$

This can be solved quickly by trial and error. Assume $R = 2$ and $B = 1$.

$$P = (\tfrac{2}{3})(\tfrac{1}{2}) = \tfrac{2}{6} = \tfrac{1}{3}$$

We can see that there must be more red balls to increase the probability. Assume $R = 3$ and $B = 1$.

$$P = (\tfrac{3}{4})(\tfrac{2}{3}) = \tfrac{6}{12} = \tfrac{1}{2}$$

This is the required probability. Thus we have a solution which satisfies the conditions, that is, 3 red balls and 1 black ball $\blacktriangleright$

Note that 6 red balls and 2 black balls is *not* another solution, as

$$P = (\tfrac{6}{8})(\tfrac{5}{7}) = \tfrac{30}{56} = 0.536 \neq 0.5$$

2-98 Wt. 3

Given the values of x and the frequency, f, with which they occur. Calculate the arithmetic mean, $\bar{x}$, and the standard deviation, σ.

x	f
2	4
4	6
7	6
12	4

Solution

The arithmetic mean, $\bar{x}$, is what we commonly call the "average" and is simply the sum of the items times their frequency divided by their number.

$$\bar{x} = \frac{\Sigma(fx)}{N}$$

x	f	fx
2	4	8
4	6	24
7	6	42
12	4	48
	$N = 20$	$122 = \Sigma(fx)$

$$\bar{x} = \frac{122}{20} = 6.1 \blacktriangleright$$

The standard deviation, σ, is a measure of the dispersion or scatter of a set of values. It is sometimes called the rms deviation, as this describes its method of calculation. First, square the deviations of individual values from the arithmetic mean. Then take the mean of these squares and extract the square root.

$$\sigma = \left[\frac{\Sigma f(x - \bar{x})^2}{N}\right]^{1/2}$$

x	f	$x - \bar{x}$	$(x - \bar{x})^2$	$f(x - \bar{x})^2$
2	4	−4.1	16.81	67.24
4	6	−2.1	4.41	26.46
7	6	0.9	0.81	4.86
12	4	5.9	34.81	139.24
	$N = 20$			$\Sigma = 237.80$

$$\sigma = \left[\frac{\Sigma f(x - \bar{x})^2}{N}\right]^{1/2} = \left(\frac{237.80}{20}\right)^{1/2} = (11.89)^{1/2} = 3.45 \blacktriangleright$$

A simplified equation, if known, can be more readily evaluated:

$$\sigma = \left[\frac{\Sigma fx^2}{N} - \left(\frac{\Sigma fx}{N} \right)^2 \right]^{1/2}$$

x	f	x^2	fx^2
2	4	4	16
4	6	16	96
7	6	49	294
12	4	144	576
	$N = 20$		$982 = \Sigma fx^2$

$\sigma = (\frac{982}{20} - 6.1^2)^{1/2} = (49.1 - 37.2)^{1/2} = (11.9)^{1/2} = 3.45$ ▶

2-99 Wt. 5

Assume a group of 9 people consists of 4 men and 5 women. Compute the probability that a committee of 3, selected at random, would consist of 2 men and 1 woman.

Solution

Here we use the notation $\binom{n}{r} = n!/(r!(n-r)!)$, which gives the total number of ways r objects can be chosen from n objects.

There are 9 people (4 men and 5 women). The total number of committees of 3 people would be equal to the total number of committees consisting of

0 men 3 women   or $\binom{4}{0} \times \binom{5}{3} = 1 \times \frac{5!}{3!\,2!} = 10$

1 man   2 women $\binom{4}{1} \times \binom{5}{2} = \frac{4!}{3!} \times \frac{5!}{2!\,3!} = 40$

2 men   1 woman $\binom{4}{2} \times \binom{5}{1} = \frac{4!}{2!\,2!} \times \frac{5!}{4!} = 30$

3 men   0 women $\binom{4}{3} \times \binom{5}{0} = \frac{4!}{3!} \times 1 \quad = 4$

Total 84

The number of committees consisting of 2 men and 1 woman is 30 out of 84 possible combinations. Hence the probability of this happening is

$$P = \tfrac{30}{84} = 0.357 \blacktriangleright$$

Alternative solution

The possible ways of selecting 2 men and 1 woman are:

man-man-woman

man-woman-man

woman-man-man

First, consider the order: man-man-woman;

Probability of selecting a man, first try $= p_1 = \dfrac{M}{M+W} = \dfrac{4}{9}$

Probability of selecting a man, second try $= p_2 = \dfrac{M-1}{M+W-1} = \dfrac{3}{8}$

Probability of selecting a woman, third try $= p_3 = \dfrac{W}{M+W-2} = \dfrac{5}{7}$

Probability of a series of events = product of probabilities of the individual events

$$P = \tfrac{4}{9} \times \tfrac{3}{8} \times \tfrac{5}{7} = \tfrac{60}{504} = 0.119$$

In both the other ways of selecting 2 men and 1 woman

man-woman-man   and   woman-man-man

the values in the numerator will be 4, 3, and 5, representing the selection of the first man, second man, and first woman. Also, the denominator will be 9, 8, and 7, representing the size of the remaining group prior to each selection.

So for man-woman-man

$$P = p_1 \times p_2 \times p_3 = \tfrac{4}{9} \times \tfrac{5}{8} \times \tfrac{3}{7} = \tfrac{60}{504} = 0.119$$

and for woman-man-man

$$P = p_1 \times p_2 \times p_3 = \tfrac{5}{9} \times \tfrac{4}{8} \times \tfrac{3}{7} = \tfrac{60}{504} = 0.119$$

Thus the probability of picking a committee of 2 men and 1 woman is

$$P = 0.119 + 0.119 + 0.119 = 0.357 \blacktriangleright$$

2-100 Wt. 1

Study the series of numbers to discover the "system" in which they are arranged. For the series 1 5 14 30 __ 91 the fifth term is

(a) 59
(b) 53
(c) 61
(d) 50
(e) 55

Solution

If we are given any series of values it is always possible to find a polynomial which will pass through the given points. So the equation of the polynomial would be the "system." Thus *any* number between 30 and 91 *could* be used. Here we derive a particular "system."

$$\begin{array}{ccccccc} 1 & & 5 & & 14 & & 30 & & x & & 91 \end{array}$$

First differences:

$$\begin{array}{ccccc} 4 & 9 & 16 & y & z \\ 2^2 & 3^2 & 4^2 & 5^2 & 6^2 \end{array}$$

We see that the first differences are successive perfect squares. Thus

$$x - 30 = y = 5^2 = 25 \qquad x = 30 + 25 = 55 \blacktriangleright$$

Answer is (e)

2-101 Wt. 1

The sum of all whole numbers from 1 to 100 inclusive is nearest to

(a) 6500
(b) 6000
(c) 5500
(d) 5050
(e) 5005

Solution

We wish to determine the sum S of an arithmetic progression. Let a be the first term, d the difference between successive terms, l the last term, and n the number of terms. Then

$$S = a + (a + d) + (a + 2d) + \cdots + [a + (n - 1)d]$$

or, in reverse order,

$$S = [a + (n - 1)d] + [a + (n - 2)d] + \cdots + a$$

Adding these two equations,
$$2S = n[2a + (n-1)d]$$
or, since $[a + (n-1)d] = l$,
$$S = \frac{n}{2}(a + l)$$

In this problem $a = 1$, $n = l = 100$, and $S = \frac{100}{2}(101)$.
$$S = 5050 \blacktriangleright$$

Answer is (d)

2-102 Wt. 5

Given: Universe = (1, 2, 3, 4, 5, 6, 7)
$A = (1, 3, 6)$
$B = (1, 2, 6, 7)$

Find:

(a) $\bar{A}$
(b) $\bar{B}$
(c) $A \cap B$
(d) $A \cap \bar{B}$
(e) $\bar{A} \cup B$

Solution

$\bar{A}$ represents non-A, or all elements of the universe which are not in set A.

(a) $\bar{A} = (2, 4, 5, 7) \blacktriangleright$

$\bar{B}$ represents non-B.

(b) $\bar{B} = (3, 4, 5) \blacktriangleright$

$A \cap B$ denotes the intersection of sets A and B, which includes all elements which are in *both* sets A and B.

(c) $A \cap B = (1, 6) \blacktriangleright$

$A \cap \bar{B}$ is the intersection of sets A and $\bar{B}$.

(d) $A \cap \bar{B} = (3) \blacktriangleright$

$\bar{A} \cup B$ represents the union of sets $\bar{A}$ and B; it includes all elements which are members of *either* $\bar{A}$ or B.

(e) $\bar{A} \cup B = (1, 2, 4, 5, 6, 7) \blacktriangleright$

Chapter 3

Statics

Statics is the subdivision of mechanics which considers the equilibrium of stationary or uniformly translating particles or rigid bodies. By this definition a unifying characteristic of *all* statics problems is the absence of any acceleration.

A variety of forces which may act on a body are considered when statics is studied. The forces may either act at specific points on the body or be distributed over a region in space; the weight of a body or a distributed pressure are examples of the latter. In connection with distributed forces it is common to discuss the determination of centroids, centers of gravity, and moments of inertia. This is done here. We defer the subject of shear and bending moment diagrams, which are related to the internal static equilibrium of a body, to Chapter 5 on Mechanics of Materials.

EQUILIBRIUM

A body is in a state of static equilibrium when no net or unbalanced resultant force **R** acts upon the body and, in addition, the forces on the body create no net tendency toward rotation or couple **M**. These requirements for equilibrium, restated, are

$$\mathbf{R} = 0 \quad \mathbf{M} = 0 \tag{3-1}$$

The sense of the vector which represents a moment is determined by use of the right-hand rule.

For the Cartesian (x, y, z) coordinate system shown in Fig. 3-1, equations (3-1) become

$$|\mathbf{R}| = [(\sum F_x)^2 + (\sum F_y)^2 + (\sum F_z)^2]^{1/2} = 0$$
$$|\mathbf{M}| = [(\sum M_x)^2 + (\sum M_y)^2 + (\sum M_z)^2]^{1/2} = 0 \tag{3-2}$$

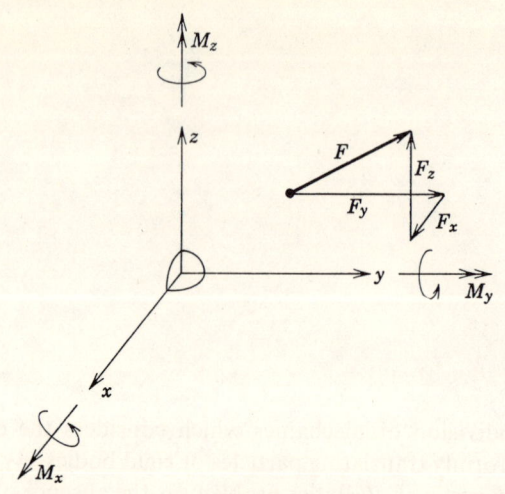

Fig. 3-1

which is equivalent to the requirements

$$\sum F_x = 0 \qquad \sum M_x = 0$$
$$\sum F_y = 0 \qquad \sum M_y = 0 \qquad (3\text{-}3)$$
$$\sum F_z = 0 \qquad \sum M_z = 0$$

For a three-dimensional statically determinate problem we thus have six independent equations for equilibrium. For the corresponding two-dimensional situation, the three equations are

$$\sum F_x = 0 \qquad \sum F_y = 0 \qquad \sum M_z = 0 \qquad (3\text{-}4)$$

Commonly encountered examples of bodies in static equilibrium are the two- and three-force bodies.

The two-force body is one that is acted upon by concentrated forces which are applied at only two points on the body. Equations (3-4) can be used to show that these two forces have the same line of action, the same magnitude, and act in opposite directions. This information is particularly useful in the solution of statically determinate truss problems.

Concentrated forces acting at three points on a body cause the body to be called a three-force body. For equilibrium these forces must either (*a*) have lines of action which all pass through the same point so that no couple acts on the body or (*b*) act along parallel lines of action.

Example 1

A 4-ft × 10-ft block which weighs 1600 lb is held in a horizontal position by a force P and hinge A, as shown in Fig. 3-2a. Determine the force P and the resultant hinge reaction R_A.

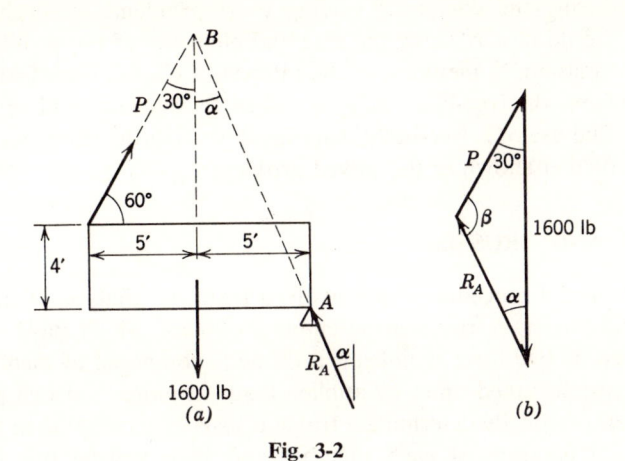

Fig. 3-2

Solution

The block, which weighs 1600 lb, is held in equilibrium by forces P and R_A, thus forming a three-force body. The point common to all three lines of action is point B, which is directly above the location of the equivalent concentrated 1600-lb weight. A force triangle, shown in Fig. 3-2b, can easily be drawn.

From Fig. 3-2a we have

$$\tan \alpha = \frac{5}{4 + 5 \tan 60°} = \frac{5}{4 + 5\sqrt{3}} = 0.396$$

$$\alpha = 21°34'$$

Hence

$$\beta = 180° - 30° - \alpha = 128°26'$$

Using the law of sines,

$$\frac{P}{\sin \alpha} = \frac{R_A}{\sin 30°} = \frac{1600}{\sin \beta} = 2040$$

$$P = 2040 \sin 21°34' = 750 \text{ lb} \blacktriangleright$$

$$R_A = 2040 \sin 30° = 1020 \text{ lb acting at an angle } \alpha \blacktriangleright$$

FREE-BODY DIAGRAM

In Fig. 3-2a we have drawn a free-body diagram of the block. Relevant dimensions and angles are also shown. It is helpful to prepare a free-body diagram during the course of solving most problems in mechanics. The diagram should clearly show the essential elements of the problem and no other information. Sometimes, in the interest of clarity, dimensions are also excluded from the free-body diagram when they would tend to clutter the drawing. The use of a free-body diagram in finding the reactions for simple bodies is further shown in the solved problems.

FRAMES AND TRUSSES

Certain useful procedures have evolved from the analysis of the forces in frames and trusses. A truss is an assemblage of pieces which may be accurately represented as two-force members; a frame is composed of members which each are usually acted upon by applied loads at more than two points. The analysis of a statically determinate frame is usually accomplished by drawing a free-body diagram of each member and then writing the appropriate equilibrium equations for each diagram. These equations are then solved for the unknown forces and reactions. The analysis of trusses, however, usually proceeds by some combination of the methods known as the method of joints or the method of sections.

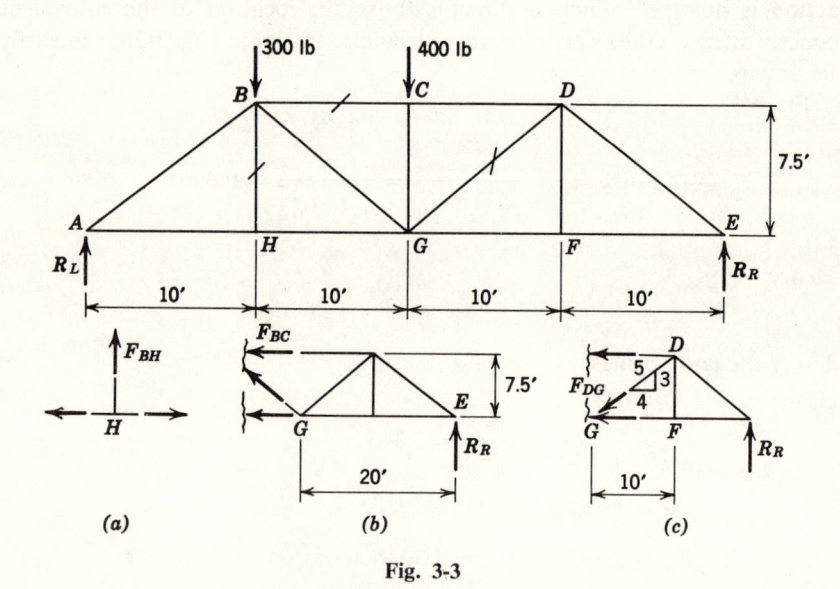

Fig. 3-3

Example 2

Determine the force in members BH, BC, and GD of the truss shown in Fig. 3-3. Note that the truss is composed of triangles 7.5 ft:10.0 ft:12.5 ft, so that they are 3:4:5 right triangles.

Solution

First we solve for the reactions R_L and R_R:

$$\sum M_E = 0 \quad 40R_L = 30(300) + 20(400)$$
$$R_L = 425 \text{ lb}$$
$$\sum F_v = 0 \quad R_R = 300 + 400 - 425 = 275 \text{ lb}$$

The method of joints considers the equilibrium of each pinned connection between members. Because of the pin, no moment can be transmitted through the joint. For a two-dimensional truss, joint equilibrium requires that $\sum F_h = 0$ and $\sum F_v = 0$ be satisfied for the forces acting at each joint. We have an especially simple case in this problem, as shown in Fig. 3-3a. All bar forces are shown to be in tension. Summing forces vertically,

$$\sum F_v = 0 = F_{BH} \blacktriangleright$$

The method of joints is most efficiently used (a) when the forces in all members of a truss are desired or (b) when special cases such as the example arise.

The method of sections is normally more efficient than the method of joints when only a few selected bar forces must be found. If the truss as a whole is in equilibrium, then any segment or section of the structure must also be in equilibrium. This principle is put to use by selecting an "appropriate" section of a structure and applying the principles of statics. In most cases the appropriate section is one which (a) severs the member of interest and (b) results in a free body acted upon by only three unknown forces. Numerous exceptions to the second requirement exist, however.

Selecting a section as shown in Fig. 3-3b, we write

$$\sum M_G = 0 \quad 7.5 F_{BC} = -20 R_R$$

$$F_{BC} = \frac{-20(275)}{7.5} = 733 \text{ lb compression} \blacktriangleright$$

Referring to Fig. 3-3c,

$$\sum F_v = 0 \quad \tfrac{3}{5} F_{DG} = R_R$$
$$F_{DG} = \tfrac{5}{3}(275) = 458 \text{ lb tension} \blacktriangleright$$

FRICTION

In many situations forces due to friction cause a body to remain in static equilibrium. In these problems it is important to recall that frictional forces always act to oppose any actual or impending motion. For most cases of dry friction the frictional force F is simply proportional to the normal force N, or $F = \mu N$ where μ is the coefficient of either static or kinetic friction. (A few more specialized friction problems, involving rolling friction or belt friction, are presented only in the statics problem section.)

Example 3

A 500-lb block rests on a 30° plane (Fig. 3-4). If the coefficient of static friction is 0.30 and the coefficient of kinetic friction is 0.20, what is the value of P needed to

(a) prevent the block from sliding down the plane?
(b) start moving the block up the plane?
(c) keep the block moving up the plane?

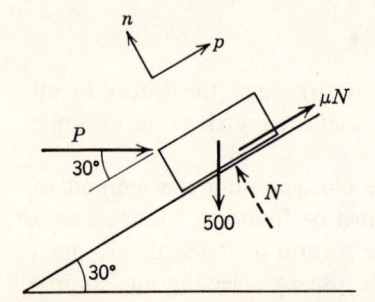

Fig. 3-4

Solution

Normal to the plane, $\sum F_n = 0$

$$N - P \sin 30° - 500 \cos 30° = 0$$

$$N = \frac{P}{2} + 433$$

(a) Parallel to the plane, $\sum F_p = 0$

$$\mu N + P \cos 30° - 500 \sin 30° = 0$$

$$0.3\left(\frac{P}{2} + 433\right) + 0.866P - 250 = 0$$

$$P = 118.2 \text{ lb} \blacktriangleright$$

(b) Here we have impending motion *up* the plane so the direction of μN in Fig. 3-4 must be directed down the plane to oppose the motion. This can be accomplished mathematically by replacing μN by $-\mu N$ in the earlier equation. Thus

$$-0.3\left(\frac{P}{2} + 433\right) + 0.866P - 250 = 0$$

$$P = 531 \text{ lb} \blacktriangleright$$

(c) The problem is unchanged from (b) except that we now consider the kinetic ($\mu = 0.2$) rather than the stationary ($\mu = 0.3$) case.

$$-0.2\left(\frac{P}{2} + 433\right) + 0.866P - 250 = 0$$

$$P = 440 \text{ lb} \blacktriangleright$$

CENTROID, CENTER OF GRAVITY, MOMENT OF INERTIA

The center of gravity (also called center of mass) and mass moment of inertia are physical properties of a body. If the density is uniform throughout the body, these properties exactly coincide with the associated, but purely geometric, properties of a body which are called the centroid and moment of inertia. The equations for the center of gravity of a body are

$$\bar{X} = \frac{1}{M} \int_V \rho x \, d\forall \qquad \bar{Y} = \frac{1}{M} \int_V \rho y \, d\forall \qquad \bar{Z} = \frac{1}{M} \int_V \rho z \, d\forall \qquad (3\text{-}5)$$

where $M = \int_V \rho \, d\forall$ is the mass of the body composed of volume elements $d\forall$ which have density ρ. These equations apply equally well for rods, plane areas, and volumes of any shape; one need only take care to select an appropriate volume element. The integrals may be replaced by finite sums for single or composite bodies whenever the component volume and the centroid or center of gravity of each individual element are already known. Note that the density ρ will cancel in equations (3-5) when it is constant throughout a body, and we then have the equations for the centroid of the body. These principles are illustrated in Example 4. Appendix A gives the centroidal coordinates for some common geometric shapes.

Example 4

Locate the X, Y, Z coordinates of the centroid for the slender rod $ABCD$ of constant density shown in Fig. 3-5. The semicircular portion ABC has a radius of π in. and lies in the Y-Z plane. The straight portion is 10 in long and lies in the X-Y plane. The point D is located at $X = 6$, $Y = 8$.

86 / Statics

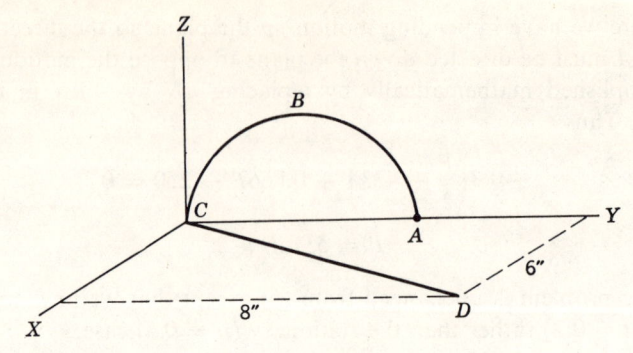

Fig. 3-5

Solution

Length $L = \pi R + \sqrt{6^2 + 8^2} = 9.86 + 10 = 19.86$ in.

$$\bar{X} = \frac{1}{L}\int x\, dL = \frac{1}{L}\sum_{i=1}^{n} x_i L_i = \frac{0 \times 9.86 + 3 \times 10}{19.86} = 1.51 \text{ in.} \blacktriangleright$$

$$\bar{Y} = \frac{1}{L}\sum_{i=1}^{n} y_i L_i = \frac{\pi \times 9.86 + 4 \times 10}{19.86} = 3.57 \text{ in.} \blacktriangleright$$

The center of gravity of a semicircular disk, measured from the diameter, is

$$\frac{4R}{3\pi} = \frac{4\pi}{3\pi} = 1.33 \text{ in.}$$

$$\bar{Z} = \frac{1}{L}\sum_{i=1}^{n} z_i L_i = \frac{1.33 \times 9.86 + 0 \times 10}{19.86} = 0.66 \text{ in.} \blacktriangleright$$

The moment of inertia (or second moment) of an area, with respect to some axis $r = 0$, is

$$I = \int_A r^2\, dA \tag{3-6}$$

In this expression r is the distance from the axis $r = 0$ to the centroid of the area element dA. Usually r is replaced by either x or y so that the moment of inertia is found with respect to the x or y coordinate axis. The mass moment of inertia is similarly defined as

$$I_m = \int_\forall r^2\, dm = \int_\forall r^2 \rho\, d\forall \tag{3-7}$$

If the moment of inertia of a body around its centroidal axis I_0 is known, the moment of inertia around any axis parallel to this centroidal axis may be found from the parallel-axis theorem

$$I = I_0 + Ad^2 \tag{3-8}$$

where A is the area and d is the distance between the two parallel axes. For mass moments of inertia, A is replaced by the mass M in equation (3-8). Appendix A gives the moment of inertia I_0 for some common areas.

Example 5

Find the moment of inertia of a rectangle of base b and height h (Fig. 3-6)

(a) about the centroidal axis
(b) about the base of the rectangle

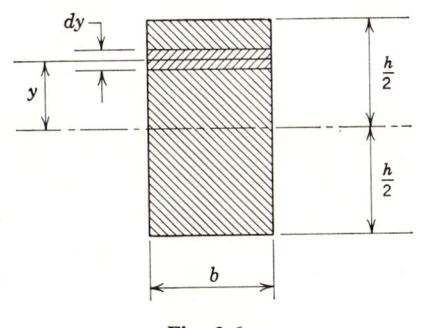

Fig. 3-6

Solution

(a) We select a differential area of width b and height dy which has a distance y to its centroid, or $dA = b\,dy$. Applying equation (3-6) with $y = 0$ at the centroidal axis,

$$I_0 = \int_{-h/2}^{h/2} y^2 (b\,dy)$$

Since b is constant,

$$I_0 = b \int_{-h/2}^{h/2} y^2\,dy = b[\tfrac{1}{3}y^3]_{-h/2}^{h/2}$$

$$= \frac{b}{3}\left(\frac{h^3}{8} + \frac{h^3}{8}\right) = \frac{b}{3}\left(\frac{h^3}{4}\right) = \frac{bh^3}{12} \blacktriangleright$$

(b) Applying equation (3-8) gives

$$I = I_0 + Ad^2 = \frac{bh^3}{12} + (bh)\left(\frac{h}{2}\right)^2$$

$$= \frac{bh^3}{3} \blacktriangleright$$

88 / Statics

3-1 Wt. 1

For the system shown (Fig. 3-7), choose the one true statement concerning the bearing reactions at A and B if the system is in equilibrium.

(a) Both reactions are vertical.
(b) Neither reaction is vertical.
(c) The left reaction at A is vertical.
(d) The right reaction at B is vertical.
(e) Since there are two unknown components at A and B, respectively, the system is statically indeterminate and the reactions cannot be described using methods of statics.

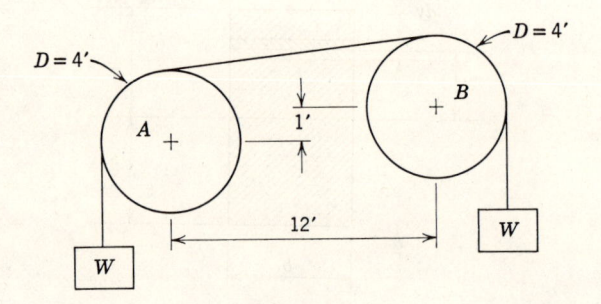

Fig. 3-7

Solution

Draw the free-body diagram:

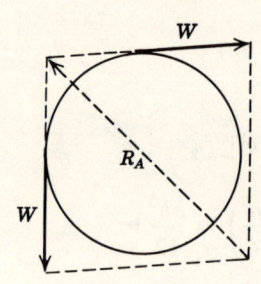

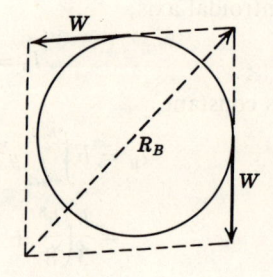

Fig. 3-8

Hence neither reaction A nor B is vertical ▶

Answer is (b)

3-2 Wt. 1

The value of reaction R (Fig. 3-9) is

(a) 9.6 kips
(b) 20.0 kips
(c) 26.4 kips
(d) 36.0 kips
(e) 44.0 kips

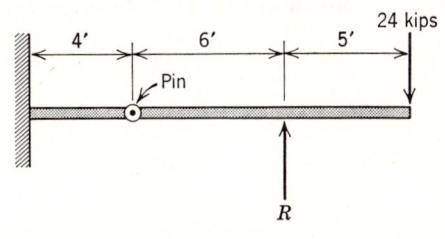

Fig. 3-9

Solution

$\sum M_{\text{pin}} = 0$

$-24 \text{ kips} \times 11 + 6R = 0 \qquad R = \dfrac{24 \times 11}{6} = 44 \text{ kips} \blacktriangleright$

Answer is (e)

3-3 Wt. 1

The moment at reaction R (Fig. 3-10) is

(a) Unknown; the structure is statically indeterminate
(b) 0
(c) Pa
(d) $Pa\left(\dfrac{b+c}{a+b+c}\right)$
(e) $P\left(\dfrac{bc}{a+b}\right)$

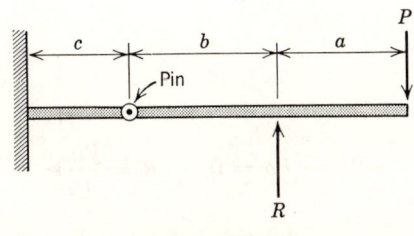

Fig. 3-10

Solution

The moment at any point on a beam may be determined by calculating the moments on either side of the point—hence $M_R = Pa$ ▶

Answer is (c)

3-4 Wt. 1

For the beam loaded as shown in Fig. 3-11, reaction R is

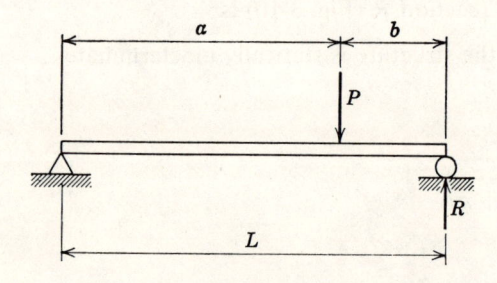

(a) $\dfrac{Pa}{L}$

(b) $\dfrac{Pb}{L}$

(c) $\dfrac{PL}{a}$

(d) $\dfrac{PL}{b}$

(e) $\dfrac{PL}{ab}$

Fig. 3-11

Solution

$\sum M_{\text{left reaction}} = 0$

$$+RL - Pa = 0 \qquad R = \frac{Pa}{L} \blacktriangleright$$

Answer is (a)

3-5 Wt. 4

Determine the beam reactions marked R in Figs. 3-12 and 3-13.

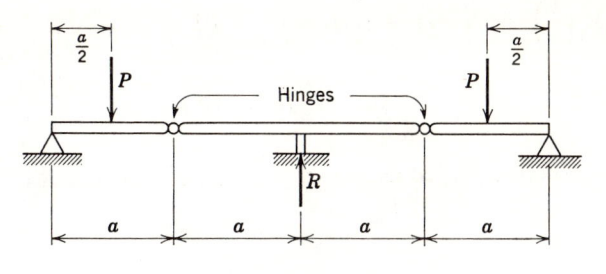

Fig. 3-12

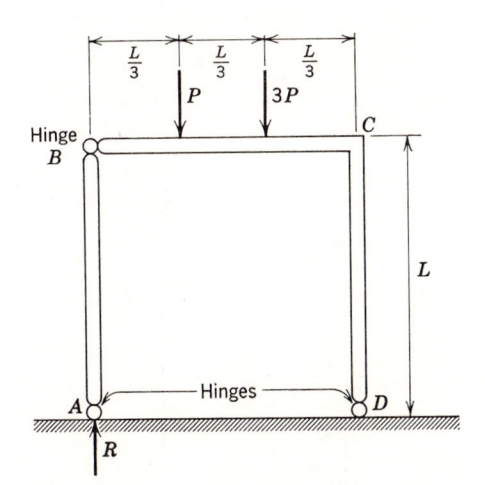

Fig. 3-13

Solution

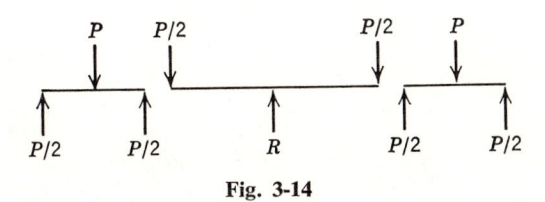

Fig. 3-14

We can see that $R = 2(P/2) = P$ ▶

92 / Statics

Since joint C in Fig. 3-13 is rigid, we have a stable rigid frame.

$$\sum M_D = 0$$

$$+3P\left(\frac{L}{3}\right) + P\left(\frac{2L}{3}\right) - R_A L = 0 \qquad PL + \tfrac{2}{3}PL - R_A L = 0$$

$$R_A = 1\tfrac{2}{3}P$$

Answer: $R = 1\tfrac{2}{3}P$ ▶

3-6 Wt. 1

For the structure loaded as shown in Fig. 3-15 reaction R is

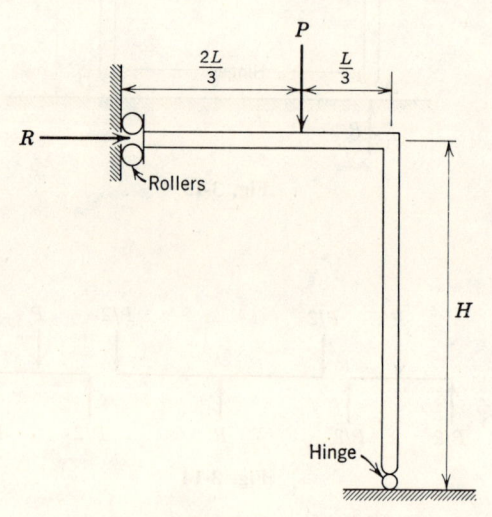

(a) $\dfrac{PL}{H}$

(b) $\dfrac{2PL}{3H}$

(c) $\dfrac{PL}{3H}$

(d) $\dfrac{3PH}{L}$

(e) $\dfrac{3PH}{2L}$

Fig. 3-15

Solution

$$\sum M_{\text{base hinge}} = 0$$

$$-RH + P\frac{L}{3} = 0 \qquad R = \frac{PL}{3H} \blacktriangleright$$

Answer is (c)

3-7 Wt. 1

In the vector diagram (Fig. 3-16) $\bar{R}$ represents

(a) $\bar{A} + \bar{B}$
(b) $\bar{A} - \bar{B}$
(c) $\bar{B} - \bar{A}$
(d) $\bar{A} \cdot \bar{B}$
(e) $\sqrt{\bar{A}^2 + \bar{B}^2}$

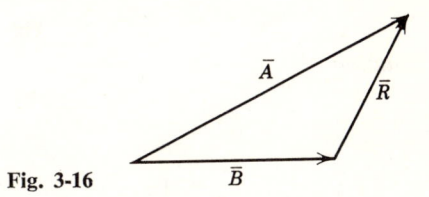

Fig. 3-16

Solution

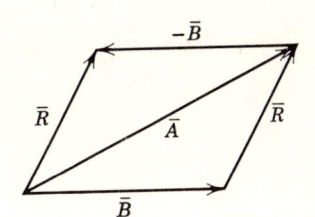

Fig. 3-17

$$\bar{B} + \bar{R} = \bar{A} \qquad \bar{R} = \bar{A} - \bar{B} \blacktriangleright$$

or

$$\bar{A} + (-\bar{B}) = \bar{R} \qquad \bar{R} = \bar{A} - \bar{B} \blacktriangleright$$

Answer is (b)

94 / Statics

3-8 Wt. 1

Given a 50-lb pulley, supported as shown (Fig. 3-18) and carrying a cable supporting an additional 50-lb load. The force the beam exerts on the pulley is

(a) 50 lb up
(b) 100 lb up
(c) 100 lb down
(d) 150 lb up
(e) 150 lb down

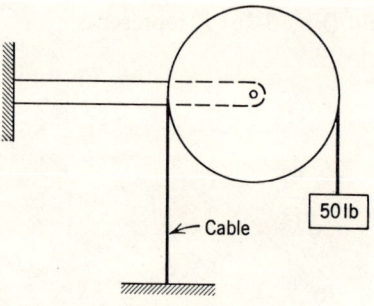

Fig. 3-18

Solution

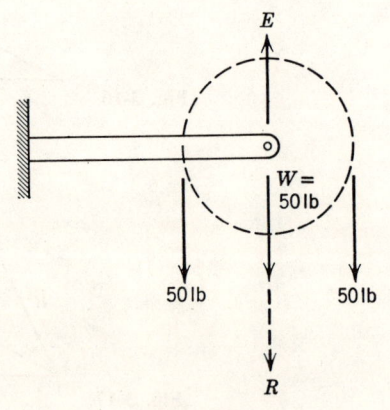

Fig. 3-19

The resultant of three forces = 50 + 50 + 50 = 150 lb down.

$\Sigma F_y = 0$

$$+E - 50 - 50 - 50 = 0$$
$$E = +150 \text{ lb equilibrant}$$

Therefore the force the beam exerts on the pulley is 150 lb up ▶

Answer is (d)

3-9 Wt. 3

A 5000-lb sphere rests against a smooth plane inclined at 45° to the horizontal and against a smooth vertical wall. What are the reactions at A and B?

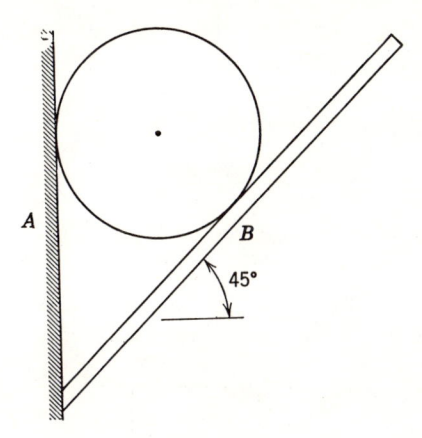

Fig. 3-20

Solution

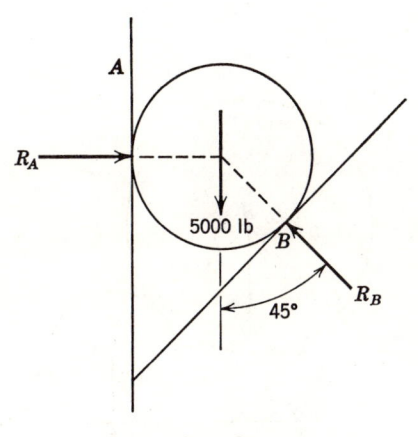

Fig. 3-21

$\sum F_V = 0$

$$-5000 + R_B \cos 45° = 0 \qquad R_B = \frac{5000}{0.707} = 7070 \text{ lb} \blacktriangleright$$

$\sum F_H = 0$

$$R_A - R_B \sin 45° = 0 \qquad R_A = 7070(0.707) = 5000 \text{ lb} \blacktriangleright$$

3-10 Wt. 6

Two cylinders lie in a trapezoidal channel as shown. Cylinder #1 weighs 500 lb, cylinder #2 weighs 1000 lb. Determine reactions R_1, R_2, and R_3. Suggestion: Try a graphical solution.

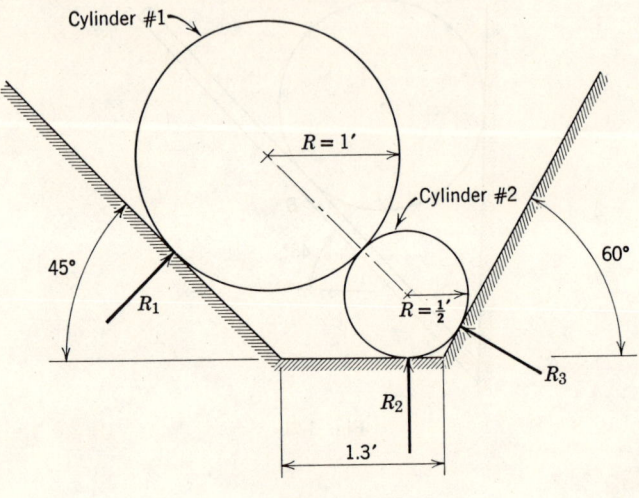

Fig. 3-22

Solution

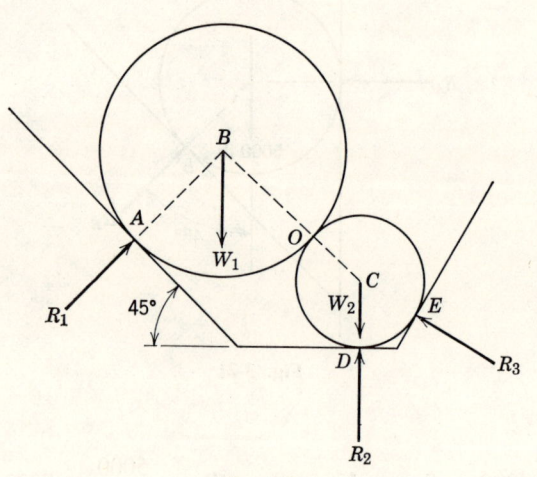

Fig. 3-23

It should be noted that the smaller cylinder is the heavier one. Lines designated as prime (′) are parallel to corresponding unprime lines. The free-body diagrams are:

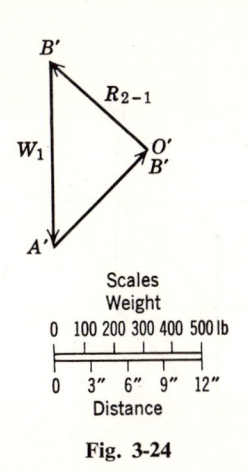

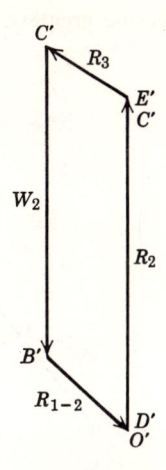

Fig. 3-24

Fig. 3-25

where R_{2-1} = reaction of cylinder #2 on cylinder #1 = 350 lb.

$$R_1 = 375 \text{ lb} \blacktriangleright$$
$$R_2 = 1100 \text{ lb} \blacktriangleright$$
$$R_3 = 320 \text{ lb} \blacktriangleright$$

By making larger free-body diagrams, greater accuracy could be obtained.

3-11 Wt. 5

The heavy, solid steel, triangular prism in Fig. 3-26 is supported by three vertical cables, attached as indicated: one at the middle of one edge with the other two at the corners of the opposite edge. All three cables are the same

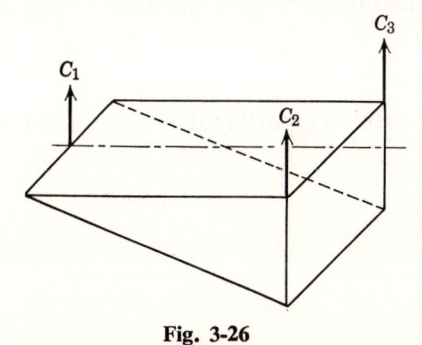

Fig. 3-26

length and are of the same size. Determine in which of the cables the tensile force will be the greatest. The steel prism weighs 1200 lb.

Solution

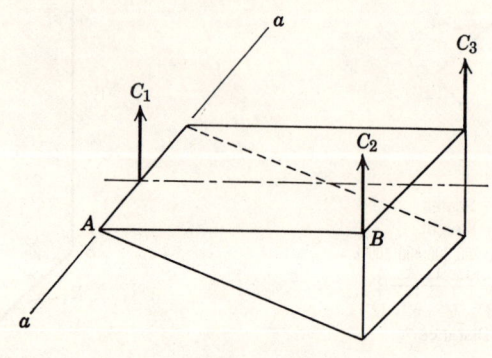

Fig. 3-27

The cables are all the same size. Thus, owing to symmetry, $C_2 = C_3$.

$$\sum M_a = 0$$

$$(C_2 + C_3)L - \tfrac{2}{3}L(1200) = 0 \qquad C_2 + C_3 = 800$$

Therefore

$$C_2 = C_3 = 400 \text{ lb}$$

$$\sum F_V = 0$$

$$400 + 400 + C_1 - 1200 = 0$$

Thus

$$C_1 = 400 \text{ lb}$$

Therefore all cable tensions are equal and have a value of 400 lb ▶

3-12 Wt. 1

The tension in the cable supporting the beam loaded as shown is approximately

(a) 5 kips
(b) 10 kips
(c) 15 kips
(d) 20 kips
(e) 25 kips

Fig. 3-28

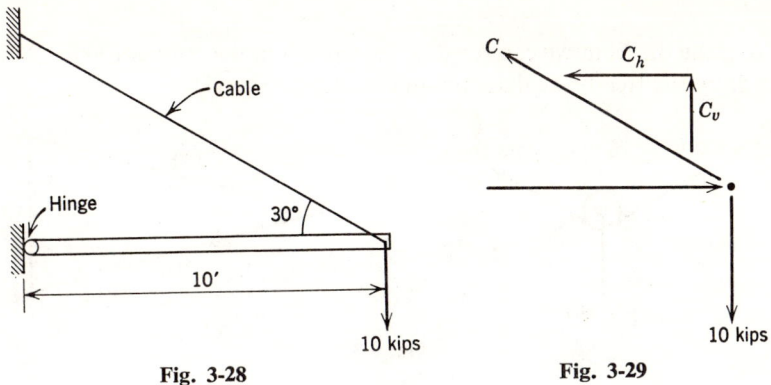

Fig. 3-29

Solution

From the free-body diagram, Fig. 3-29, we can see that for $\sum F_y = 0$

$$C_v - 10 \text{ kips} = 0 \qquad C_v = 10 \text{ kips}$$

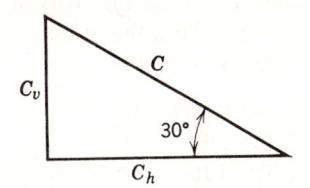

Fig. 3-30

$$\sin 30° = \frac{C_v}{C} = \frac{1}{2} \qquad C = 2C_v = 2 \times 10 \text{ kips} = 20 \text{ kips}$$

Answer is (*d*)

3-13 Wt. 3

Two weights are suspended on an inextensible cord as shown. What is the angle θ at equilibrium?

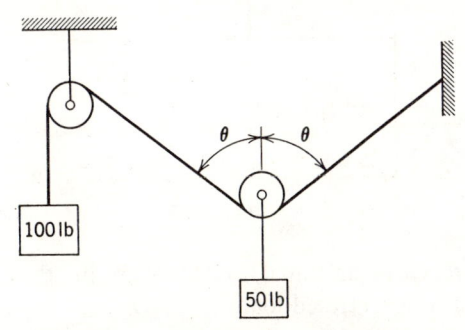

Fig. 3-31

Solution

From the diagram we can see that the tension in the cord is 100 lb. We can then draw the free-body diagram and the force triangle.

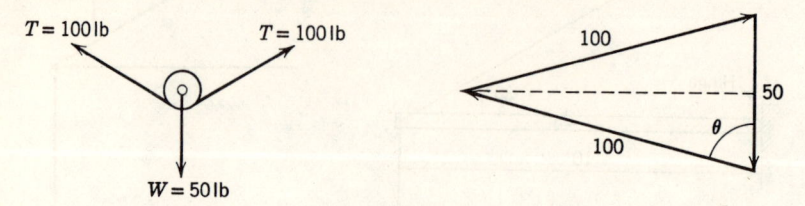

Fig. 3-32

$$\theta = \cos^{-1} \tfrac{25}{100} = \cos^{-1} 0.250 = 75°30' \blacktriangleright$$

3-14 Wt. 1

The center of gravity of a log 10 ft long and weighing 100 lb is 4 ft from one end of the log. It is to be carried by two men. If one is at the heavy end, how far from the other end does the second man have to hold the log if each is to carry 50 lb?

(a) at the end
(b) 2 ft
(c) 4 ft
(d) 5 ft
(e) 6 ft

Solution

This problem can be treated as a beam with a concentrated load of 100 lb 4 ft from one end.

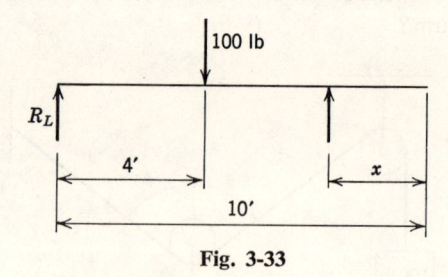

Fig. 3-33

For both men to carry an equal weight of 50 lb, they must be the same distance from the concentrated load.

The second man is $10 - 4 - 4 = 2$ ft from the other end $\blacktriangleright$

Alternative solution

Taking moments about R_L, we know that $\sum M_{R_L} = 0$

$$+4 \times 100 - 50(10 - x) = 0$$
$$400 - 500 + 50x = 0$$
$$50x = 100 \qquad x = 2 \text{ ft} \blacktriangleright$$

Answer is (b)

3-15 Wt. 2

Determine the theoretical mechanical advantage of the system shown in the figure.

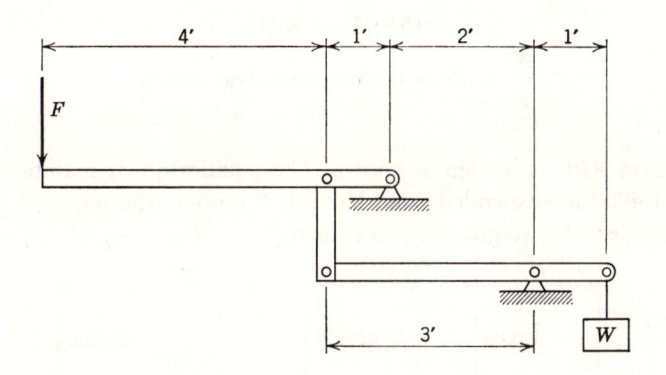

Fig. 3-34

Solution

The problem can be solved by considering it as two separate problems:

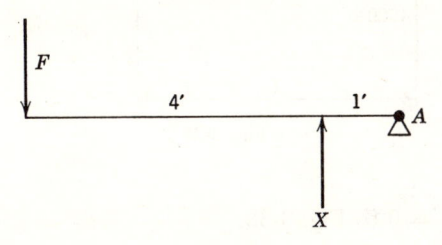

Fig. 3-35

$\sum M_A = 0$

$$X = \frac{5F}{1} = 5F$$

102 / Statics

(Thus the mechanical advantage for this section is 5.)

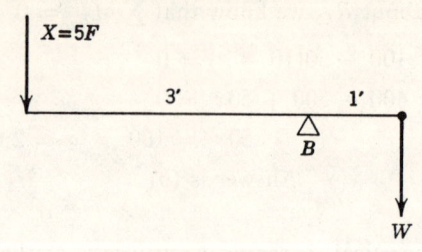

Fig. 3-36

$\sum M_B = 0$

$$(5F)(3) = W(1)$$

$$\frac{W}{F} = 15 = \text{mechanical advantage} \blacktriangleright$$

3-16 Wt. 3

The beam *ABC* is loaded as shown. The equilibrium is maintained by a weight of 4000 lb suspended from bar *DE*. Compute the required length *L* of *DE*. Neglect the weight of the members.

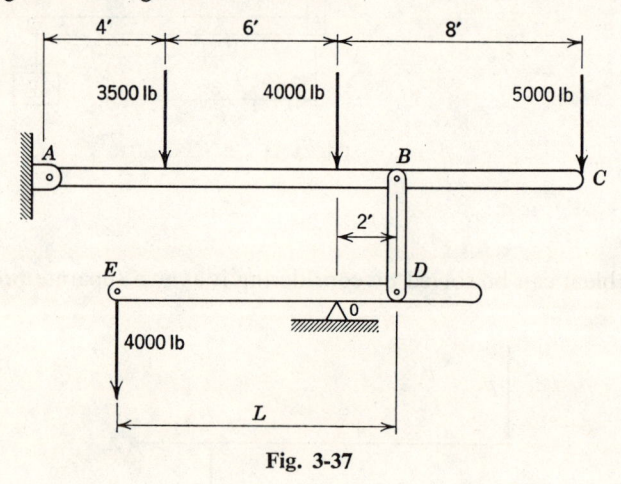

Fig. 3-37

Solution

Applying $\sum M_A = 0$ to Fig. 3-38,

$$\frac{(4 \times 3500) + (10 \times 4000)}{12} + \frac{(18 \times 5000)}{12} = \text{force at } B = 12{,}000 \text{ lb}$$

From Fig. 3-39, $\sum M_0 = 0$

$$12{,}000 \times 2 = 4000(L - 2) \qquad L = 8 \text{ ft} \blacktriangleright$$

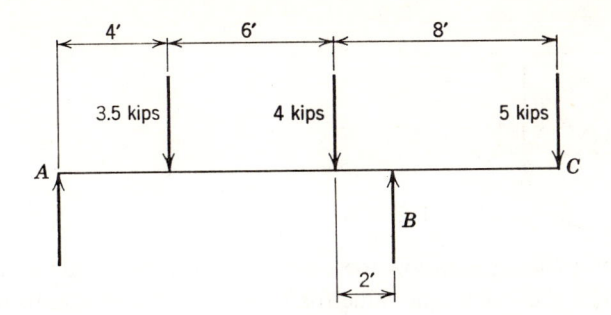

Fig. 3-38

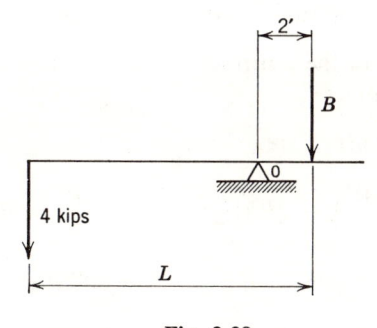

Fig. 3-39

3-17 Wt. 2

A homogeneous body is composed of a semicylinder and a rectangular parallelepiped as shown. Find the maximum value of h such that the system will be in a *stable* equilibrium on a horizontal plane. Assume no sliding between plane and cylinder.

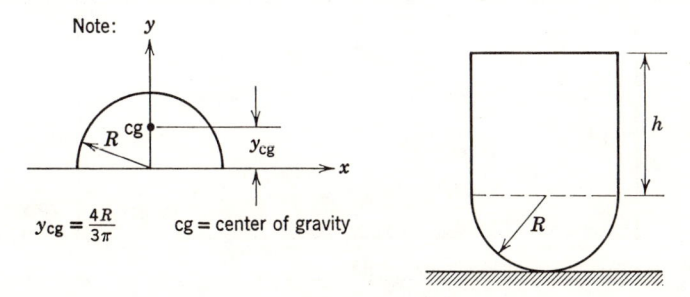

Note: $y_{cg} = \dfrac{4R}{3\pi}$ cg = center of gravity

Fig. 3-40

Solution

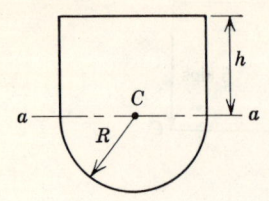

Fig. 3-41

To have stable equilibrium the centroid has to be in the semicylindrical portion; thus the maximum value for h occurs when the centroid is on line *a-a*. One of the properties of the centroid is that the first moment of area about it is equal to zero. Hence

$$0 = \text{moment of semicylinder} - \text{moment of rectangle}$$

Having the coordinates of the centroid of a semicircle given in the problem, the equation above becomes:

$$\overset{\text{area} \cdot \text{arm}}{\frac{\pi R^2}{2}\left(\frac{4R}{3\pi}\right)} - \overset{\text{area} \cdot \text{arm}}{2Rh\left(\frac{h}{2}\right)} = 0 \qquad \frac{2R^3}{3} = Rh^2$$

$$h^2 = \tfrac{2}{3}R^2 \qquad h = R\sqrt{\tfrac{2}{3}} = 0.82R \blacktriangleright$$

3-18 Wt. 6

A three-hinged frame must support a vertical monorail load of 12 kips and a lateral load of 10 kips in addition to the usual dead loads. What are the reactions at A and E owing to the loads shown? Neglect the dead load of the frame, and give both magnitude and direction for the reactions.

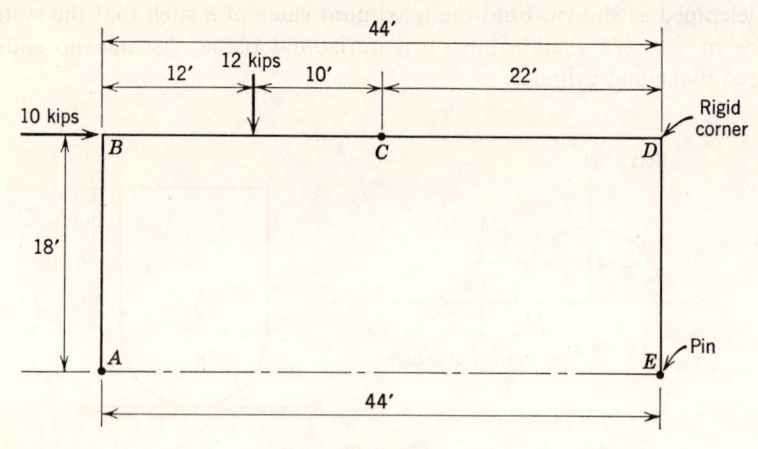

Fig. 3-42

Solution

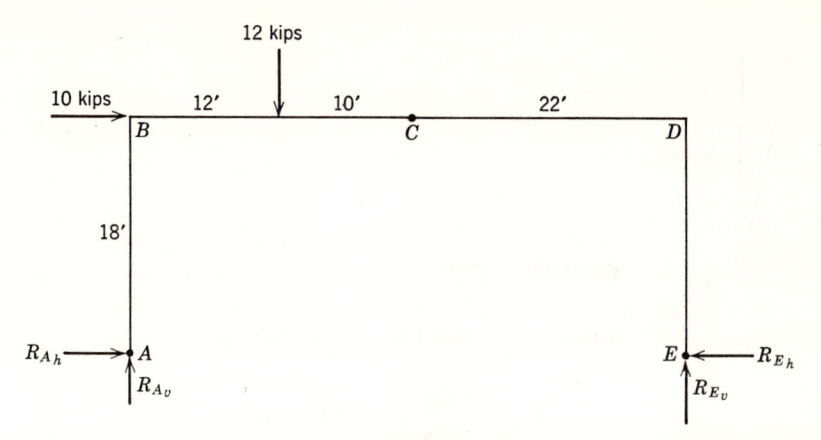

Fig. 3-43

$\sum M_C = 0$, $\sum M_E = 0$. Therefore the reactions at C and at E are parallel to a line connecting C and E.

$$11 R_{E_v} = 9 R_{E_h}$$

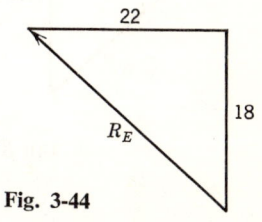

Fig. 3-44

The reactions at E on member CDE are equal to the reactions at C on member ABC. Therefore

$$11 R_{C_v} = 9 R_{C_h} \quad \text{or} \quad R_{C_v} = \tfrac{9}{11} R_{C_h}$$

$\overset{\frown}{+} \sum M_A = 0$ $\quad$ 10 kips × 18 ft + 12 kips × 12 ft − R_{C_v} × 22 ft
$\qquad\qquad\qquad - R_{C_h} \times 18 \text{ ft} = 0$
$\qquad\qquad\qquad 180 + 144 - 22(\tfrac{9}{11} R_{C_h}) - 18 R_{C_h} = 0$
$\qquad\qquad\qquad 324 - 18 R_{C_h} - 18 R_{C_h} = 0$
$\qquad\qquad\qquad R_{C_h} = \tfrac{324}{36} = 9 \text{ kips} = R_{E_h}$

$+ \rightarrow \sum F_X = 0$ $\quad$ 10 kips $- R_{C_h} + R_{A_h} = 10$ kips $- 9$ kips $+ R_{A_h} = 0$
$\qquad\qquad\qquad R_{A_h} = -1$ kip

$+ \uparrow \sum F_Y = 0$ $\quad$ $R_{A_v} - 12$ kips $+ R_{C_v} = R_{A_v} - 12$ kips $+ 7.36 = 0$
$\qquad\qquad\qquad R_{A_v} = 4.64$ kips

106 / Statics

Reaction A:

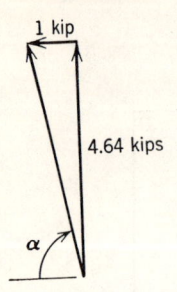

Fig. 3-45

$$\tan \alpha = \frac{4.64}{1} = 4.64 \qquad \alpha = 77.8° \blacktriangleright$$

$$R_A = \frac{4.64}{\sin \alpha} = 4.75 \text{ kips} \blacktriangleright$$

Reaction E:

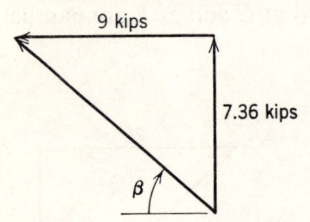

Fig. 3-46

$$\tan \beta = \frac{7.36}{9} = 0.818 \qquad \beta = 39.3° \blacktriangleright$$

$$R_E = \frac{7.36}{\sin \beta} = 11.63 \text{ kips} \blacktriangleright$$

3-19 Wt. 6

Given the pin-connected frame shown below with a 10-kip load at point D and a 5-kip load at point F. Determine the magnitude and direction of the resulting reactions at A, B, and C.

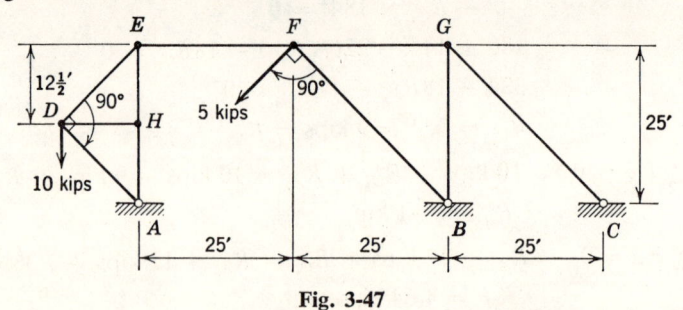

Fig. 3-47

Solution

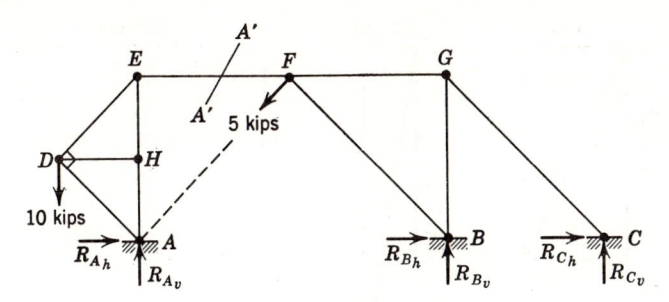

Fig. 3-48

Section $A' - A'$:

Left

$\sum M_E = 0 \;\; +\rangle \qquad 12.5(10) + R_{A_h}(25) = 0$

$\qquad\qquad\qquad\qquad R_{A_h} = -5 \text{ kips} \quad \text{or} \quad 5 \text{ kips}\leftarrow$

$\sum F_v = 0 \;\; \uparrow^+ \qquad R_{A_v} - 10 \text{ kips} = 0$

$\qquad\qquad\qquad\qquad R_{A_v} = 10 \text{ kips} \;\;\uparrow$

$\sum M_A = 0 \;\; +\rangle \qquad 10(12.5) - 25 F_{EF} = 0$

$\qquad\qquad\qquad\qquad F_{EF} = 5 \text{ kips (tension)}$

Right

$\sum M_B = 0 \;\; +\rangle \qquad 5 \text{ kips}(25)\sqrt{2} + 5(25) + 25 R_{C_v} = 0$

$\qquad\qquad\qquad\qquad R_{C_v} = -12.07 \text{ kips} = 12.07 \text{ kips} \;\;\downarrow$

$\sum F_v = 0 \;\; \uparrow^+ \qquad -12.07 \text{ kips} + R_{B_v} - \dfrac{5}{\sqrt{2}} = 0$

$\qquad\qquad\qquad\qquad R_{B_v} = 15.61 \text{ kips} \;\;\uparrow$

At joint C:

$\sum F_h = 0 \;\; \xrightarrow{+} \qquad -\dfrac{F_{GC}}{\sqrt{2}} + R_{C_h} = 0$

$\sum F_v = 0 \;\; \uparrow^+ \qquad \dfrac{F_{GC}}{\sqrt{2}} = 12.07 \text{ kips}$

$\qquad\qquad\qquad\qquad R_{C_h} = 12.07 \text{ kips} \;\;\rightarrow$

Total structure:

$\sum F_h = 0 \;\; \xrightarrow{+} \qquad -5 + R_{B_h} + 12.07 \text{ kips} - \dfrac{5}{\sqrt{2}} = 0$

$\qquad\qquad\qquad\qquad R_{B_h} = -3.53 \text{ kips} = 3.53 \text{ kips} \;\;\leftarrow$

Reactions:

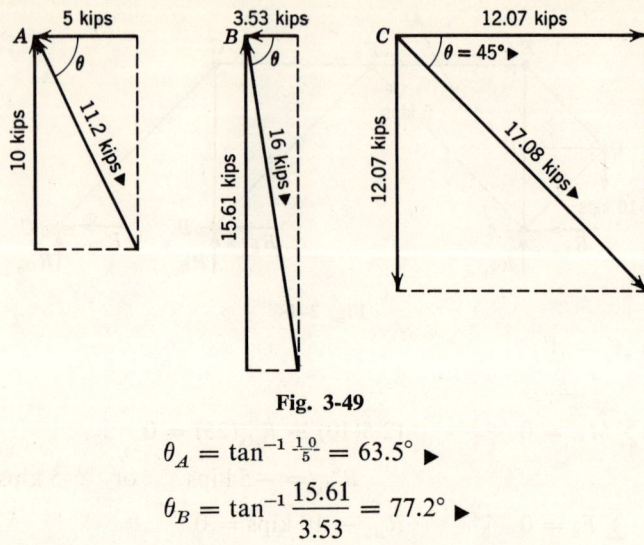

Fig. 3-49

$$\theta_A = \tan^{-1} \tfrac{10}{5} = 63.5° \blacktriangleright$$

$$\theta_B = \tan^{-1} \frac{15.61}{3.53} = 77.2° \blacktriangleright$$

3-20 Wt. 5

A tripod whose legs are each 10 ft long supports a load of 1000 lb. The feet of the tripod are at the vertices of a horizontal isosceles triangle whose base is 12 ft and whose altitude is 8 ft. Determine the total load in each leg.

Solution

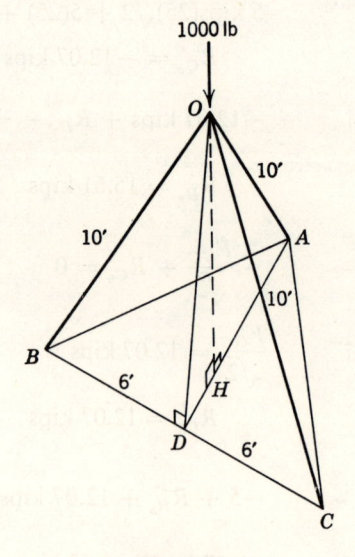

Fig. 3-50

From the figure,
$$OD = \sqrt{10^2 - 6^2} = \sqrt{64} = 8$$

Consider triangle ODA separately.

$$\overline{OH}^2 + \overline{DH}^2 = 8^2 \tag{1}$$

$$\overline{OH}^2 + (8 - DH)^2 = 10^2 \tag{2}$$

$$\underline{-\overline{OH}^2 - \overline{DH}^2 = -8^2} \quad -(1)$$

$$(8 - DH)^2 - \overline{DH}^2 = 36$$

$$64 - 16DH + \overline{DH}^2 - \overline{DH}^2 = 36$$

$$DH = \frac{64 - 36}{16} = 1.75 \text{ ft}$$

$$HA = 8 - 1.75 = 6.25 \text{ ft}$$

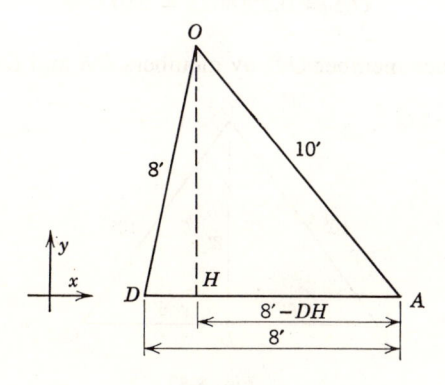

Fig. 3-51

Solving for the distance OH,

$$\overline{OH}^2 + \overline{DH}^2 = \overline{OD}^2$$

$$\overline{OH}^2 + 1.75^2 = 8^2$$

$$\overline{OH}^2 = 64 - 3.06 = 60.94$$

$$OH = 7.8 \text{ ft}$$

110 / Statics

With the dimensions calculated we can now solve triangle ODA, assuming that members OB and OC have been replaced by the single member OD.

$$\sum F_x = 0$$

$$OD_x - OA_x = 0 \qquad \frac{1.75}{8}OD = \frac{6.25}{10}OA \qquad OA = \frac{10 \times 1.75}{8 \times 6.25}OD$$

$$OA = 0.35\,OD$$

$$\sum F_y = 0$$

$$OD_y + OA_y - 1000 = 0 \qquad \frac{7.8}{8}OD + \frac{7.8}{10}OA - 1000 = 0$$

$$0.975\,OD + 0.78(0.35\,OD) = 1000$$

$$0.975\,OD + 0.273\,OD = 1000$$

$$OD = \frac{1000}{1.248} = 802 \text{ lb.}$$

$$OA = 0.35(802) = 280 \text{ lb} \blacktriangleright$$

Now we can replace member OD by members OB and OC.

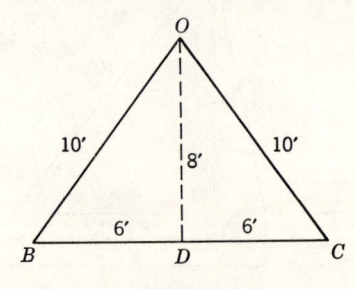

Fig. 3-52

$$\sum F_y = 0$$

$$OB_y + OC_y = OD$$

Since $OB = OC$ and $OD = 802$ lb,

$$2\,OB_y = 802 \qquad 2(\tfrac{8}{10})OB = 802$$

$$OB = OC = \frac{802 \times 10}{2 \times 8} = 501 \text{ lb} \blacktriangleright$$

(All legs are in compression.)

3-21 Wt. 6

The three-hinged arch ABC is loaded as shown. Determine the hinge force at B and the reactions at A and C (neglect weight of arch).

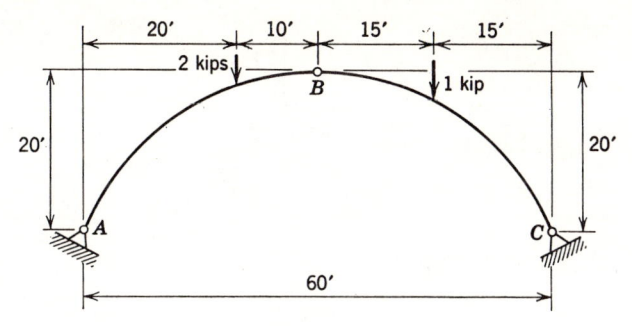

Fig. 3-53

Solution

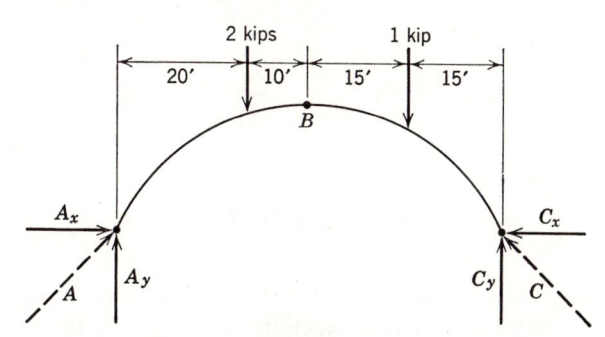

Fig. 3-54

$\sum M_A = 0 \quad (+ \; -)$

$$-2 \times 20 - 1 \times 45 + 60C_y = 0$$
$$C_y = \tfrac{85}{60} = 1.42 \text{ kips}$$

$\sum F_y = 0$

$$A_y - 2 - 1 + 1.42 = 0$$
$$A_y = 1.58 \text{ kips}$$

$\sum M_B = 0$ (left side)

$$2 \times 10 + 20A_x - 30(1.58) = 0$$
$$A_x = \frac{47.4 - 20}{20} = 1.37 \text{ kips}$$

$\sum F_x = 0$

$$A_x = C_x \qquad C_x = 1.37 \text{ kips}$$

112 / Statics

Reactions:

$$A = (1.37^2 + 1.58^2)^{1/2} = 2.10 \text{ kips} \blacktriangleright$$
$$C = (1.37^2 + 1.42^2)^{1/2} = 1.98 \text{ kips} \blacktriangleright$$

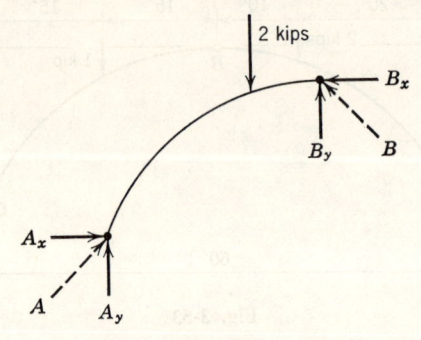

Fig. 3-55

Shear at hinge:

$$1.58 - 2 + B_y = 0 \qquad B_y = 0.42 \text{ kips}$$
$$B_x = A_x = 1.37 \quad \text{kips}$$

Reaction at hinge:

$$B = (1.37^2 + 0.42^2)^{1/2} = 1.43 \text{ kips} \blacktriangleright$$

3-22 Wt. 5

A contractor placed an unbalanced roof rafter as shown in the figure. Determine the components of the reactions at A and C. Do not give the

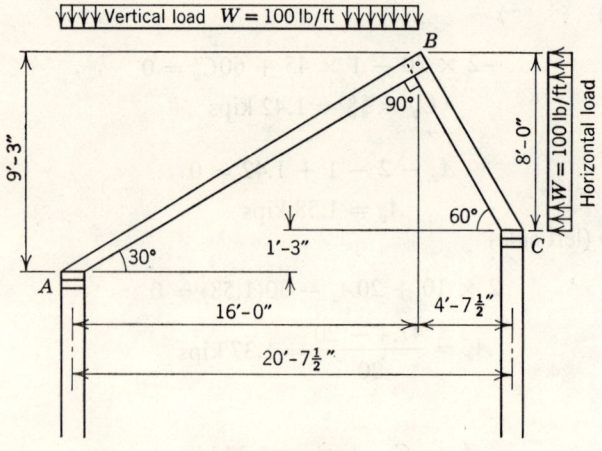

Fig. 3-56

final resultant reaction. The structure is assumed to be stable by other factors which are not part of the problem, and the dead load of the members may be neglected.

Solution

Free-body diagrams

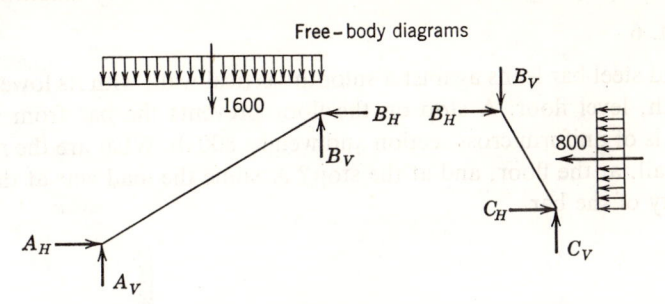

Fig. 3-57

$\sum M_A = 0$

$$16B_V + 9.25B_H - 8(1600) = 0$$

$$B_V = \frac{8(1600) - 9.25B_H}{16} = 800 - 0.578B_H \qquad (1)$$

$\sum M_C = 0$

$$4.625B_V - 8B_H + 4(800) = 0$$

Substituting in the value of B_V from equation (1)

$$4.625(800 - 0.578B_H) - 8B_H + 3200 = 0$$

$$3700 - 2.68B_H - 8B_H + 3200 = 0 \qquad B_H = \frac{6900}{10.68} = 646$$

$$B_V = 800 - 0.578(646) = 426$$

$\sum F_X = 0$

Member AB
$A_H - B_H = 0$
$A_H = B_H = 646$ ▶

Member BC
$B_H - 800 + C_H = 0$
$646 - 800 + C_H = 0$
$C_H = 154$ ▶

$\sum F_Y = 0$

Member AB
$A_V - 1600 + B_V = 0$
$A_V - 1600 + 426 = 0$
$A_V = 1174$ ▶

Member BC
$C_V - B_V = 0$
$C_V = B_V = 426$ ▶

114 / Statics

Answers:

$$A_H = 646 \blacktriangleright \quad A_V = 1174 \blacktriangleright$$
$$C_H = 154 \blacktriangleright \quad C_V = 426 \blacktriangleright$$

Magnitudes in pounds. Direction as indicated on free-body diagrams.

3-23 Wt. 6

A solid steel bar leans against a smooth vertical wall, with its lower end on a smooth, level floor. A stop on the floor prevents the bar from slipping. The bar is of uniform cross section and weighs 800 lb. What are the reactions at the wall, at the floor, and at the stop? Assume the load acts at the center of gravity of the bar.

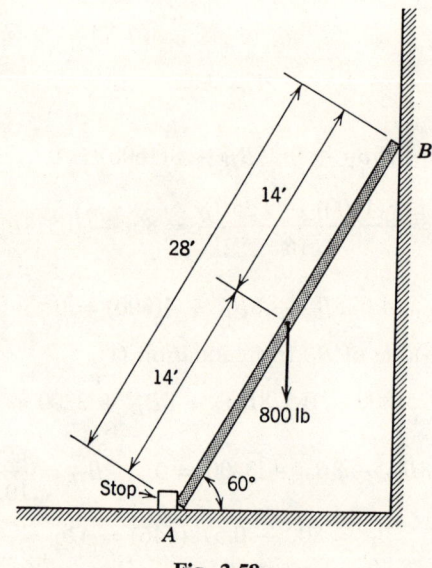

Fig. 3-58

Solution

The wall is said to be smooth, hence there is no friction, so $B_v = 0$ ▶
Referring to Fig. 3-59,

$$\sum M_A = 0 \; \curvearrowleft +$$

$$-800(14 \cos 60°) + B_h(28 \sin 60°) = 0$$

$$B_h = \tfrac{5600}{24.25} = 231 \text{ lb} \blacktriangleright$$

$$\sum F_x = 0$$

$$A_h - B_h = 0 \quad A_h - 231 = 0 \quad A_h = 231 \text{ lb} \blacktriangleright$$

$$\sum F_y = 0$$

$$A_v - 800 = 0 \quad A_v = 800 \text{ lb} \blacktriangleright$$

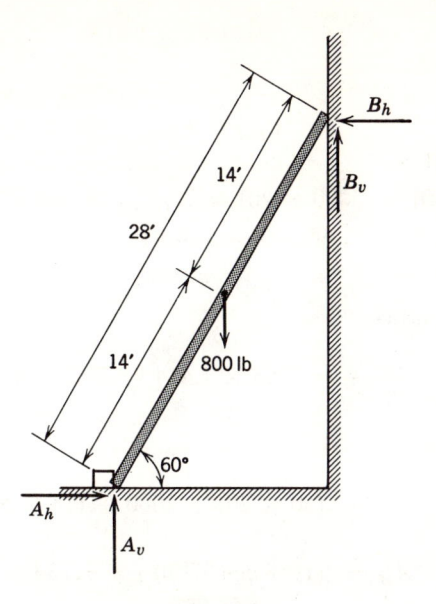

Fig. 3-59

3-24 Wt. 4

The boom and mast of the crane weigh 50 lb/lin ft. Neglect the weights of the other members. Solve for the reactions at F, G, and A when the plane of the boom bisects the angle GAF. A and B are hinge connections.

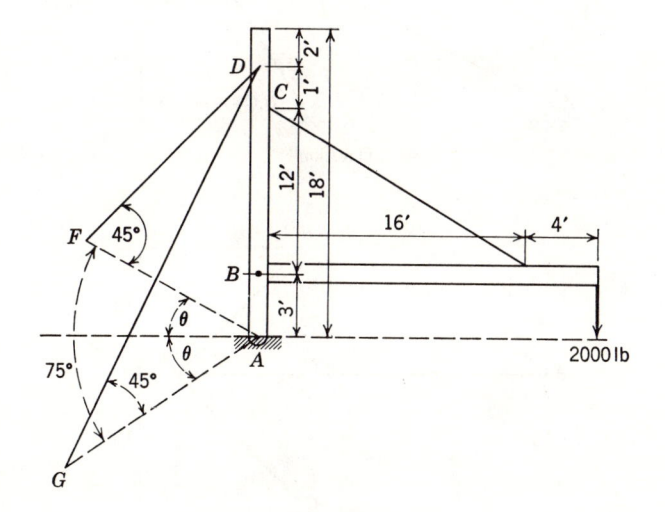

Fig. 3-60

116 / Statics

Solution

Reactions at F and G:
$$F_{DG} = F_{DF}$$
Taking moments at A:
$$(20 \times 50 \times 10) + (2000 \times 20) = 2F_{(DG)_H} \times \cos 37°30' \times 16$$
$$50{,}000 = 25.4 F_{(DG)_H}$$
$$F_{(DG)_H} = F_{(DF)_H} = 1968 \text{ (tension)}$$

Reactions at F and G:
$$\frac{F_{(DG)_H}}{\cos 45°} = 1968 \times 1.414 = 2782 \text{ lb} \blacktriangleright$$

Reactions at A:
$$\sum F_v = 0$$
$$A_v = 1968 + 1968 + (50 \times 20) + 2000 + (18 \times 50) = 7836 \text{ lb}$$
$$\sum F_H = 0$$
$$A_H = 2(1968 \cos 37°30') = 3123 \text{ lb}$$
$$\text{Reaction } A = \sqrt{3123^2 + 7836^2} = 8423 \text{ lb} \blacktriangleright$$

3-25 Wt. 4

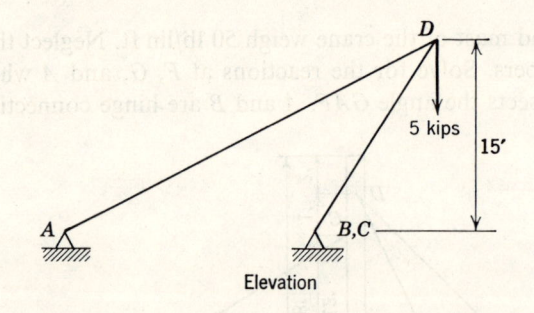

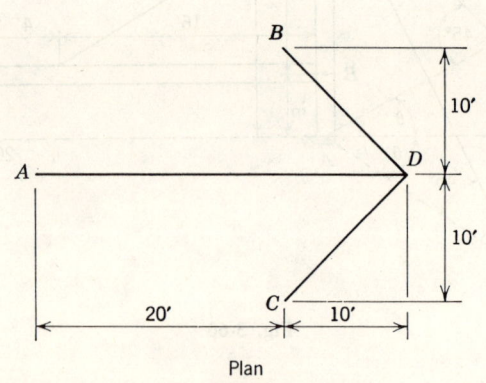

Fig. 3-61

Compute the force in each of the three members (*AD*, *BD*, and *CD*) of the space frame shown in the plan and elevation views in Fig. 3-61.

Solution

The easiest method of solution would be to determine the force in the frame assuming members *BD* and *CD* are replaced by a single member *ED*.
Length of members:

$$DE = (15^2 + 10^2)^{1/2} = (325)^{1/2} = 18.0 \text{ ft}$$
$$AD = (15^2 + 30^2)^{1/2} = (1125)^{1/2} = 33.6 \text{ ft}$$

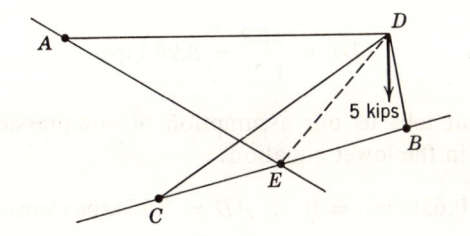

Fig. 3-62

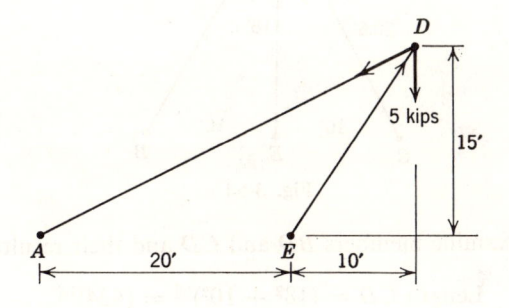

Fig. 3-63

$\sum F_y = 0$ (assume *DE* in compression and *AD* in tension)

$$-\frac{15}{33.6} AD + \frac{15}{18} DE - 5 = 0$$

$$-0.446 AD + 0.834 DE - 5 = 0$$
$$-AD + 1.87 DE - 11.2 = 0 \tag{1}$$

$\sum F_x = 0$

$$-\frac{30}{33.6} AD + \frac{10}{18} DE = 0$$

$$-0.893 AD + 0.556 DE = 0$$

$$-AD + 0.623 DE = 0 \qquad (2)$$

Subtracting (2) from (1):

$$-AD + 1.87 DE - 11.2 = 0$$
$$\underline{AD - 0.62 DE \qquad\quad = 0}$$
$$1.25 DE - 11.2 = 0$$

$$DE = \frac{11.2}{1.25} = 8.96 \text{ kips}$$

(The positive result tells us our assumption of compression was correct.) Substituting back in the lower equation:

$$AD - 0.62(8.96) = 0 \qquad AD = 5.55 \text{ kips (tension)} \blacktriangleright$$

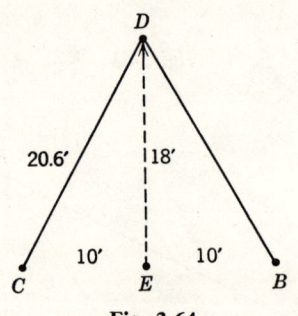

Fig. 3-64

Now we can examine members BD and CD and their resultant ED.

$$\text{Length } CD = (18^2 + 10^2)^{1/2} = (424)^{1/2}$$
$$= 20.6 \text{ ft}$$

By symmetry we can see that CD and BD have equal forces. Therefore, the "vertical" component of CD along $DE = \frac{1}{2} DE$.

$$BD_{\text{along } DE} = CD_{\text{along } DE} = \frac{1}{2}(8.96) = 4.48 \text{ kips}$$

$$BD = CD = \frac{20.6}{18}(4.48) = 5.13 \text{ kips} \blacktriangleright$$

Recap:
$$BD = CD = 5.13 \text{ kips compression} \blacktriangleright$$
$$AD = 5.55 \text{ kips tension} \blacktriangleright$$

3-26 Wt. 3

Given the pin-jointed structure shown. What is the force in member CE?

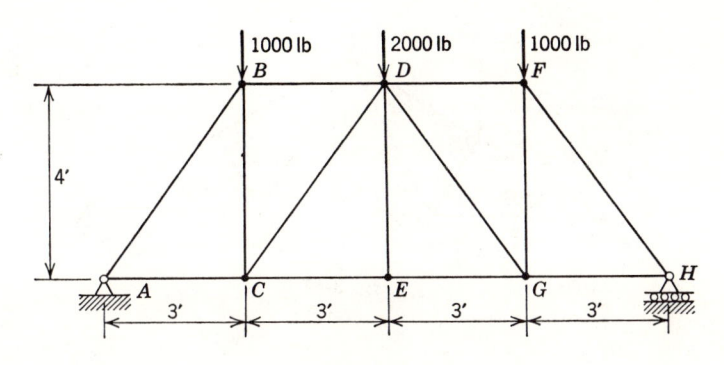

Fig. 3-65

Solution

It will be noted that the structure and the loading pattern are symmetrical. Therefore
$$R_L = R_R = \tfrac{4000}{2} = 2000 \text{ lb}$$
To find the force in member CE, take a section $X - X$ and analyze either side.

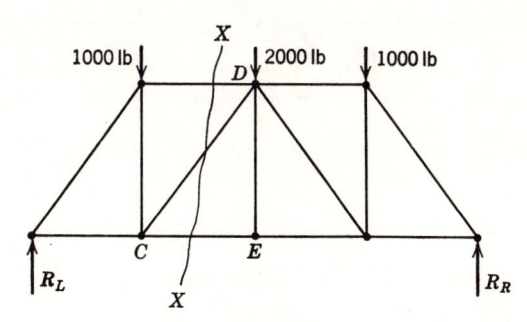

Fig. 3-66

$\Sigma M_D = 0$

$$2 \text{ kips} \times 6 \text{ ft} - 1 \text{ kip} \times 3 \text{ ft} - F_{CE} \times 4 \text{ ft} = 0$$

$$F_{CE} = \frac{12 - 3}{4} = 2.25 \text{ kips (tension)} \blacktriangleright$$

120 / Statics

3-27 Wt. 6

A pin-jointed arch truss is loaded as shown. Determine the forces in members *a* and *b*. (Member *b* is horizontal.)

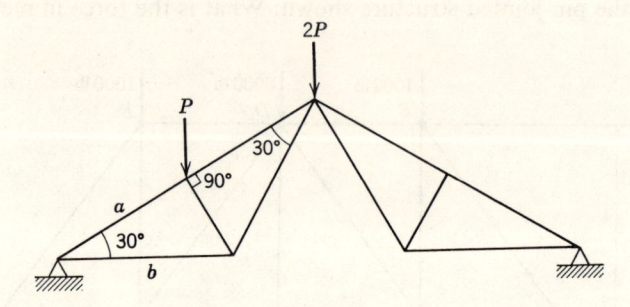

Fig. 3-67

Solution

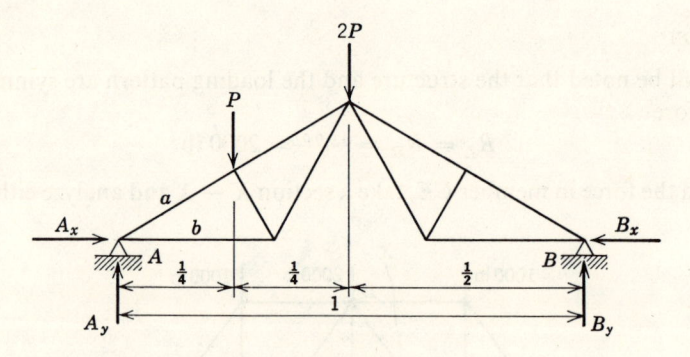

Fig. 3-68

$\sum M_A = 0 \quad (+\,-)$

$\qquad +1B_y - \frac{1}{2}(2P) - \frac{1}{4}(P) = 0 \qquad B_y = 1.25P$

$\sum F_y = 0 \quad +\uparrow\, -\downarrow$

$\qquad -P - 2P + 1.25P + A_y = 0 \qquad A_y = 1.75P$

$\sum F_x = 0$

$\qquad A_x - B_x = 0 \qquad A_x = B_x$

At joint *D* we see there is no force in member *e*. As a result there can be no vertical component of *g*, hence neither *f* nor *g* carry any load.

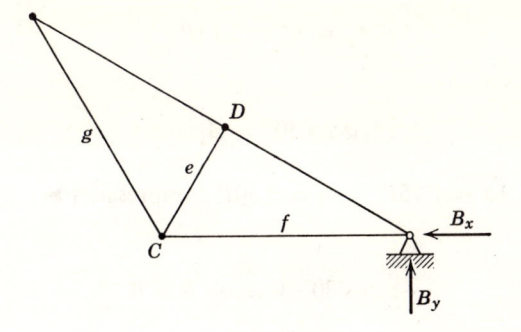

Fig. 3-69

Joint *B* becomes

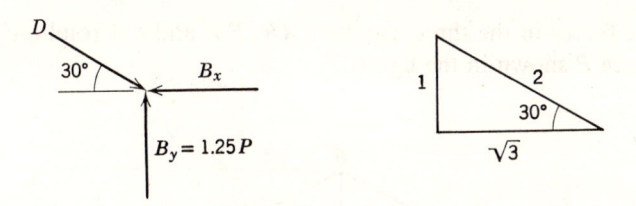

Fig. 3-70

$\sum V = 0$
$$D_v = B_y = 1.25P \qquad D = 2 \times 1.25P = 2.50P$$
$\sum H = 0$
$$B_x = D_H = \frac{\sqrt{3}}{2} \times 2.50P = 2.17P$$

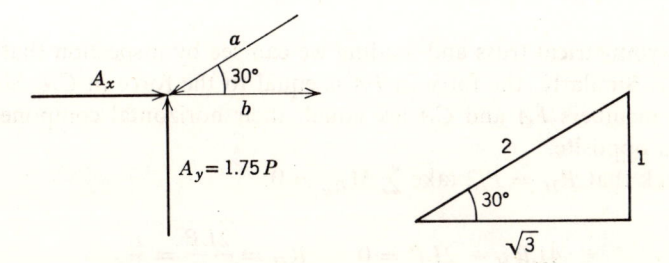

Fig. 3-71

Joint A:

$$A_x = B_x = 2.17P$$

$\sum V = 0$

$$a \sin 30° = A_y$$

$$\tfrac{1}{2}a = 1.75P \qquad a = 3.50P \text{ compression} \blacktriangleright$$

$\sum H = 0$

$$-a \cos 30° + A_x + b = 0$$

$$-3.50P(0.866) + 2.17P + b = 0$$

$$b = 3.03P - 2.17P = 0.86P \text{ tension} \blacktriangleright$$

3-28 Wt. 5

Find the forces in the three members AB, FA, and CA resulting from the vertical force P shown in the figure.

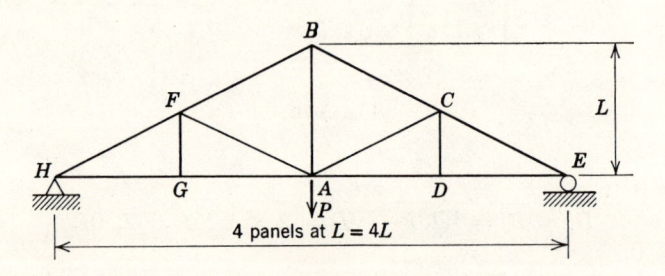

Fig. 3-72

Solution

For a symmetrical truss and loading we can see by inspection that $R_H = R_E = P/2$. Similarly, the force in FA is equal to the force in CA. Since the forces in members FA and CA are equal, their horizontal components are equal and opposite.

To check that $R_H = P/2$ take $\sum M_{R_E} = 0$:

$$4LR_H - 2LP = 0 \qquad R_H = \frac{2LP}{4L} = \frac{P}{2}$$

Method of joints:

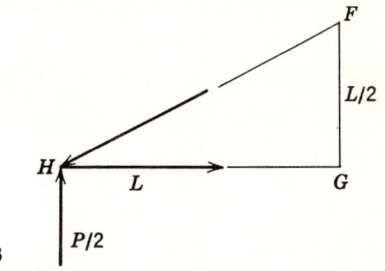

Fig. 3-73

Joint H. The vertical component of FH must equal $P/2$. The horizontal component will equal P, hence $HG = P$ (tension).

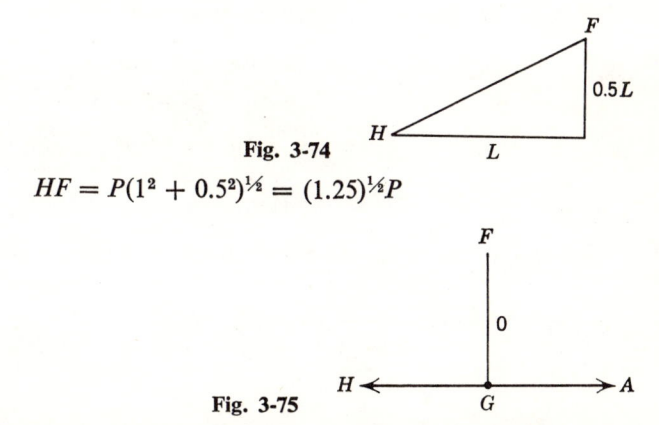

Fig. 3-74

$$HF = P(1^2 + 0.5^2)^{1/2} = (1.25)^{1/2} P$$

Fig. 3-75

Joint G. $GH = GA$. FG must be equal to zero. At joint H we established that HG was equal to P, hence GA must also be equal to P.

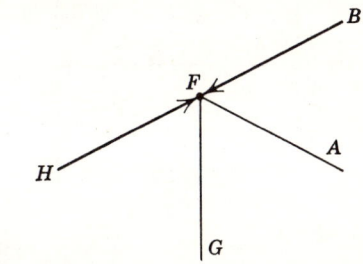

Fig. 3-76

Joint F. Since FG is zero, FH must equal FB and FA has no component perpendicular to FH, hence is equal to zero.

Joint A. From Fig. 3-77,

$$GA = AD = +P$$
$$FA = CA = 0$$
$$AB = P$$

124 / Statics

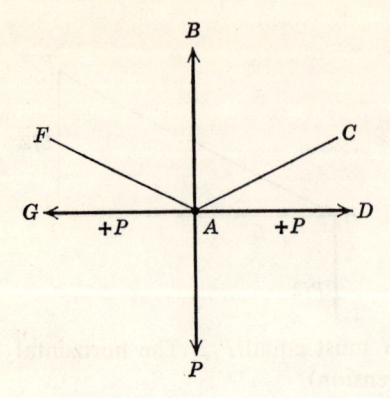

Fig. 3-77

We can now answer the question presented:

Member	Force
AB	$+P$ ▶
FA	0 ▶
CA	0 ▶

3-29 Wt. 5

Find the forces in the various members of the truss as shown.

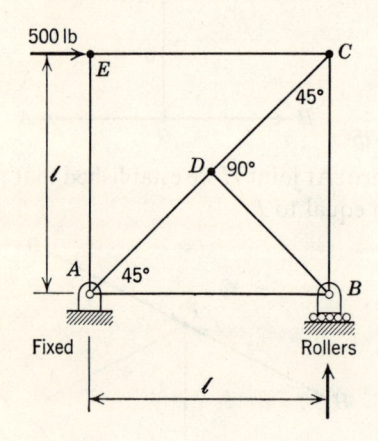

Fig. 3-78

Solution

$\sum M_A = 0$

$\qquad B_v \times \ell = 500 \times \ell \qquad B_v = 500 \text{ lb} \uparrow$ ▶

$\sum V = 0$

$\qquad A_v = 500 \text{ lb} \downarrow$ ▶

$\sum H = 0$

$\qquad A_h = 500 \text{ lb} \leftarrow$ ▶

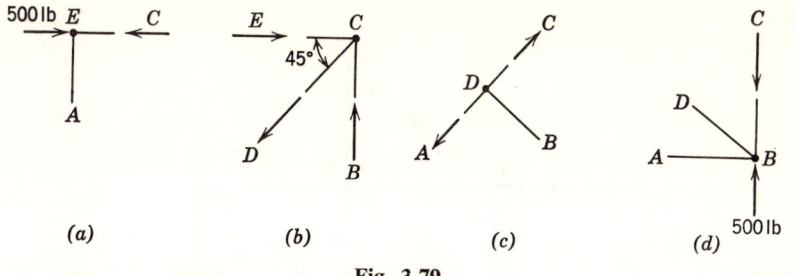

Fig. 3-79

By the method of joints:

(a) force in EC = 500 lb compression ▶
   force in AE = 0 lbs ▶
(b) H comp CD = 500 lb   force in CD = 707 lb tension
   V comp BC = V comp CD force in BC = 500 lb compression
(c) force in BD = 0 ▶
   force in AD = CD = 707 lb tension ▶
(d) force in AB = H comp BD = 0
   force in AB = 0 ▶
   force in BC = 500 lb compression ▶

3-30 Wt. 6

A pin-connected truss has a horizontal load of 2000 lb and a vertical load of 1200 lb as shown. Determine the total axial force in members CD, AD, and FD. Give the magnitude and state whether the force is tension or compression.

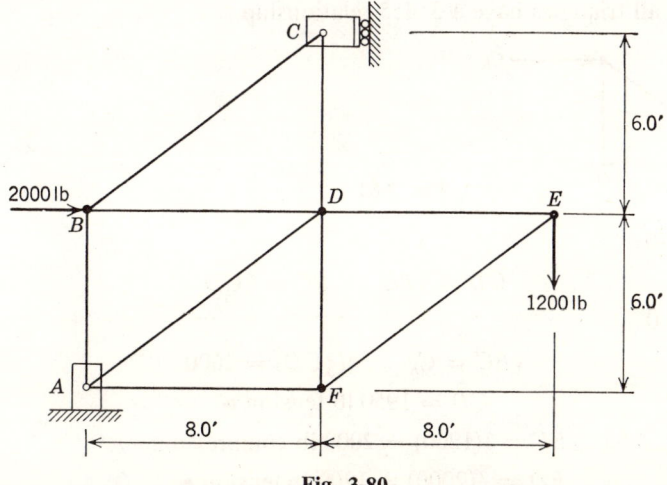

Fig. 3-80

126 / Statics

Solution

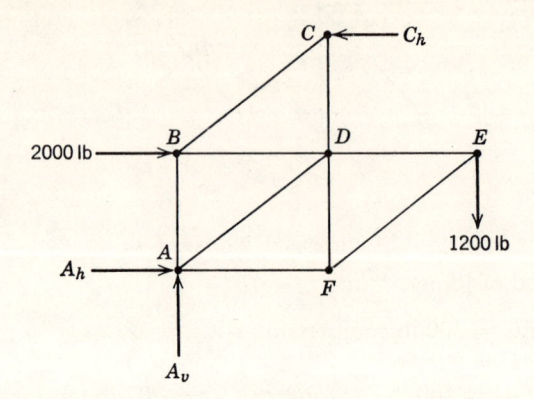

Fig. 3-81

$\sum F_y = 0$

$$A_v - 1200 = 0 \qquad A_v = 1200 \text{ lb}$$

$\sum M_A = 0 \; (+$

$$-1200 \times 16 + C_h \times 12 - 2000 \times 6 = 0$$

$$C_h = \frac{19{,}200 + 12{,}000}{12} = 2600 \text{ lb}$$

$\sum F_x = 0$

$$A_h + 2000 - C_h = 0$$

$$A_h + 2000 - 2600 = 0 \qquad A_h = 600 \text{ lb}$$

Note that all triangles have a 3:4:5 relationship.

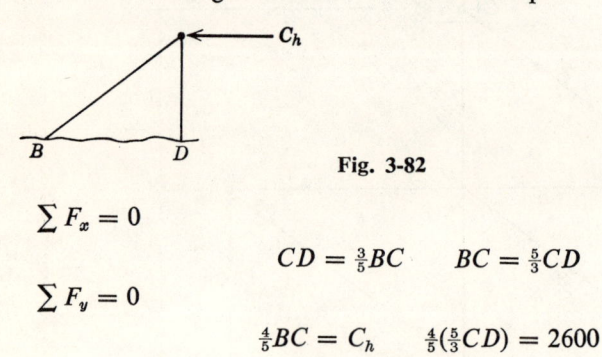

Fig. 3-82

$\sum F_x = 0$

$$CD = \tfrac{3}{5} BC \qquad BC = \tfrac{5}{3} CD$$

$\sum F_y = 0$

$$\tfrac{4}{5} BC = C_h \qquad \tfrac{4}{5}(\tfrac{5}{3} CD) = 2600$$

$$CD = 1950 \text{ lb tension} \blacktriangleright$$

From Fig. 3-83, $EF = \tfrac{5}{3}(1200) = 2000 \text{ lb compression}$

$$FD = \tfrac{3}{5}(2000) = 1200 \text{ lb tension} \blacktriangleright$$

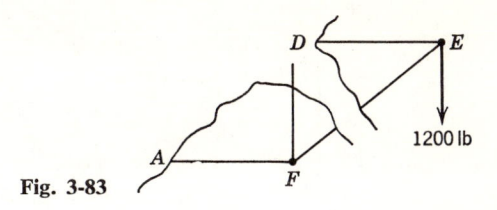

PROBLEMS / 127

Fig. 3-83

At joint D $\quad \sum F_y = 0$

$$+1950 - 1200 - \tfrac{3}{5}AD = 0$$

$$AD = \tfrac{5}{3}(750) = 1250 \text{ lb tension} \blacktriangleright$$

3-31 Wt. 6

Determine graphically the force in each member of the truss for the loads and reactions shown. Use a scale of 2 kips/in.

Note: No credit will be given for an algebraic solution

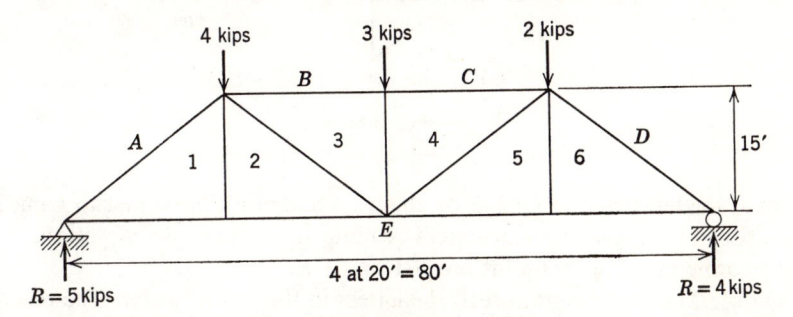

Fig. 3-84

Solution

The graphical solution desired (only a graphical solution is acceptable) is a *Maxwell diagram*. To construct a Maxwell diagram, start by drawing the force polygon for the truss as a free body (Fig. 3-85a). Work in clockwise order. Then we can examine the joints individually. For joint E-A-1 we have Fig. 3-85b. We do the same for the rest of the joints until we obtain a complete Maxwell diagram, as in Fig. 3-85c.

The Maxwell diagram appears to be properly drawn. Looking again at the truss we see that the force in members 1-2 and 5-6 must be zero. As an additional check, we see that A-1 has a greater load than D-6.

Now the magnitude of the force in every member can be obtained by scaling the length of the corresponding line on the Maxwell diagram. The direction of the force can also be determined from the diagram. Consider the

128 / Statics

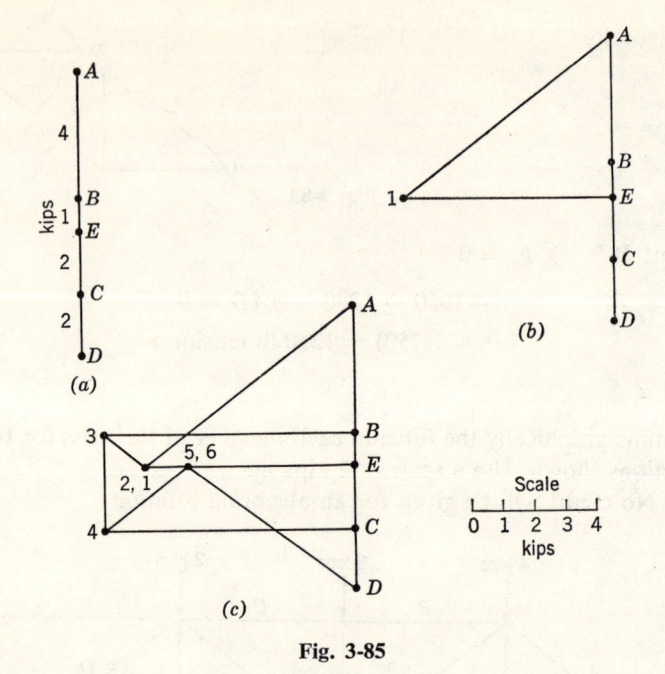

Fig. 3-85

joint at the left reaction (*A*-1-*E* on the truss diagram). Now we can name the members by assigning them letters reading in a clockwise direction. Thus the two members at this joint are *A*-1 and 1-*E*.

On the Maxwell diagram read the letters in their same order. The reaction of the member upon the joint is in the same direction in which the force is read. Reading the diagonal member (*A*-1) on the Maxwell diagram shows that *A*-1 acts downward or toward the joint—hence *A*-1 is in compression. Reading the horizontal member (1-*E*) on the Maxwell diagram shows that 1-*E* goes from left to right or away from joint *A*, hence member 1-*E* is in tension.

The forces in the members, scaled from the diagram, are:

A-1	8.3 kips compression ▶	4-3	3.0 kips compression ▶
1-*E*	6.7 kips tension ▶	*D*-6	6.7 kips compression ▶
1-2	0 ▶	6-5	0 ▶
B-3	8.0 kips compression ▶	5-4	3.4 kips tension ▶
3-2	1.6 kips tension ▶	*E*-5	5.2 kips tension ▶
2-*E*	6.7 kips tension ▶	6-*E*	5.2 kips tension ▶
C-4	8.0 kips compression ▶		

3-32 Wt. 5

The two blocks shown are connected by a cord passing over a smooth pulley. Will the system move if the coefficient of friction under the 40-lb block is 0.25 and that under the 100-lb block is 0.4? Show the calculations that justify your answer.

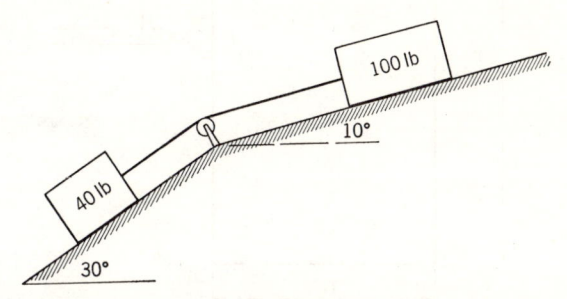

Fig. 3-86

Solution

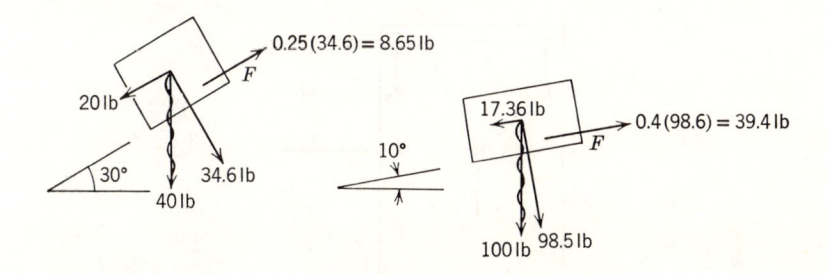

Fig. 3-87

Forces down planes $17.36 + 20.0 = 37.36$

Friction forces $8.65 + 39.4 = 48.05$

System will *not* move ▶

3-33 Wt. 6

A hangar door weighs 5 psf and is supported on two rollers. The rollers have rusted, causing the door to slide along the track when moved. If the door must be opened in an emergency by a materials-handling machine pushing horizontally at point P, what is the maximum distance d that will not cause one roller to leave the track? Assume that the coefficient of friction is equal to 0.40.

130 / Statics

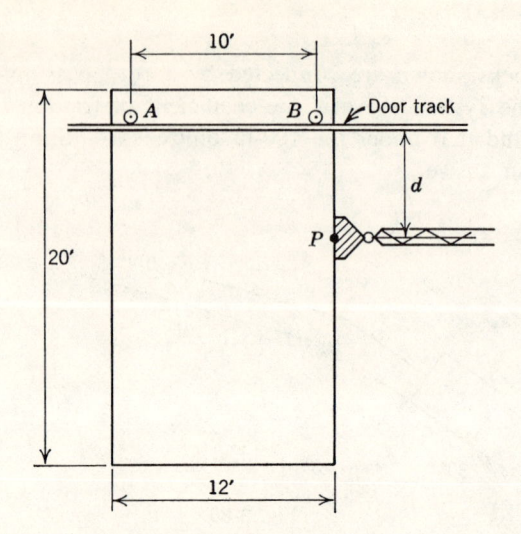

Fig. 3-88

Solution

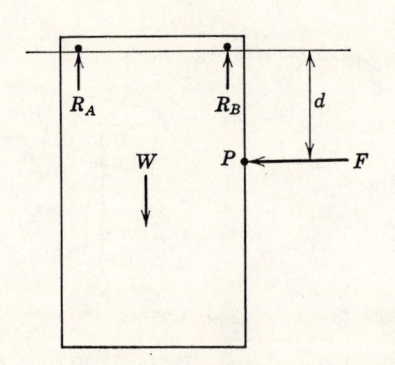

Fig. 3-89

$$\text{Total weight of door} = 20 \text{ ft} \times 12 \text{ ft} \times 5 \text{ psf}$$
$$W = 1200 \text{ lb}$$

located midway between A and B.

In order to slide the door along the track a force $F = fN$ must be applied

$$F = 0.40 \times 1200 = 480 \text{ lb}$$

Taking moments about roller B we see that roller A will lift off the track only when the moment Fd is greater than $5W$

$$Fd = 5W \qquad 480d = 5 \times 1200 \qquad d_{\max} = \tfrac{6000}{480} = 12.5 \text{ ft} \blacktriangleright$$

3-34 Wt. 3

A solid block 5 in. × 5 in. × 8 in. weighs 200 lb. What force P is required to cause the block to slide if $\mu = 0.25$?

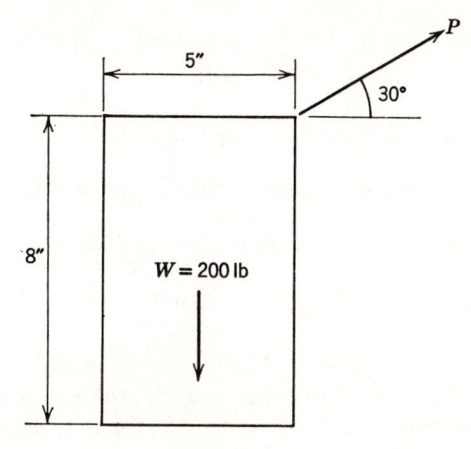

Fig. 3-90

Solution

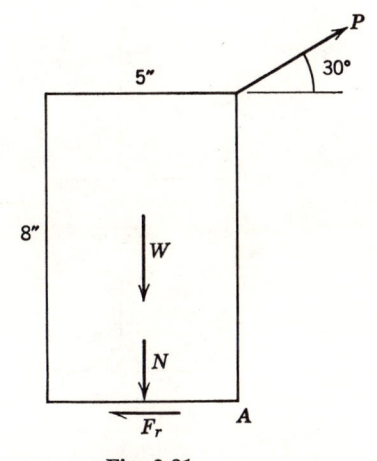

Fig. 3-91

$$W = 200 \text{ lb} \qquad \mu = 0.25$$

$$\Sigma F_H = 0 \qquad F_r = P \cos 30°$$

$$\Sigma F_V = 0 \qquad N = W - \frac{P}{2}$$

132 / Statics

Friction relation $F_r = \mu N$

$$P\frac{\sqrt{3}}{2} = \frac{1}{4}\left(200 - \frac{P}{2}\right) \quad P\left(\frac{\sqrt{3}}{2} + \frac{1}{8}\right) = 50$$

$$P = 50\left(\frac{8}{7.93}\right) = 50.5 \text{ lb} \blacktriangleright$$

The block will slide at $P = 50.5$ lb if it does not overturn. Around point A,

Overturning moment = $50.5 \cos 30°(8) = 350$ in-lb

Righting moment = $200(\frac{5}{2}) = 500$ in-lb

Therefore at $P = 50.5$ lb the block will not overturn; it will slide.

3-35 Wt. 5

The cam arrangement shown was designed to develop large friction forces upon a cable which is subjected to a tension force. If the tension force on the cable is 800 lb and the coefficient of friction is 0.3, complete the calculations to demonstrate whether the cable will or will not slip.

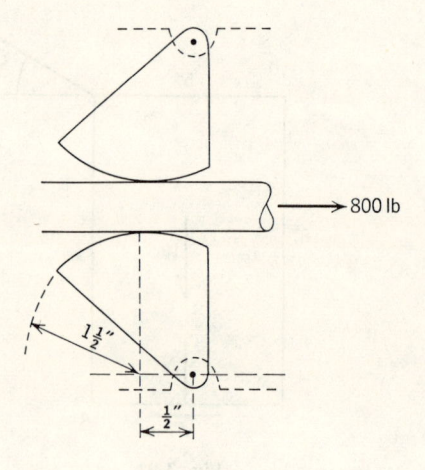

Fig. 3-92

Solution

From the cable in Fig. 3-93, $2F = 800$:

$$F = 400 \text{ lb}$$

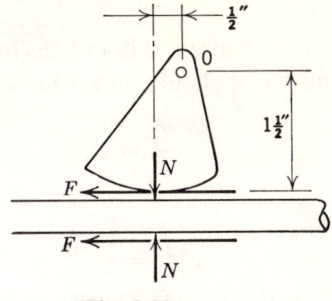

Fig. 3-93

From the cam, $\sum M_0 = 0$:

$$\tfrac{1}{2}N = 1.5F \qquad N = 3F$$

$$N = 3 \times 400 = 1200 \text{ lb}$$

The maximum friction possible $F = 0.3 \times 1200 = 360$ lb. $2F = 720$, which is insufficient to hold 800 lb. Therefore the cable will slip ▶

3-36 Wt. 5

For a belt passing over a pulley, the ratio of the forces is given as

$$\frac{T_2}{T_1} = e^{\mu\beta}$$

where μ is the coefficient of friction and β is the angle of contact. For the pulley shown, calculate the ratio T_2/T_1 for $\mu = 0.30$.

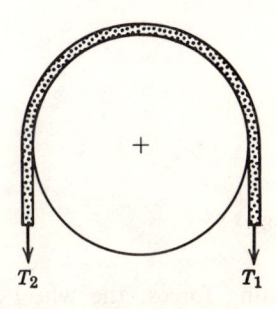

Fig. 3-94

Solution

The angle of contact β is expressed in radians in this equation. Since there are 2π rad in 360° and the angle of contact in this case is 180°, $\beta = \pi$ rad.

$$\frac{T_2}{T_1} = e^{0.3\pi}$$

Slide rule solution

$$e^{0.3\pi} = e^{0.943}$$

Set indicator to 0.943 on the D scale. On the LL2 scale read the value under the indicator (2.57). Therefore

$$\frac{T_2}{T_1} = 2.57 \blacktriangleright$$

3-37 Wt. 1

In a problem involving rolling friction, the value of it would be given as

(a) an angle ϕ
(b) μ_s, the tangent F_s/N of an angle ϕ (F_s = static friction)
(c) μ_k, the tangent F_k/N of an angle ϕ (F_k = kinetic friction)
(d) r_f, the radius of the friction circle ($r_f = r \sin \phi$)
(e) b, a deformation, as a linear dimension

Solution

In rolling friction we encounter a situation where the "ideal" situation fails to give us any appreciation of the actual situation. If we were to place a rigid wheel on a rigid smooth surface and set the wheel in motion, the forces would appear as in Fig. 3-95.

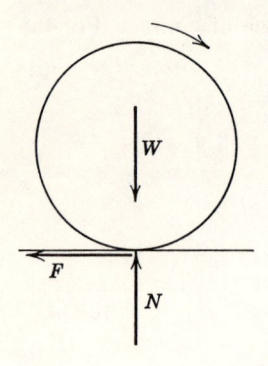

Fig. 3-95

In the absence of retarding forces, the wheel would theoretically roll forever. In the actual case there is deflection of the surface, or of the wheel, or

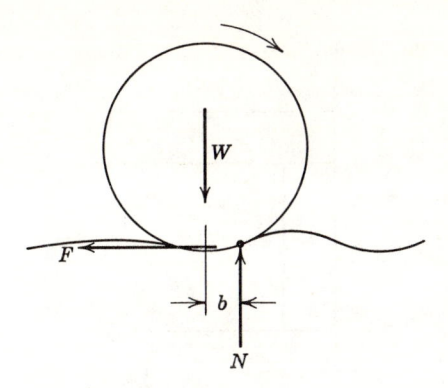

Fig. 3-96

more likely both. The result is that the normal force N acts ahead of the line of action of the weight W. This small distance (b) is called the coefficient of rolling friction and is a linear dimension ▶

Answer is (e)

3-38 Wt. 3

Locate the centroid of the figure shown.

Suggestion: Locate the centroid by dimensions from the left side and bottom of the figure.

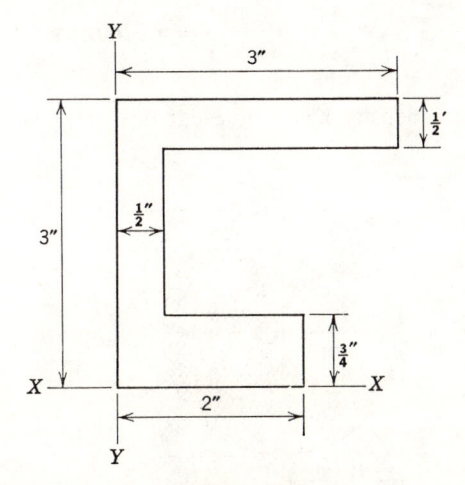

Fig. 3-97

136 / Statics

Solution

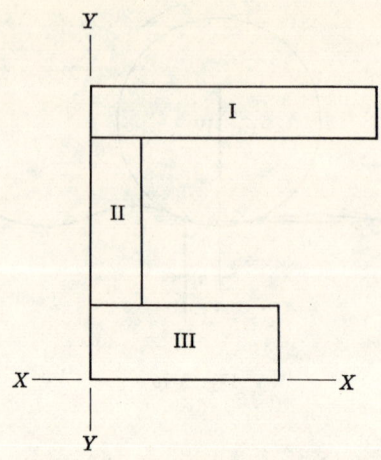

Fig. 3-98

Section	b (in.)	h (in.)	A (in.²)	X (in.)	Y (in.)	AX (in.³)	AY (in.³)
I	3.0	0.5	1.5	1.5	2.75	2.25	4.125
II	0.5	1.75	0.875	0.25	1.625	0.22	1.422
III	2.0	0.75	1.5	1.0	0.375	1.5	0.562
			3.875			3.97	6.109

$$\bar{X} = \frac{\sum AX}{\sum A} = \frac{3.97}{3.875} = 1.025 \text{ in.} \blacktriangleright$$

$$\bar{Y} = \frac{\sum AY}{\sum A} = \frac{6.109}{3.875} = 1.576 \text{ in.} \blacktriangleright$$

3-39 Wt. 3

A disk of uniform density has a hole cut out of it, as shown. Find the center of mass.

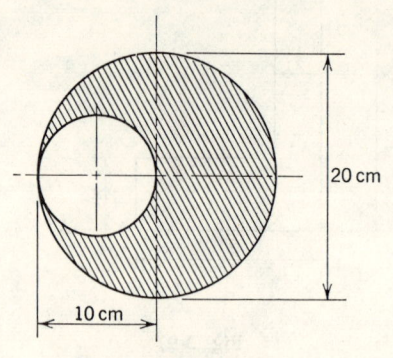

Fig. 3-99

Solution

The center of mass, or center of gravity as it probably is more frequently called, is the point through which the resultant of the total weight of the object will pass regardless of the orientation of the object.

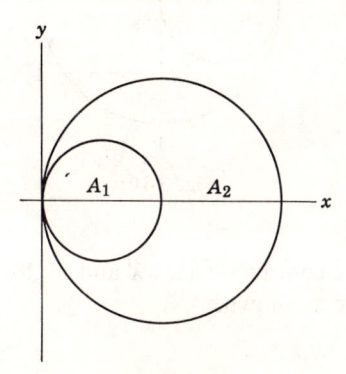

Fig. 3-100

Since the disk has an axis of symmetry, the center of mass is on that axis and $\bar{y} = 0$. Also

$$\bar{x} = \frac{\int x\, dA}{\int dA} = \frac{\Sigma Ax}{\Sigma A}$$

$$A_1 = \frac{\pi}{4} 10^2 = 25\pi$$

A_2 (total area of disk incl. A_1) $= \dfrac{\pi}{4} 20^2 = 100\pi$

$$\bar{x} = \frac{A_2 x_2 - A_1 x_1}{A_2 - A_1} = \frac{100\pi \times 10 - 25\pi \times 5}{100\pi - 25\pi}$$

$$= \frac{1000\pi - 125\pi}{75\pi} = \frac{875\pi}{75\pi} = 11\tfrac{2}{3} \text{ cm}$$

$$\bar{x} = 11\tfrac{2}{3} \text{ cm} \blacktriangleright$$

$$\bar{y} = 0 \qquad \blacktriangleright$$

138 / Statics

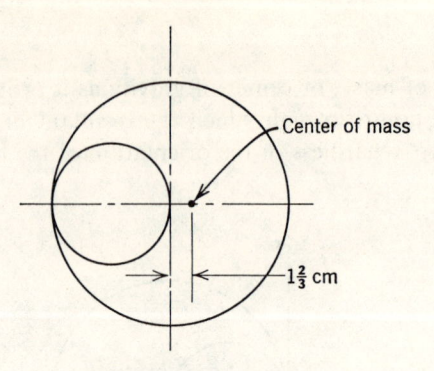

Fig. 3-101

3-40 Wt. 4

With reference to the coordinate axes X and Y, locate the centroid of the area of the plane figure as shown.

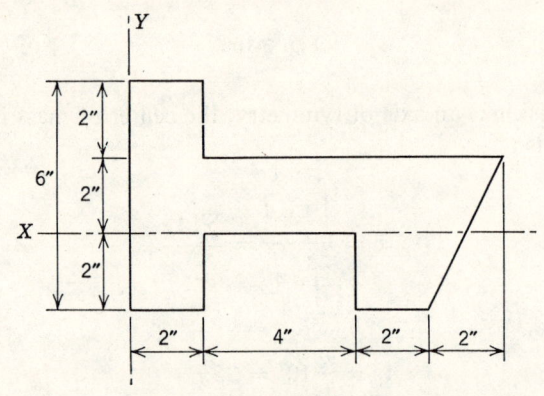

Fig. 3-102

Solution

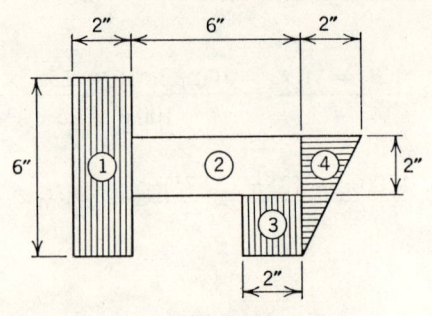

Fig. 3-103

$$\bar{Y} = \frac{\sum\limits_{i=1}^{n} A_i \bar{y}_i}{\sum\limits_{i=1}^{n} A_i} \quad \frac{12(1) + 12(1) + 4(-1) + 4(2 - \frac{4}{3})}{2(19) + 2(4)} = \frac{22.67}{32} = 0.71 \text{ in.} \blacktriangleright$$

$$\bar{X} = \frac{\sum\limits_{i=1}^{n} A_i \bar{x}_i}{\sum\limits_{i=1}^{n} A_i} = \frac{12(1) + 12(5) + 4(7) + 4(8 + \frac{2}{3})}{32} = \frac{134.7}{32} = 4.21 \text{ in.} \blacktriangleright$$

3-41 Wt. 1

The formula $I = \int y^2 \, dA$ represents the

(a) product of inertia
(b) section modulus
(c) area of cross section
(d) moment of inertia
(e) modulus of elasticity

Solution

The formula represents the moment of inertia $\blacktriangleright$

Answer is (d)

3-42 Wt. 1

The term $I = bh^3/12$ refers to the

(a) radius of gyration
(b) section modulus
(c) instantaneous center
(d) moment of inertia
(e) product of inertia

Solution

$I = bh^3/12$ is the equation for the moment of inertia for a rectangular cross section $\blacktriangleright$

Answer is (d)

140 / Statics

3-43 Wt. 2

The following data are taken from the AISC Handbook for an angle 5 in. × 3 in. × $\frac{1}{2}$ in.:

$$\begin{aligned}
\text{Wt/ft} &= 12.8 \text{ lb} & \text{Area} &= 3.75 \text{ in.}^2 \\
I_{xx} &= 9.5 \text{ in.}^4 & I_{yy} &= 2.6 \text{ in.}^4 \\
r_{xx} &= 1.59 \text{ in.} & r_{yy} &= 0.83 \text{ in.} \\
y &= 1.75 \text{ in.} & x &= 0.75 \text{ in.} \\
r_{zz} &= 0.65 \text{ in.} & \tan \alpha &= 0.357
\end{aligned}$$

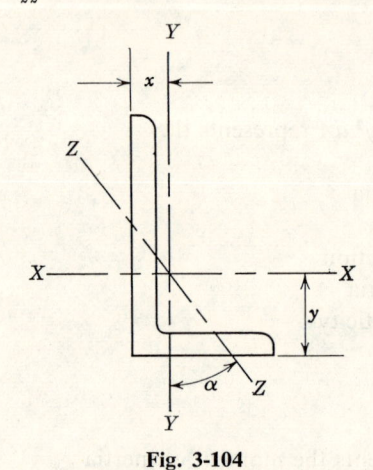

Fig. 3-104

1. The polar moment of inertia is nearest

(a) 5 in.^{4}
(b) 7 in.^{4}
(c) 12 in.^{4}
(d) 19 in.^{4}
(e) 25 in.^{4}

2. The minimum moment of inertia for the angle is nearest

(a) 0.9 in.^{4}
(b) 1.6 in.^{4}
(c) 2.4 in.^{4}
(d) 2.6 in.^{4}
(e) 5.4 in.^{4}

Solution

1. The polar moment of inertia for an area with respect to an axis perpendicular to its plane is equal to the sum of the moments of inertia about

any two mutually perpendicular axes in its plane which intersect on the polar axis. Thus

$$J_z = I_{xx} + I_{yy} = 9.5 + 2.6 = 12.1 \text{ in.}^4 \blacktriangleright$$

Answer is (c)

2. In the AISC Handbook r_z is equal to the least radius of gyration. The radius of gyration is defined as $(I/A)^{1/2}$; thus $r_z = \sqrt{I_{min}/A}$.

$$I_{min} = r_z^2 A = 0.65^2 \times 3.75 = 1.58 \text{ in.}^4 \blacktriangleright$$

Answer is (b)

3-44 Wt. 1

The moment of inertia of the area shown in the figure about the x-x axis is

(a) 28 in.⁴
(b) 40 in.⁴
(c) 64 in.⁴
(d) 110 in.⁴
(e) 256 in.⁴

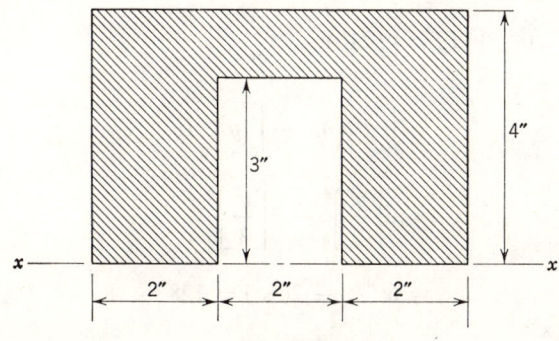

Fig. 3-105

Solution

The moment of inertia $I_x = \int y^2 \, dA$. Thus the moment of inertia of a cross-sectional area is equal to the sum of the differential areas dA multiplied by the square of their moment arms about the reference axis.

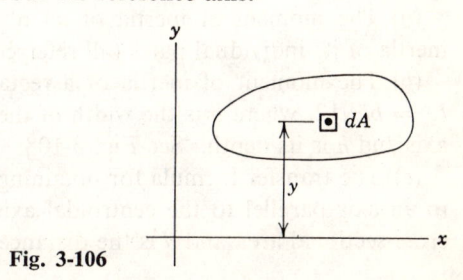

Fig. 3-106

142 / Statics

This problem can be solved in two ways:

1. Integration of the differential areas

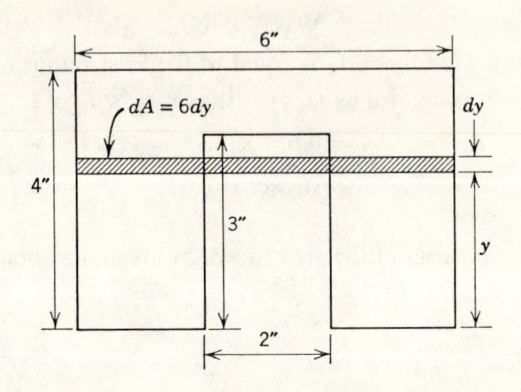

Fig. 3-107

$$I_x = \int y^2 \, dA$$

$$= \int_0^4 y^2 6 \, dy - \int_0^3 y^2 2 \, dy$$

$$= 6\left[\frac{y^3}{3}\right]_0^4 - 2\left[\frac{y^3}{3}\right]_0^3$$

$$= 6(\tfrac{64}{3}) - 2(\tfrac{27}{3}) = 128 - 18$$

$$= 110 \text{ in.}^4 \blacktriangleright$$

2. Transfer of the moment of inertia, with respect to the centroid, to a parallel axis. In this method three facts are utilized:

(a) The moment of inertia of an object is the sum of the moments of inertia of its individual parts (all referred to the same axis).

(b) The moment of inertia of a rectangle with respect to its centroid is $I_{x_0} = bh^3/12$, where b is the width of the rectangle parallel to the centroidal axis and h is its depth. See Fig. 3-108.

(c) The transfer formula for obtaining the moment of inertia with respect to an axis parallel to the centroidal axis is $I_x = I_{x_0} + Ad^2$, where A is the cross-sectional area and d is the distance between the axes.

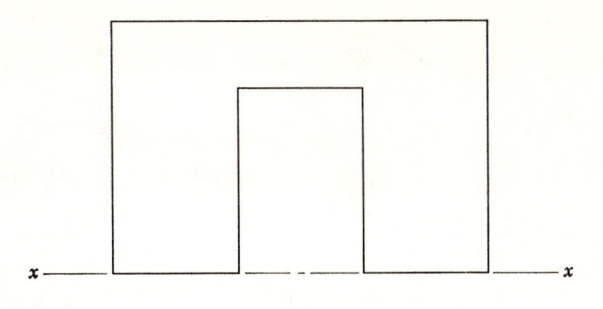

Fig. 3-108

For a rectangle

$$I_x = \frac{bh^3}{12} + bh\left(\frac{h}{2}\right)^2 = \frac{bh^3}{3}$$

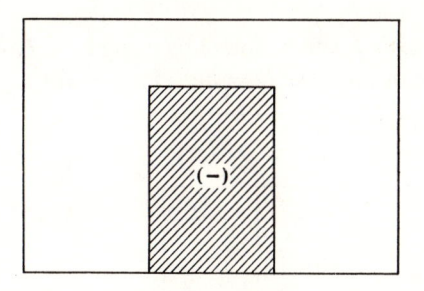

Fig. 3-109

Large rectangle $\dfrac{bh^3}{3} = \dfrac{6(4)^3}{3} = 128$ in.^{4}

minus

Small rectangle $\dfrac{bh^3}{3} = \dfrac{2(3)^3}{3} = 18$ in.^{4}

$$I_x = 110 \text{ in.}^4 \blacktriangleright$$

Answer is (d)

3-45 Wt. 3

What is the moment of inertia of this figure about the axis X-X?

144 / Statics

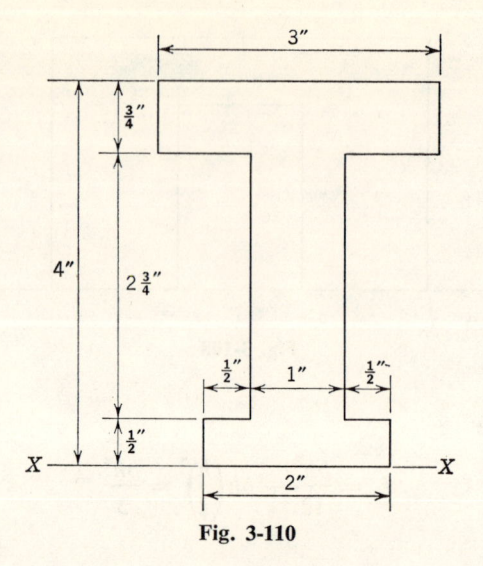

Fig. 3-110

Solution

The moment of inertia I_x can be found by using $I_x = I_0 + Ay^2$, the parallel-axis theorem, where in this case $I_0 = bh^3/12$.

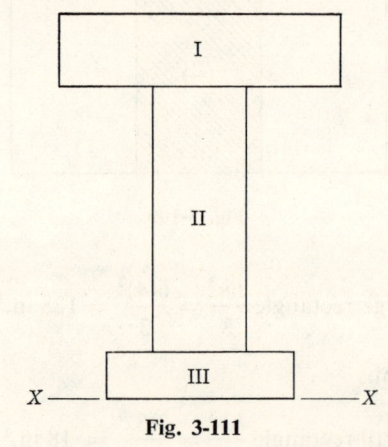

Fig. 3-111

Section	b (in.)	h (in.)	h³ (in.³)	A (in.²)	y (in.)	y² (in.²)	$\frac{bh^3}{12}$ (in.⁴)	Ay² (in.⁴)	I_section (in.⁴)
I	3.0	0.75	0.422	2.25	3.625	13.14	0.106	29.60	29.706
II	1.0	2.75	20.80	2.75	1.875	3.52	1.733	9.68	11.413
III	2.0	0.50	0.125	1.00	0.25	0.062	0.021	0.06	0.081
									41.2

$$I_x = \sum I_{\text{section}} = 41.2 \text{ in.}^4 \blacktriangleright$$

3-46 Wt. 1

The moment of inertia of a rectangle with respect to an axis passing through its base is

(a) $\dfrac{bh^3}{3}$

(b) $\dfrac{bh^3}{12}$

(c) $\dfrac{bh^2}{3}$

(d) $\dfrac{bh}{2}$

(e) none of the above

Solution

Using the transfer formula for the moment of inertia $I_x = I_0 + Ad^2$, where I_0 = moment of inertia about the centroid, A = area = bh, and $d = h/2$, we have

$$I_x = \frac{bh^3}{12} + bh\left(\frac{h}{2}\right)^2 = \frac{bh^3}{12} + \frac{bh^3}{4} = \frac{bh^3}{3} \blacktriangleright$$

Solution using calculus: The general equation for the moment of inertia is $I = \int y^2 \, dA$.

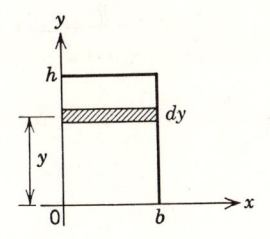

Fig. 3-112

$$dA = b \, dy$$
$$dI_x = y^2(b \, dy) = by^2 \, dy$$
$$I_x = b \int_0^h y^2 \, dy = b\left[\frac{y^3}{3}\right]_0^h = \frac{bh^3}{3} \blacktriangleright$$

Answer is (a)

146 / Statics

3-47 Wt. 5

Derive the formula for computing the moment of inertia of an equilateral triangle about one of its sides.

Solution

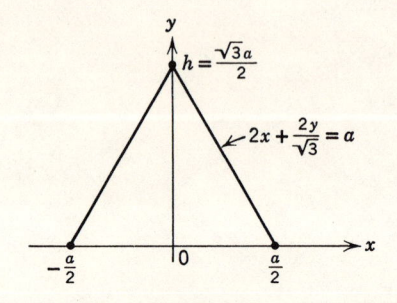

Fig. 3-113

The altitude h is

$$\left(\frac{a}{2}\right)^2 + h^2 = a^2 \qquad h^2 = a^2 - \frac{a^2}{4} = \frac{3a^2}{4}$$

$$h = \frac{\sqrt{3}\,a}{2}$$

The equation of the side in terms of the x and y equation of a straight line (intercept form) is

$$\frac{x}{a/2} + \frac{y}{\sqrt{3}\,a/2} = 1$$

$$2x + \frac{2y}{\sqrt{3}} = a \qquad 2x = a - \frac{2y}{\sqrt{3}}$$

The moment of inertia is

$$I_x = \int y^2\, dA = \int y^2 \left(a - \frac{2}{\sqrt{3}} y\right) dy = \int_0^{\sqrt{3}a/2} \left(ay^2 - \frac{2}{\sqrt{3}} y^3\right) dy$$

$$= \left[\frac{ay^3}{3} - \frac{2y^4}{4\sqrt{3}}\right]_0^{\sqrt{3}a/2}$$

$$= \frac{\sqrt{3}\,a^4}{8} - \frac{3\sqrt{3}\,a^4}{32} = \frac{\sqrt{3}}{32} a^4 \blacktriangleright$$

3-48 Wt. 1

12"

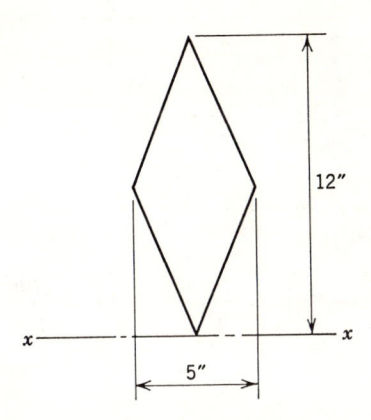

5"

Fig. 3-114

The moment of inertia about the x-x axis is

(a)   420 in.⁴
(b) 1230 in.⁴
(c) 1260 in.⁴
(d) 1380 in.⁴
(e) 2460 in.⁴

Solution

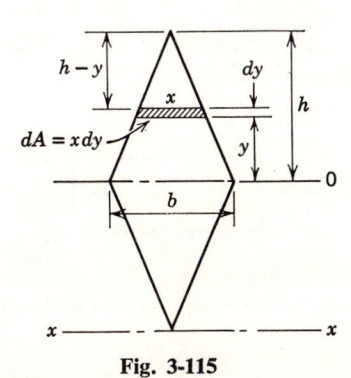

$h - y$ dy h
$dA = x\,dy$ y
b

Fig. 3-115

From similar triangles in Fig. 3-115,

$$\frac{b}{h} = \frac{x}{h-y} \qquad x = \frac{b}{h}(h-y)$$

$$I_0 = \int y^2 \, dA = 2\int_0^h y^2 x \, dy = 2\int_0^h y^2 \frac{b}{h}(h-y) \, dy$$

$$= \frac{2b}{h}\left(\int_0^h hy^2 \, dy - \int_0^h y^3 \, dy\right)$$

$$= \frac{2b}{h}\left[\frac{hy^3}{3} - \frac{y^4}{4}\right]_0^h = \frac{2b}{h}\left(\frac{h^4}{3} - \frac{h^4}{4}\right) = \frac{2bh^4}{12h}$$

$$I_0 = \frac{bh^3}{6}$$

But we want I_x.

Using the transfer formula $I_x = I_0 + Ad^2$, where $A = 2(b/2)h = bh$ and $d = h$, we have

$$I_x = \frac{bh^3}{6} + bh(h^2) = \frac{7bh^3}{6}$$

In our case $h = 6$ and $b = 5$. Therefore

$$I_x = \frac{7 \times 5 \times 6^3}{6} = 1260 \text{ in.}^4 \blacktriangleright$$

Answer is (c)

3-49 Wt. 3

The moment of inertia of the rectangle shown below is 128 in.^{4} about its base. Compute its moment of inertia about the X'-X' axis.

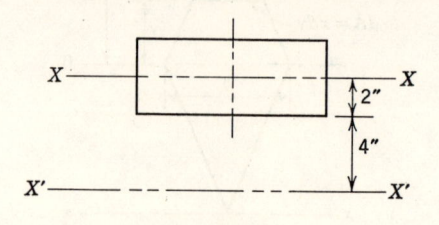

Fig. 3-116

Solution

Parallel-axis theorem: $I_z = I_x + Ad^2$, where x is located at the centroid of the piece and z is some distance (d) away. For a rectangle $I_x = bd^3/12$. Then we know that

$$I_{\text{base}} = \frac{bd^3}{12} + bd(2^2) = 128 \text{ in.}^4$$

By inspection the depth of the piece $d = 4$ in. So

$$\frac{b \times 4^3}{12} + b(4)(4) = 128 \text{ in.}^4 \qquad 5.33b + 16b = 128$$

$$b = \frac{128}{21.33} = 6 \text{ in.}$$

Now we can calculate $I_{X'}$:

$$I_{X'} = I_x + Ad^2 = \frac{6 \times 4^3}{12} + 6(4)(6^2) = 32 + 864 = 896 \text{ in.}^4 \blacktriangleright$$

3-50 Wt. 4

Find the moment of inertia of the angle section shown about the X_0 axis and the product of inertia about the Y_0-X_0 axes.

Since the parallel-axis theorems may be applied to the whole area or its component parts, checking methods are available. Very little credit will be given if wrong numerical answers are obtained.

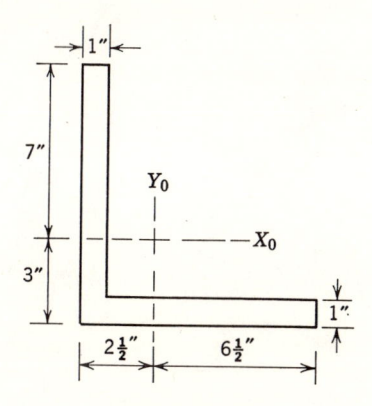

Fig. 3-117

Solution

$$I_x = \sum (I_0 + Ad^2) = \frac{1(10)^3}{12} + 1(10)(2)^2 + \frac{8(1)^3}{12} + 8(1)(2.5)^2$$

$$= 83.3 + 40 + 0.7 + 50 = 174 \text{ in.}^4 \blacktriangleright$$

$$I_{xy} = \sum (I_{xy_0} - Aab)$$

Here a and b are the x and y distances, respectively, between the new axes and the centroids of the component areas. Hence

$$I_{xy} = 0 + (1)(10)(-2)(2) + 0 + (8)(1)(2.5)(-2.5)$$
$$= -40 - 50 = -90 \text{ in.}^4 \blacktriangleright$$

Chapter 4

Dynamics

Dynamics is the study of the motion of nondeformable bodies. This study is normally divided into *kinematics*, the study of acceleration-velocity-displacement relations, and *kinetics*, which relates these motions and the forces causing, and caused by, them. Statics may be regarded as merely a special (but important) subdivision of dynamics where forces are in equilibrium and no accelerations are present.

KINEMATICS

Kinematics ignores the forces causing motion and considers only the motion itself. The acceleration a, velocity v, and displacement x of a body are related by the very basic expressions

$$a = \frac{dv}{dt} \qquad v = \frac{dx}{dt} \qquad (4\text{-}1)$$

where t is time. When the acceleration is a known function of time, velocity, or displacement, equations (4-1) may be integrated to give direct velocity-time or displacement-time relations. For the important case of rectilinear motion beginning at a point x_0 with a constant acceleration a_0 and initial velocity v_0, the displacement at any later time, by integration, is

$$x(t) = x_0 + v_0 t + \tfrac{1}{2} a_0 t^2 \qquad (4\text{-}2)$$

Since we also have

$$a_0 = \frac{dv}{dt} = \frac{dv}{dx}\frac{dx}{dt} = v\frac{dv}{dx} = \frac{d}{dx}\!\left(\frac{v^2}{2}\right)$$

we also find by integration that the velocity and displacement are related by

$$\tfrac{1}{2}(v^2 - v_0^2) = a_0(x - x_0) \qquad (4\text{-}3)$$

Example 1

A driver sees a stoplight when his car is traveling at 55 mph. If it takes him 0.6 sec to apply the brakes, and the brakes produce a deceleration of 15 ft/sec² in the car, how many feet does the car travel before coming to a stop?

Solution

Using the conversion factor 88 fps = 60 mph, we find the initial speed $v_0 = 55(88/60) = 80.7$ fps. During the 0.6-sec reaction period the car moves a distance $s_1 = v_0 t = (80.7)(0.6) = 48.4$ ft.

During the deceleration period $a_0 = -15$ ft/sec² while the velocity decreases from v_0 to zero. From equation (4-3) the distance traveled $s_2 = x - x_0$ is

$$s_2 = \frac{1}{2a_0}(v^2 - v_0^2) = \frac{1}{2(-15)}[0 - (80.7)^2] = 217 \text{ ft}$$

Thus the total distance traveled is

$$s = s_1 + s_2$$
$$s = 48.4 + 217 = 265 \text{ ft} \blacktriangleright$$

In addition to rectilinear or straight-line motions there are curvilinear motions. Equations (4-1) apply in a vectorial sense. For two- or three-dimensional motion, the trajectory and velocity of the body are often expressed by a set of parametric equations, usually with time t as the parameter. The trajectory of a body in a gravity field is a good example. Choosing x and y to be the horizontal and vertical displacements, respectively,

$$x = x_0 + v_0 \cos \theta_0 \, t$$
$$y = y_0 + v_0 \sin \theta_0 \, t - \tfrac{1}{2} g t^2 \tag{4-4}$$

when the body is released at $t = 0$ from the point (x_0, y_0) with initial velocity v_0 and orientation θ_0 from the horizon. Here the location of the body is defined in terms of the time parameter.

The accelerations of a body experiencing curvilinear motion are often split into tangential and normal components. For the case of circular motion of radius r at velocity v, the tangential acceleration a_t and normal acceleration a_n are

$$a_t = \frac{dv}{dt} \qquad a_n = \frac{v^2}{r} \tag{4-5}$$

If the speed of the body is unchanging, $a_t = 0$ but a normal acceleration still exists. In terms of the angular velocity ω, $v = \omega r$ and $a_n = \omega^2 r$.

KINETICS

Here we consider the kinetics of bodies whose dynamic behavior is adequately described by reference to only one body property, the mass, which is assumed to be concentrated at a point. Mass is the proportionality constant that relates the acceleration of a body to the net force which acts upon it. The Newtonian law expresses this as

$$\Sigma \mathbf{F} = m\mathbf{a} \tag{4-6}$$

for a body of constant mass m. This vector equation is usually written in component scalar form for use in obtaining numerical answers, however.

Example 2

The blocks A and B, of weights $W_A = 500$ lb and $W_B = 100$ lb, are connected by a rope which passes over a small pulley (Fig. 4-1). No friction

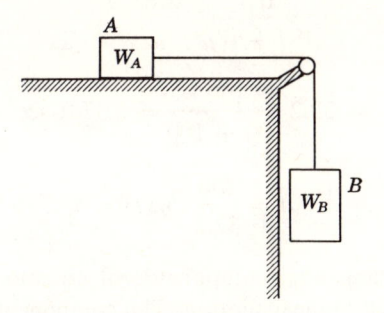

Fig. 4-1

is present. When the system is released from rest, what is the acceleration a of the bodies and the tension T in the rope?

Solution

Drawing a diagram of each block (Fig. 4-2) shows that the blocks are clearly not in equilibrium but will accelerate to the right (block A) and

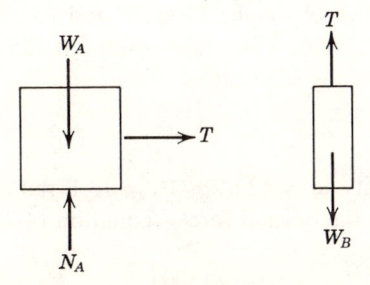

Fig. 4-2

downward (block B) at the same rate a. Applying the Newtonian equation, we have for block A

$$\sum F_x = ma$$

$$T = \frac{W_A}{g} a$$

and for block B

$$\sum F_y = ma$$

$$W_B - T = \frac{W_B}{g} a$$

Hence

$$W_B - \frac{W_A}{g} a = \frac{W_B}{g} a$$

$$a = g \frac{W_B}{W_A + W_B}$$

$$= 32.2 \frac{100}{500 + 100} = 5.37 \text{ ft/sec}^2 \blacktriangleright$$

$$T = \frac{W_A}{g} a = \frac{500}{32.2} (5.37) = 83.3 \text{ lb} \blacktriangleright$$

The normal and tangential components of equation (4-6) are useful in situations involving curvilinear motion. The components are

$$\sum F_t = m \frac{dv}{dt} \qquad \sum F_n = m \frac{v^2}{r} \qquad (4\text{-}7)$$

The normal force component is also called the centrifugal force.

Example 3

A segment of a flywheel weighs 1300 lb, and its center of gravity is 6 ft from the center of the shaft. The wheel rotates at 125 rpm. What is the pull on the arm supporting the segment?

Solution

The angular velocity $\omega = 125(2\pi/60) = 13.1$ rad/sec. The speed of the segment is $v = \omega r$ so the normal force—equation (4-7)—is

$$F_n = m \frac{v^2}{r} = mr\omega^2 = \left(\frac{1300}{32.2}\right)(6)(13.1)^2 = 41{,}500 \text{ lb} \blacktriangleright$$

ENERGY CONSERVATION

If Newton's law of motion is rearranged and integrated with respect to distance between two points, we find that the work done in moving a body from one point to the other is equal to the change in kinetic energy *KE* of that body. If the amount of work done depends only on the end points of the path, that is, if a conservative force produces the work, then the work can be expressed as a change in potential energy *PE*. Common conservative forces include the weight of a body and the force due to an elastic spring. A common *non*conservative force is that due to friction; it must be excluded from uses of the energy method. For conservative forces we may then say that the total mechanical energy of a system is conserved, or

$$PE + KE = \text{constant} \tag{4-8}$$

for a process. This may be restated in terms of changes of kinetic and potential energy as

$$\Delta(KE) = -\Delta(PE) \tag{4-9}$$

Example 4

A block of mass *m* is released from rest on an inclined frictionless plane, as in Fig. 4-3. What is the block's speed when it has dropped a vertical distance of 4 ft?

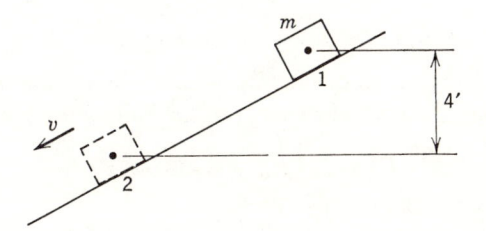

Fig. 4-3

Solution

Here we apply equation (4-9):

$$\Delta(KE) = -\Delta(PE)$$
$$KE_2 - KE_1 = -(PE_2 - PE_1)$$
$$\tfrac{1}{2}mv^2 - 0 = -(0 - mgh)$$
$$v^2 = 2gh$$
$$v^2 = 2(32.2)(4)$$
$$v = 16.05 \text{ fps} \blacktriangleright$$

Example 5

The spring of a spring gun has an uncompressed length of 8 in. The modulus of the spring $k = 1$ lb/in. The spring is compressed to a length of 4 in., and a ball weighing 1 oz is put in the barrel against the compressed spring, as in Fig. 4-4. If the spring is then released, find the velocity with which the ball leaves the gun. Neglect friction.

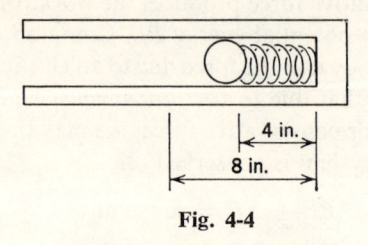

Fig. 4-4

Solution

The force involved in the compression of an elastic spring is a conservative force whose magnitude is proportional to the distance x that the spring is compressed, that is, $F = kx$, where k is the spring constant. The work done on the spring is

$$W = \int_0^x F\,dx = \int_0^x kx\,dx = \tfrac{1}{2}kx^2$$

In our case

$$W = \tfrac{1}{2}(1 \text{ lb/in.})(4 \text{ in.})^2 = 8 \text{ in-lb} = \tfrac{2}{3} \text{ ft-lb}$$

This work represents stored or potential energy. When the spring is released, this is converted into the kinetic energy of the ball, which was initially at rest.

$$\Delta(KE) = \Delta(PE)$$

$$\tfrac{1}{2}mv^2 = W$$

$$v = \left(\frac{2W}{m}\right)^{1/2}$$

$$v = \left[\frac{2(\tfrac{2}{3})}{(\tfrac{1}{16})/32.2}\right]^{1/2} = 26.2 \text{ fps} \blacktriangleright$$

MOMENTUM CONSERVATION

Newton's law of motion, when viewed another way, yields a useful principle which relates impulse and momentum. Expressing equation (4-6) as

$$\sum \mathbf{F} = \frac{d}{dt}(m\mathbf{v}) \qquad (4\text{-}10)$$

we may integrate with respect to time to obtain

$$\sum \int F \, dt = (mv)_2 - (mv)_1 \tag{4-11}$$

The left term represents an external impulse which changes the momentum mv of a body from that of state 1 to that of state 2. In the absence of impulsive external forces, we note that momentum is conserved for the body. Extended to a system of several bodies, we may state again, if the net external impulsive force on the system is zero, that system momentum is conserved, or

$$(\sum mv)_1 = (\sum mv)_2 \tag{4-12}$$

This is true even though individual bodies within a system may impact with one another; the reason is that the impulsive forces are internal, not external, to the system.

The nature of the impact which occurs between two bodies is important even when no external impulsive forces act. An index of the kind of impact which occurs between two bodies is the coefficient of restitution e, which is the ratio of the relative velocity between the two bodies after impact to the relative velocity before impact. Two cases are of particular interest:

(a) $e = 1$ represents elastic impact where the relative velocities before and after impact are equal. This is the only impact situation where energy is conserved; all other impact cases involve a change in total mechanical energy.

(b) $e = 0$ represents inelastic or plastic impact. After impact the two bodies move together with a common velocity. Energy is not conserved.

Different parts of a single problem commonly use the principles of momentum and energy conservation and the equation of motion together to achieve a solution.

Example 6

A bullet of mass m, traveling at velocity v_1, makes an inelastic impact with a simple pendulum composed of a mass M on the end of a flexible cord. If the impact occurs a distance L below the pendulum's suspension point, as illustrated in Fig. 4-5, through what angle θ will the pendulum move? (Assume the impact occurs at a right angle to the vertical.)

Solution

By the principle of conservation of momentum for an inelastic impact, the sum of the momenta of the bullet and the pendulum before impact equals the momentum of the combination after impact, or

$$mv_1 + 0 = (m + M)V_c$$

158 / Dynamics

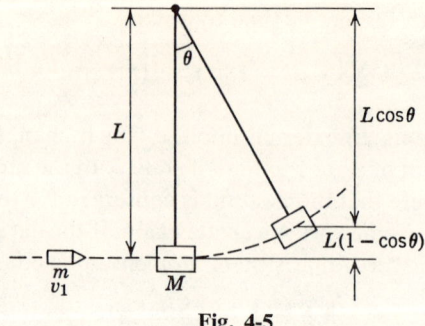

Fig. 4-5

where V_c is the velocity of the combination immediately after impact. Thus

$$V_c = \frac{mv_1}{m + M}$$

Using the energy conservation principle for the motion after the impact, we write

$$Wh = \tfrac{1}{2}mV^2$$

which in this case gives

$$g(m + M)L(1 - \cos \theta) = \tfrac{1}{2}(m + M)V_c^2 = \tfrac{1}{2}(m + M)\left(\frac{mv_1}{m + M}\right)^2$$

Solving for θ,

$$\theta = \cos^{-1}\left[1 - \frac{1}{2gL}\left(\frac{mv_1}{m + M}\right)^2\right] \blacktriangleright$$

RIGID BODY DYNAMICS

The motion of some bodies cannot be properly analyzed by assuming the mass of the body is concentrated at a point and analyzing it as a particle. In particular this is true when bodies execute rotational motion. We consider briefly the analysis of such motions now.

A general plane motion can always be regarded as the sum of a translational motion and a rotational motion. We have already examined translational motion. For rotational motion the velocity v of a point which is a distance r from the center of rotation on a body rotating with an angular velocity ω is $v = \omega r$, and the normal and tangential acceleration components are $a_n = \omega^2 r$, $a_t = \alpha r$, where $\alpha = d\omega/dt$ is the angular acceleration of the body. These quantities are often easier to calculate when one first finds the instantaneous center of rotation, the point around which the body appears to rotate at a given instant.

Example 7

A 5-in.-radius cylinder rolls to the right at a constant rate of 10 in./sec. For the instant shown in Fig. 4-6, what is the

(a) velocity at point A?
(b) acceleration of point A?

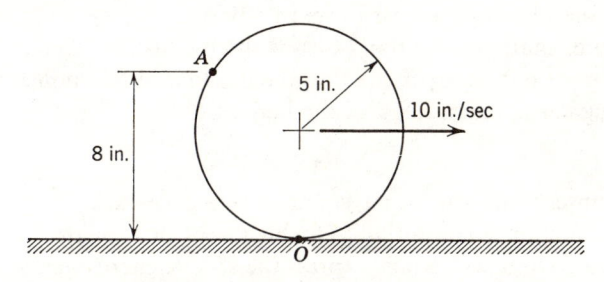

Fig. 4-6

Solution

We first note that the point on the cylinder which is in contact with the plane, point O, is the instantaneous center of rotation. At a given instant it is stationary, and all other points are executing a purely rotational motion about point O, see Fig. 4-7.

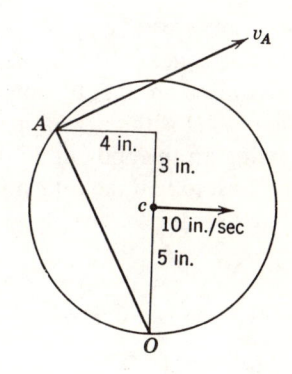

Fig. 4-7

The angular velocity ω of the body is

$$\omega = \frac{v}{r} = \frac{10}{5} = 2 \text{ rad/sec.}$$

Also the distance $OA = d = (8^2 + 4^2)^{1/2} = 8.94$ in.

(a) The velocity $v_A = \omega d = (2)(8.94) = 17.88$ in./sec. ▶
(b) Since ω is constant, the angular acceleration $\alpha = d\omega/dt = 0$, and the

tangential acceleration $a_t = 0$ also. The normal acceleration

$$a_n = \omega^2 d = (2)^2(8.94) = 35.8 \text{ in./sec}^2 \blacktriangleright$$

The kinetics of the plane motion of rigid bodies is quite similar to the kinetics of particles. The differences arise because we are now analyzing the motion of distributed masses rather than point masses.

To study the plane motion of a rigid body directly, we must supplement the equation of motion, equation (4-6), with an equation which says the net external moment on a body is equal to the product of the moment of inertia I and the angular acceleration α of the body, or

$$\sum M_0 = I_0 \alpha \qquad (4\text{-}13)$$

where the subscript denotes some common reference axis.

For plane motion equation (4-9), which says that energy is conserved, remains valid so long as we now express the kinetic energy as the sum of the translational kinetic energy $\frac{1}{2}mV^2$ and the rotational kinetic energy $\frac{1}{2}I_0\omega^2$.

The momentum of a body may be treated as the sum of linear momentum and angular momentum just as a general motion is the sum of a translational and a rotational motion. The treatment of linear momentum conservation is unchanged from the presentation given in equations (4-10)–(4-12). Conservation of angular momentum may be expressed as

$$I_0\omega_1 + \int M_0 \, dt = I_0\omega_2 \qquad (4\text{-}14)$$

In this equation the $I_0\omega$ terms are the angular momentum of the body before and after an angular impulse, given by the middle term, is applied to the body. M_0 is the net moment, computed with respect to axis O, of the forces acting on the body at an instant. One then integrates over the time period of interest to find the total angular impulse. In the absence of an applied angular impulse, the angular momentum $I_0\omega$ is constant.

4-1 Wt. 1

A dyne is a unit of

(a) weight
(b) force
(c) mass
(d) pressure
(e) energy

Solution

The dyne is the magnitude of the force which imparts an acceleration of 1 cm/sec² to a mass of 1 g ▶

Using $F = ma$,
$$1 \text{ dyne} = (1 \text{ g}) \times (1 \text{ cm/sec}^2)$$

Answer is (b)

4-2 Wt. 1

A newton is the force required to give

(a) 1 kg an acceleration of 1 m/sec^2
(b) 1 kg a velocity of 1 m/sec
(c) 1 kg a velocity of 1 cm/sec
(d) 1 g an acceleration of 1 cm/sec^2
(e) 1 g a velocity of 1 cm/sec

Solution

One newton of force will accelerate a 1-kg mass at 1 m/sec^2. Again using $F = ma$,
$$1 \text{ newton} = (1 \text{ kg}) \times (1 \text{ m/sec}^2) = 10^5 \text{ dynes} \blacktriangleright$$

Answer is (a)

4-3 Wt. 1

In a vacuum near the earth's surface a body falls with a uniform

(a) acceleration
(b) velocity
(c) energy
(d) momentum
(e) inertia

Solution

A body will fall with a uniform acceleration of approximately 32.2 ft/sec^2 when it is in a vacuum near the surface of the earth $\blacktriangleright$

Answer is (a)

4-4 Wt. 4

A pebble is dropped in a well, and it is found that 4.25 sec elapse after release of the pebble before the splash is heard. If the velocity of sound in the well is 1030 fps, what is the depth to the water surface?

Solution

The pebble is initially at rest ($V_0 = 0$) and thereafter falls with a constant acceleration due to gravity g so that
$$V(t) = \frac{dy}{dt} = gt$$

162 / Dynamics

The fall distance is then $y = \frac{1}{2}gt_1^2$ by integration, where t_1 is the time for the pebble to reach the water. The sound wave will return the same distance y at a constant velocity so that $y = 1030t_2$, where t_2 is the time for the sound to travel from the base to the top of the well. By the problem statement,

$$t_1 + t_2 = 4.25 \text{ sec}$$

and we also have

$$\tfrac{1}{2}gt_1^2 = 1030t_2$$

Solving these equations simultaneously,

$$\tfrac{1}{2}(32.2)t_1^2 - 1030t_2 = 0$$

$$\underline{1030t_1 + 1030t_2 = 4.25(1030)}$$

$$16.1t_1^2 + 1030t_1 - 4377.5 = 0$$

or

$$t_1^2 + 64.0t_1 - 272 = 0$$

$$t_1 = \tfrac{1}{2}[-64.0 \pm \sqrt{64.0^2 + 4(272)}] = \tfrac{1}{2}(-64.0 \pm 72.0)$$

Since t_1 must be positive, $t_1 = 4.0$ sec, and

$$y = \tfrac{1}{2}gt_1^2 = \tfrac{1}{2}(32.2)(4.0)^2 = 258 \text{ ft} \blacktriangleright$$

4-5 Wt. 4

A body moves so that the x component of acceleration is given by the equation $6 - t$ and the y component of acceleration is given by $6 + t$. If the initial x and y components of velocity are both 2, what is the speed of the body at the end of 2 sec?

Solution

$$a_x = \frac{d^2x}{dt^2} = 6 - t \qquad a_y = \frac{d^2y}{dt^2} = 6 + t$$

Integrating these expressions once,

$$V_x = \frac{dx}{dt} = 6t - \frac{1}{2}t^2 + C_1 \qquad V_y = \frac{dy}{dt} = 6t + \frac{1}{2}t^2 + C_2$$

At $t = 0$, $V_x = 2$ and $V_y = 2$ so $C_1 = C_2 = 2$ by substitution into the velocity expressions. At the end of 2 sec we have

$$V_x = 6(2) - \tfrac{1}{2}(2)^2 + 2 \qquad V_y = 6(2) + \tfrac{1}{2}(2)^2 + 2$$

$$V_x = 12 \qquad V_y = 16$$

The speed

$$V = (V_x^2 + V_y^2)^{1/2} = (12^2 + 16^2)^{1/2}$$

$$V = 20 \blacktriangleright$$

4-6 Wt. 4

A fast-moving train, traveling at a velocity V_1, rounds a turn onto a straightaway; at this moment the engineer observes at a distance d a slow-moving train on the same track going in the same direction at velocity V_2. The engineer instantly applies his brakes, which causes the fast train to decelerate at a constant rate $-a$. Determine the minimum value of d so that no rear-end collision will occur.

Solution

To avoid the impending collision, the fast-moving train must in time t decelerate from V_1 to V_2 in a maximum distance of $d + V_2 t$. For a constant deceleration, the final velocity minus the initial velocity must equal the deceleration times the time interval, or

$$V_2 - V_1 = (-a)t$$

giving

$$t = \frac{V_1 - V_2}{a}$$

In general, the distance traveled s is given by

$$s = V_0 t + \tfrac{1}{2} a t^2$$

Here

$$s = d + V_2 t$$

so that

$$d + V_2 t = V_1 t + \tfrac{1}{2}(-a)t^2$$

Solving for d and substituting for t results in

$$d = (V_1 - V_2)t - \tfrac{1}{2} a t^2$$

$$d = (V_1 - V_2)\left(\frac{V_1 - V_2}{a}\right) - \frac{1}{2} a \left(\frac{V_1 - V_2}{a}\right)^2$$

Finally,

$$d = \frac{(V_1 - V_2)^2}{2a} \blacktriangleright$$

4-7 Wt. 4

The driver of a car traveling 30 mph suddenly applies the brakes and skids 60 ft before coming to a stop. If the car weighs 1600 lb and a constant rate of deceleration is assumed, compute

(a) the rate of deceleration
(b) the average coefficient of sliding friction

Solution

The general displacement equation for rectilinear motion can be obtained by integrating the expression $d^2x/dt^2 = a$ twice. The result is

$$x = x_0 + V_0 t + \tfrac{1}{2}at^2$$

where x_0 and V_0 are the initial displacement and velocity, respectively, and a is the (constant) acceleration. If t is measured from the moment the brakes are applied and the initial location of the car is $x_0 = 0$, then $V_0 = 30$ mph $= 44$ fps. When the car stops at time t later, $x = 60$ ft, so

$$60 = 44t + \tfrac{1}{2}at^2$$

or, solving for a,

$$a = \frac{120}{t^2} - \frac{88}{t}$$

Since the deceleration is to be constant,

$$\frac{da}{dt} = 0 = \frac{120(-2)}{t^3} - \frac{88(-1)}{t^2}$$

yielding $t = 240/88 = 2.73$ sec as the time to stop. Hence

$$a = \frac{120}{(2.73)^2} - \frac{88}{2.73} = -16.1 \text{ ft/sec}^2 \blacktriangleright$$

(The minus sign denotes deceleration.)

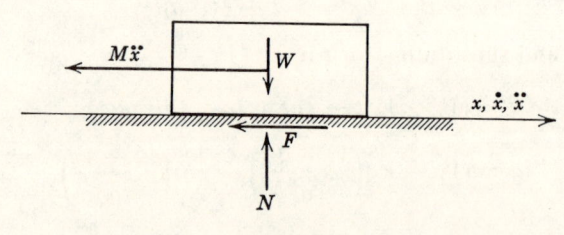

Fig. 4-8

For sliding,

$$F = -\mu N = -1600\mu$$

Also

$$F = M\ddot{x} = Ma$$

Equating these expressions,

$$-1600\mu = \left(\frac{1600}{32.2}\right)(-16.1) \quad \text{and} \quad \mu = 0.5 \blacktriangleright$$

4-8 Wt. 3

An aircraft begins its take-off run with an acceleration of 12 ft/sec², which then decreases uniformly to zero in 15 sec, at which time the craft becomes airborne.

(a) What is its take-off speed, in feet per second?
(b) How long (in feet) is its take-off run?

Solution

The aircraft's acceleration, as a function of time, is

$$a = \frac{d^2x}{dt^2} = 12 - \frac{12}{15}t = 4\left(3 - \frac{t}{5}\right)$$

At the initial instant when $t = 0$, $V = 0$ and $x = 0$. By integration, the velocity is

$$V = \frac{dx}{dt} = 4\left(3t - \frac{t^2}{10}\right) + C_1$$

At $t = 0$, $V = 0$ so $C_1 = 0$. At $t = 15$ sec the velocity is then

$$V = 4\left(3t - \frac{t^2}{10}\right)\bigg|_{t=15}$$

$$V = 4\left[3(15) - \frac{(15)^2}{10}\right] = 4[45 - 22.5]$$

$$V = 90 \text{ fps} \blacktriangleright$$

Integrating once more, the position $x(t)$ is

$$x = 4\left(\frac{3}{2}t^2 - \frac{t^3}{30}\right) + C_2$$

At $t = 0$, $x = 0$, hence $C_2 = 0$. At $t = 15$ sec the length of the take-off run is

$$x = 4\left[\frac{3}{2}(15)^2 - \frac{(15)^3}{30}\right] = 4[337.5 - 112.5]$$

$$x = 900 \text{ ft} \blacktriangleright$$

4-9 Wt. 4

A body moves in rectilinear motion with an acceleration of $a = 24 - t - 6t^2$. Starting with an initial velocity of 4 fps to the left and an initial position at the origin, what is the velocity of the body at the end of 4 sec and where is it located?

166 / Dynamics

Solution

In this problem we assume that motion to the right is positive.

$$a = \frac{d^2x}{dt^2} = 24 - t - 6t^2$$

Integrate once,

$$v = \frac{dx}{dt} = 24t - \frac{1}{2}t^2 - 2t^3 + C_1$$

At $t = 0$, $v = -4$ fps (motion to the left is negative), hence $C_1 = -4$, and

$$v = \frac{dx}{dt} = 24t - \frac{1}{2}t^2 - 2t^3 - 4$$

Integrate again,

$$x = 12t^2 - \tfrac{1}{6}t^3 - \tfrac{1}{2}t^4 - 4t + C_2$$

At $t = 0$, $x = 0$ so $C_2 = 0$. At $t = 4$ sec,

$v = 24(4) - \tfrac{1}{2}(4)^2 - 2(4)^3 - 4 = -44$ fps
$v = 44$ fps ←▶
$x = 12(4)^2 - \tfrac{1}{6}(4)^3 - \tfrac{1}{2}(4)^4 - 4(4) = +37.33$ ft
$x = 37.33$ ft ▶

to the right of the origin.

4-10 Wt. 5

A body experiences an acceleration a given by the expression $a = At - Bt^2$, where A and B are constants and t is time. If, at time $t = 0$, the body has zero displacement and velocity, at what next value of time does the body again have zero displacement?

Solution

$$a = \frac{d^2x}{dt^2} = At - Bt^2$$

Integrating once to obtain the velocity,

$$v = \frac{dx}{dt} = A\frac{t^2}{2} - B\frac{t^3}{3} + C_1$$

At $t = 0$, $v = 0$ and therefore $C_1 = 0$. To obtain the displacement $x(t)$ we integrate once more to get

$$x = A\frac{t^3}{6} - B\frac{t^4}{12} + C_2$$

Using the initial condition $x = 0$ at $t = 0$, $C_2 = 0$. Hence

$$x = \frac{t^3}{6}\left(A - B\frac{t}{2}\right)$$

For $t > 0$, x is again zero when $t = 2A/B$ ▶

4-11 Wt. 6

A body weighing 322 lb is subjected to an acceleration (in the positive x direction) which is a linearly decreasing function of the velocity. The body is stationary at $x = 0$ when $t = 0$, and the force acting on the body at this instant is 100 lb. The acceleration is zero when the velocity reaches 100 fps.

(a) Write a differential equation which expresses this situation mathematically.
(b) Find the displacement $x(t)$.
(c) Find the velocity $v(t)$.

Solution

Using $F = Ma$, the initial acceleration

$$a(0) = \ddot{x} = \frac{F}{M} = \frac{F}{W/g} = \frac{100(32.2)}{322} = 10 \text{ ft/sec}^2$$

The other initial conditions are $\dot{x}(0) = 0$, $x(0) = 0$. The acceleration as a function of time is of the form

$$\ddot{x} = -K\dot{x} + C \qquad K > 0$$

At $t = 0$, $\ddot{x} = 10$ when $\dot{x} = 0$ so $C = 10$. Also when $\dot{x} = 100$, $\ddot{x} = 0$, or

$$0 = -100K + 10 \qquad K = 0.1$$

Hence

$$\ddot{x} + 0.1\dot{x} - 10 = 0 \blacktriangleright$$

plus initial conditions, is the solution to part (a).

We seek the general solution to the differential equation for x, which is the sum of a particular solution and the complementary solution to

$$\ddot{x} + 0.1\dot{x} = 10$$

By inspection, a particular solution is $x = 100t$. To find the complementary solution, that is, the solution to $\ddot{x} + 0.1\dot{x} = 0$, assume a solution of the form

$$x = Ae^{\alpha t} + B$$

Then

$$\dot{x} = A\alpha e^{\alpha t}$$

and
$$\ddot{x} = A\alpha^2 e^{\alpha t}$$

Substituting into the homogeneous differential equation, we obtain
$$A\alpha^2 e^{\alpha t} + 0.1 A\alpha e^{\alpha t} = 0$$
or
$$\alpha(\alpha + 0.1) = 0$$

The nontrivial solution is $\alpha = -0.1$. The general solution is then
$$x = Ae^{-0.1t} + B + 100t$$
subject to the initial conditions $\dot{x} = x = 0$ at $t = 0$.
$$x = 0 = A + B$$
$$\dot{x} = 0 = -0.1A + 100$$

The solution is $A = 1000$, $B = -1000$. Hence
$$x(t) = 1000(e^{-0.1t} - 1) + 100t \blacktriangleright$$

The velocity $v(t) = \dot{x}(t)$,
$$v(t) = -100e^{-0.1t} + 100 \blacktriangleright$$

4-12 Wt. 4

A truck moves horizontally to the left at a steady 10 fps. At the instant shown, what is

(a) the velocity of the weight?
(b) the acceleration of the weight?

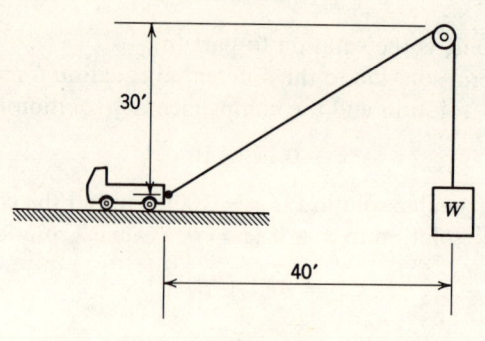

Fig. 4-9

Solution

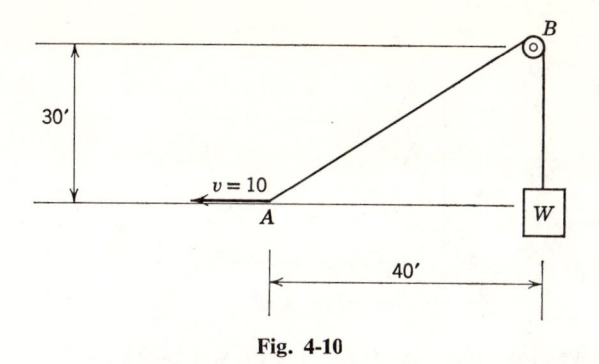

Fig. 4-10

Select the instant shown to be $t = 0$, and let $x(t)$ be the distance AB. At $t = 0$, $x(0) = 50$ ft. More generally,

$$x(t) = [(40 + 10t)^2 + 30^2]^{1/2}$$
$$= 10(t^2 + 8t + 25)^{1/2}$$

$$v(t) = \frac{dx}{dt} = 10\left(\frac{1}{2}\right)\frac{2t + 8}{(t^2 + 8t + 25)^{1/2}}$$

Evaluating $v(t)$ at $t = 0$ gives $v = 8$ fps ▶

$$a(t) = \frac{d^2x}{dt^2} = 5\frac{2}{(t^2 + 8t + 25)^{1/2}} - \frac{5}{2}\frac{(2t+8)^2}{(t^2 + 8t + 25)^{3/2}}$$

At $t = 0$

$$a(0) = \frac{10}{5} - \frac{5}{2}\frac{(8)^2}{(25)^{3/2}} \quad \text{or} \quad a = 0.72 \text{ ft/sec}^2 ▶$$

4-13 Wt. 4

A weight is attached to one end of a 53-ft rope passing over a small pulley 29 ft above the ground. A man whose hand is 5 ft above the ground grasps the other end of the rope and walks away at the rate of 5 fps. How fast is the weight rising when the man is 7 ft from a point directly under the pulley?

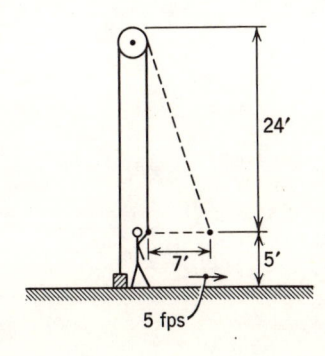

Fig. 4-11

170 / Dynamics

Solution

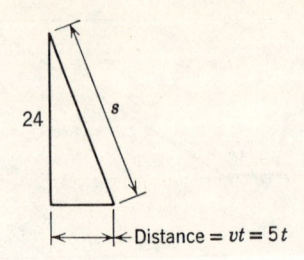

Fig. 4-12

At $t = 0$ we note that the rope is taut since $29 \text{ ft} + 24 \text{ ft} = 53 \text{ ft}$. Let s be the distance between the pulley and the man's hand:

$$s = [24^2 + (5t)^2]^{1/2} = (576 + 25t^2)^{1/2}$$

$$v = \frac{ds}{dt} = \frac{1}{2}(576 + 25t^2)^{-1/2} \times 50t$$

When the horizontal distance $5t = 7$, $t = 1.4$ sec, and

$$v = \tfrac{1}{2}[576 + 25(1.4)^2]^{-1/2} \times 50(1.4) = 1.4 \text{ fps}$$

The rope is therefore moving at 1.4 fps, and consequently the weight is rising at this same velocity of 1.4 fps ▶

4-14 Wt. 1

A river flows north with a speed of 3 mph. A man rows a boat across the river. His speed relative to the water is 4 mph. What is his velocity relative to the earth?

(*a*) 4 mph
(*b*) 3 mph
(*c*) 7 mph
(*d*) 1 mph
(*e*) 5 mph

Solution

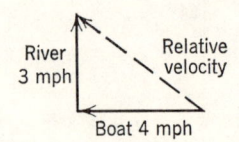

Fig. 4-13

$$V = (3^2 + 4^2)^{1/2} = 5 \text{ mph} \blacktriangleright$$

Answer is (e)

4-15 Wt. 6

Boat A is chasing boat B and wishes to fire its gun so that a hit may be scored. Boats A and B are traveling in the same direction at constant velocities V_A and V_B respectively. The gun on boat A has a muzzle velocity of V_1 and is inclined an angle θ to the horizontal. What must be the separation of the boats, L, in order for the hit to be scored?

Obtain the distance L in terms of the variables V_A, V_B, V_1, and θ. Neglect air friction, and assume that the shell leaves the gun at the same elevation as the point of impact on boat B.

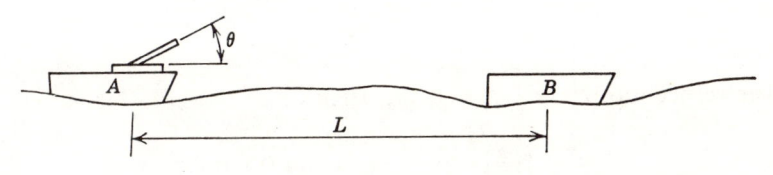

Fig. 4-14

Solution

The initial velocity of the shell is the vector sum of the muzzle velocity V_1 and the boat velocity V_A. The components of this velocity are

$$\text{Horizontal component} = V_1 \cos \theta + V_A$$
$$\text{Vertical component} = V_1 \sin \theta$$

The required range of the shell is

$$L + V_B t = (V_1 \cos \theta + V_A)t$$

Since there is no difference in elevation between boats A and B, the net vertical distance traveled by the shell is zero, or the vertical distance traveled is

$$y = (V_1 \sin \theta)t - \tfrac{1}{2}gt^2 = 0$$

Hence

$$t = \frac{2}{g} V_1 \sin \theta$$

From the range equation,

$$L = (V_1 \cos \theta + V_A - V_B)t$$

and

$$L = \frac{2}{g} V_1 \sin \theta (V_1 \cos \theta + V_A - V_B) \blacktriangleright$$

172 / Dynamics

4-16 Wt. 3

A car jumps across a 10-ft-wide ditch with a constant velocity V. The ditch is 6 in. lower on the far side. What is the minimum velocity (in miles per hour) that will keep the car from falling into the ditch?

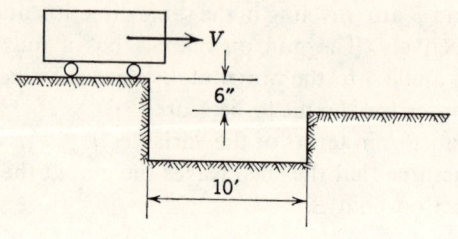

Fig. 4-15

Solution

The vertical drop $y = \tfrac{1}{2}gt^2 = 0.5$ ft. Hence

$$t = \left(\frac{2y}{g}\right)^{1/2} = \left[\frac{2(0.5)}{32.2}\right]^{1/2} = 0.176 \text{ sec}$$

The horizontal motion is not accelerated and is

$$x = Vt = 10 \text{ ft}$$

$$V = \frac{10}{t} = \frac{10}{0.176} = 56.8 \text{ fps}$$

Since 60 mph = 88 fps,

$$V = \left(\frac{60}{88}\right)(56.8) = 38.8 \text{ mph} \blacktriangleright$$

4-17 Wt. 5

A batted baseball leaves the bat at an angle of 30° above the horizontal and is caught by an outfielder 400 ft from home plate. What is the initial velocity of the ball? Assume the ball's height when hit is the same as when caught and that air resistance is negligible.

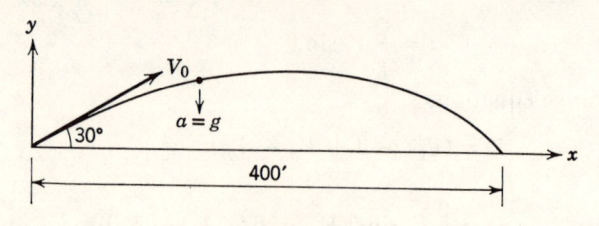

Fig. 4-16

Solution

Since we neglect air resistance, the only force acting on the ball is due to its weight. Hence the acceleration of the ball at all times is g directed downward.

The horizontal distance traversed is $x = V_x t$, where $V_x = V_0 \cos 30°$ and $x = 400$ ft.

$$400 = V_0 \frac{\sqrt{3}}{2} t \quad \text{or} \quad t = \frac{800}{V_0 \sqrt{3}} \sec$$

The displacement vertically is

$$y = V_y t - \tfrac{1}{2}gt^2$$

or

$$0 = V_0 \sin 30° \, t - \tfrac{1}{2}gt^2$$

Substituting for t,

$$0 = \frac{800}{\sqrt{3}} \sin 30° - \frac{1}{2} g \left(\frac{800}{V_0\sqrt{3}}\right)^2$$

$$V_0 = 32.2 \left(\frac{800}{\sqrt{3}}\right) \qquad V_0 = 122 \text{ fps} \blacktriangleright$$

4-18 Wt. 3

A projectile from an antiaircraft gun is designed to rise 10 miles into the air when the gun is fired at an angle of 60° from horizontal. What must its muzzle velocity be in feet per second if air friction is neglected?

Solution

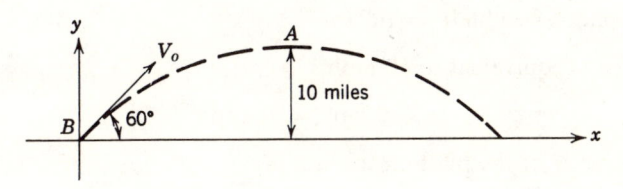

Fig. 4-17

The vertical displacement is $y = V_{0y} t - \tfrac{1}{2}gt^2$, where $V_{0y} = V_0 \sin 60°$. The vertical velocity is

$$V_y = \frac{dy}{dt} = V_{0y} - gt$$

At A, 10 miles in the air, $V_y = 0$, yielding $t = V_{0y}/g$ at that instant. By substitution, y is then 10 miles $= y = V_{0y}^2/2g$, and

$$V_{0y} = [2(32.2)(10)(5280)]^{1/2} = 1844 \text{ fps}$$

174 / Dynamics

Hence
$$V_0 = V_{0y}/\sin 60° = 1844\left(\frac{2}{\sqrt{3}}\right)$$
$$V_0 = 2130 \text{ fps} \blacktriangleright$$

4-19 Wt. 2

How much force does a 160-lb pilot exert on his plane seat at the top of a vertical loop of radius 1000 ft if the plane's speed is 180 mph?

Solution

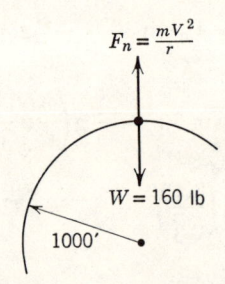

Fig. 4-18

The two normal forces acting on the pilot are:

(*a*) the centrifugal force due to his motion, which is directed upward and has a magnitude $F_n = mV^2/r$; and
(*b*) the pilot's weight $W = 160$ lb.

Since 88 fps is equivalent to 60 mph,
$$180 \text{ mph} = 264 \text{ fps}$$

The net force N of the pilot on the seat is

$$N = \frac{mV^2}{r} - W = \left(\frac{160}{32.2}\right)\frac{(264)^2}{1000} - 160$$

$$= 346 - 160 = 186 \text{ lb upward} \blacktriangleright$$

4-20 Wt. 3

Calculate the speed of a satellite traveling in a perfectly circular orbit around the earth at an altitude of 1000 miles. Assume the earth is a perfect sphere with a radius of 4000 miles and that the force of the earth's gravity g at the height of the satellite is 16.3 ft/sec².

Solution

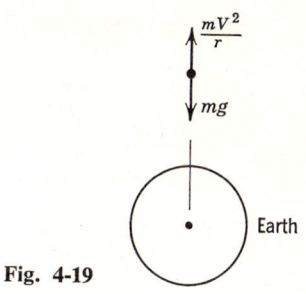

The two forces acting on the satellite are:

(a) the attractive force of the earth mg
(b) the centrifugal force mV^2/r

If the satellite is to remain a constant distance from earth, these two forces must be in equilibrium. Hence

$$\frac{mV^2}{r} = mg$$

$$V = (rg)^{1/2}$$
$$= [(4000 + 1000) \text{ miles} \times (5280 \text{ ft/mile})(16.3 \text{ ft/sec}^2)]^{1/2}$$
$$= 2.08 \times 10^4 \text{ fps} = 14{,}150 \text{ mph} \blacktriangleright$$

4-21 Wt. 5

The distance of the planet Neptune from the sun is 30 times that of the earth from the sun. Calculate the approximate period of Neptune's revolution about the sun. (The force of attraction between two bodies is directly proportional to the product of their masses and inversely proportional to the square of the distance between them.)

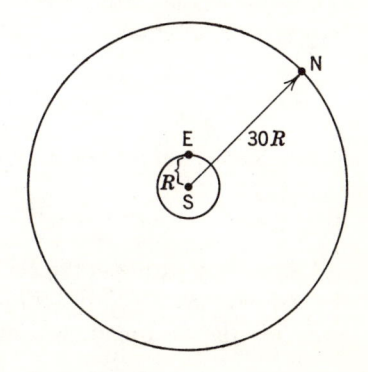

Fig. 4-20

176 / Dynamics

Solution

Mathematically, the attractive force f is

$$f = k \frac{MM'}{R^2}$$

Between the sun and earth it is

$$f = k \frac{M_s M_e}{R^2}$$

and between the sun and Neptune

$$f = k \frac{M_s M_n}{(30R)^2}$$

For uniform circular motion the force f causes a centripetal acceleration $a = V^2/R$, where $V = 2\pi R/T$ and T is the period of revolution. Hence

$$a_e = \frac{4\pi^2 R}{T_e^2} \quad \text{and} \quad a_n = \frac{4\pi^2(30R)}{T_n^2}$$

Using $f = ma$,

$$k\frac{M_s M_e}{R^2} = \frac{M_e 4\pi^2 R}{T_e^2} \quad \text{and} \quad k\frac{M_s M_n}{(30R)^2} = \frac{M_n 4\pi^2(30R)}{T_n^2}$$

or

$$\frac{R^3}{T_e^2} = \frac{kM_s}{4\pi^2} = \frac{(30R)^3}{T_n^2}$$

The period of the earth's revolution about the sun T_e is 1 yr, so the period for Neptune's revolution is

$$T_n = (30)^{3/2} \text{ yr} = 164.3 \text{ yr} \blacktriangleright$$

(Using Kepler's third law of planetary motion yields the same result.)

4-22 Wt. 3

A small boy sitting in the back of a pickup truck holds a weight of 4 lb on the end of a flexible cord 3 ft long. If the truck accelerates at 10 ft/sec², what angle will the cord make with the vertical? What is the tension in the cord?

Solution

A free-body diagram of the weight and the corresponding force triangle are shown below.

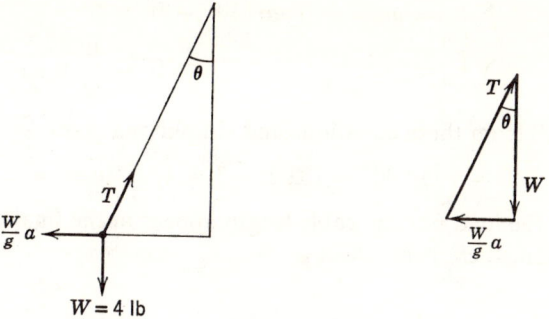

Fig. 4-21

From the force triangle,

$$\tan \theta = \frac{(W/g)a}{W} = \frac{a}{g} = \frac{10}{32.2} = 0.311$$

$$\theta = 17°15' \blacktriangleright$$

Also from the force triangle, $\cos \theta = W/T$ or

$$T = \frac{W}{\cos \theta} = \frac{4 \text{ lb}}{\cos 17°15'} = \frac{4 \text{ lb}}{0.955}$$

$$T = 4.19 \text{ lb} \blacktriangleright$$

4-23 Wt. 4

A traveling crane lifts a 1000-lb load on a 20-ft hoisting cable.

(a) What is the maximum horizontal acceleration that the crane may have without producing a deviation of the cable of more than 30° from the vertical?

(b) What will be the effect on the acceleration of increasing the load to 2000 lb and shortening the cable to 15 ft for the same maximum angular deviation of the cable?

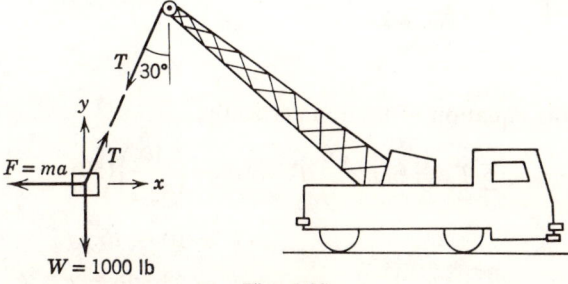

Fig. 4-22

178 / Dynamics

Solution

Writing Newton's equation of motion in the x and y directions for the weight

$$\sum F_y = ma_y \qquad T \cos 30° - W = 0$$

$$\sum F_x = ma_x \qquad T \sin 30° = \frac{W}{g} a$$

Eliminating W from these equations and simplifying gives

$$a = g \tan 30° = (32.2)/\sqrt{3} = 18.6 \text{ ft/sec}^2 \blacktriangleright$$

Since neither the load nor the cable length appear in the final expression for the acceleration, there is no effect ▶

4-24 Wt. 2

A test rocket weighing 10 lb is to be launched vertically from a stationary platform. What thrust T, in pounds, must the rocket motor develop to accelerate the rocket at 50 ft/sec²? Neglect the loss of weight due to the burning fuel.

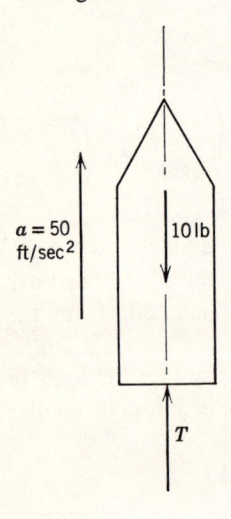

Fig. 4-23

Solution

Applying the equation of motion vertically,

$$\sum F_v = ma_v \qquad T - 10 = \left(\frac{10}{32.2}\right)(50)$$

$$T = 10 + 15.5$$

$$T = 25.5 \text{ lb} \blacktriangleright$$

4-25 Wt. 3

An elevator, which with its load weighs 8 tons, is descending with a speed of 900 fpm. If the load on the cable must not exceed 14 tons, find the shortest distance in which the elevator should be stopped.

Solution

Since the elevator weighs 8 tons and the cable load cannot exceed 14 tons, the maximum allowable inertia force to stop the elevator is 6 tons. Writing the inertia force as

$$F = -ma = -\frac{W}{g}a$$

$$6(2000) = -\frac{8(2000)}{32.2}a$$

and the maximum allowable deceleration is $a = -24.2$ ft/sec². Since

$$a = \frac{dv}{dt} = \frac{dv}{ds}\frac{ds}{dt} = \frac{d}{ds}\left(\frac{v^2}{2}\right)$$

$$as = \tfrac{1}{2}(V_2^2 - V_1^2)$$

The initial elevator velocity $V_1 = (900 \text{ fpm})(\tfrac{1}{60} \text{ min/sec}) = 15$ fps, and the final velocity $V_2 = 0$. Hence the minimum stopping distance is

$$s = \frac{1}{2(-24.2)}(-15^2) = 4.65 \text{ ft} \blacktriangleright$$

4-26 Wt. 4

The system of pulleys carries two loads as shown in Fig. 4-24. Find the acceleration of the 8-lb weight.

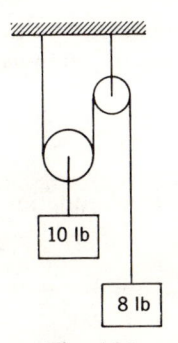

Fig. 4-24

180 / Dynamics

Solution

The pulleys are assumed to be weightless and frictionless. We will select the downward direction to be positive in each free-body diagram. For the movable pulley,

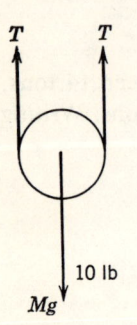

Fig. 4-25

$$\sum F = -2T + Mg = Ma_1 \qquad (1)$$

and for the 8-lb weight

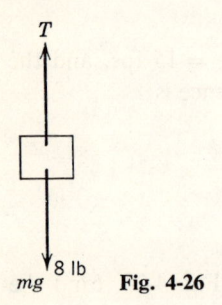

Fig. 4-26

$$\sum F = -T + mg = ma_2 \qquad (2)$$

The accelerations a_1 and a_2 are related by the pulley geometry since the 8-lb weight will rise twice the distance the 10-lb weight will fall; hence

$$a_1 = \frac{-a_2}{2} \qquad (3)$$

Equations (1), (2), and (3) are three equations for the unknowns T, a_1, and a_2. First eliminate a_1:

$$2T - Mg = \frac{Ma_2}{2}$$

$$-2T + 2mg = 2ma_2$$

Now add:

$$a_2\left(2m + \frac{M}{2}\right) = (2m - M)g$$

$$a_2 = \left(\frac{2mg - Mg}{2mg + Mg/2}\right)g$$

$$a_2 = \left[\frac{2(8) - 10}{2(8) + 10/2}\right](32.2)$$

Downward acceleration $= a_2 = 9.20 \text{ ft/sec}^2$ ▶

4-27 Wt. 6

In the following pulley arrangement, what should be the value of the weight W so that the 100-lb weight will remain stationary? What is the resulting acceleration of W? Neglect the friction and inertia of the pulleys.

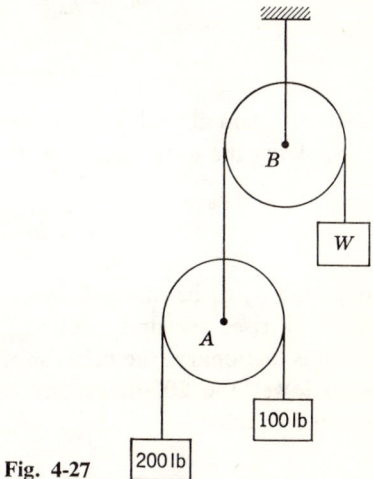

Fig. 4-27

Solution

This problem is concerned with the kinetics of rectilinear translation. The basic equation is

$$\Sigma F_v = \frac{W}{g}a$$

First consider pulley A in Fig. 4-28.
Since the 100-lb weight is to be stationary, $T_1 = 100$ lb. Summing moments about the pulley axis gives $T_2 = 100$ lb. Considering next the 200-lb weight

182 / Dynamics

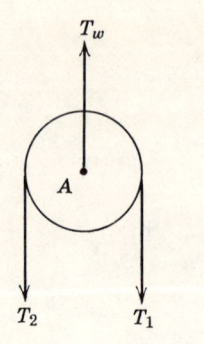

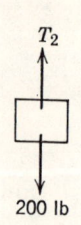

Fig. 4-28

(Fig. 4-28), our basic equation gives

$$-200 + T_2 = -100 = \frac{200}{g} a_2$$

$$a_2 = -0.5g \text{ ft/sec}^2 \text{ (downward)}$$

Looking again at pulley A, summing forces vertically gives $T_w = 200$ lb.

Applying the basic equation to weight W,

$$T_w - W = \frac{W}{g} a_w$$

assuming a_w to be upward. We must now examine the pulley arrangement. When W rises one unit, pulley A falls by one unit. Since the 100-lb weight remains stationary, the cable supporting it shortens one unit. The net effect is to lower the 200-lb weight two units. The accelerations are similarly affected so that

$$a_w = -\tfrac{1}{2} a_2 = 0.25g \text{ ft/sec}^2 \text{ upward}$$

or

$$a_w = 0.25(32.2) = 8.05 \text{ ft/sec}^2 \text{ upward} \blacktriangleright$$

Using

$$T_w - W = \frac{W}{g} a_w$$

$$200 - W = 0.25W$$

$$1.25W = 200$$

$$W = 160 \text{ lb} \blacktriangleright$$

4-28 Wt. 1

A 16-lb weight and an 8-lb weight resting on a horizontal frictionless surface are connected by a cord A and are pulled along the surface with a uniform acceleration of 4 ft/sec² by a second cord attached to the 16-lb weight. The tension in cord A is closest to

(a) 4 lb
(b) 1 lb
(c) 2 lb
(d) 3 lb
(e) 32 lb

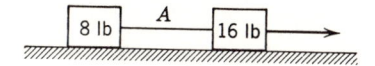

Fig. 4-29

Solution

The presence of the 16-lb weight does not affect the tension in A. The tensile force is that needed to accelerate the 8-lb weight.

$$F = ma = \frac{W}{g}a = \left(\frac{8}{32.2}\right)(4) \approx 1 \text{ lb} \blacktriangleright$$

Answer is (b)

4-29 Wt. 4

In Fig. 4-30 block A weighs 96.6 lb and block B weighs 64.4 lb. The coefficient of friction under A is 0.20 and under B is 0.25. $P = 200$ lb. Determine the acceleration of A and B and the tension T in the connecting rope.

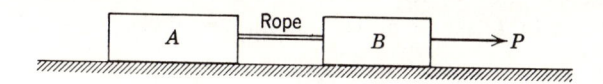

Fig. 4-30

Solution

Applying the equations of motion $\sum F = ma$ to the two free-body diagrams,

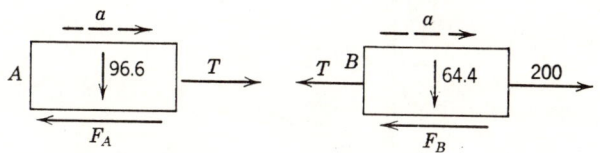

Fig. 4-31

184 / Dynamics

we obtain

$$\Sigma F = T - F_A = \frac{96.6}{32.2} a \qquad (1)$$

$$\Sigma F = 200 - T - F_B = \frac{64.4}{32.2} a \qquad (2)$$

Using the friction relation $F = \mu N$,

$$F_A = 0.2(96.6) = 19.32 \qquad F_B = 0.25(64.4) = 16.1$$

We now solve the two simultaneous equations for a and T Solve equation (1) for T

$$T = 19.32 + 3a \qquad (3)$$

and substitute this relation into equation (2):

$$200 - T - 16.1 = 2a$$

$$200 - 19.32 - 3a - 16.1 = 2a$$

$$5a = 164.58 \qquad a = 32.92 \text{ ft/sec}^2 \blacktriangleright$$

From equation (3),

$$T = 19.32 + 3(32.92)$$

$$T = 118.1 \text{ lb} \blacktriangleright$$

4-30 Wt. 4

Find the acceleration of the masses and the tension in the rope for the arrangement shown. (Assume no friction and weightless pulleys.)

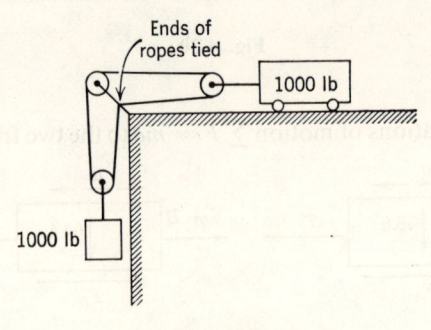

Fig. 4-32

Solution

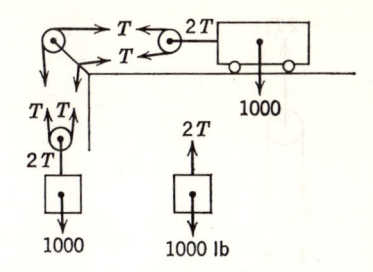

Fig. 4-33

For the suspended weight, $\sum F_y = ma$

$$1000 - 2T = \frac{1000}{32.2} a \tag{1}$$

For the cart, $\sum F_x = ma$

$$2T = \frac{1000}{32.2} a \tag{2}$$

Solving equations (1) and (2) for T and a,

$$1000 - 2T = 2T \qquad T = 250 \text{ lb} \blacktriangleright$$

From (2),

$$2(250) = \frac{1000}{32.2} a$$

$$a = \frac{500}{1000} (32.2) = 16.1 \text{ ft/sec}^2 \blacktriangleright$$

4-31 Wt. 4

Two 3-lb weights are connected by a massless string hanging over a smooth, frictionless peg. If a third weight of 3 lb is added to one of the weights and the system is released, by how much is the force on the peg increased?

Solution

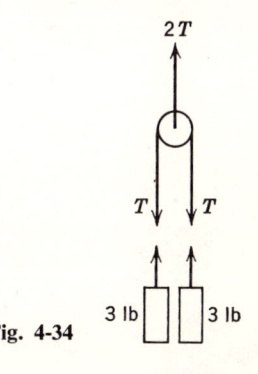

Fig. 4-34

186 / Dynamics

The initial system is in the static equilibrium with $T = 3$ lb, $2T = 6$ lb.

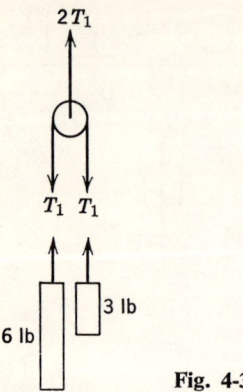

Fig. 4-35

The additional weight is then added. Applying $\sum F = ma$ successively to the 6- and 3-lb weights, we obtain

$$6 - T_1 = \frac{6}{g} a \qquad (1)$$

$$T_1 - 3 = \frac{3}{g} a \qquad (2)$$

The increase in force on the peg is represented by the quantity $2T_1 - 2T$, so $2T_1$ must be found. Doubling equation (2) and comparing it with equation (1),

$$6 - T_1 = 2(T_1 - 3)$$

$$3T_1 = 12 \quad \text{and} \quad 2T_1 = 8$$

Hence the increase in force on the peg $= 8 - 6 = 2$ lb ▶

4-32 Wt. 4

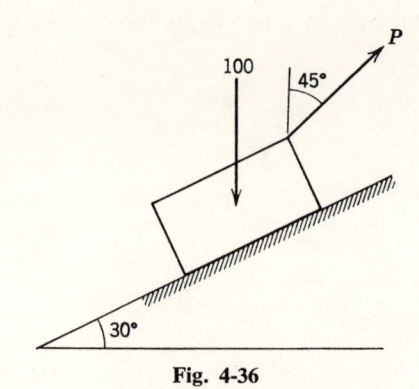

Fig. 4-36

Compute the force P which will be required to give the 100-lb block a velocity V of 10 fps up the plane in a time interval of 5 sec. The system is initially at rest.

Solution

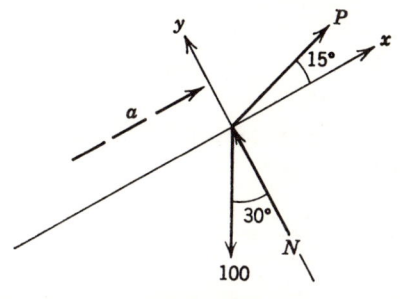

Fig. 4-37

Since no coefficient of friction is mentioned, let us assume the plane to be smooth. Considering the block to be a particle, we choose the coordinate system to be as shown and write the equation of motion

$$\sum F_x = \frac{W}{g} a_x$$

$$P \cos 15° - 100 \sin 30° = \frac{100}{32.2} a_x$$

To determine a_x, $V = V_0 + at$ with $V_0 = 0$. At $t = 5$ sec,

$$10 \text{ fps} = a(5 \text{ sec})$$

$$a = 2 \text{ ft/sec}^2$$

Hence

$$P = \left[\frac{100}{32.2}(2) + 100(0.500)\right] \Big/ 0.966$$

from the earlier equation, or $P = 116$ lb ▶

4-33 Wt. 4

A locomotive weighing 120 tons is coupled to, and pulls, a car weighing 40 tons. The resistances to motion on a level track are $\frac{1}{100}$th of its weight for the locomotive and $\frac{1}{160}$th of its weight for the car. The tractive force exerted by the locomotive is 8000 lb. Find the tension T in the coupling.

188 / Dynamics

Solution

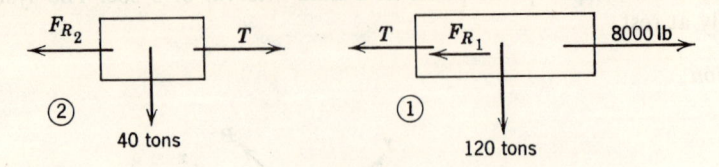

```
  FR₂ ┌──────┐  T      T  ┌── FR₁ ──┐  8000 lb
 ←────│      │────→   ←───│  ←──    │─────→
      └──────┘            └─────────┘
         │                      │
         ▼                      ▼
       40 tons               120 tons
        ②                       ①
```

Fig. 4-38

The diagram on the right represents the locomotive; the car is to the left. The forces resisting motion are

$$F_{R_1} = \frac{W_1}{100} = \frac{120(2000)}{100} = 2400 \text{ lb}$$

$$F_{R_2} = \frac{W_2}{160} = \frac{40(2000)}{160} = 500 \text{ lb}$$

Successively applying $\sum F_x = ma_x$ to the two bodies,

$$8000 - 2400 - T = \frac{120(2000)}{g}a \qquad (1)$$

$$T - 500 = \frac{40(2000)}{g}a \qquad (2)$$

We wish to solve for the tension T. Since the common factor $2000a/g$ appears on the right side of equations (1) and (2), we can immediately write

$$\frac{5600 - T}{120} = \frac{T - 500}{40}$$

$$T = 1775 \text{ lb} \blacktriangleright$$

4-34 Wt. 5

If the coefficient of friction between the 50-lb weight and the 300-lb weight is 0.5, determine the acceleration of each weight for $P = 16$ lb. The 300-lb weight is free to roll, and the weight and friction for the pulley are negligible.

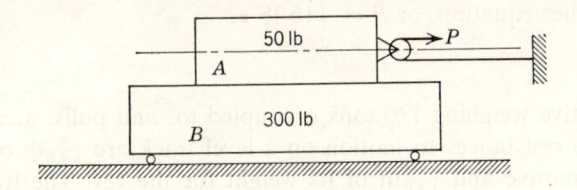

Fig. 4-39

Solution

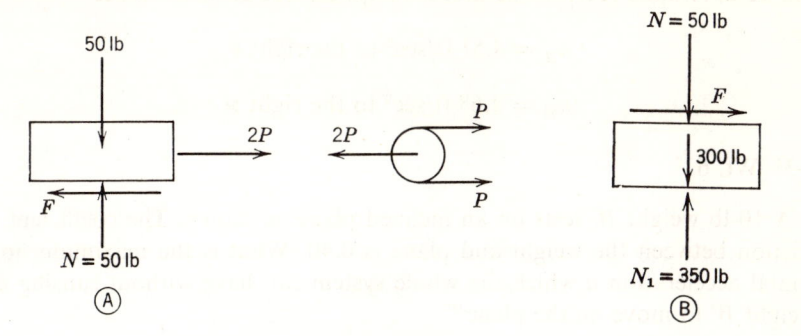

Fig. 4-40

Let us first assume that slipping occurs between the blocks. Then the frictional force developed is

$$F = \mu N = 0.5(50) = 25 \text{ lb}$$

Now apply $\sum F_x = ma$ to each block. For block A

$$2P - F = \frac{50}{g} a_A$$

$$a_A = \frac{g}{50}(32 - 25) = 4.51 \text{ ft/sec}^2 \text{ to the right}$$

and for block B,

$$25 = \frac{300}{g} a_B$$

$$a_B = \frac{25}{300} g = 2.68 \text{ ft/sec}^2 \text{ to the right}$$

To check our original assumption, assume the blocks move together. For the two blocks, $\sum F_x = ma$

$$32 = 2P = \frac{350}{g} a \qquad a = 2.94 \text{ ft/sec}^2$$

For block B alone, this acceleration requires a frictional force

$$F = \frac{300}{g}(2.94) = 27.4 \text{ lb}$$

190 / Dynamics

However, since this force is greater than the maximum frictional force that can be developed (25 lb), the first assumption was correct. Hence

$$a_A = 4.51 \text{ ft/sec}^2 \text{ to the right} \blacktriangleright$$

$$a_B = 2.68 \text{ ft/sec}^2 \text{ to the right} \blacktriangleright$$

4-35 Wt. 6

A 10-lb weight W rests on an inclined plane as shown. The coefficient of friction between the weight and plane is 0.40. What is the maximum horizontal acceleration a which the whole system can have without causing the weight W to move on the plane?

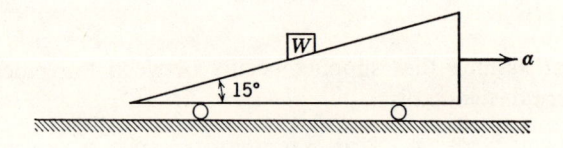

Fig. 4-41

Solution

We first draw a diagram of the weight W:

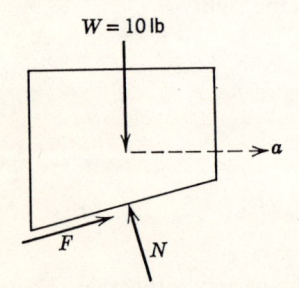

Fig. 4-42

For impending slip, $F = \mu N = 0.4N$. Summing forces vertically,

$$N \cos 15° + 0.4N \sin 15° = 10$$

$$N = \frac{10}{\cos 15° + 0.4 \sin 15°} = \frac{10}{0.966 + 0.4(0.259)}$$

$$N = 9.34 \text{ lb}$$

To find the maximum allowable horizontal acceleration a (when slip is impending),

$$\sum F_x = ma_x$$

$$0.4N \cos 15° - N \sin 15° = \frac{10}{32.2} \times a$$

$$a = \frac{32.2}{10} \times 9.34(0.4 \times 0.966 - 0.259)$$

$$a = 3.84 \text{ ft/sec}^2 \blacktriangleright$$

4-36 Wt. 2

A weight W is suspended by two strings, AB and AC, as shown. What is the ratio of the forces in AC (1) just before; to (2) just after the instant that AB is cut?

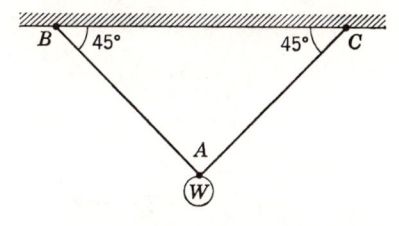

Fig. 4-43

Solution

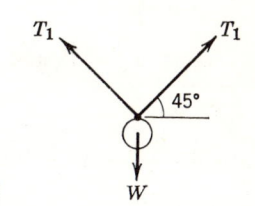

Fig. 4-44

Before string AB is cut, the tensile forces in strings AB and AC are equal because of symmetry. Summing forces vertically,

$$2T_1 \sin 45° = W \qquad T_1 = \frac{W}{\sqrt{2}} \blacktriangleright$$

At the initial instant after string AB is cut, the body possesses only a tangential acceleration; the radial acceleration becomes nonzero only when the

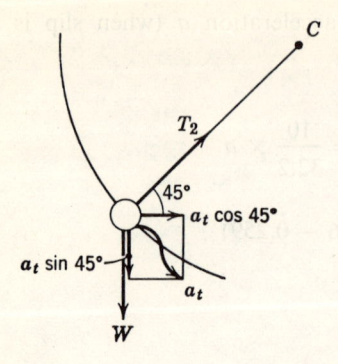

Fig. 4-45

body is moving. Writing the equations of motion,

$\sum F_y = ma_y$

$$W - T_2 \sin 45° = ma_t \sin 45°$$

$\sum F_x = ma_x$

$$T_2 \cos 45° = ma_t \cos 45°$$

$$T_2 = ma_t$$

Inserting this relation to eliminate the acceleration term in the first equation, we obtain

$$W - T_2 \sin 45° = T_2 \sin 45°$$

$$W = 2T_2 \sin 45°$$

$$T_2 = \frac{W}{\sqrt{2}}$$

Hence $T_1 = T_2$ and the ratio $T_1/T_2 = 1$ ▶

4-37 Wt. 1

Kinetic energy is a function of

(a) mass and position
(b) moment of inertia, momentum, and speed
(c) mass and velocity
(d) moment of inertia and acceleration
(e) mass and acceleration

Solution

Kinetic energy may be expressed as $mV^2/2$ and consequently is a function of mass and velocity ▶

Answer is (c)

4-38 Wt. 1

The kinetic energy of a 2400-lb automobile traveling at 100 fps is closest to

(a) 3.75×10^5 ft-lb
(b) 7.5×10^5 ft-lb
(c) 24×10^6 ft-lb
(d) 12×10^6 ft-lb
(e) 3.75×10^6 ft-lb

Solution

$$KE = \tfrac{1}{2}mV^2 = \frac{1}{2}\frac{W}{g}V^2 \approx \frac{1}{2}\left(\frac{2400}{32}\right)(100)^2 = 375{,}000 \text{ ft-lb} \blacktriangleright$$

Answer is (a)

4-39 Wt. 3

An automobile of mass m travels on a straight level highway at velocity V_0. At time $t = 0$ the brakes lock, and the automobile skids to a stop. The length of the skid is L, and the coefficient of sliding friction between the tires and the highway is μ. Derive an expression for V_0 in terms of m, L, and μ.

Solution

The initial kinetic energy of the automobile is $\tfrac{1}{2}mV_0^2$. Since the car comes to rest, this is also the change in kinetic energy. This change in kinetic energy must equal the work done by the automobile while skidding:

$$\tfrac{1}{2}mV_0^2 = FL = (\mu N)L = \mu(mg)L$$

Hence

$$V_0^2 = 2\mu g L$$

and

$$V_0 = (2\mu g L)^{\frac{1}{2}} = (64.4\mu L)^{\frac{1}{2}} \blacktriangleright$$

4-40 Wt. 4

A 9-oz ball is thrown horizontally with a kinetic energy of 36 ft-lb from a vertical cliff 64 ft high.

(a) What is the ball's kinetic energy when it strikes the ground, level with the base of the cliff?

(b) What is the distance from the foot of the cliff to the point where the ball strikes the ground?

194 / Dynamics

Solution

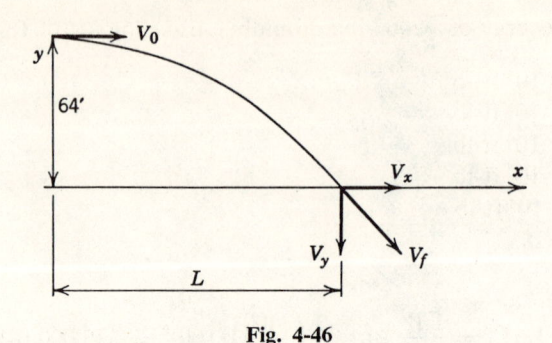

Fig. 4-46

The mechanical energy, which is the sum of kinetic and potential energy, of the ball remains constant while the ball falls. If the base of the cliff is chosen as a datum, then relative to this datum the ball will possess only kinetic energy when it hits the ground. Hence

$$\text{Ground kinetic energy} = \text{initial kinetic energy} + \text{change in potential energy}$$

$$= 36 + Wh$$

$$= 36 + \tfrac{9}{16}(64) = 72 \text{ ft-lb} \blacktriangleright$$

Initially,

$$KE = 36 = \tfrac{1}{2}mV_0^2$$

$$V_0^2 = \frac{72}{m} = \frac{72g}{W} = \frac{72(32.2)}{\tfrac{9}{16}} = 4120$$

$$V_0 = V_x = 64.3 \text{ fps}$$

The ball falls a vertical distance of 64 ft, so

$$64 = \tfrac{1}{2}gt^2 \quad \text{or} \quad t = \left(\frac{128}{g}\right)^{1/2} = 1.99 \text{ sec}$$

Finally,

$$L = V_0 t = (64.3)(1.99) = 128 \text{ ft} \blacktriangleright$$

4-41 Wt. 4

A static weight W produces a static deflection of 2 in. in a spring having a spring constant k. Neglecting the mass of the spring, what will be

(*a*) the deflection δ when the weight W is dropped on the spring from a point 6 in. above the free position at the spring?

(*b*) the maximum velocity of the falling weight?

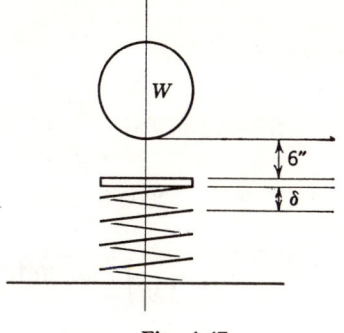

Fig. 4-47

Solution

Potential energy must be conserved here; consequently the decrease in potential energy of the weight during its fall is equal to the gain in potential energy of the spring. Thus

$$W(6 + \delta) = -\tfrac{1}{2}k\delta^2$$

Since the spring constant relates applied force to the resulting deflection of the spring $F = -kx$, or in this case $W = -2k$, and $k = -W/2$. Now

$$W(6 + \delta) = -\frac{1}{2}\left(-\frac{W}{2}\right)\delta^2$$

$$\delta^2 - 4\delta - 24 = 0$$

$$\delta = \tfrac{1}{2}[+4 \pm \sqrt{16 + 4(24)}]$$

$$\delta = 2 \pm \sqrt{28} = 2 \pm 5.3$$

Only the positive root is relevant here: $\delta = 7.3$ in. ▶

The maximum velocity will occur just before the weight begins to decelerate, that is, when $kx/m = g$.

$$mgh = \tfrac{1}{2}mV^2 + \tfrac{1}{2}kx^2 \quad \text{with } h = 6'' = 0.5' \text{ and } x = 2'' = 0.167'$$

$$k = \frac{mg}{x} = \frac{W}{0.167}$$

$$W(0.5 + 0.167) = \frac{1}{2}\frac{W}{g}V^2 + \frac{1}{2}\left[\frac{W}{0.167}\right](0.167)^2$$

$$0.667 = \frac{V^2}{2(32.2)} + \frac{0.167}{2} \qquad V = 6.13 \text{ fps } ▶$$

4-42 Wt. 4

A 50-ton car moving on a level track at 3 mph strikes a bumping post equipped with a spring whose constant is 60,000 lb/in. Assuming no energy loss, how far will the spring be compressed?

196 / Dynamics

Solution

The initial kinetic energy of the car is

$$KE = \tfrac{1}{2}mV^2 = \frac{1}{2}\frac{W}{g}V^2$$

Using the conversion factors 1 ton = 2000 lb and 60 mph = 88 fps,

$$KE = \frac{1}{2}\frac{50(2000)}{32.2}\left[3\left(\frac{88}{60}\right)\right]^2 = 30,100 \text{ ft-lb}$$

When the spring is fully compressed, all of this kinetic energy will have been converted into additional potential energy which is stored in the spring. For a deflection x and spring constant k, the gain in potential energy equals $Fx = \tfrac{1}{2}kx^2$. Hence

$$\tfrac{1}{2}kx^2 = KE$$

$$x = \left(\frac{2KE}{k}\right)^{1/2} = \left[\frac{2(30,100 \text{ ft-lb})}{(60,000 \text{ lb/in.})(12 \text{ in./ft})}\right]^{1/2}$$

$$x = 0.29 \text{ ft} = 3.47 \text{ in.} \blacktriangleright$$

4-43 Wt. 3

A 300,000-lb aircraft descends from 36,000 ft to 6000 ft in 10 min without changing speed. At what average rate (horsepower) is energy being dissipated? If the aircraft drops to sea level in 5 more minutes, at what rate must energy be dissipated?

Solution

The average rate of energy loss is equal to the change in potential energy divided by the corresponding time interval:

(a) $$\frac{\Delta(PE)}{\Delta t} = \frac{W\,\Delta h}{\Delta t} = \frac{(300,000)(36,000 - 6000)}{(10)(60)}$$

$$= \frac{1.5 \times 10^7 \text{ ft-lb/sec}}{550 \text{ ft-lb/sec/hp}}$$

$$= 27,300 \text{ hp} \blacktriangleright$$

(b) $$\frac{\Delta(PE)}{\Delta t} = \frac{W\,\Delta h}{\Delta t} = \frac{(300,000)(6000)}{5(60)(550)}$$

$$= 10,900 \text{ hp} \blacktriangleright$$

4-44 Wt. 1

An impulse is the product of

(a) force and displacement
(b) force and time
(c) force and velocity
(d) mass and acceleration
(e) mass and velocity

Solution

The product of force and the time over which it acts is known as impulse. The time period is usually short, as in impact ▶

$$\text{Answer is } (b)$$

4-45 Wt. 2

A projectile weighing 100 lb strikes the concrete wall of a fort with an impact velocity of 1200 fps. The projectile comes to rest in 0.01 sec, having penetrated the 8-ft-thick wall to a distance of 6 ft. What is the average force exerted on the wall by the projectile?

Solution

Knowing that the impulse exerted on the wall is equal to the change of momentum of the projectile, one may write

$$\int F \, dt = m \, \Delta V$$

If F is assumed to be the average force, then it is a constant, and the equation becomes

$$F \, \Delta t = \frac{W}{g} \Delta V$$

$$F(0.01) = \frac{100}{32.2}(1200)$$

$$F = 3.73(10^5) \text{ lb } ▶$$

4-46 Wt. 6

The initial positions of two elastic balls A and B, each attached to 20-ft-long strings, are as shown below. The mass of ball B is M and of ball A is $3M$. If ball B is released, determine the velocity of each ball just after impact. (Assume elastic impact, and that the center of gravity of each ball is at the same level at the instant of impact.)

198 / Dynamics

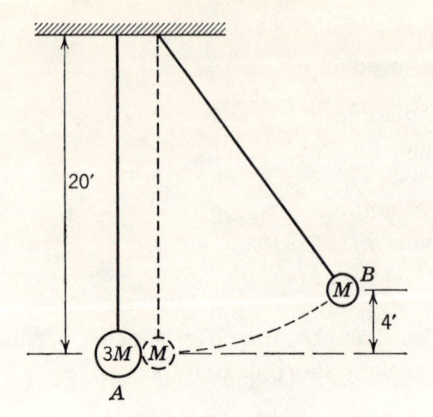

Fig. 4-48

Solution

By the principle of conservation of momentum, the momentum before impact equals the momentum after impact, or

$$MV_B = 3MV_{2A} + MV_{2B}$$
$$V_B = 3V_{2A} + V_{2B} \tag{1}$$

Writing the work-energy relation for ball B before impact,

$$Wh = \tfrac{1}{2}MV_B^2$$
$$Mg \times 4 = \tfrac{1}{2}MV_B^2 \qquad V_B = (8g)^{1/2} = 16.1 \text{ fps}$$

For an elastic impact, energy is conserved in addition to momentum. Writing the equation which states that the energy before impact equals the energy after impact,

$$\tfrac{1}{2}MV_B^2 = \tfrac{1}{2}(3M)V_{2A}^2 + \tfrac{1}{2}MV_{2B}^2$$
$$V_B^2 = 3V_{2A}^2 + V_{2B}^2 \tag{2}$$

Equations (1) and (2) are two equations for the unknowns V_{2A} and V_{2B}. Solving equation (1) for V_{2B} and substituting this expression into (2) gives

$$V_B^2 = 3V_{2A}^2 + (V_B - 3V_{2A})^2$$
$$\cancel{V_B^2} = \cancel{V_B^2} - 6V_{2A}V_B + 12V_{2A}^2$$
$$V_{2A} = \frac{V_B}{2} = 8.05 \text{ fps} \blacktriangleright$$

Now from equation (1)

$$V_B = 3\left(\frac{V_B}{2}\right) + V_{2B}$$

$$V_{2B} = \frac{-V_B}{2} = -8.05 \text{ fps} \blacktriangleright$$

4-47 Wt 5

A 10-lb block which is suspended by a long cord is at rest when a 0.05-lb bullet traveling horizontally to the left strikes and is embedded in it. The impact causes the block to swing upward 0.5 ft measured vertically from its lowest position. Determine

(a) the velocity of the bullet just before it strikes the block
(b) the loss of kinetic energy of the system during impact

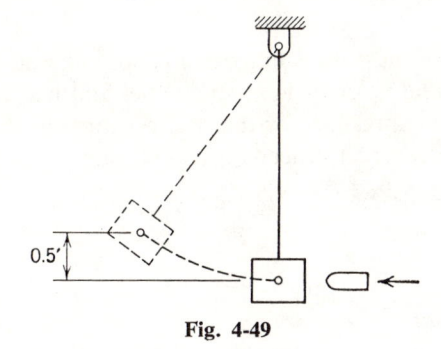

Fig. 4-49

Solution
Let m = mass of bullet
M = mass of block
V_b = velocity of bullet before impact
V = system velocity immediately after impact

Energy is conserved for the motion of the system after impact so that

$$\Delta(KE) = \Delta(PE)$$
$$\tfrac{1}{2}(M + m)V^2 = (M + m)gh$$
$$V^2 = 2gh = 2g(0.5)$$
$$V = g^{1/2} \text{ fps}$$

Also momentum is conserved during impact.

$$mV_b = (M + m)V$$
$$V_b = \left(\frac{M}{m} + 1\right)V$$
$$V_b = \left(\frac{10}{0.05} + 1\right)(32.2)^{1/2} = 1140 \text{ fps} \blacktriangleright$$

200 / Dynamics

The initial kinetic energy of the bullet is therefore

$$\frac{1}{2} m V_b^2 = \frac{1}{2}\left(\frac{0.05}{32.2}\right)(1140)^2 = 1010 \text{ ft-lb}$$

The kinetic energy immediately after impact is

$$\frac{1}{2}(M + m)V^2 = \frac{1}{2}(M + m)g$$
$$= \frac{1}{2}(10 + 0.05) = 5.025 \text{ ft-lb}$$

The system loss of kinetic energy is the difference between these two values, or

$$1010 - 5.025 = 1005 \text{ ft-lb} \blacktriangleright$$

4-48 Wt. 6

A soft lead ball of mass m_2 is hung as a pendulum and is struck by a lead bullet of mass m_1 and velocity V_1. If the bullet and ball stick together after impact, develop an expression for the rise in temperature ΔT of the system in terms of the specific heat of lead c_p, m_1, m_2, and V_1.

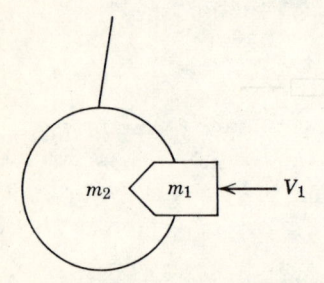

Fig. 4-50

Solution

The solution of this problem combines momentum and energy relations with some elementary concepts from thermodynamics. Let the velocity after impact be V_2. Momentum is conserved during the impact, so

$$m_1 V_1 = (m_1 + m_2)V_2 \quad \text{or} \quad V_2 = \frac{m_1}{m_1 + m_2} V_1$$

The rise in temperature ΔT is related to the heat generated by the impact; the amount of heat produced must equal the change in kinetic energy that occurs because of the collision.

Before the impact,

$$KE_1 = \frac{1}{2} m_1 V_1^2$$

After the impact,
$$KE_2 = \tfrac{1}{2}(m_1 + m_2)V_2^2$$
$$= \frac{1}{2}\frac{m_1^2}{m_1 + m_2}V_1^2$$

The change in kinetic energy is
$$\Delta(KE) = KE_1 - KE_2 = JQ$$
where Q = the amount of heat generated
J = mechanical equivalent of heat (a conversion factor)
The specific heat c_p has units of heat per unit mass per degree of temperature so that
$$Q = c_p(m_1 + m_2)\,\Delta T$$
Hence,
$$\tfrac{1}{2}m_1V_1^2 - \tfrac{1}{2}\left(\frac{m_1^2}{m_1 + m_2}\right)V_1^2 = Jc_p(m_1 + m_2)\,\Delta T$$
and by rearrangement
$$\Delta T = \frac{m_1 m_2 V_1^2}{2Jc_p(m_1 + m_2)^2} \blacktriangleright$$

This derivation will be valid for any unit system so long as that unit system is internally consistent.

4-49 Wt. 4

A 5-lb block of wood slides down a frictionless inclined plane at an angle of 45° with the horizontal and lands on a 10-lb cart with frictionless wheels. The slant length of the plane is $16\sqrt{2}$ ft. If the block sticks to the cart, at what speed will the cart and the block move away from the bottom of the inclined plane?

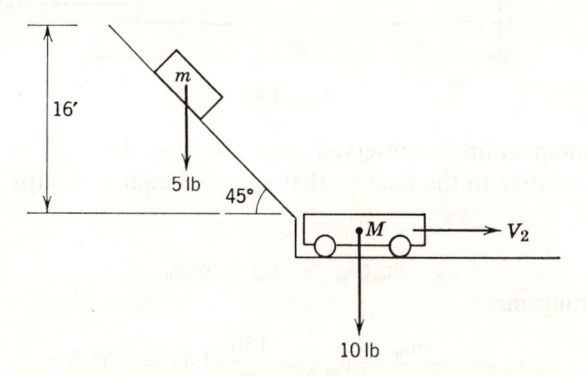

Fig. 4-51

202 / Dynamics

Solution

Since the plane is frictionless, the fall of the block is unimpeded. Equating the changes in kinetic and potential energy between the top and bottom of the plane,

$$\tfrac{1}{2}mV^2 = mgh$$
$$V^2 = 2gh = 2(32.2)(16)$$

Because of the 45° slope, V_x and V_y are equal, or

$$V_x^2 = V_y^2 = \tfrac{1}{2}V^2 = (32.2)(16)$$
$$V_x = 22.7 \text{ fps}$$

Momentum is conserved when the block and cart collide:

$$mV_x = (m + M)V_2$$
$$V_2 = \frac{m}{m+M}V_x = \left(\frac{1}{1+M/m}\right)V_x$$
$$= \left(\frac{1}{1+2}\right)(22.7)$$
$$V_2 = 7.56 \text{ fps} \blacktriangleright$$

4-50 Wt. 3

A 150-lb man stands at the rear of a 250-lb boat. The distance from the man to the pier is 30 ft, and the length of the boat is 16 ft. What is the distance of the man from the pier after he walks to the front of the boat at a velocity of 3 mph? (Assume no friction between the boat and the water.)

Solution

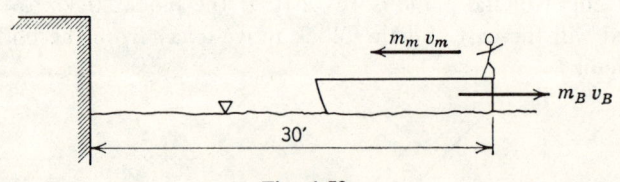

Fig. 4-52

So that momentum is conserved, $m_m v_m = m_b v_b$. Let $v_{m/b}$ be the velocity of the man relative to the boat so that $v_{m/b} = 3 \text{ mph} = 4.4 \text{ fps}$. Since

$$v_m = v_{m/b} - v_b$$
$$m_m(v_{m/b} - v_b) = m_b v_b$$

or, by rearrangement,

$$v_b = \frac{m_m}{m_m + m_b} v_{m/b} = \frac{150}{400}(4.4) = 1.65 \text{ fps}$$

The man walks to the front of the boat in

$$t = \frac{s_m}{v_m} = \frac{16.0}{4.4} = 3.64 \text{ sec}$$

In this time the boat will move away from the pier a distance

$$s_b = v_b t = (1.65)(3.64) = 6.0 \text{ ft}$$

The rear of the boat is now $30 + 6 = 36$ ft from the pier, and the man is

$$36 - 16 = 20 \text{ ft from the pier} \blacktriangleright$$

4-51 Wt. 5

A pile weighing 2000 lb is driven vertically into the ground by a "monkey" (pile hammer) weighing 6000 lb which falls freely from rest through a height of 16 ft to the head of the pile. The impact of the monkey on the pile is assumed to be inelastic. The resistance of the ground can be assumed uniform and equivalent to a force of 150,000 lb. Find how far the pile penetrates into the ground at each blow.

Solution

For conservation of momentum,

$$M_h V_1 + M_p V_{1_p} = M_h V_{2_h} + M_p V_{2_p}$$

Here $V_{1_p} = 0$ and $V_{2_h} = V_{2_p} = V_2$ for inelastic impact. The equation is now

$$M_h V_1 = (M_h + M_p) V_2$$

From the work-energy relation for a freely falling body (the hammer),

$$V_1 = (2gh)^{\frac{1}{2}} = [2(32.2)(16)]^{\frac{1}{2}} = 32.1 \text{ fps}$$

Hence

$$V_2 = \frac{M_h}{M_h + M_p} V_1 = \frac{6000}{8000}(32.1) = 24.1 \text{ fps}$$

Since the ground resistance F is uniform, the deceleration of the pile-hammer combination will be constant.

$$+8000 - 150,000 = F = (M_h + M_p)a = \frac{8000}{32.2}a$$

or the deceleration $a = -571 \text{ ft/sec}^2$.

The pile displacement $x = x_0 + V_0 t + \frac{1}{2}at^2$, where $x_0 = 0$, $V_0 = V_2$, and $a = -571 \text{ ft/sec}^2$ in this case. The pile stops moving when $dx/dt = 0$.

$$0 = \frac{dx}{dt} = V_2 + at$$

$$t = -\frac{V_2}{a} = -\frac{24.1}{-571} = 0.0422 \text{ sec}$$

The set $x = 24.1(0.0422) - \frac{1}{2}(571)(0.0422)^2$.

Set $x = 0.51$ ft/blow ▶

4-52 Wt. 4

The crankshaft of an engine is turning at the rate of 20 rps. The connecting rod is 6 in. long, and the radius is 2 in. At what rate is the piston moving when the angle θ is

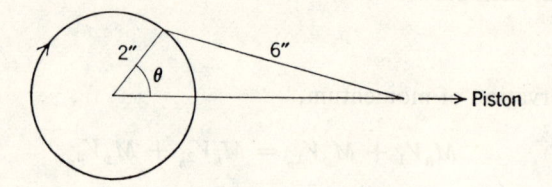

(a) 30°?
(b) 90°?

Fig. 4-53

Solution

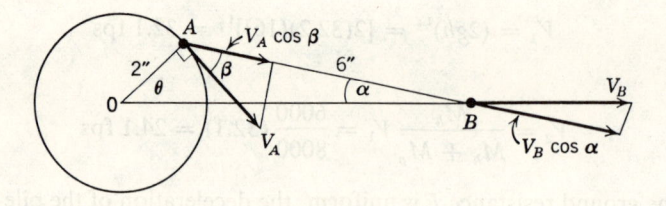

Fig. 4-54

The velocity at point A is $V_A = 20(2\pi)2 = 80\pi$ in./sec. The velocity of the piston is V_B. From the diagram $V_B \cos \alpha = V_A \cos \beta$, and β is a function of the given angles α and θ.

Using the law of sines for triangle OAB,

$$\frac{6}{\sin \theta} = \frac{2}{\sin \alpha} \qquad \sin \alpha = \tfrac{1}{3} \sin \theta$$

Also from triangle OAB,

$$\alpha + \theta + 90° + \beta = 180°$$

(a) For $\theta = 30°$,

$$\sin \alpha = \tfrac{1}{3} \sin 30° \qquad \alpha = 9°36'$$

$$\beta = 180° - 90° - 30° - 9°36' = 50°24'$$

and

$$V_B = V_A \frac{\cos \beta}{\cos \alpha} = 80\pi \frac{\cos 50°24'}{\cos 9°36'} = 80\pi \frac{0.637}{0.986}$$

$$V_B = 162 \text{ in./sec} = 13.5 \text{ fps}$$

(b) For $\theta = 90°$, we find from the angle relation for triangle OAB that $\beta = -\alpha$. Since $\cos \beta = \cos \alpha$, $V_B = V_A$ and

$$V_B = 251 \text{ in./sec} = 21.0 \text{ fps} \blacktriangleright$$

4-53 Wt. 5

A block weighs 6.44 lb. It is being slid along the plane by a pin on the rotating arm OP, which moves in a smooth vertical slot. In the 45° position shown, the 6-in.-long arm OP is rotating clockwise at a constant angular velocity of 5 rad/sec. The coefficient of friction between the block and the plane is 0.2. Determine the torque exerted by the arm OP.

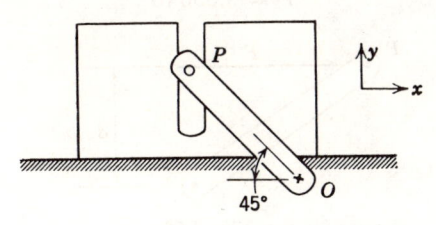

Fig. 4-55

Solution

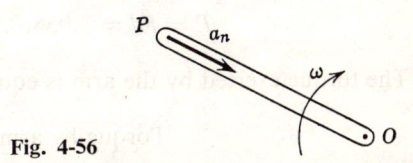

Fig. 4-56

Consider first the rotating arm. The normal acceleration of point P is

$$a_n = r\omega^2 = (0.5)(5)^2 = 12.5 \text{ ft/sec}^2$$

For the position shown, the component in the x direction is $(a_n)_x = 12.5 \sin 45° = 8.84 \text{ ft/sec}^2$.

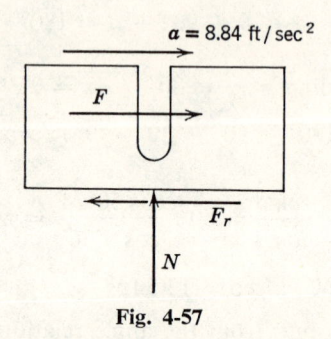

Fig. 4-57

The block slides along the plane, creating a frictional force

$$F_r = \mu N = \mu W = 0.2(6.44) = 1.288 \text{ lb}$$

Writing the equation of motion in the x direction, $\sum F_x = ma_x$ or

$$F - F_r = ma$$

$$F - 1.288 = \frac{6.44}{32.2}(8.84) = 1.768$$

$$F = 3.056 \text{ lb}$$

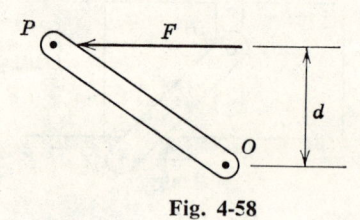

Fig. 4-58

The torque on the arm T is

$$T = Fd = 3.056(0.5 \sin 45°) = 1.08 \text{ lb-ft}$$

The torque exerted by the arm is equal and opposite to this torque.

$$\text{Torque by arm} = 1.08 \text{ lb-ft} \; \rangle \; \blacktriangleright$$

4-54 Wt. 6

A homogeneous cylinder begins to roll down the inclined plane without slipping.

(*a*) Derive the value of the acceleration of the center O as a function of the one variable θ.

(*b*) What is the minimum coefficient of friction μ to prevent the cylinder from slipping?

Note: The moment of inertia of the cylinder about the axis through O is $I_0 = \tfrac{1}{2}mr^2$.

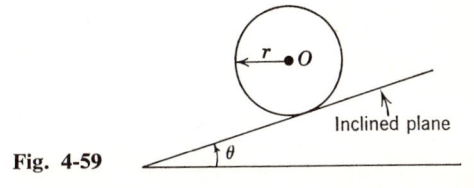

Fig. 4-59

Solution

The cylinder rotates about the contact point A between the plane and the cylinder in a case of noncentroidal rotation.

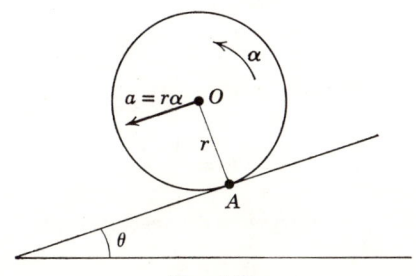

Fig. 4-60

The governing equation is $\sum M_A = I_A \alpha$. The only force that can produce a nonzero moment about A is the weight W of the cylinder. This moment is equal to rH, where H is the component of W parallel to the inclined plane.

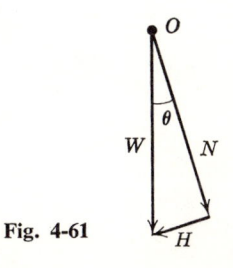

Fig. 4-61

Note that $H = W \sin \theta = mg \sin \theta$. The moment of inertia I_A, by the parallel-axis theorem, is

$$I_A = I_0 + md^2$$
$$= \tfrac{1}{2}mr^2 + mr^2$$
$$= \tfrac{3}{2}mr^2$$

We now have $rmg \sin \theta = \tfrac{3}{2}mr^2\alpha$. Using $a = r\alpha$ and simplifying,

$$a = \tfrac{2}{3}g \sin \theta = 21.5 \sin \theta \blacktriangleright$$

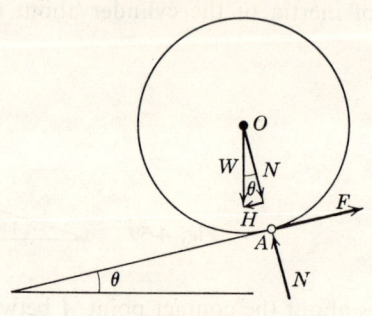

Fig. 4-62

Parallel to the inclined plane, $\sum F = ma$

$$H - F = ma$$
$$W \sin \theta - F = \frac{W}{g}\left(\frac{2}{3}g \sin \theta\right)$$
$$F = \frac{W}{3} \sin \theta$$

For impending slipping $F = \mu N = \mu W \cos \theta$. Hence

$$\mu = \frac{F}{W \cos \theta} = \frac{W \sin \theta}{3W \cos \theta} \qquad \mu = \frac{1}{3} \tan \theta \blacktriangleright$$

4-55 Wt. 5

A homogeneous solid cylinder of mass m and radius R has a string wound around it. One end of the string is fastened to a fixed point, and the cylinder is allowed to fall as shown.

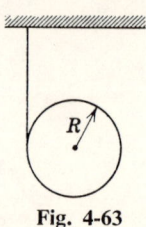

Fig. 4-63

Develop the expression for the angular acceleration of the cylinder in terms of the given parameters. What is the tension in the string? The moment of inertia of a cylinder about its axis is $\frac{1}{2}mR^2$.

Solution

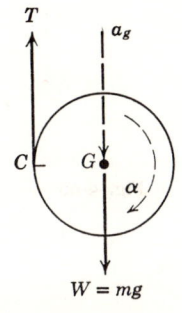

Fig. 4-64

For dynamic equilibrium vertically, $\sum F_y = ma_y$ or

$$mg - T = ma_g$$

Since the contact point C has no vertical acceleration, $a_g = R\alpha$ and

$$mg - T = mR\alpha \qquad (1)$$

Also, about point G, $\sum M_G = I_G \alpha$ so that

$$TR = \tfrac{1}{2}mR^2\alpha \quad \text{or} \quad T = \tfrac{1}{2}mR\alpha \qquad (2)$$

Substituting equation (2) for T into equation (1),

$$\alpha = \frac{2}{3}\frac{g}{R} \blacktriangleright$$

From equation (2), $T = \dfrac{1}{2}mR\left(\dfrac{2}{3}\dfrac{g}{R}\right)$

$$T = \tfrac{1}{3}mg \blacktriangleright$$

4-56 Wt. 5

If the system shown in Fig. 4-65 is released from rest, compute the angular acceleration of the circular drum.

210 / Dynamics

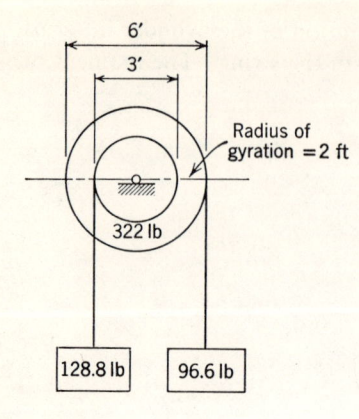

Fig. 4-65

Solution

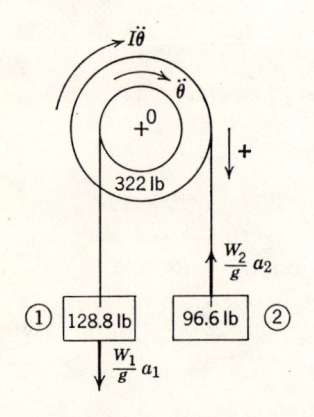

Fig. 4-66

For dynamic equilibrium $\sum M = I\ddot{\theta}$. The moment of inertia of the drum is $I = (W/g)k^2$, where k is the radius of gyration of the drum.

$$I = \frac{322}{32.2}(2)^2 = 40 \text{ lb-ft-sec}^2$$

The acceleration of each weight is $a = r\alpha = r\ddot{\theta}$. Therefore $a_1 = 1.5\ddot{\theta}$ and $a_2 = 3\ddot{\theta}$. Now summing moments about 0 gives

$$3\left(W_2 - \frac{W_2}{g}a_2\right) - 1.5\left(W_1 + \frac{W_1}{g}a_1\right) = I\ddot{\theta}$$

$$3\left[96.6 - \frac{96.6}{32.2}(3\ddot{\theta})\right] - 1.5\left[128.8 + \frac{128.8}{32.2}(1.5\ddot{\theta})\right] = 40\ddot{\theta}$$

$$289.8 - 27\ddot{\theta} - 193.2 - 9\ddot{\theta} = 40\ddot{\theta}$$

$$\ddot{\theta} = \frac{96.6}{76} = 1.27 \text{ rad/sec}^2 \blacktriangleright$$

4-57 Wt. 3

A figure skater is spinning with arms extended. Explain why his speed of rotation increases as he brings his arms in toward his body.

Solution

As the skater spins there is no significant outside force present which would tend to change the skater's angular momentum. Hence the skater's angular momentum, which may be expressed as $I\omega$, is constant during this action. When the arms are drawn inward the radius of gyration, and hence the moment of inertia I, is decreased. The angular velocity ω must then increase proportionally to conserve angular momentum ▶

4-58 Wt. 3

A putty ball of mass m, moving with a velocity v, as shown, makes inelastic impact with the end of a long thin bar of length $2L$ which has a moment of inertia I_0 about its center. The bar is frictionlessly pivoted at its center. Find the expression for the angular velocity of the bar just after the impact occurs.

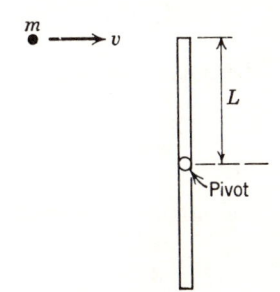

Fig. 4-67

Solution

In this problem we employ the principle of conservation of angular momentum (moment of momentum). This may be expressed as $H_{p1} = H_{p2}$, where H_p is the moment of momentum at any given instant about the pivot.

Before impact, the amount of momentum of the putty ball is $H_{p1} = mvL$. After impact, the moment of momentum of the ball-bar combination is $H_{p2} = mv_2L + I_0\omega$, where $v_2 = L\omega$. Hence

$$mvL = mL^2\omega + I_0\omega$$

$$\omega = \frac{mvL}{mL^2 + I_0} \blacktriangleright$$

4-59 Wt. 4

A homogeneous cylindrical wheel, of 3-ft radius, weighing 400 lb carries two symmetrically placed weights, each weighing 64.4 lb, attached 2 ft from its center. If each weight moves radially outward 1 ft when the wheel is rotating at 2000 rpm, determine the change, in radians per second, which will occur in the angular velocity of the wheel.

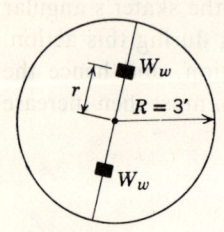

Fig. 4-68

Solution

Angular momentum is conserved for the wheel-weight combination. Expressed mathematically, $I_1\omega_1 = I_2\omega_2$. The initial angular velocity ω_1 is

$$\omega_1 = (2000 \text{ rpm})(\tfrac{1}{60} \text{ min/sec})(2\pi \text{ rad/rev}) = 210 \text{ rad/sec}$$

The polar moment of inertia of the wheel is

$$I_0 = \frac{1}{2}mR^2 = \frac{1}{2}\frac{W}{g}R^2 = \frac{1}{2}\left(\frac{400}{32.2}\right)(3)^2 = 56.0 \text{ lb-ft-sec}^2$$

The moment of inertia of the combination is then

$$I = I_0 + 2\frac{W_w}{g}r$$

For $r = 2$ ft,

$$I_1 = 56.0 + 2\left(\frac{64.4}{32.2}\right)(2)^2 = 72.0 \text{ lb-ft-sec}^2$$

For $r = 3$ ft,

$$I_2 = 56.0 + 2\left(\frac{64.4}{32.2}\right)(3)^2 = 92.0 \text{ lb-ft-sec}^2$$

For conservation of angular momentum,

$$I_1\omega_1 = (72.0)(210) = 92.0\omega_2$$

$$\omega_2 = 164 \text{ rad/sec}$$

Therefore the *decrease* in angular velocity is $210 - 164 = 46$ rad/sec ▶

4-60 Wt. 6

In Fig. 4-69 a cylinder weighing 80 lb is caused to roll on a horizontal plane by means of a cord passing over a weightless and frictionless pulley and attached to a 10-lb weight. Determine the acceleration of W_2 and the tension T in the cord.

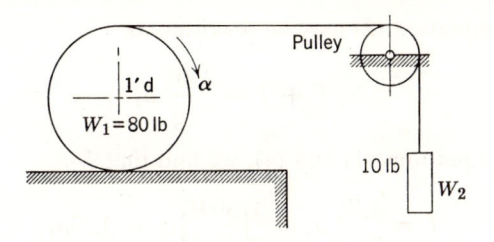

Fig. 4-69

Solution

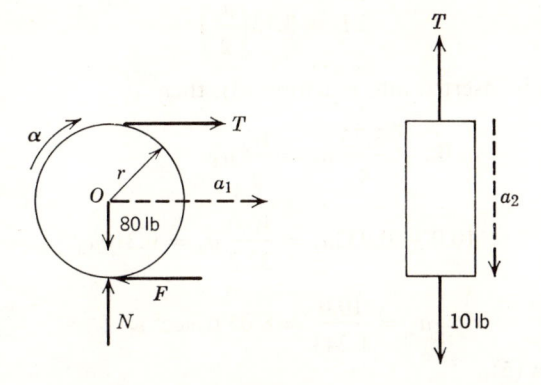

Fig. 4-70

For the state of motion of the cylinder, we can write

$$\sum F_h = ma$$

or

$$T - F = \frac{W_1}{g} a_1 \quad (1)$$

and

$$\sum M_0 = I_0 \alpha$$

or

$$(F + T)r = \frac{1}{2}\left(\frac{W_1}{g}\right) r^2 \alpha \quad (2)$$

Looking next at the weight W_2, we may write

$$\Sigma F_v = ma_2$$

or

$$W_2 - T = \frac{W_2}{g} a_2 \qquad (3)$$

Since $a_1 = \alpha r$, equation (2) may be rewritten as

$$F + T = \frac{1}{2} \frac{W_1}{g} a_1 \qquad (4)$$

If we now add equations (1) and (4), we find that

$$2T = \frac{3}{2} \frac{W_1}{g} a_1 = \frac{3}{2}\left(\frac{80.0}{32.2}\right) a_1 = 3.73 a_1$$

If we assume that the cylinder does not slip on the plane surface, then $a_2 = a_1 + r\alpha = 2a_1$, or

$$2T = 3.73\left(\frac{a_2}{2}\right) \qquad (5)$$

If this relation is inserted into equation (3), then

$$W_2 - \frac{3.73}{4} a_2 = \frac{W_2}{g} a_2$$

$$10.0 - 0.933 a_2 = \frac{10.0}{32.2} a_2 = 0.310 a_2$$

$$a_2 = \frac{10.0}{1.243} = 8.05 \text{ ft/sec}^2 \blacktriangleright$$

From equation (5),

$$T = \frac{3.73}{4} a_2 = \frac{3.73}{4}(8.05)$$

$$T = 7.5 \text{ lb} \blacktriangleright$$

4-61 Wt. 5

An unbalanced flywheel has its center of mass 4.00 in. from the axis of rotation. The radius of gyration of the flywheel with respect to an axis through the center of mass parallel to the axis of rotation is 16.00 in. The flywheel, which weighs 145.0 lb, is rotating clockwise about its axis at an angular speed of 3600 rpm when a counterclockwise torque $T = 18.0 t^2$ is applied, where T is in pound-feet and t is in seconds. Neglecting friction, determine the angular speed in revolutions per minute of the flywheel when t is 10.00 sec.

Solution

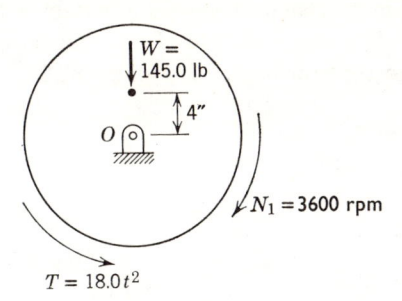

Fig. 4-71

By the parallel-axis theorem, the moment of inertia about axis O is

$$I_0 = \bar{I} + Md^2 = Mk^2 + Md^2 = M(k^2 + d^2)$$

where M = flywheel mass = $145/32.2$ = 4.50 slugs
k = radius of gyration = 16 in.
d = distance from O to center of mass = 4 in.

Thus

$$I_0 = 4.5[(\tfrac{16}{12})^2 + (\tfrac{4}{12})^2] = 8.50 \text{ lb-ft-sec}^2$$

In this problem angular momentum is not constant, but instead the change in angular momentum is equal to the angular impulse caused by the torque T.

$$\text{Angular impulse} = \int T\, dt = \int_0^{10} 18.0 t^2\, dt = 6.0 t^3 \Big|_0^{10} = 6000 \text{ lb-ft-sec}$$

This is now equated to the change in angular momentum, so

$$6000 = I_0(\omega_2 - \omega_1) = 8.50(\omega_2 - \omega_1)$$

$$\omega_2 = \omega_1 + \frac{6000}{8.50} = \omega_1 + 706$$

Hence

$$N_2 = N_1 + \frac{60}{2\pi}(706)$$

$$= -3600 + 6750 = 3150 \text{ rpm counterclockwise} \blacktriangleright$$

4-62 Wt. 1

A simple spring-mass system possesses a certain natural frequency. If the mass is quadrupled in value, the ratio of the new period of oscillation to the original value is closest to which of the following?

(a) 4
(b) $\frac{1}{4}$
(c) 2
(d) $\frac{1}{2}$
(e) 16

Solution

The period T of a spring may be written as

$$T = 2\pi \left(\frac{m}{k}\right)^{1/2}$$

where m is the mass and k is the spring constant. Hence the ratio

$$\frac{\text{New } T}{\text{Old } T} = \left(\frac{4m}{m}\right)^{1/2} = 2 \blacktriangleright$$

Answer is (c)

Chapter 5

Mechanics of Materials

This field of mechanics considers the equilibrium behavior of *deformable* solid bodies. It differs from statics because statics is primarily concerned with the action of external forces on *rigid* bodies, whereas mechanics of materials is primarily concerned with the stress-deformation behavior of nonrigid solid bodies subjected to these external force systems.

In this chapter we consider some fundamental relations which describe the behavior of a solid body which is subjected to axial tension or compression, or to twisting or torsion, or to transverse bending or flexure. These forces or force combinations create internal stresses in the body and cause related material displacements and deformations. A knowledge of these relations is valuable in the analysis and design of structures and machinery.

AXIAL STRESS AND STRAIN

The axial stress s is the axial force per unit area which acts on a member. If P is the magnitude of the force and A is the cross-sectional area, then the axial stress acting on that area is

$$s = \frac{P}{A} \tag{5-1}$$

When the axial force P acts upon a deformable solid bar, the bar changes length. The elongation (or shortening) per unit length is the strain ϵ, or

$$\epsilon = \frac{\delta}{L} \tag{5-2}$$

where L is the initial length of the bar and δ is the change in that length. For elastic materials, that is, materials that completely return to their initially undeformed state when an applied stress is removed, it is found that

218 / Mechanics of Materials

stress is proportional to strain

$$s = E\epsilon \tag{5-3}$$

This is Hooke's law; within limits (below the yield point of the material) many common engineering materials, such as steel, wood, and concrete, follow this law. E is called the modulus of elasticity.

Application of the foregoing equations to an axially loaded bar segment of differential length dx which elongates a distance $d\delta$ shows

$$\delta = \int_0^\delta d\delta = \int_0^L \frac{P}{EA} \, dx \tag{5-4}$$

when integrated over the entire bar length L. If P, E, and A are constant for the bar, equation (5-4) is easily evaluated to give $\delta = PL/EA$.

When a bar is stretched axially in tension, it is found that the cross-sectional area of the bar is decreased by a small amount. A measure of the behavior is Poisson's ratio μ, sometimes called v, which is formed by dividing the lateral strain by the axial strain. The range of Poisson's ratio is $0 < \mu < 0.5$, although 0.25–0.30 is typical of most metals.

These very basic stress-deformation relations can be used to solve many statically indeterminate problems which could not be solved by statics alone. Here we use the observation that all body members will deform when stressed and require that the deformations be *geometrically consistent*.

Example 1

A steel rod containing a turnbuckle has its ends attached to rigid walls and is tightened by the turnbuckle in summer when the temperature is 90°F to give a stress of 2000 psi. What is the stress in the rod in winter when its temperature is $-20°F$? (For steel, $E = 30 \times 10^6$ psi and $\alpha = 6.5 \times 10^{-6}$.)

Solution

The strain ϵ_t induced in an equivalent unrestrained rod by the temperature change is

$$\epsilon_t = \alpha \, \Delta T = 6.5(10^{-6})[90 - (-20)] = 7.15 \times 10^{-4}$$

This shortening, however, cannot occur, and instead the stress in the rod increases. The thermally induced additional tensile stress is

$$s_t = \epsilon_t E = (7.15 \times 10^{-4})(30 \times 10^6) = 21{,}450 \text{ psi}$$

The total stress s in the rod is

$$s = 2000 + 21{,}450 = 23{,}450 \text{ psi tension} \blacktriangleright$$

Example 2

A short column is made of a $\frac{1}{2}$-in.-thick pipe (8-in. I.D.) filled with concrete and capped with a rigid plate in contact with both the steel and the concrete. What load P can be carried by the column if the maximum allowable compressive stress is 15,000 psi in the steel and 900 psi in the concrete? Assume $E_s = 30(10^6)$ psi for steel and $E_c = 2(10^6)$ psi for concrete.

Solution

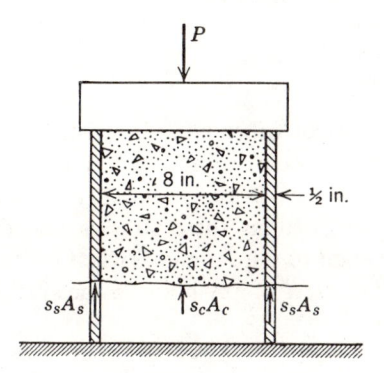

Fig. 5-1

Figure 5-1 shows a free-body diagram of the column. Summing forces vertically,

$$P = s_c A_c + s_s A_s$$

In addition, each material must compress or shorten by the same amount since the rigid plate is in contact with both materials. Thus our consistent deformation requirement requires equal strains $\epsilon_c = \epsilon_s$. In terms of stress

$$\frac{s_s}{E_s} = \frac{s_c}{E_c}$$

or

$$s_s = s_c \left(\frac{E_s}{E_c}\right) = s_c \left[\frac{30(10^6)}{2(10^6)}\right] = 15 s_c$$

If the maximum allowable steel stress of 15,000 psi is attained, the resulting concrete stress would be

$$\frac{15,000}{15} = 1000 \text{ psi}$$

which is greater than the allowable stress. Hence the concrete stress governs, and

$$P = s_c A_c + 15 s_c A_s$$

$$P = (900)\frac{\pi}{4}(8)^2 + 15(900)\frac{\pi}{4}(9^2 - 8^2)$$

$$P = 45{,}200 + 180{,}000$$

$$P = 225{,}200 \text{ lb} \blacktriangleright$$

TORSION

When a circular shaft or other body is twisted, as in cases where shafts transmit power from one point to another by rotational motion, the body is in a state of torsion; shear stresses τ and angular rotations ϕ are the result. A moment or torque T causes these stresses and deformations. When a circular shaft is in a state of pure torsion (when no axial or bending stresses are present), the shear stress at any point in the shaft is

$$\tau = \frac{Tr}{J} \tag{5-5}$$

Thus the stress increases in direct proportion to the distance r from the shaft axis. For a hollow circular shaft of outer diameter d_o and inner diameter d_i, the polar moment of inertia J of the cross section is

$$J = \frac{\pi}{32}(d_o^4 - d_i^4) \tag{5-6}$$

For solid shafts we set $d_i = 0$.

The effect of this torque and resulting shear stress is to cause a relative rotation between different shaft sections. For a shaft section of length L the total angle of twist is

$$\phi = \frac{TL}{GJ} \tag{5-7}$$

where G is the shearing modulus of the material. In these relations the shear equivalent of Hooke's law $\tau = G\gamma$ has been used, where γ is the shear strain.

Example 3

A solid steel shaft 8 ft long is to transmit a torque $T = 20{,}000$ ft-lb. The shear modulus of the material is $G = 12 \times 10^6$ psi, and the allowable shearing stress is 10,000 psi. Compute

(a) the required shaft diameter
(b) the angle of twist between the two ends of the shaft

Solution

(a) Since the shear stress is largest when $r = R$, the radius of the shaft, we have

$$\tau_{max} = \frac{TR}{J}$$

For a solid shaft,

$$J = \frac{\pi}{32} d_o^4 = \frac{\pi}{2} R^4$$

and

$$R^3 = \frac{2T}{\pi \tau} = \frac{2(20,000)}{\pi(10,000)(144)} = 0.00884 \text{ ft}^2$$

$$R = 0.206 \text{ ft} = 2.46 \text{ in.}$$

The required shaft diameter is $2R = 4.92$ in. or, practically, 5 in. ▶

(b) The twist is

$$\phi = \frac{TL}{GJ}$$

$$\phi = \frac{TL}{G\left(\frac{\pi}{2} R^4\right)}$$

$$\phi = \frac{(20,000)(8)}{12(10^6)(144) \frac{\pi}{2}(0.206)^4}$$

$$\phi = 0.0325 \text{ rad} = 0.0325 \left(\frac{180}{\pi}\right) = 1.86° \blacktriangleright$$

BEAM EQUILIBRIUM

The principles of statics are adequate to analyze the external equilibrium of any statically determinate beam for any combination of concentrated or distributed loads. Here, however, we wish to determine the shear force V and bending moment M which exist at a point *in* the beam so that the beam is internally in equilibrium. With the aid of a systematic sign convention, the same statics principles may still be directly used. This information is commonly displayed in diagrams which give V and M for every cross section of the beam. This information can then be used to determine stresses and deformations.

The shear force V, bending moment M, and distributed beam loading w (directed downward) are related by the expressions

$$\frac{dV}{dx} = -w \qquad \frac{dM}{dx} = V \qquad (5\text{-}8)$$

where w, V, and M are all functions of the distance along the beam x. Here we have adopted the sign convention shown in Fig. 5-2 for positive shear and moment.

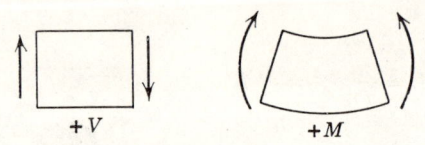

Fig. 5-2

The shear diagram can be constructed a bit more easily when one proceeds from left to right along the beam, for then the loads on the beam act in the same direction as the change in the shear ordinate. From equations (5-8) the slope in the shear diagram is equal to minus the loading intensity at that point. Wherever concentrated loads are applied, the shear ordinate abruptly changes by that amount. Also from equations (5-8) a positive shear gives a positive slope to the moment diagram, and a zero shear occurs where the moment M is a maximum or minimum, since the ordinate of the shear diagram equals the slope of the moment diagram. Finally, if equations (5-8) are integrated, we find that (a) minus the area under the load curve between two points equals the change in shear between these points, and (b) the area under the shear curve between two points equals the change in the bending moment between these same two points.

Example 4

The loading diagram for a beam is shown in Fig. 5-3. The beam is 6 in. × 12 in. and is placed on edge with respect to the loads. Determine the reactions R_1 and R_2 and construct the shear and bending moment diagrams.

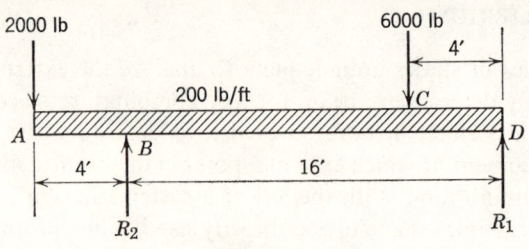

Fig. 5-3

Solution

To find R_2 we sum moments around point D.

$$\sum M_D = 0 = 2000(20) + 200(20)(10) + 6000(4) - 16R_2$$

$$R_2 = \tfrac{1}{16}(40{,}000 + 40{,}000 + 24{,}000)$$

$$R_2 = 6500 \text{ lb} \blacktriangleright$$

Summing forces vertically,

$$\sum F_v = 0$$

$$R_1 = 2000 + 6000 + 200(20) - 6500$$

$$R_1 = 5500 \text{ lb} \blacktriangleright$$

Diagrams for the shear force V and bending moment M are given in Fig. 5-4.

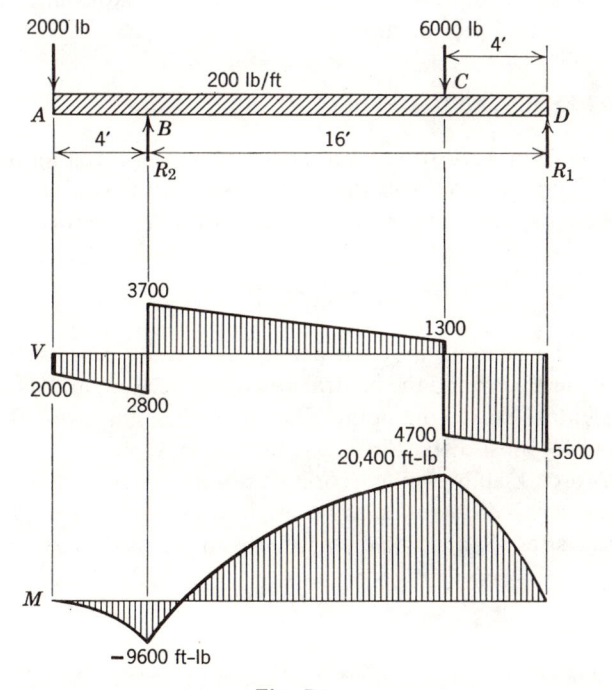

Fig. 5-4

The shear curve drops continuously down to the right at a slope of 200 lb/ft, the magnitude of the distributed loading. At points A, B, and C the shear ordinate jumps by the amount, and in the direction, of the concentrated loads at those points. At D the 5500-lb reaction returns the shear ordinate to zero.

224 / Mechanics of Materials

The moment diagram M can be constructed directly from the shear diagram since no concentrated couples act on the beam. The change in the moment ordinate equals the area under the shear curve. From A to B we find

$$M_B = -\tfrac{1}{2}(4)(2000 + 2800) = -9600 \text{ ft-lb}$$

since $M_A = 0$. From B to C the shear is positive, giving

$$M_C = M_B + \tfrac{1}{2}(12)(3700 + 1300)$$
$$= -9600 + 30,000 = 20,400 \text{ ft-lb}$$

From C to D the shear is negative, which causes the moment to decrease to zero at D. The slope of the moment diagram is steepest where the ordinate of the shear diagram is largest. Finally, we note that the maximum positive moment is at C where the shear curve passes through zero at the support. [As an alternative to this procedure, we could mathematically integrate equations (5-8) and plot them.]

BEAM STRESSES AND DEFLECTIONS

The existence of a shearing force or bending moment at an internal beam cross section creates axial and shearing stresses at the section.

The bending stress f_s at a point is given by the flexure formula

$$f_s = \frac{My}{I} \tag{5-9}$$

where M is the bending moment acting on a beam cross section, I is the moment of inertia around the neutral axis of the section, and y is the distance from the neutral axis to the point. The largest bending stress thus occurs in the beam fibers most distant from the neutral axis.

A shear force V acting on a section causes an internal longitudinal (horizontal) shear stress distribution to be set up across the section. The magnitude of this shear stress S_s at some distance y from the neutral axis is

$$S_s = \frac{VQ}{Ib} \tag{5-10}$$

Here I is again the section moment of inertia and b is the section width at height y. The quantity Q is the first moment of the area of a portion of the beam cross section. The area is that portion of the section which lies outside the distance y from the neutral axis. If c denotes the extreme fiber in the section, then Q is

$$Q = \int_y^c y \, dA \tag{5-11}$$

This equation shows immediately that the extreme fibers in a beam carry zero shear stress since $Q = 0$ when $y = c$.

Example 5

For the beam in Example 4 determine

(a) the location and magnitude of the maximum bending stress
(b) the location and magnitude of the maximum shear stress

Solution

(a) The maximum bending moment occurs at point C and is $M_C = 20{,}400$ ft-lb. The moment of inertia of the 6 in. × 12 in. section is $I = bh^3/12 = (6)(12)^3/12 = 864$ in.^{4}. The maximum bending stress occurs farthest from the neutral axis where $y = h/2 = 6$ in. Equation (5-9) now gives the maximum stress as

$$f_s = \frac{My}{I} = \frac{(20{,}400)(12)(6)}{864} = 1700 \text{ psi} \blacktriangleright$$

(b) Since cross-sectional properties are constant throughout the length of the beam, the maximum shear stress will occur at some point in the section where the shear force V is a maximum. From the shear diagram in Example 4 the shear is a maximum at the right support where $V = 5500$ lb. (Notice here that the maximum shear ordinate does not occur at the point of application of the largest concentrated force.) The maximum stress occurs at the point in this section where Q is maximized. First we evaluate equation (5-11) for any point y on the rectangular beam section of height h and width b:

$$Q = \int_y^{h/2} y(b\,dy) = \frac{b}{2}\left(\frac{h^2}{4} - y^2\right)$$

Thus Q, and consequently S_s, is a maximum at the neutral axis $y = 0$.

$$S_s = \frac{VQ}{Ib} = \frac{V(bh^2/8)}{(bh^3/12)b} = \frac{3V}{2bh} = \frac{3V}{2A}$$

is the maximum shear stress in a rectangular beam. Hence

$$S_s = \frac{3(5500)}{2(6)(12)} = 114.5 \text{ psi} \blacktriangleright$$

For small deflections of beams the basic relation between the beam deflection y, curvature $1/\rho$, and moment M at a point is

$$\frac{d^2y}{dx^2} = \frac{1}{\rho} = \frac{M}{EI} \tag{5-12}$$

226 / Mechanics of Materials

Here we assume that y is positive upward when M is positive according to the earlier sign convention for moment diagrams. If M is written as a function of x, equation (5-12) may be integrated twice. The constants of integration are evaluated by noting the slope and deflection constraints for the particular beam.

The moment-area method is a rapid, efficient alternative method of interpreting and using the basic deflection equation. Integrated once between points A and B on the beam, equation (5-12) gives

$$\theta_B - \theta_A = \int_A^B \frac{M}{EI} dx \tag{5-13}$$

which shows that the change in slope between A and B equals the area of the M/EI diagram between A and B. It can also be shown that the deflection δ of point B from the tangent to point A is

$$\delta = \int_A^B \frac{M}{EI} x \, dx \tag{5-14}$$

The integral is the first moment of the M/EI diagram about point B. In applying the moment-area principles, one should treat positive and negative sections of the moment diagram separately.

Since equation (5-12) is a linear equation, the principle of linear superposition is valid for finding slopes and deflections for a beam which is simultaneously acted upon by several loads. This principle is therefore a useful tool in the solution of statically indeterminate beam problems.

Example 6

A propped cantilever beam 12 ft long carries a uniformly distributed load of 2000 lb/ft (2 kips/ft), as shown in Fig. 5-5. Calculate the reactions.

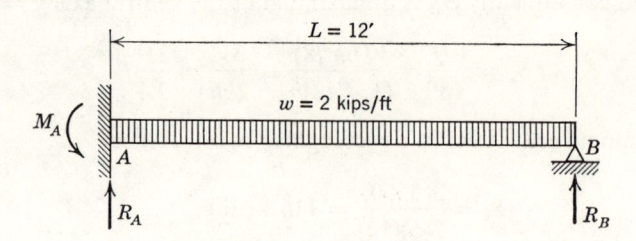

Fig. 5-5

Solution

The beam is statically indeterminate to the first degree. We consider the problem as the sum of two statically determinate problems and require the

BEAM STRESSES AND DEFLECTIONS / 227

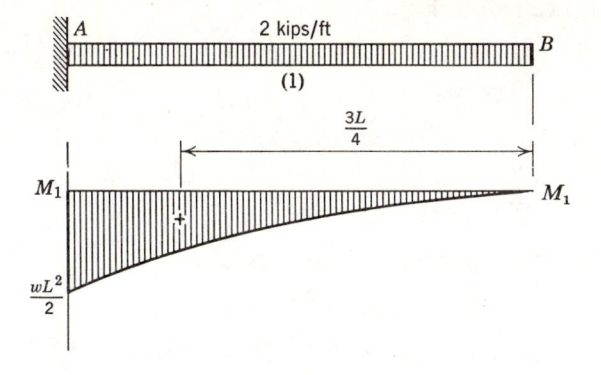

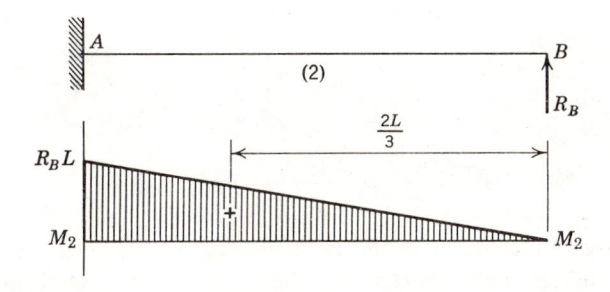

Fig. 5-6

net deflection of the propped end (point B) to be zero. Shown beneath each statically determinate cantilever in Fig. 5-6 is its associated bending moment diagram; also shown is the distance to the centroid of the moment diagram from the free end of the beam. For this problem we note that EI is constant throughout the beam. According to equation (5-14), the deflection at B for case 1 is

$$\delta_{B_1} = \frac{1}{EI} (\text{area of } M_1 \text{ diagram}) \bar{x}_1$$

$$\delta_{B_1} = \frac{1}{EI}\left(\frac{-wL^3}{6}\right)\frac{3L}{4} = \frac{-wL^4}{8EI}$$

For case 2,

$$\delta_{B_2} = \frac{1}{EI}\left(\frac{+R_B L^2}{2}\right)\frac{2L}{3} = \frac{+R_B L^3}{3EI}$$

Since point B cannot deflect,

$$\delta_B = \delta_{B_1} + \delta_{B_2} = 0$$

$$\frac{R_B L^3}{3EI} - \frac{wL^4}{8EI} = 0$$

$$R_B = \frac{+3}{8} wL = \frac{+3}{8}(2)(12) = +9 \text{ kips} \blacktriangleright$$

Now, by using the equations of statics, $\sum F_v = 0$

$$R_A + R_B = wL$$

$$R_A = \tfrac{5}{8} wL = \tfrac{5}{8}(2)(12) = +15 \text{ kips} \blacktriangleright$$

At point A, $\sum M_A = 0$

$$M_A + R_B L - \frac{wL^2}{2} = 0$$

$$M_A = \frac{wL^2}{2} - \frac{3}{8} wL^2$$

$$M_A = \frac{wL^2}{8} = \frac{2(12)^2}{8} = +36 \text{ kip-ft} \blacktriangleright$$

The plus sign indicates that the moment acts in the indicated direction. The shear and moment diagrams for the propped cantilever beam can now be constructed directly, if desired.

EULER COLUMN BUCKLING

Columns are slender members which carry primarily axial loads. The critical load P_{cr} which will cause lateral buckling, an instability phenomenon, in long slender columns is given by the Euler column formula, which may be expressed as

$$P_{cr} = \frac{\pi^2 EI}{L_1^2} \tag{5-15}$$

where L_1 is related to the length of the column. The moment of inertia may be written $I = Ar^2$, r being the radius of gyration. Euler's formula is applicable only when the modified slenderness ratio L_1/r is greater than 100, approximately. Shorter columns do not fail by buckling which is described by the Euler equation. The relation between L_1 and the column length L depends on the end conditions of the column. For both ends fixed $L_1/L = \tfrac{1}{2}$; for one end fixed and one end pinned $L_1/L = 0.7$; for both ends pinned $L_1/L = 1$; and for one end fixed and one end free $L_1/L = 2$.

Example 7

A $\frac{3}{4}$-in.-diameter solid steel rod ($E = 30 \times 10^6$ psi) is 5 ft long and is pin-connected at its ends. How large a compressive load can be applied before buckling occurs?

Solution

The moment of inertia for a circular rod of diameter d is

$$I = \frac{\pi}{64} d^4 = Ar^2$$

The radius of gyration is

$$r = \left(\frac{\pi}{64} d^4 \Big/ \frac{\pi}{4} d^2\right)^{1/2} = \frac{d}{4} = \frac{3}{16} \text{ in.}$$

For pinned ends $L_1 = L$ so

$$\frac{L_1}{r} = \frac{5(12)}{\frac{3}{16}} = 320 > 100$$

and the Euler formula applies. The critical load is

$$P_{cr} = \frac{\pi^2 E I}{L^2}$$

$$P_{cr} = \frac{\pi^2 E}{L^2} \left(\frac{\pi}{64} d^4\right)$$

$$P_{cr} = \frac{\pi^3 (30)(10^6)(0.75)^4}{64[5(12)]^2}$$

$$P_{cr} = 1280 \text{ lb} \blacktriangleright$$

5-1 Wt. 1

The only strength property of stress grade lumber that can be precisely determined without breaking the piece is

(a) rigidity
(b) factor of safety
(c) modulus of rupture
(d) stiffness
(e) horizontal shear

Solution

Stiffness is a measure of the wood's ability to resist deformation or bending. It is expressed in terms of the *modulus of elasticity* and applies only within the proportional limit. It can therefore be determined without breaking the piece ▶

Answer is (d)

5-2 Wt. 1

The ultimate strength divided by the allowable stress is the

(a) yield point
(b) percentage of elongation
(c) percentage of reduction in area
(d) working stress
(e) factor of safety

Solution

The ultimate strength divided by the allowable stress is the factor of safety ▶

Answer is (e)

5-3 Wt. 1

If an engineering structure is to be designed against rupture of any of its members under steady load, the factor of safety would be a ratio based on working stress and

(a) elastic strength
(b) ultimate strength
(c) toughness
(d) resilience
(e) endurance limit

Solution

Where a structure is designed against rupture, with a steady load, the governing situation would be ultimate strength. Therefore, the factor of safety would be the ratio of ultimate strength to working stress ▶

Answer is (b)

5-4 Wt. 1

A body having the same elastic properties in all directions is

(a) isotropic
(b) homogeneous
(c) isothermal
(d) orthotropic
(e) inhomogeneous

Solution

A body is said to be isotropic if its elastic properties are the same in all directions ▶

Answer is (a)

5-5 Wt. 1

The coefficient of thermal expansion of steel is approximately what percent of the coefficient of thermal expansion of concrete?

(a) 50%
(b) 75%
(c) 100%
(d) 125%
(e) 150%

Solution

Coefficient of thermal expansion of steel = 0.0000065 in./in./°F
Coefficient of thermal expansion of concrete = 0.000006 in./in./°F

$$\frac{6.5 \times 10^{-6}}{6.0 \times 10^{-6}} = 1.08 = 108\% \blacktriangleright$$

Answer is (c)

5-6 Wt. 1

The designations A36, A441, and T1 are used to indicate

(a) concrete
(b) wood
(c) brick masonry
(d) steel
(e) plastic

Solution

These are ASTM designations for various types of steel ▶

Answer is (d)

5-7 Wt. 1

Structural steel designated as A36 has a minimum yield strength close to

(a) 32,000 psi
(b) 33,000 psi
(c) 36,000 psi
(d) 42,000 psi
(e) 46,000 psi

232 / Mechanics of Materials

Solution

A36 refers to ASTM Designation A36. ASTM A36-61T sets the required minimum yield point at 36,000 psi ▶

Answer is (*c*)

5-8 Wt. 1

The function of connectors in composite construction is to

(*a*) transfer vertical shear from the slab to the beam
(*b*) transfer moment from the slab to the beam
(*c*) transfer diagonal tension from the slab to the beam
(*d*) transfer horizontal shear from the slab to the beam
(*e*) help balance the dead loads from top to bottom of the beam

Solution

Connectors are used to transfer horizontal shear from the slab to the beam

Answer is (*d*)

5-9 Wt. 1

Pan joists are generally used in the construction of

(*a*) railroad bridges
(*b*) timber wharves
(*c*) roof trusses
(*d*) concrete flooring
(*e*) wood frame ceilings

Solution

Concrete joist construction consists of joists and a top slab of concrete, the whole formed by creating longitudinal void spaces by means of either permanent or removable forms of steel (or removable wood forms). The metal forms are called pans, hence *pan joists* refer to this system of concrete construction ▶

Answer is (*d*)

5-10 Wt. 1

Split rings are commonly used in the construction of

(*a*) bearing piles
(*b*) concrete pipes
(*c*) steel joists
(*d*) timber trusses
(*e*) earth fill dams

Solution

Split rings are timber connectors used to transmit forces between wood members or between wood and metal members ▶

Answer is (*d*)

5-11 Wt. 1

The unit lateral deformation of a body under stress divided by the unit longitudinal deformation is known as _____ ratio.

Solution

$$\text{Poisson's ratio } \mu = \frac{\text{unit lateral deformation (lateral strain)}}{\text{unit longitudinal deformation (axial strain)}}$$

Answer is Poisson's ratio ▶

5-12 Wt. 1

A thin wire is subject to a tensile stress. If the temperature is constant, the electric resistance of the stressed wire with respect to the electric resistance of the unstressed wire will

(*a*) increase
(*b*) decrease
(*c*) remain the same
(*d*) become negative as the elastic limit of the material is exceeded
(*e*) any of the above, depending on material

Solution

The resistance of wire varies inversely with the area. Since the stressed wire has a reduced cross-sectional area and increased length, its resistance is increased compared to the unstressed wire. This property is utilized in strain gages ▶

Answer is (*a*)

5-13 Wt. 1

The "offset method" is used to find a property of metals which do not have a well defined stress-strain curve, such as steel. This property is called the

(*a*) yield point
(*b*) modulus of elasticity
(*c*) section modulus
(*d*) proportional limit
(*e*) moment of inertia

Solution

The "offset method" is used to determine the yield point ▶

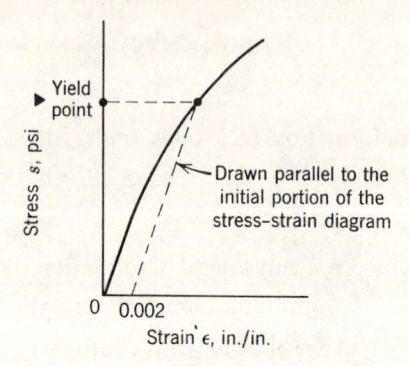

Fig. 5-7

Answer is (a)

5-14 Wt. 5

The rod in Fig. 5-8a of a material having the stress-strain curve shown in Fig. 5-8b has a spring attached at one end. The spring constant of this spring is 20,000 lb/in. The rod has a cross-sectional area of 1.00 in.² and is 20.00 in. long. The load F is increased until the spring has elongated 0.75 in. and then decreased to zero. What is the length of the rod after the load is removed?

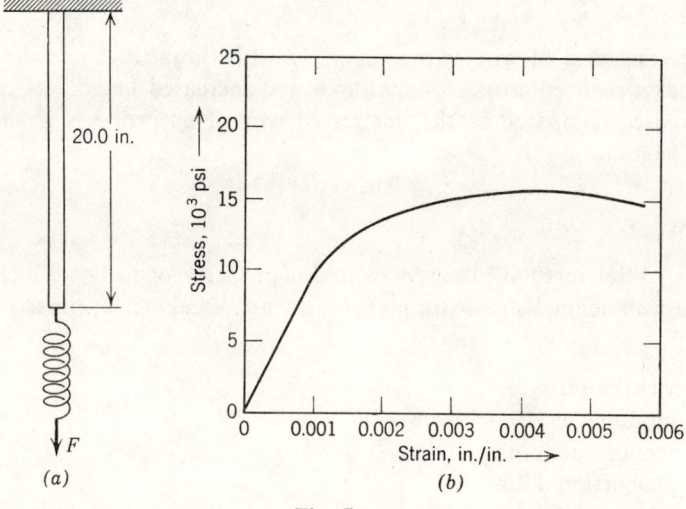

Fig. 5-8

Solution

The spring tells us the magnitude of the load applied. The load F is increased to $0.75 \times 20{,}000 = 15{,}000$ lb, then removed. For a rod of 1.00 in.² the stress applied is

$$f_s = \frac{P}{A} = \frac{15{,}000}{1.00} = 15{,}000 \text{ psi}$$

Up to 10,000 psi on the stress-strain diagram there is a linear relationship between stress and strain (Hooke's law applies). For stresses above the elastic limit (10,000 psi) there will be a permanent deformation. As the load is increased, the stress-strain relation is as represented by the curve. Beyond the elastic limit the stress-strain curve will not be retraced as the load is removed but instead will decrease with a slope equal to that below the elastic limit.

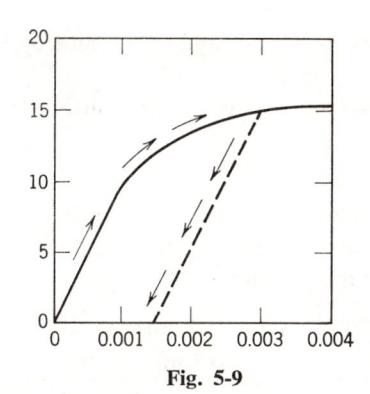

Fig. 5-9

The result is a permanent elongation of the rod.

$$\Delta L = \varepsilon L = 0.0015 \times 20.00 = 0.03 \text{ in.}$$

Thus the new rod length will be

$$20.00 + 0.03 = 20.03 \text{ in.} \blacktriangleright$$

5-15 Wt. 3

The rails of a tramway are welded together at $+50°$F. What stress will be produced in these rails when heated by the sun to $+100°$F?

(Assume the coefficient of linear expansion of rails is 70×10^{-7} in./in.-°F and the modulus of elasticity is 30×10^6 lb/in.².)

Solution

Stress s = unit strain (ε) × modulus of elasticity (E)

If the rails were free to expand, then their increase per unit length due to temperature would be

$$\Delta L = \text{coefficient of expansion } (\alpha) \times \text{length } (L) \times \text{change of temperature } (\Delta T)$$
$$= 70 \times 10^{-7} \times 1 \times (100 - 50) = 350 \times 10^{-6} \text{ in./in.}$$

Since ΔL is found per unit length of the rail, it is the unit strain, or $\Delta L = \varepsilon$. Thus the stress developed in the rail is given by

$$s = \varepsilon E = 350 \times 10^{-6} \times 30 \times 10^6 = 10{,}500 \text{ psi} \blacktriangleright$$

5-16 Wt. 6

A $\frac{1}{2}$-in. O.D. brass rod is 100 ft 5 in. long and a 4-in. nominal cast-iron pipe (4.80 in. O.D., 0.38 in. wall thickness) is 100 ft $5\frac{1}{2}$ in. long when both are at the same temperature (60°F). At what temperature will both be the same length?

Given: C_e brass $= 10 \times 10^{-6}$ units per unit length per degree Fahrenheit

C_e cast iron $= 6 \times 10^{-6}$ units per unit length per degree Fahrenheit

Solution

Since brass has a greater coefficient of expansion than cast iron and the brass piece is presently shorter than the cast-iron one, they will be the same length at some elevated temperature.

$$\Delta T (C_{e \text{ brass}})(\text{length}_{\text{brass}}) = 0.5 \text{ in.} + \Delta T (C_{e \text{ c.i.}})(\text{length}_{\text{c.i.}})$$

$$\Delta T = \frac{0.5 \text{ in.}}{(C_{e \text{ brass}})(\text{length}_{\text{brass}}) - (C_{e \text{ c.i.}})(\text{length}_{\text{c.i.}})}$$

$$\Delta T = \frac{0.5 \text{ in.}}{(10 \times 10^{-6})(1205 \text{ in.}) - (6 \times 10^{-6})(1205.5 \text{ in.})}$$

$$\Delta T = \frac{0.5}{0.01205 - 0.00723} = \frac{0.5}{0.00482} = 103.7°F$$

Therefore the temperature at which both pieces will be the same length $= 60 + 103.7 = 163.7°F \blacktriangleright$

5-17 Wt. 5

A $\frac{1}{2}$-in.-diameter steel tie rod 18 ft in length is joined to two rigid walls in such a way that an axial tensile stress of 20,000 psi is induced in the rod. Determine

(a) the change in the diameter of the rod due to the application of this tensile load

(b) the temperature change which would nearly reduce the stress to zero

Use Poisson's ratio of $\frac{1}{4}$, a temperature coefficient of 6.5×10^{-6}, and $E = 30 \times 10^6$ psi.

Solution

$$\text{Unit strain } \varepsilon = \frac{S}{E} = \frac{20{,}000}{30 \times 10^6} = \frac{2}{3} \times 10^{-3}$$

(a) Unit change in diameter = $\mu\varepsilon = \frac{1}{4}\varepsilon = \frac{1}{6} \times 10^{-3}$
Total change in diameter = $d\mu\varepsilon = \frac{1}{2} \times \frac{1}{6} \times 10^{-3}$
$= 8.33 \times 10^{-5}$ in. ▶

(b) The temperature-induced strain must be equal to the initial strain $\varepsilon = \alpha\Delta T$:

$$\Delta T = \frac{\varepsilon}{\alpha} = \frac{\frac{2}{3} \times 10^{-3}}{6.5 \times 10^{-6}} = 102.6°F \text{ temperature rise} \blacktriangleright$$

5-18 Wt. 1

The stress in an elastic material is

(a) inversely proportional to the material's yield strength
(b) inversely proportional to the force acting
(c) proportional to the displacement of the material acted upon by the force
(d) inversely proportional to the strain
(e) proportional to the length of the material subject to the force

Solution

According to Hooke's law, stress is directly proportional to strain. Strain is the deformation (or displacement) of the material per unit length. Thus we can say: Stress is proportional to the displacement of the material acted upon by the force ▶

Answer is (c)

5-19 Wt. 1

What load must be applied to a 1-in. round steel bar 8 ft long ($E = 30{,}000{,}000$ psi) to stretch the bar 0.05 in.?

(a) 7,200 lb
(b) 9,850 lb
(c) 8,600 lb
(d) 12,250 lb
(e) 15,000 lb

Solution

$$\delta = \frac{PL}{AE} \qquad P = \frac{AE\delta}{L} = \frac{\pi}{4}(1)^2 \frac{30{,}000{,}000(0.05)}{96} = 12{,}250 \text{ lb} \blacktriangleright$$

Answer is (*d*)

5-20 Wt. 6

A 2-in. diameter bolt must be stressed to a total load of 70,000 lb. If the initial length of the bolt is 22.250 in., what is the final length after the load is applied?

Solution

$$\delta = \frac{PL}{AE} \qquad A = \pi R^2 = \pi(1)^2 = 3.14 \text{ in.}^2$$

$$\delta = \frac{(70 \text{ kips})(22.250 \text{ in.})}{(3.14 \text{ in.}^2)(30{,}000 \text{ kips/in.}^2)} = 0.0165 \text{ in.}$$

Final length = 22.250 + 0.0165 = 22.266 in. ▶

5-21 Wt. 5

A steel bar having a 1-in.² cross section is 150 in. long when lying on a horizontal surface. What is its length when suspended vertically from one end? Use $E = 30 \times 10^6$ and $W = 0.283$ lb/in.³.

Solution

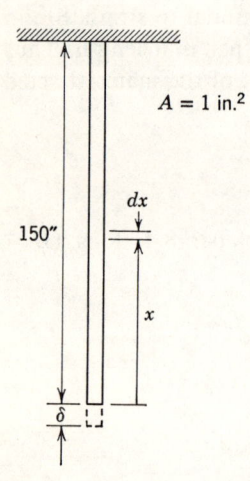

Fig. 5-10

$$d\delta = \frac{P(x)\,dx}{AE} \quad \text{where } P(x) = WAx$$

$$\int_0^\delta d\delta = \int_0^L \frac{WAx\,dx}{AE}$$

$$\int_0^\delta d\delta = \frac{W}{E}\int_0^L x\,dx$$

$$\delta = \frac{W}{E}\frac{L^2}{2}$$

$$\delta = \frac{0.283\text{ lb/in.}^3}{30\times 10^6\text{ lb/in.}^2} \times \frac{150^2\text{ in.}^2}{2}$$

$$\delta = 1.06 \times 10^{-4}\text{ in.}$$

Suspended length = 150 in. + 1.06×10^{-4} in. ▶

5-22 Wt. 1

A steel test specimen is $\frac{5}{8}$ in. in diameter at the root of the thread. It is to be stressed to 50,000 psi tension. What load must be applied?

(a) 12,790 lb
(b) 15,340 lb
(c) 16,320 lb
(d) 25,600 lb
(e) 31,250 lb

Solution

Stress = P/A. Here $A = (\pi/4)(\frac{5}{8})^2$ and stress = 50,000 psi. The load P is

$$P = 50{,}000\,\frac{\pi}{4}\left(\frac{5}{8}\right)^2 = 15{,}340\text{ lb} \blacktriangleright$$

Answer is (b)

5-23 Wt. 6

Given a rigid bar hanging from three wires of length L, modulus of elasticity E, cross-sectional area A, spaced a distance a apart as shown in the figure. Calculate the force exerted by each wire when the bar is subjected to a force P located a distance $a/2$ from one wire. Neglect the weight of the bar.

240 / Mechanics of Materials

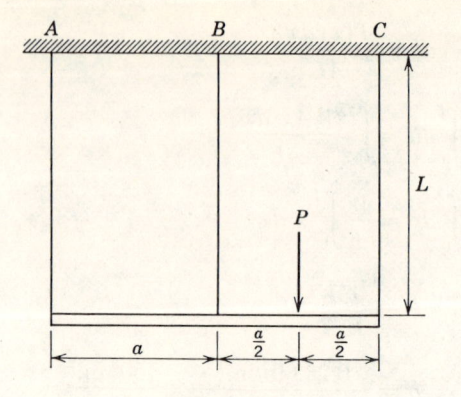

Fig. 5-11

Solution

When load P is applied, we know that $P = P_A + P_B + P_C$, and since A, B, and C are wires, none are in compression. Further, we know that each wire will elongate; hence

$$\text{length } A = L + \Delta_A$$
$$\text{length } B = L + \Delta_B$$
$$\text{length } C = L + \Delta_C$$

As the bar is rigid, B will have a deflection intermediate to that of A and C, or

$$\Delta_B = \frac{\Delta_A + \Delta_C}{2}$$

But since all wires are identical, load is proportional to deflection, so

$$P_B = \frac{P_A + P_C}{2}$$

Taking moments about wire A, $\sum M_A = 0$

$$aP_B - 1.5aP + 2aP_C = 0$$

Dividing by a,

$$P_B - 1.5P + 2P_C = 0$$

Thus we have three equations in three unknowns:

$$P = P_A + P_B + P_C \tag{1}$$
$$0 = P_A - 2P_B + P_C \tag{2}$$
$$1.5P = + P_B + 2P_C \tag{3}$$

Solving simultaneously,

$$-(1) \quad -P = -P_A - P_B - P_C$$
$$(2) \quad 0 = P_A - 2P_B + P_C$$
$$\overline{-P = \quad -3P_B} \qquad P_B = \frac{P}{3} \blacktriangleright$$

$$(3) \quad 1.5P = +\frac{P}{3} + 2P_C \qquad P_C = \frac{1.5P - P/3}{2} = \frac{7P}{12} \blacktriangleright$$

$$(1) \quad P = P_A + \frac{P}{3} + \frac{7P}{12} \qquad P_A = P - \frac{4P + 7P}{12} = \frac{P}{12} \blacktriangleright$$

5-24 Wt. 4

Three 1-in.-diameter rods AD, BD, and CD support the 2000-lb load as shown. What is the load in each rod?

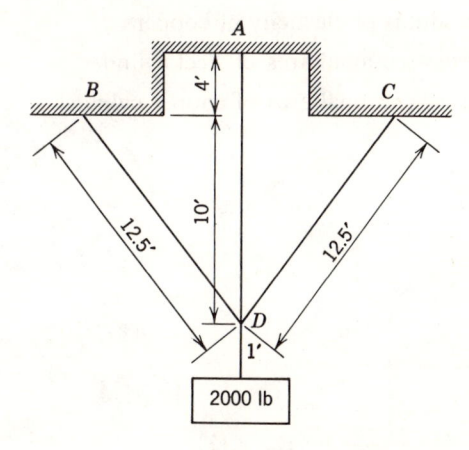

Fig. 5-12

Solution

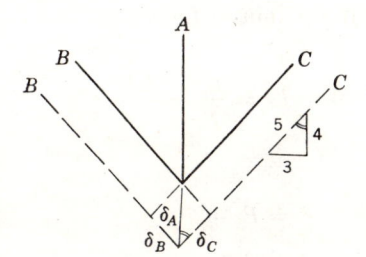

Fig. 5-13

By geometry, $\delta_B = \delta_C = \frac{4}{5}\delta_A$. By symmetry, $P_B = P_C$. Using $\delta = PL/EA$,

242 / Mechanics of Materials

$$\frac{4}{5}\frac{P_A(14)}{AE} = \frac{P_B(12.5)}{AE} \qquad P_A = 1.12P_B = 1.12P_C$$

$$\sum F_v = 0 \qquad \tfrac{4}{5}P_B + P_A + \tfrac{4}{5}P_C = 2000$$

$$\tfrac{4}{5}P_B + 1.12P_B + \tfrac{4}{5}P_B = 2000$$

$$P_B = 736 \text{ lb tension} \blacktriangleright$$

$$P_C = 736 \text{ lb tension} \blacktriangleright$$

$$P_A = 824 \text{ lb tension} \blacktriangleright$$

5-25 Wt. 5

Two concentric cylinders of length L are loaded between rigid smooth plates as shown. If the inner cylinder is copper and the outer cylinder is steel, what is the algebraic formula for stress in the copper cylinder?

Assume E_s = modulus of elasticity of steel
E_c = modulus of elasticity of copper
A_s = cross-sectional area of steel cylinder
A_c = cross-sectional area of copper cylinder

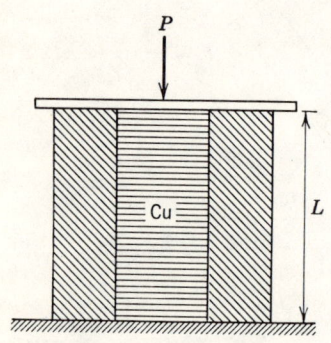

Fig. 5-14

Solution

$$\text{Stress} = \frac{\text{force}}{\text{area}} = \frac{P}{A} \qquad \text{Deflection} = \text{length} \times \text{strain} = L\varepsilon = \Delta$$

$$\text{Modulus of elasticity} = \frac{\text{stress}}{\text{strain}} = \frac{s}{\varepsilon} \qquad \delta = L\varepsilon = \frac{Ls}{E} = \frac{LP}{AE}$$

But $\delta_s = \delta_c$. Therefore

$$\frac{LP_s}{A_s E_s} = \frac{LP_c}{A_c E_c} \qquad P_s = \frac{P_c A_s E_s}{A_c E_c} \qquad P = P_s + P_c$$

$$P = \frac{P_c A_s E_s}{A_c E_c} + P_c \qquad P = P_c\left(1 + \frac{A_s E_s}{A_c E_c}\right)$$

Therefore
$$P_c = \frac{A_c E_c P}{A_c E_c + A_s E_s}$$
But
$$s_c = \frac{P_c}{A_c} \qquad s_c = \frac{E_c P}{A_c E_c + A_s E_s} \blacktriangleright$$

5-26 Wt. 1

Normally the diameter of a rivet is $\frac{1}{16}$ in. less than that of the hole in which it is to fit. This is to insure easy entrance of a hot rivet. In boiler calculations, the rivet is assumed to

(a) remain the same size after driving
(b) shrink an additional $\frac{1}{16}$ in. on cooling
(c) shrink on cooling in proportion to its cross-sectional area
(d) attain a size found by tables
(e) fill the hole after driving

Solution

The ASME Boiler and Pressure Vessel Code, among others, requires that all rivets be driven to fill the rivet holes completely ▶

Answer is (e)

5-27 Wt. 5

Shown in the figure is an eccentrically loaded riveted connection consisting of six $\frac{3}{4}$-in. rivets. If the maximum allowable shear stress is 15,000 psi, what is the maximum load P permissible?

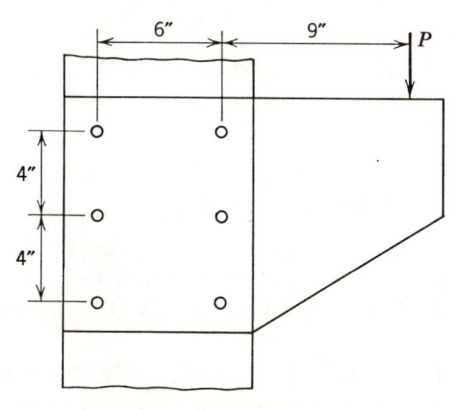

Fig. 5-15

244 / Mechanics of Materials

Solution

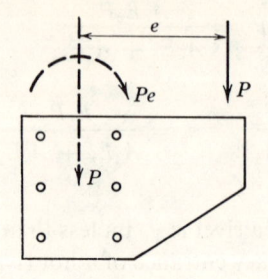

Fig. 5-16

In the solution of this type of problem we use an equivalent loading. The load P is equivalent to the sum of a parallel and equal force P passing through the centroid of the rivets and a moment Pe. The centroidal force P is assumed to load all rivets equally; hence force/rivet $= P/n$. The moment Pe tends to rotate the bracket clockwise about the centroid of the rivet areas. The force F developed by each rivet in resisting the moment Pe is proportional to the distance of that rivet from the centroid of the rivet areas. The sum of the moments of these forces must equal Pe. The total force on any rivet is found by vectorially adding the axial shearing stress P/n to force F.

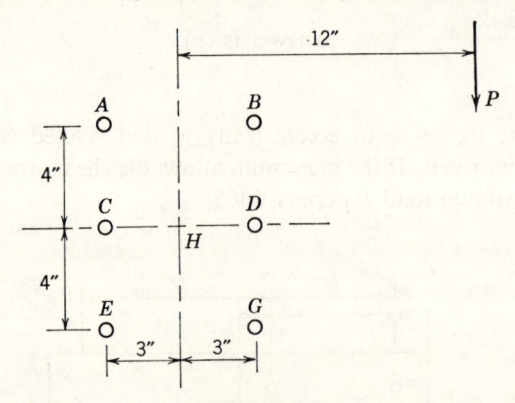

Fig. 5-17

In this case the centroid is at H. Thus the distance from the centroid to each of the corner rivets is maximum and equal. The distance is 5 in., assuming uniform stress over the rivet cross section. From H to C or D the distance is 3 in., so we know the force F for these two rivets is three-fifths the force in the corner rivets.

For $\frac{3}{4}$-in. rivets $A = (\pi/4)(0.75)^2 = 0.442$ in.^{2}, and for a maximum shear stress of 15,000 psi the maximum force in a rivet may be

$$S_s A = 15{,}000 \times 0.442 = 6630 \text{ lb}$$

Taking moments about H, $\sum M_H = 0$,
$$-12P + 4(F \times \tfrac{4}{5}) + 2(\tfrac{3}{5}F \times 3) = 0$$
$$23.6F = 12P \qquad F = 0.508P$$

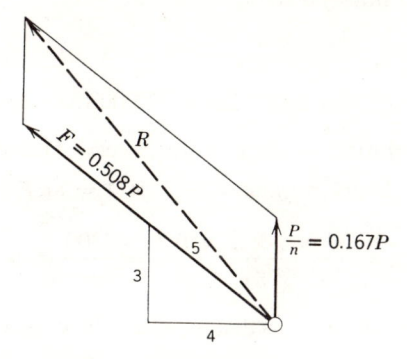

Fig. 5-18

The resulting force R is
$$R_v = 0.167P + \tfrac{3}{5} \times 0.508P = 0.472P$$
$$R_h = \tfrac{4}{5} \times 0.508P = 0.406P$$
$$R = \sqrt{(0.472P)^2 + (0.406P)^2} = 0.622P$$

The maximum force in a rivet is 6630 lb, so the maximum load is
$$P = \frac{6630}{0.622} = 10{,}650 \text{ lb} \blacktriangleright$$

5-28 Wt. 4

A $\tfrac{1}{4}$-in.-diameter shear pin, as shown in the figure, is used on the screw conveyor of a coal furnace feed to protect the mechanism when jamming is

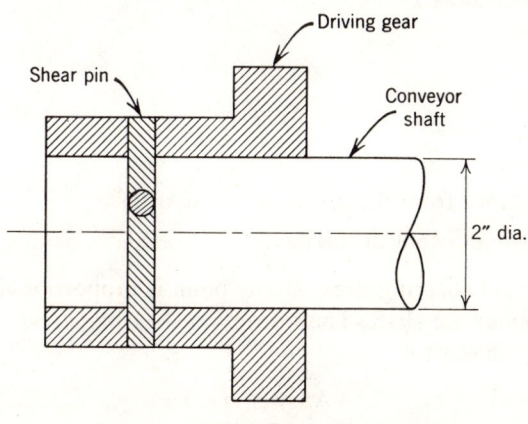

Fig. 5-19

caused by tramp metal. The screw conveyor revolves at 4.0 rpm and is driven by a $\frac{1}{8}$-hp motor through a reduction gearing having an efficiency of 80%. What will be the unit shearing stress on the shear pin when the motor is delivering its rated horsepower?

Solution

T = torque in pound-feet and 1 hp = 33,000 ft-lb/min.

$$\text{Work/revolution} = \text{force} \times \text{distance} = F(2\pi R) = 2\pi T$$
$$\text{Work/minute} = 2\pi T \times 4 = 8\pi T$$

$$\text{Torque} = \frac{33,000 \times \frac{1}{8} \times 0.8}{8\pi} = \frac{3300}{8\pi} \text{ lb-ft}$$

$$\text{Torque} = \text{force} \times \text{distance} = S_s A \times \text{lever arm}$$

$$= S_s \times \frac{2\pi}{4}\left(\frac{1}{4}\right)^2 \times 1 = \frac{3300}{8\pi} \times 12$$

$$S_s = \frac{(3300/8\pi) \times 12}{\pi/32} = \frac{3300 \times 32 \times 12}{8\pi^2} = 16,000 \text{ psi} \blacktriangleright$$

5-29 Wt. 1

The maximum shear stress in a solid round shaft subjected only to torsion occurs

(a) on principal planes
(b) on planes containing the axis of the shaft
(c) on the surface of the shaft
(d) only on planes perpendicular to the axis of the shaft
(e) at the neutral axis

Solution

$$S_s = \frac{Tc}{J}$$

where T = torque
c = distance from the center of the shaft
J = polar moment of inertia

Thus the torsional shearing stress at any point is proportional to its distance from the center of the shaft. The maximum shearing stress, therefore, is at the surface of the shaft $\blacktriangleright$

Answer is (c)

5-30 Wt. 6

A shaft coupling is to be designed, using 1-in.-diameter bolts at a distance of 6 in. from the center of the shaft. If the shaft is to transmit 5400 hp at a speed of 1200 rpm, how many bolts should be used in the connection? Allowable shearing stress for the bolts is 15,000 psi.

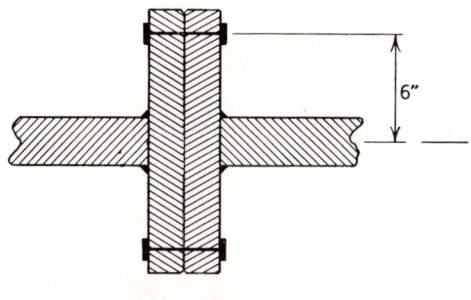

Fig. 5-20

Solution

$$\text{Work/revolution} = \text{force} \times \text{distance} = F(2\pi R) = 2\pi T$$

where T = torque in pound-feet. At 1200 rpm

$$\text{Work/minute} = 1200 \times 2\pi T \text{ lb-ft/min}$$

Since 1 hp = 33,000 ft-lb/min,

$$\text{hp} = \frac{1200 \times 2\pi T}{33,000} = 5400$$

$$\text{Torque} = \frac{33,000 \times 5400}{1200 \times 2\pi} = 23,600 \text{ lb-ft}$$

Assuming that the shearing stress is uniform over the bolt cross section, we can determine the torsional resistance per bolt.

$$\text{Torque} = \text{force} \times \text{lever arm} = S_s A \times \text{lever arm}$$

$$= 15,000 \text{ psi} \times \frac{\pi}{4} 1^2 \text{ in.}^2 \times 0.5 \text{ ft}$$

$$= 5880 \text{ lb-ft}$$

$$\text{No. of bolts required} = \frac{23,600}{5880} = 4.02$$

Use four bolts ▶

5-31 Wt. 6

A multistage vertical turbine pump is installed in a sump as shown in the figure. To facilitate installation and removal of the pump, a slip-type coupling is used to connect the pump discharge which passes through the concrete wall. The discharge valve is normally open when the pump is running, but it may be inadvertently closed. With the data given, determine the diameter of the tie bolts required to contain the lateral thrust produced by the pump. Assume there are two tie bolts connected to lugs on each side of the slip coupling. Also assume that none of the thrust is taken at the motor base plate.

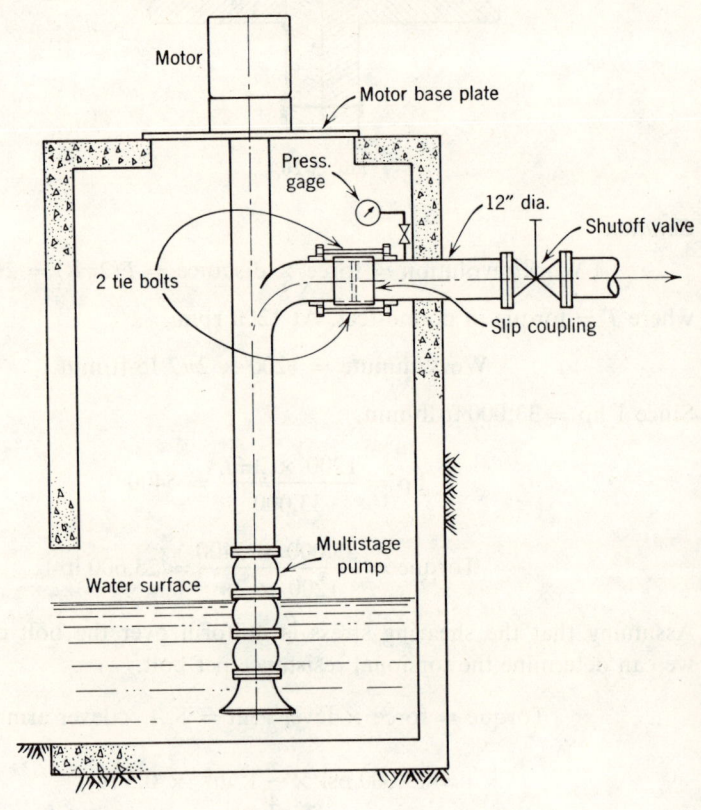

Fig. 5-21

Given: Pressure gage readings

(a) Normal operation 530 ft
(b) Shutoff head 30% above normal
(c) Working stress 14,000 psi

Solution

$$\text{Shutoff head} = 530 \times 1.3 = 689 \text{ ft of water}$$

Since the weight of a 1-ft column of water is 62.4 lb/ft³/144 in.²/ft² = 0.433 psi,

$$P_{equiv} = 0.433 \times 689 = 298 \text{ psi}$$

$$\text{Thrust} = PA = 298 \times \frac{\pi}{4} \times 12^2$$

$$= 10{,}720\pi \text{ lb}$$

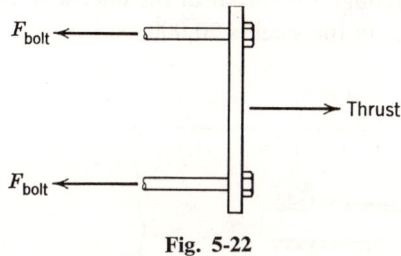

Fig. 5-22

$$\text{Force/bolt} = 5360\pi \text{ lb}$$

$$\text{Stress/bolt} = \frac{\text{Force}}{\text{Area/bolt}} = \frac{5360\pi}{(\pi/4)d^2}$$

$$= 14{,}000 \text{ psi}$$

$$\text{Net bolt diameter } d = \left(\frac{5360 \times 4}{14{,}000}\right)^{1/2}$$

$$= (1.53)^{1/2}$$

$$= 1.24 \text{ in.}$$

Assuming the tie bolts are threaded, then d is not the bolt diameter but the net root diameter at the threaded portion of the bolt. The necessary diameter of the tie bolts is $1\frac{1}{2}$ in. to obtain the necessary net thread root diameter of 1.24 in. ▶

5-32 Wt. 1

Intermittent welds which are generally used to hold material in place temporarily are called

(*a*) double vee
(*b*) tack
(*c*) fillet
(*d*) groove
(*e*) butt

250 / Mechanics of Materials

Solution

A tack weld is a temporary or auxiliary weld. The other alternatives listed are types of structural welds ▶

Answer is (b)

5-33 Wt. 6

Two steel angles 3 in. × 3 in. × $\frac{1}{2}$ in. are to be welded to a steel gusset plate as shown. Determine the minimum lengths of the $\frac{3}{8}$-in. fillet welds *a* and *b* which will develop the full strength of the angles in tension. Assume the allowable shear through the throat of the fillet weld is 13,600 psi and the allowable tensile stress in the steel is 20,000 psi.

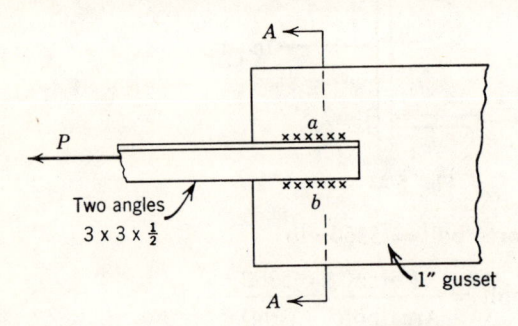

Fig. 5-23

Solution

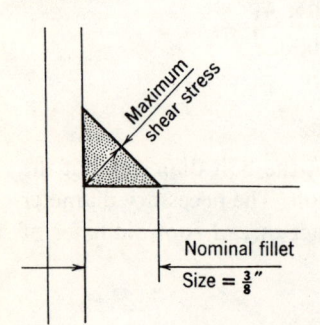

Fig. 5-24

$$\text{Area through throat/inch weld} = 0.707 \times \tfrac{3}{8}$$
$$= 0.265 \text{ in.}^2$$
$$\text{Allowable load/inch weld} = 13,600 \times 0.265$$
$$= 3600 \text{ lb}$$
$$\text{Total force } P = f_s A = 20,000 \, (2.75)$$
$$= 55,000 \text{ lb/angle}$$

Here the cross-sectional area of the angle is

$$A = 3 \times 0.5 + 2.5 \times 0.5 = 2.75 \text{ in.}^2$$

$$\text{Minimum no. of inches of weld} = \frac{55{,}000}{3{,}600}$$

$$= 15.28 \text{ in./angle}$$

To proportion properly the lengths of welds a and b, we must determine the centroid of the cross-section of the angle.

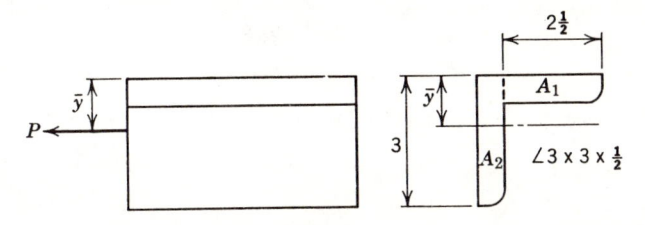

Fig. 5-25

$$A\bar{y} = \sum ay \qquad 2.75\bar{y} = 3 \times 0.5 \times 1.5 + 2.5 \times 0.5 \times 0.25$$

$$2.75\bar{y} = 2.56$$

$$\bar{y} = \frac{2.56}{2.75} = 0.93 \text{ in.}$$

For the load per linear inch of weld to be the same in all parts of the weld, the resultant force exerted by the welds must be collinear with force P.

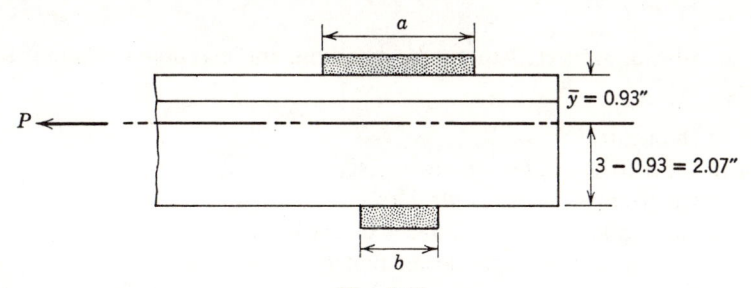

Fig. 5-26

Taking moments about weld b,

$$2.07P - a(\text{load/in. weld})(3) = 0$$

$$a = \frac{2.07 \times 55{,}000}{3 \times 3{,}600} = 10.52 \text{ in.} \blacktriangleright$$

252 / Mechanics of Materials

And taking moments about weld a,

$$-0.93P + b(\text{load/in. weld})(3) = 0$$

$$b = \frac{0.93 \times 55,000}{3 \times 3,600} = 4.76 \text{ in.} \blacktriangleright$$

Thus $a + b = 10.52 + 4.76 = 15.28$ in. This agrees with our earlier calculation.

$$a = 10.52 \text{ in.} \blacktriangleright$$
$$b = 4.76 \text{ in.} \blacktriangleright$$

5-34 Wt. 1

The allowable flexural stress for Douglas fir beams is nearest

(*a*)   15 psi
(*b*)   150 psi
(*c*)   1,500 psi
(*d*) 15,000 psi
(*e*) 150,000 psi

Solution

Depending upon the quality of the beam and the conditions under which it is to be utilized, the allowable flexural stress for Douglas fir ranges from about 875 to 1800 psi ▶

Answer is (*c*)

5-35 Wt. 1

In an I-beam subjected to simple bending, the maximum bending stress occurs

(*a*) at the neutral axis
(*b*) in the web above the neutral axis
(*c*) in the web below the neutral axis
(*d*) at the top and bottom surfaces of the beam
(*e*) where the web joins the lower flange

Solution

In simple bending the stress at each point in the cross-section of a beam is directly proportional to that point's distance from the neutral axis. For a symmetrical section, like an I-beam, the maximum bending stress will occur at both the top and bottom surfaces of the beam ▶

Answer is (*d*)

5-36 Wt. 1

The bending moment of a beam

(a) depends on the modulus of elasticity of the beam
(b) is minimum where the shear is zero
(c) is maximum at the free end of a cantilever
(d) is plotted as a straight line for a simple beam with a uniformly distributed load
(e) may be determined from the area of the shear diagram

Solution

The shear at any point along the beam is

$$V = \frac{dM}{dx}$$

so

$$M = \int V \, dx$$

or the bending moment is the area of the shear diagram ▶

Answer is (e)

5-37 Wt. 1

The moment curve for a simple beam with a concentrated load at mid-span takes the shape of a

(a) triangle
(b) semicircle
(c) semiellipse
(d) parabola
(e) rectangle

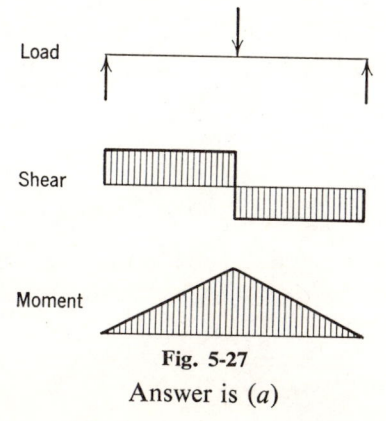

Solution

Fig. 5-27

Answer is (a)

254 / Mechanics of Materials

5-38 Wt. 1

The moment diagram for a beam uniformly loaded with a concentrated load in the center is the sum of

(a) two triangles
(b) a rectangle and a triangle
(c) a parabola and a rectangle
(d) a parabola and a triangle
(e) a rectangle and a trapezoid

Solution

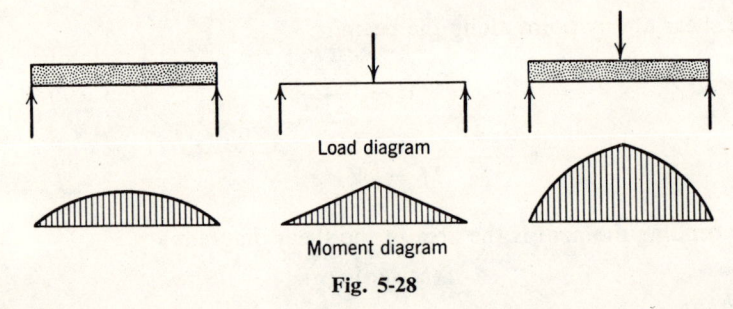

Load diagram

Moment diagram

Fig. 5-28

The moment diagram for a beam with a uniform load is a parabola, and the moment diagram for a concentrated load is a triangle. By superposition, the combined diagram is the sum of the two individual diagrams ▶

Answer is (d)

5-39 Wt. 1

A beam is fixed at both ends. What is the fixed end moment if the load is uniformly distributed?

(a) $\dfrac{wL^2}{8}$

(b) $\dfrac{wL^2}{10}$

(c) $\dfrac{wL^2}{12}$

(d) $\dfrac{wL^2}{16}$

(e) $\dfrac{wL^2}{24}$

Solution

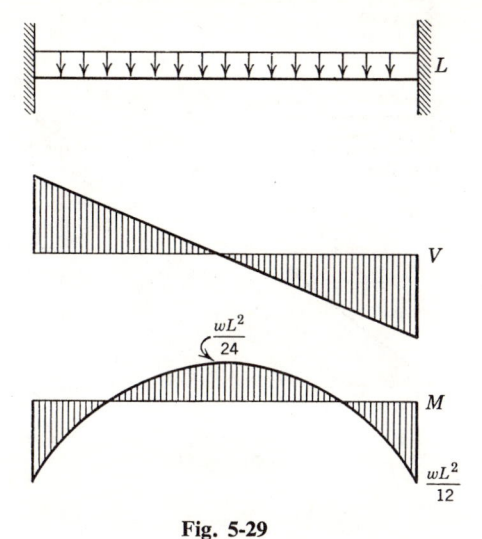

Fig. 5-29

This problem presents a very basic statically indeterminate beam loading. The load, shear, and bending moment diagrams for the problem are given here, and one can observe that the fixed end moment is $wL^2/12$. The answer is a standard result; its derivation, however, requires the use of statically indeterminate stress analysis and is not given here ▶

Answer is (c)

5-40 Wt. 6

Determine the reactions at A and C and construct the moment and shear diagrams for the beam AD.

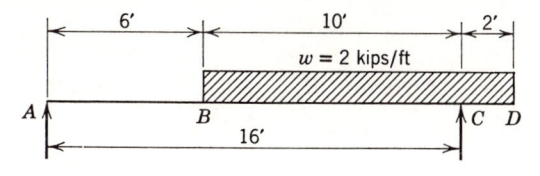

Fig. 5-30

Solution

$\sum M_A = 0$

$+16R_C - (12 \times 2)(12) = 0 \qquad R_C = \dfrac{24 \times 12}{16} = 18 \text{ kips} \blacktriangleright$

$\sum F_y = 0$

$R_A + 18 - (12 \times 2) = 0 \qquad R_A = 6 \text{ kips} \blacktriangleright$

256 / Mechanics of Materials

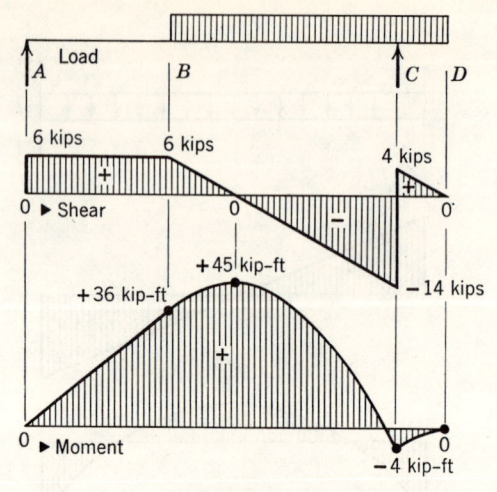

Fig. 5-31

5-41 Wt. 6

A simple beam is loaded as shown. Where is the point of maximum moment and what is the value of the maximum moment?

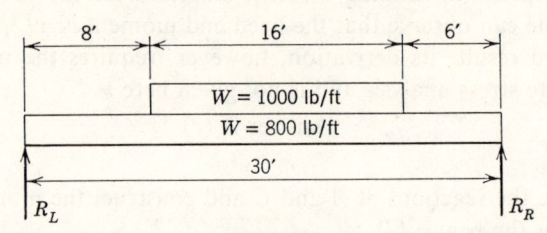

Fig. 5-32

Solution

Solve first for the left reaction R_L and the right reaction R_R.

$\sum M_{R_L} = 0$

$$+30R_R - 800 \times 30 \times 15 - 1000 \times 16 \times 16 = 0$$

$$+30R_R - 360{,}000 - 256{,}000 = 0 \qquad R_R = \frac{616{,}000}{30} = 20{,}533 \text{ lb}$$

$\sum F_y = 0$

$$R_L + R_R - 800 \times 30 - 1000 \times 16 = 0$$

$$R_L + 20{,}533 - 24{,}000 - 16{,}000 = 0 \qquad R_L = 19{,}467 \text{ lb}$$

Then draw the shear diagram:

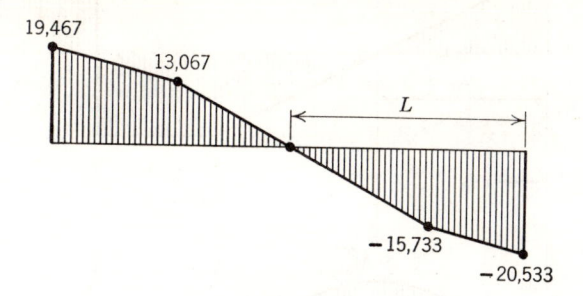

Fig. 5-33

$$L = 6 + \frac{15{,}733}{1800} = 14.74 \text{ ft}$$

The maximum moment occurs at the point where the shear is zero, or 14.74 ft from the right end of the beam ▶

$$M_{\max} = +14.74 \times 20{,}533 - 14.74 \times 800 \times \frac{14.74}{2} - 8.74 \times 1000 \times \frac{8.74}{2}$$
$$= 302{,}656 - 86{,}907 - 38{,}194 = 177{,}555 \text{ lb-ft} \blacktriangleright$$

5-42 Wt. 6

Draw the shear and moment diagrams for the beam loaded as shown. Locate the point of maximum moment and give the value of the maximum moment in kip-feet.

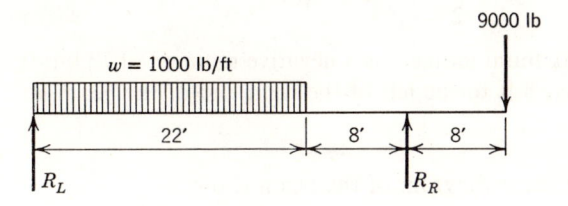

Fig. 5-34

Solution

$\sum M_{R_L} = 0$

$$22 \times 1 \text{ kip} \times 11 + 9 \text{ kips} \times 38 - 30R_R = 0$$

$$+242 + 342 - 30R_R = 0 \qquad R_R = \frac{242 + 342}{30} = 19.47 \text{ kips}$$

$\sum F_y = 0$

$$R_L + 19.47 - 22 \times 1 \text{ kip} - 9 \text{ kips} = 0 \qquad R_L = 11.53 \text{ kips}$$

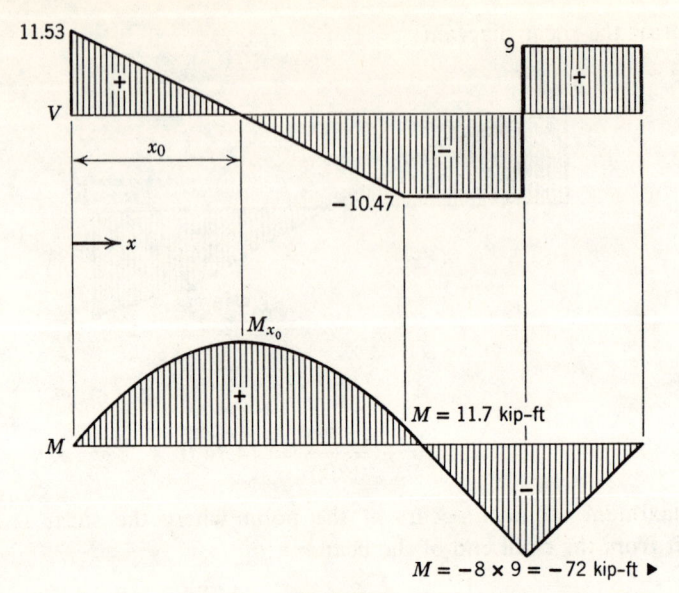

Fig. 5-35

The moment is a maximum at the point where the shear is zero. Expressing the shear as a function of x from the left end, $V(x) = 11.53$ kips $- x$ kips $= 0$ when $x = x_0$. $x_0 = 11.53$ ft.

$$M(x_0) = M_{x_0} = x_0 R_L - \frac{w x_0^2}{2} = 11.53(11.53) - \frac{1(11.53)^2}{2}$$

$$= \frac{11.53^2}{2} = 66.5 \text{ kip-ft}$$

Thus the maximum moment is a negative moment of 72 kip-ft, located at the right support, 8 ft to the left of the right end of the beam ▶

5-43 Wt. 6

Given the shear diagram of the beam shown.

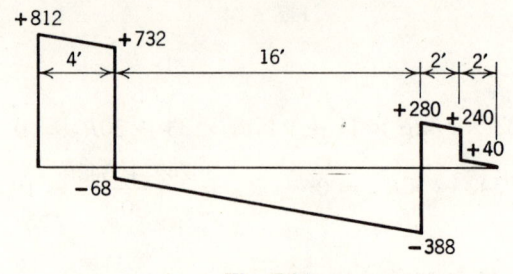

Fig. 5-36

(a) Draw the beam showing its loads, reactions, and dimensions.
(b) Draw the moment diagram of the beam showing dimensions and magnitudes at points of maximum positive and negative moments.

Solution

Assuming the units on the shear diagram are pounds, then:

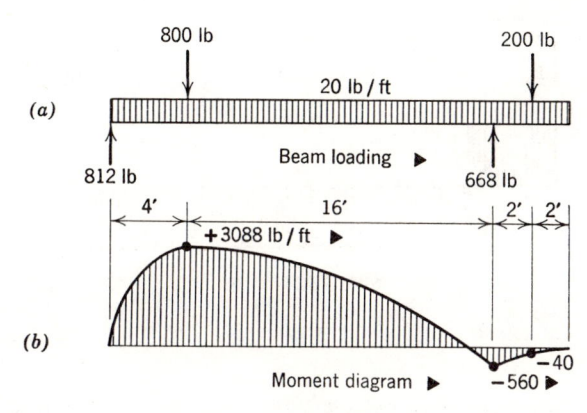

Fig. 5-37

5-44 Wt. 6

Calculate the magnitude and direction of the reactions of beam AB loaded as shown. Draw the shear and moment diagrams, approximately to scale, showing the magnitudes of the shears and moments at all discontinuities.

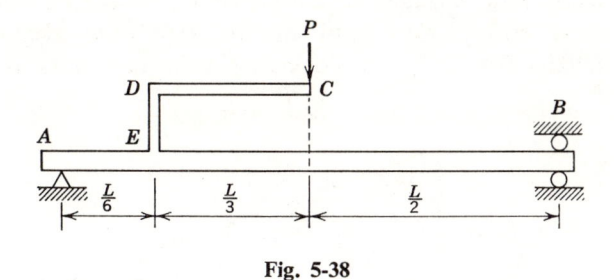

Fig. 5-38

Solution

Solve for the left reaction R_A and the right reaction R_B.

$\Sigma M_{R_A} = 0$

$$-P\left(\frac{L}{6} + \frac{L}{3}\right) + LR_B = 0 \qquad R_B = \frac{0.5LP}{L} = 0.5P \blacktriangleright$$

$\Sigma F_y = 0$

$$-P + 0.5P + R_A = 0 \qquad R_A = 0.5P \blacktriangleright$$

260 / Mechanics of Materials

The load P can be resolved into a shear of P plus a moment $PL/3$, each applied at $L/6$ from the left end of the beam.

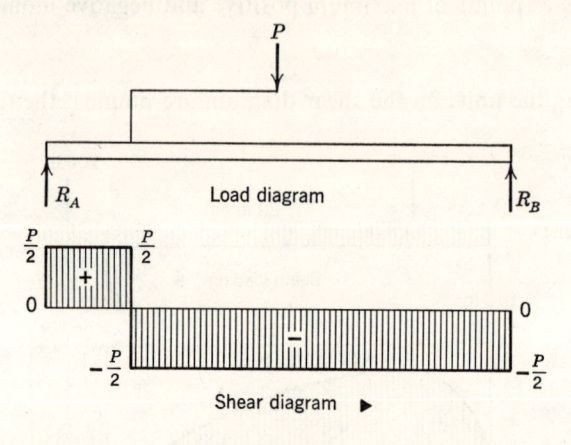

Fig. 5-39

The change in bending moment between two sections of a beam is equal to the area of the shear diagram between the sections, taking proper account of the sign. Beginning at the left end, the bending moment is zero. At $L/6$ the moment increases at a constant rate to $(L/6) \times (P/2) = +PL/12$. At $L/6$ there is a concentrated moment $PL/3$; hence the moment diagram increases by this amount to $PL/3 + PL/12 = 5PL/12$. To the right of this point the shear is a constant negative amount, so the moment diagram declines at a steady rate until at the right end the moment is $5PL/12 - (P/2)(L/2 + L/3) = 0$. The moment diagram looks like this:

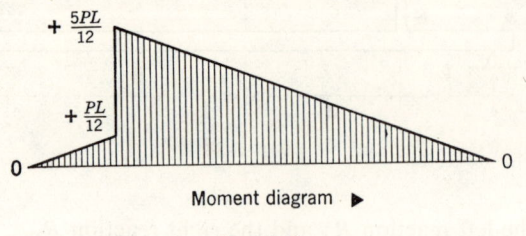

Fig. 5-40

Shear and moment diagrams of part CDE:

1. The shear in segment CD is constant and equal to P.
2. There is no shear in DE.

3. The moment at point C is zero (it is a free end), and at D the moment is $P(L/3)$.

4. The moment in segment DE is constant and equal to $-P(L/3)$.

The shear and moment diagrams are as follows:

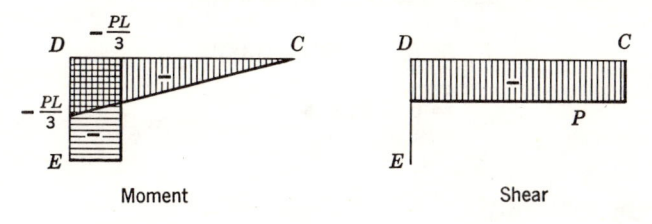

Fig. 5-41

5-45 Wt. 6

Calculate the magnitude and direction of the reactions for the beam AC loaded as shown. The load on the left 6 ft of the beam varies uniformly in intensity from 0 kips/ft at A to 4 kips/ft. Draw the shear and moment diagrams and indicate the magnitude of the shears and moments at all discontinuities.

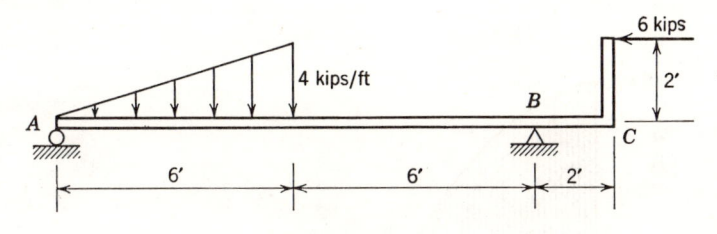

Fig. 5-42

Solution

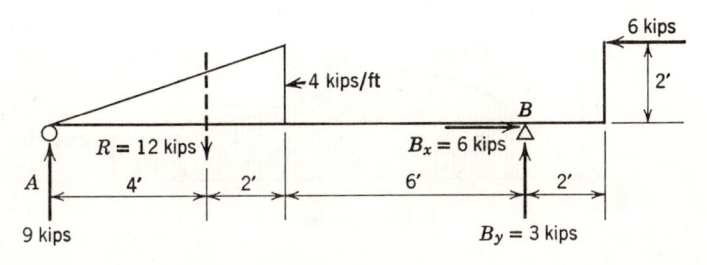

Fig. 5-43

262 / Mechanics of Materials

$\sum M_A = 0$

$$-4(12 \text{ kips}) + 12B_y + 2(6 \text{ kips}) = 0 \qquad B_y = 3 \text{ kips}$$

$\sum F_H = 0$

$$B_x - 6 = 0 \qquad B_x = 6 \text{ kips}$$

$$R_B = (6^2 + 3^2)^{1/2} = (45)^{1/2}$$

$$= 6.71 \text{ kips} \blacktriangleright$$

Fig 5-44

$$\tan \alpha = \tfrac{3}{6} = 0.5 \qquad \alpha = 26.6° \blacktriangleright$$

$\sum M_B = 0$

$$-12A + 8(12 \text{ kips}) + 2(6 \text{ kips}) = 0 \qquad R_A = A = 9 \text{ kips} \blacktriangleright$$

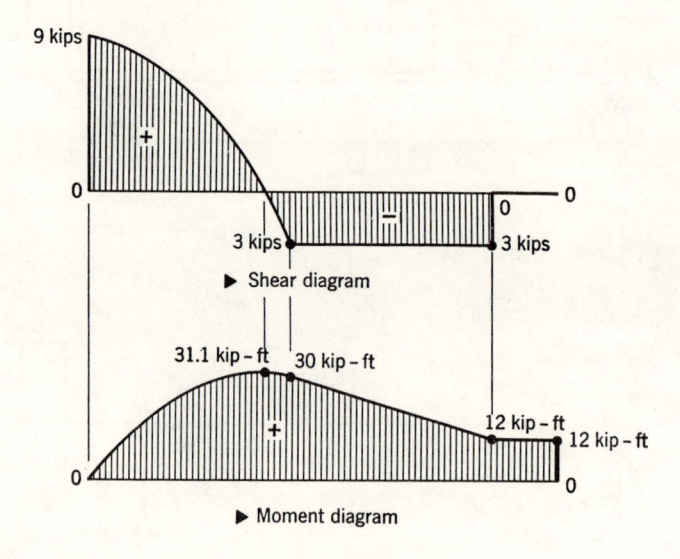

Fig. 5-45

5-46 Wt. 1

Which is the proper shape of the moment diagram for the cantilever beam AB loaded as shown?

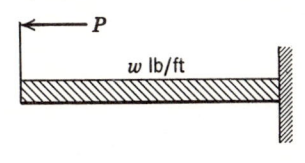

Fig. 5-46

Solution

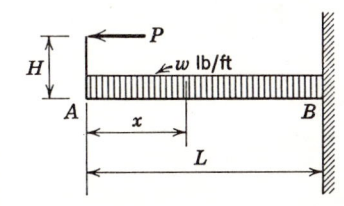

Fig. 5-47

The force P contributes a constant moment $-PH$ from A to B. The uniform load w, at any point x, produces a moment $-wx^2/2$. The total moment diagram is the sum of the moment diagrams of the component loads.

264 / Mechanics of Materials

Fig. 5-48

The moment diagram, therefore, takes the shape of (c) ▶

5-47 Wt. 5

A cantilever beam 10 ft long is loaded with a 1000-lb concentrated load at its free end, a uniform load of 30 lb/ft, and a load which varies from zero at the free end to 300 lb at the fixed end. For this cantilever beam determine

(*a*) the shear equation
(*b*) the moment equation

(Note: Answers are to be equations in terms of constants given and any other mathematical symbols required.)

Solution

Although unstated, we will assume the load varies uniformly from zero at the free end to 300 lb at the fixed end.

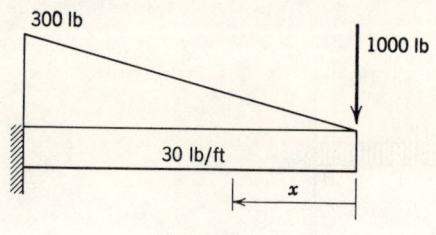

Fig. 5-49

$$V(x) = 1000 + 30x + 30x\left(\frac{x}{2}\right) = 15x^2 + 30x + 1000 \blacktriangleright$$

$$M(x) = -1000x - 30x\left(\frac{x}{2}\right) - 30x\left(\frac{x}{2}\right)\left(\frac{x}{3}\right) = -5x^3 - 15x^2 - 1000x \blacktriangleright$$

5-48 Wt. 6

A cantilever beam supports a uniformly varying load as shown. Neglect the weight of the beam.

(*a*) Draw the shear and moment diagrams.
(*b*) Compute the maximum bending stress.

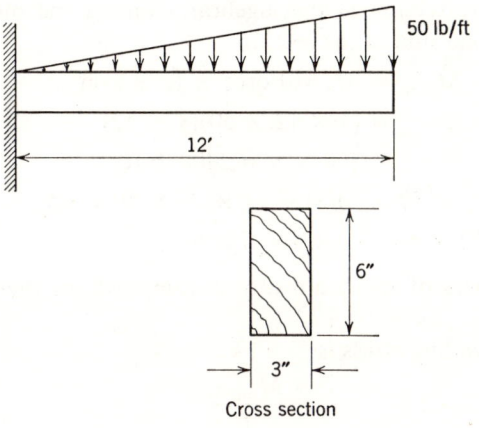

Cross section
Fig. 5-50

Solution

Load diagram

Shear diagram: $V_0 = 300$ lb, $V_6 = 225$ lb, $V_{12} = 0$

Moment diagram: -750 lb-ft, -2400 lb-ft

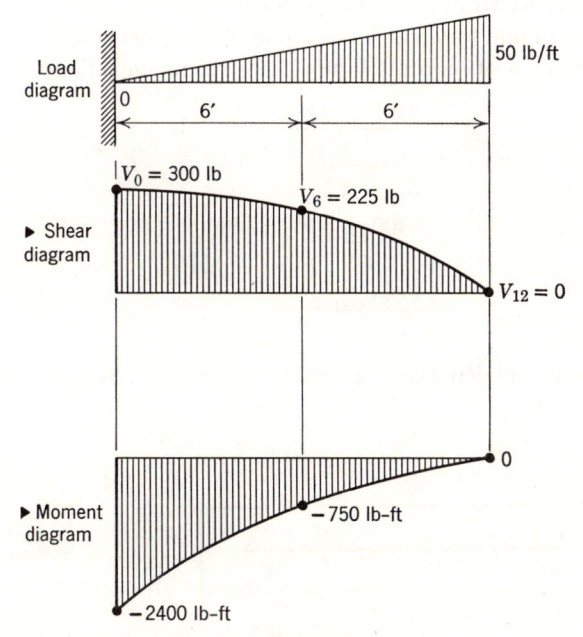

Fig. 5-51

266 / Mechanics of Materials

The shear at any section is equal to the net load on one side of the section.

$$V_{max} = V_0 = \tfrac{1}{2}(12)(50) = 300 \text{ lb}$$
$$V_6 = \tfrac{6}{12}(50)(6) + \tfrac{1}{2} \times \tfrac{6}{12} \times 50 \times 6$$
$$= 225 \text{ lb} \qquad V_{12} = 0$$

The moment at any section is the algebraic sum of the moments of all external forces on one side of the section.

$$M_{max} = M_0 = \text{Force} \times \text{lever arm}$$
$$= (\tfrac{1}{2} \times 12 \times 50)(\tfrac{2}{3} \times 12)$$
$$= 300 \times 8 = 2400 \text{ lb-ft}$$
$$M_6 = (\tfrac{6}{12} \times 50 \times 6)(3) + (\tfrac{1}{2} \times \tfrac{6}{12} \times 50 \times 6)(\tfrac{2}{3} \times 6)$$
$$= 150 \times 3 + 75 \times 4 = 750 \text{ lb-ft}$$

Since the lower fibers of the beam are in compression, the moments are negative.

The maximum bending stress is

$$f_s = \frac{M_{max} c}{I}$$

where f_s = bending stress in pounds per square inch
  c = distance from neutral axis to extreme fiber = 3 in.
M_{max} = maximum moment in inch-pounds = 2400×12
  I = moment of inertia ($= bh^3/12$ for rectangular beams)

$$f_s = \frac{2400 \times 12 \times 3}{\frac{3 \times 6^3}{12}}$$
$$= \frac{2400 \times 12^2}{6^3} = \frac{2400 \times 144}{216}$$
$$= 1600 \text{ psi} \blacktriangleright$$

5-49 Wt. 5

A rectangular wooden beam is loaded as shown. (Neglect the weight of the beam.)

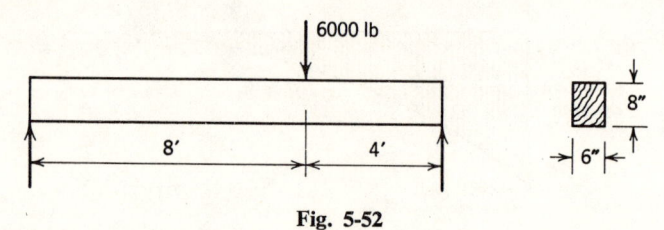

Fig. 5-52

(a) Is the beam overloaded if the maximum allowable bending stress is 2000 psi? Show proof of your answer.

(b) Is the beam overloaded in horizontal shear if the maximum allowable horizontal shearing stress is 100 psi? Show proof of your answer.

Solution

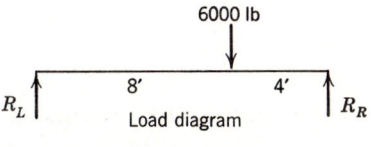

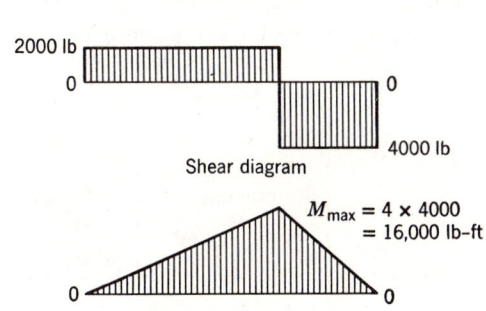

Fig. 5-53

$\sum M_{R_L} = 0$

$12 R_R - 8(6000) = 0 \qquad R_R = 4000 \text{ lb}$

$\sum F_y = 0$

$-6000 + 4000 + R_L = 0 \qquad R_L = 2000 \text{ lb}$

(a) The flexure formula for bending stress is

$$f_s = \frac{Mc}{I}$$

where c = distance from neutral axis to extreme fiber, and

$$I = \frac{bh^3}{12} = \frac{6 \times 8^3}{12} = 256 \text{ in.}^4$$

$$f_s = \frac{16{,}000 \text{ lb-ft} \times 4 \text{ in.} \times 12 \text{ in./ft}}{256 \text{ in.}^4}$$

$f_s = 3000 \text{ psi} > 2000 \text{ psi}$ ▶

Therefore the beam is overloaded in bending ▶

268 / Mechanics of Materials

(b) $S_s = VQ/Ib$; for a rectangular section $S_s = 3V/2A$ since then $Q = bh^2/2$ and $I = bh^3/12$.

$$S_s = \frac{3 \times 4000}{2 \times 6 \times 8} = \frac{12{,}000}{96} = 125 \text{ psi} > 100 \text{ psi} \blacktriangleright$$

Therefore the horizontal shearing stress exceeds the allowable shearing stress ▶

5-50 Wt. 4

A solid steel block, which weighs 49 kips, rests on a level concrete slab. If a horizontal force of 6 kips is applied to the top, what are the maximum and minimum pressures under the base in pounds per square foot?

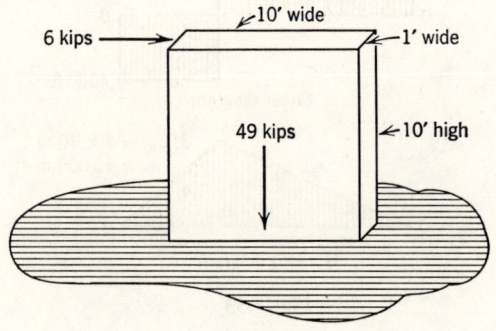

Fig. 5-54

Solution

$$S_{\max} = \frac{P}{A} + \frac{My}{I} = \frac{49{,}000}{10 \times 1} + \frac{6000 \times 10 \times 5}{\frac{1}{12} \times 1 \times 10^3} = 4900 + 3600 = 8500 \text{ psf} \blacktriangleright$$

$$S_{\min} = \frac{P}{A} - \frac{My}{I} = 4900 - 3600 = 1300 \text{ psf} \blacktriangleright$$

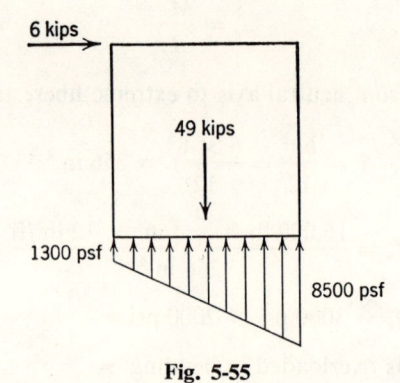

Fig. 5-55

5-51 Wt. 2

For the structure loaded as shown, sketch the elastic curve and indicate on the sketch the points of inflection.

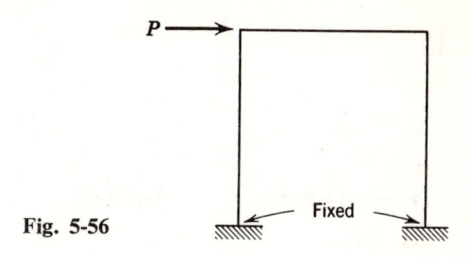

Fig. 5-56

Solution

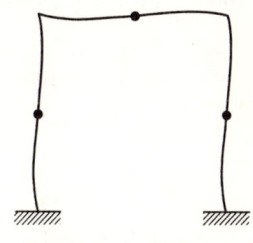

Fig. 5-57   • Point of inflection

5-52 Wt. 2

For the structure loaded as shown, sketch the elastic curve and indicate on the sketch the points of inflection.

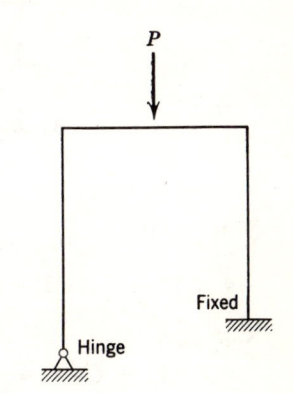

Fig. 5-58

270 / Mechanics of Materials

Solution

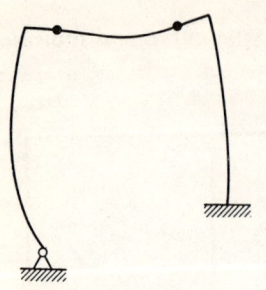

• Point of inflection   **Fig. 5-59**

5-53 Wt. 6

Two steel beams are fabricated to form a three-hinged structure as shown. Determine the horizontal and vertical components of the reactions at A and C and the maximum moment in the beam AB.

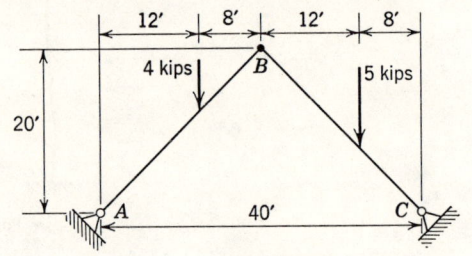

Fig. 5-60

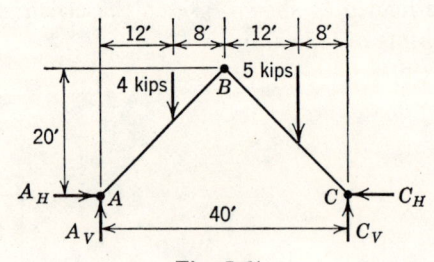

Solution

Fig. 5-61

$\sum M_A = 0$

$-4 \times 12 - 5 \times 32 + C_V \times 40 = 0 \qquad C_V = \dfrac{48 + 160}{40} = 5.2 \text{ kips} \blacktriangleright$

$\sum F_y = 0$

$\qquad A_V + C_V - 4 - 5 = 0 \qquad A_V = 9 - 5.2 = 3.8 \text{ kips} \blacktriangleright$

In member AB, $\sum M_B = 0$ since B is a hinge.

$$-A_V \times 20 + 4 \times 8 + A_H \times 20 = 0$$

$$A_H = \frac{76 - 32}{20} = 2.2 \text{ kips} \blacktriangleright$$

$$C_H = A_H = 2.2 \text{ kips} \blacktriangleright$$

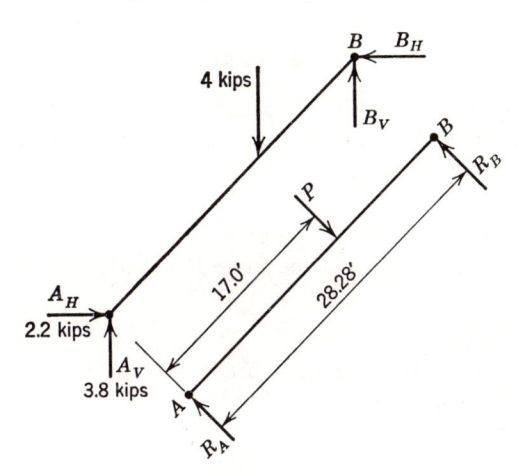

Fig. 5-62

$$B_V = 4 - 3.8 = 0.2 \text{ kips}$$

$$\text{Length } AB = 20 \times \sqrt{2} = 28.28 \text{ ft}$$

$$R_A = 3.8(0.707) - 2.2(0.707) = 1.13 \text{ kips}$$

$$R_B = 2.2(0.707) + 0.2(0.707) = 1.70 \text{ kips}$$

$$\text{Length } AP = 12 \times \sqrt{2} = 17.0 \text{ ft}$$

$$P = 4(0.707) = 2.83 \text{ kips}$$

Since axial forces along AB do not contribute to the moment, we need not consider them.

The maximum moment occurs under P and is equal to $(R_A)(L_{AP})$:

$$M_{\max} = 1.13 \times 17.0 = 19.2 \text{ kip-ft} \blacktriangleright$$

Alternatively, we observe that AB is a simply supported beam with additional axial forces at the ends, so $M_{\max}$ must occur at the 4-kip load and

$$M_{\max} = 3.8(12) - 2.2(12) = 19.2 \text{ kip-ft} \blacktriangleright$$

5-54 Wt. 5

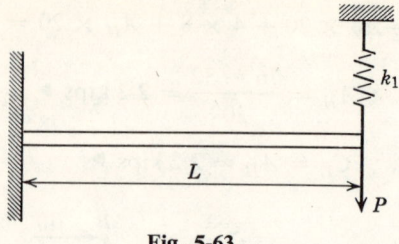

Fig. 5-63

Shown in the figure is a cantilever beam of length L loaded at one end by a force P and restrained by a spring having a spring constant equal to k_1. (k_1 is defined in units of pounds per inch.) The deflection of the end of the beam is 0.003 in.

When the spring is replaced by another spring having a constant k_2, the deflection of the beam because of the force P is 0.002 in. If $k_2 = 2k_1$, determine the ratio k_1/P.

Assume E = modulus of elasticity = 20×10^6 psi
A = cross-sectional area of beam = 6 in.^{2}
I = moment of inertia = 2 in.^{4}
L = 30 in.

Solution

To solve this problem, we must first derive the equation relating P and beam deflection. The second moment-area proposition says that the vertical displacement Δ of point A from the tangent to the elastic curve at B equals the moment (with respect to A) of the area of the bending moment diagram between A and B divided by EI.

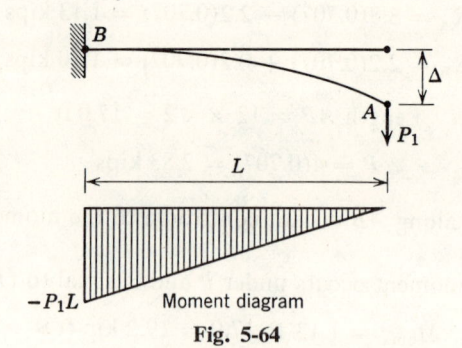

Fig. 5-64

$$\Delta = \frac{\text{area of } M \text{ diagram} \times \text{lever arm}}{EI}$$

$$= \frac{-P_1 L(L/2) \times \tfrac{2}{3} L}{EI} = \frac{-P_1 L^3}{3EI}$$

For $\Delta = 0.003$,

$$\frac{P_1 \times 30^3}{3(20 \times 10^6) \times 2} = 0.003$$

$$P_1 = \frac{0.003 \times 2 \times 3 \times 20 \times 10^6}{30^3} = 13.33 \text{ lb}$$

With the spring,

$$P = 13.33 + 0.003k_1$$

When the first spring is replaced by the second one,

$$P = \tfrac{2}{3}(13.33) + 0.002(2k_1) = 8.9 + 0.004k_1$$

Equating the two expressions for P,

$$13.33 + 0.003k_1 = 8.9 + 0.004k_1$$

$$k_1 = \frac{4.43}{0.001} = 4430 \text{ lb/in.}$$

$$P = 13.33 + 0.003(4430) = 26.62$$

Thus the ratio $k_1/P = 4430/26.62 = 166$ ▶

5-55 Wt. 6

A cantilever beam with a rectangular cross section supports a concentrated load at the end which deflects the beam 2.0 in. It is proposed to replace this beam with another beam of rectangular cross section, of the same material, but with a width of only 0.6 of that of the original beam. The deflection with the new beam must be limited to 0.5 in. What will be the depth of the new beam in relation to the depth of the original beam?

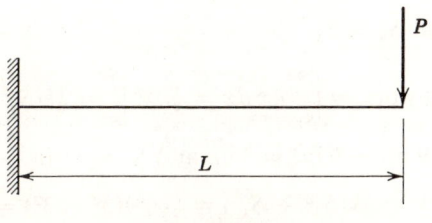

Fig. 5-65

Solution

From Problem 5-54,

$$\Delta = \frac{PL^3}{3EI} \quad (P, L, \text{ and } E \text{ are constants here})$$

$\Delta_1 = 2.0$ in. $\quad \Delta_2 = 0.5$ in.

$b_1 = b$ $\quad b_2 = 0.6b$

$h_1 = h$ $\quad h_2 = ah$

$I_1 = \dfrac{bh^3}{12}$ $\quad I_2 = \dfrac{0.6b(ah)^3}{12}$

$\dfrac{I_2}{I_1} = \dfrac{0.6a^3}{1} \quad \dfrac{\Delta_2}{\Delta_1} = \dfrac{0.5}{2.0} = 0.25 \quad \dfrac{I_2}{I_1} = \dfrac{\Delta_1}{\Delta_2}$

$$0.6a^3 = \frac{1}{0.25} \quad \text{or} \quad 0.6a^3 = 4$$

$$a^3 = \frac{4}{0.6} = 6.67 \quad a = 1.88 \blacktriangleright$$

The new depth is 1.88 times the original depth.

5-56 Wt. 3

A helical spring has a natural length of 6 in. It requires a force of 20 lb to hold it extended to a length of 12 in. How much work in inch-pounds does it take to stretch the spring from a total length of 9 in. to 11 in.? Assume the spring does not exceed its elastic limit.

Solution

$$\text{Spring constant } k = \frac{\text{force}}{\text{deflection}} = \frac{20 \text{ lb}}{(12 - 6) \text{ in.}} = \frac{20 \text{ lb}}{6 \text{ in.}}$$

The energy required to extend a spring a distance dx is $F\, dx$.

Force F = spring constant $k \times$ distance extended x

For a displacement $S_2 - S_1$,

$$\text{Total work} = \int_{S_1}^{S_2} kx\, dx = \tfrac{1}{2}kx^2 \Big|_{S_1}^{S_2} = \tfrac{1}{2}k(S_2^2 - S_1^2)$$

In this case $S_1 = 9$ in. $- 6$ in. $= 3$ in. and $S_2 = 11$ in. $- 6$ in. $= 5$ in.

$$\text{Total work} = \tfrac{1}{2}k(S_2^2 - S_1^2) = \tfrac{1}{2}(\tfrac{20}{6})(5^2 - 3^2) = 26.7 \text{ in-lb} \blacktriangleright$$

5-57 Wt. 6

A 3-in.-diameter solid steel shaft 10 ft long is subjected to a constant torque of 100,000 in-lb at each end, together with an axial tensile load of 70,000 lb also applied at each end. Determine the maximum tensile, compressive, and shearing stresses in the shaft under this loading.

Solution

$$\sigma_x = \sigma_{\text{tension}} = \frac{P}{A} = \frac{70,000(4)}{\pi 3^2} = 9900 \text{ psi}$$

$$\sigma_y = 0$$

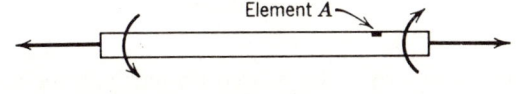

Fig. 5-66

Torsion

$$\tau_{xy} = \frac{16T}{\pi d^3} = \frac{16(100,000)}{\pi(27)} = 18,900 \text{ psi}$$

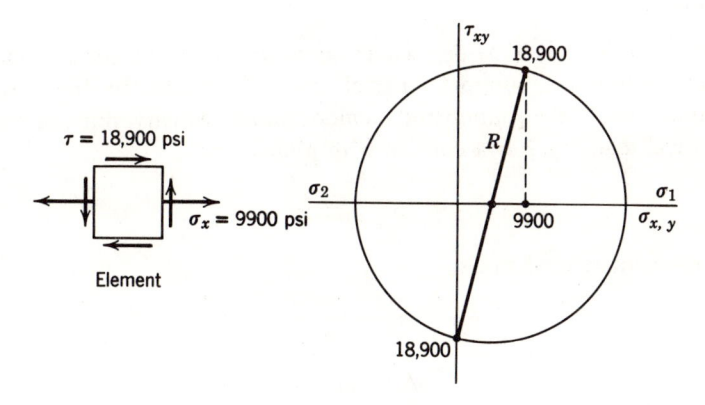

Fig. 5-67

$$R = (18,900^2 + 4950^2)^{1/2} = 19,600$$

$$(\tau_{xy})_{\max} = R = 19,600 \text{ psi} \blacktriangleright$$

$$\sigma_1 = 4950 + R = 24,550 \text{ psi tension} \blacktriangleright$$

$$\sigma_2 = 4950 - R = -14,650 \text{ psi compression} \blacktriangleright$$

276 / Mechanics of Materials

5-58 Wt. 1

The "least radius of gyration" is required in the design of

(a) shaft couplings
(b) columns
(c) helical springs
(d) cantilevered beams
(e) riveted joints

Solution

The ratio L/r of the column length to the least radius of gyration is the column slenderness ratio ▶

Answer is (b)

5-59 Wt. 1

The formula developed by Euler is used primarily in the design of

(a) beams
(b) slabs
(c) footings
(d) columns
(e) girders

Solution

Leonhard Euler in 1757 devised an equation for the maximum load that a slender column will support (critical load). Knowing the dimensions and end conditions of the column and its modulus of elasticity, one can determine the critical load P_{cr}. For a column with pinned ends,

$$P_{cr} = \frac{\pi^2 EI}{L^2}$$

or in terms of critical stress,

$$\sigma_{cr} = \frac{\pi^2 E}{(L/r)^2} \blacktriangleright$$

Answer is (d)

5-60 Wt. 1

The "slenderness ratio" of a column is generally defined as the ratio of its

(a) length to its minimum width
(b) unsupported length to its maximum radius of gyration
(c) length to its moment of inertia
(d) unsupported length to its least radius of gyration
(e) unsupported length to its minimum cross-sectional area

Solution

In Euler's equation

$$\sigma_{cr} = \frac{\pi^2 E}{(L/r)^2}$$

where L is the unsupported length and r is the least radius of gyration, the ratio L/r is called the column *slenderness ratio* ▶

Answer is (d)

5-61 Wt. 5

A cylindrical steel tank 11 ft in diameter and 14 ft high contains a brine solution (brine weighs 10 lb/gal). The thickness of the steel shell is $\frac{1}{4}$ in., and the tank is full, with no cover. What is the unit stress in the steel for a 1-ft-wide strip, the center of which is 3 ft above the concrete pad?

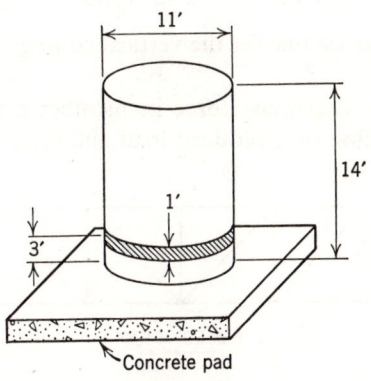

Fig. 5-68

Solution

Consider the ring of diameter D, thickness t, and length L which is part of the tank having internal pressure R. Considering half this ring in equilibrium, the force H equals the unit stress in the steel S times tL. The force F due to the internal pressure equals the pressure R times diameter D times length L.

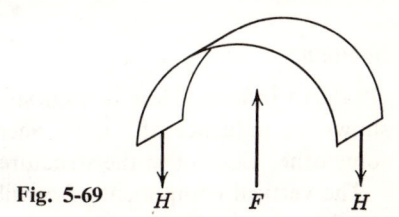

Fig. 5-69

$\sum F_y = 0$

$2H = F \qquad 2StL = RDL \qquad S = \dfrac{RD}{2t}$

278 / Mechanics of Materials

In this problem

$$R = 10 \text{ lb/gal} \times 7.48 \text{ gal/ft}^3 \times \tfrac{1}{144} \text{ ft}^2/\text{in.}^2 \times 11 \text{ ft}$$
$$= 5.71 \text{ psi}$$
$$D = 11 \times 12 = 132 \text{ in.}$$
$$t = 0.25 \text{ in.}$$
$$S = \frac{5.71 \times 132}{2 \times 0.25} = 1510 \text{ psi} \blacktriangleright$$

5-62 Wt. 6

In the Pratt truss shown, loads are applied at the lower panel points through a system of floor beams and stringers.

(a) Draw the influence line for the vertical component of force in member a.

(b) Determine the maximum force in member a for a vehicle with the following wheel loading (neglect dead load and impact):

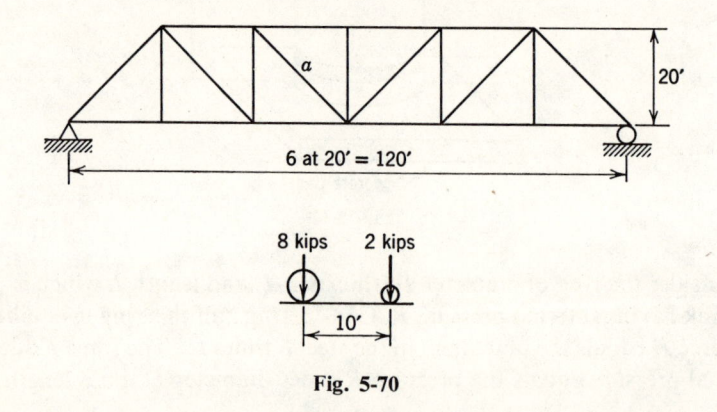

Fig. 5-70

Solution

(a) An influence line is a curve drawn so that the ordinate at any point shows the influence of a unit concentrated load, placed at any point, upon some other element in the structure—in this case member a.

The vertical component of a will be equal to the shear in the panel. The panel must transmit the shear to the reactions, so when the unit load is to the left of the panel the shear in the panel is equal to the right reaction. Similarly, when the unit load is to the right of the panel, the shear in the panel is equal to the left reaction.

▶ Influence line for the vertical component of force in member a:

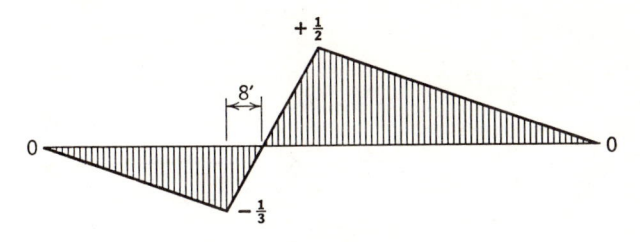

Fig. 5-71

(b) By inspection of the influence line, we can see that the maximum force in member a will occur when the 8-kip load is at the middle panel point and the 2-kip load is 10 ft to the right.

Under these conditions the shear in the panel would be:

$$V_{max} = 8 \text{ kips} \times 0.5 + 2 \text{ kips}(\tfrac{5}{6})(0.5) = 4 + 0.83 = +4.83 \text{ kips}$$

Since the panel is square (20 × 20), the tensile force in member a will be $\sqrt{2}$ × shear in the panel:

$$\text{Tensile force in } a = \sqrt{2} \times 4.83 \text{ kips} = +6.84 \text{ kips} \blacktriangleright$$

5-63 Wt. 3

Derive the equivalent spring constant K' for three springs hooked in series as shown, where K_1, K_2, and K_3 are the spring constants for the individual springs.

Fig. 5-72

Solution

Total displacement S = sum of displacements of individual springs

$$= S_1 + S_2 + S_3$$

Also,

$$\text{Displacement} = \frac{\text{Load}}{\text{Spring constant}}$$

$$S = \frac{P}{K'} = \frac{P}{K_1} + \frac{P}{K_2} + \frac{P}{K_3}$$

$$\frac{1}{K'} = \frac{1}{K_1} + \frac{1}{K_2} + \frac{1}{K_3} \qquad K' = \frac{1}{\frac{1}{K_1} + \frac{1}{K_2} + \frac{1}{K_3}} \blacktriangleright$$

5-64 Wt. 6

A reinforced concrete beam 12 in. wide and 30 in. deep is simply supported on a span of 20 ft center-to-center of the supports. The beam is reinforced with four #7 bars ($A_s = 2.4$ in.^{2}, $\Sigma_0 = 11$ in.) placed 3 in. from the bottom of the beam. The concrete beam has a compressive strength of 3000 psi at 28 days. The beam supports its own weight and 2.0 kips/lin ft. Use concrete weight $= 150$ lb/ft^3, $n = 10$. Determine the

(a) maximum bending moment in foot-pounds
(b) maximum shear in pounds
(c) location of the neutral axis based on the transformed section

Solution

$$\text{Weight of beam per foot} = (1)(2.5)(150) = 375 \text{ lb/ft}$$

$$W_{\text{total}} = 2375 \text{ lb/ft}$$

(a)

$$M_{\max} = \frac{wL^2}{8} = \frac{(2375)(20)^2}{8} = 118{,}750 \text{ lb-ft} \blacktriangleright$$

(b)

$$V_{\max} = \frac{wL}{2} = \frac{(2375)(20)}{2} = 23{,}750 \text{ lb} \blacktriangleright$$

(c)

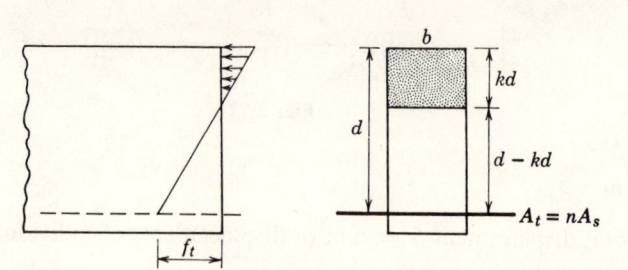

Fig. 5-73

From similar triangles, $\dfrac{f_t}{d-kd} = \dfrac{f_c}{kd}$.

Compressive force $= \tfrac{1}{2}f_c kd b$ Tensile force $= f_t n A_s$

For force equilibrium, $C = T$.

$$\tfrac{1}{2}f_c kd b = f_t n A_s = f_c \dfrac{d-kd}{kd} n A_s = f_c \dfrac{1-k}{k} n A_s$$

$$\dfrac{k^2 bd}{2} = (1-k)(nA_s)$$

$$\dfrac{k^2(12)(27)}{2} = (1-k)(10)(2.4)$$

$$k^2 + 0.148k - 0.148 = 0$$

$$k = \dfrac{-0.148 \pm \sqrt{0.0219 + 0.592}}{2} = \dfrac{-0.148 + 0.783}{2} = 0.318$$

Therefore

$$kd = 27(0.318) = 8.6 \text{ in.} \blacktriangleright$$

The neutral axis is located 8.6 in. below the top of the beam.

5-65 Wt. 5

For the reinforced concrete beam shown below, determine

(a) the distance kd

(b) the maximum resisting moment of the beam with the conditions given (Assume a balanced design, using straight-line theory.)

Data: $f_c = 1350$ psi
     $f_s = 20{,}000$ psi
     $A_s = 2.18$ in.^{2}
     $n = 10$

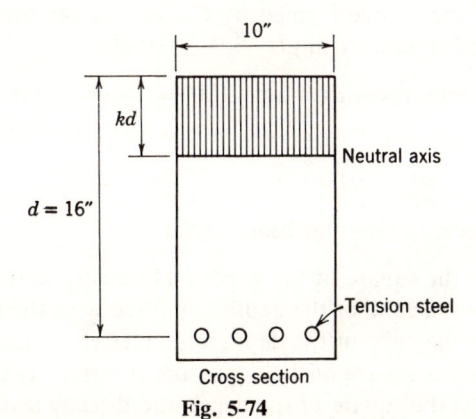

Cross section
Fig. 5-74

Solution

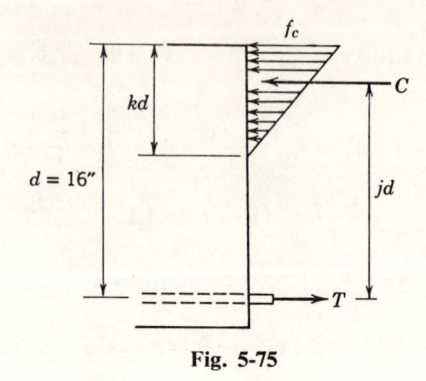

Fig. 5-75

Width $b = 10$ in.
$$C = \tfrac{1}{2}f_c kdb = \tfrac{1}{2}(1350)(kd)(10) = 6750kd$$
$$T = A_s f_s = 2.18(20{,}000) = 43{,}600 \text{ lb}$$

In a *balanced design* the maximum allowable concrete stress f_c and allowable steel stress f_s are simultaneously developed.

(a) $\sum F_x = 0$
$$T = C$$
$$43{,}600 = 6750\, kd \qquad kd = \frac{43{,}600}{6750} = 6.46 \text{ in.} \blacktriangleright$$

(b)
$$jd = d - \frac{kd}{3} = 16 - \frac{6.46}{3} = 16 - 2.15 = 13.85 \text{ in.}$$

The moment of the couple formed by C and T is the product of the force and the distance between them jd.

Maximum resisting moment $M = Tjd = 43{,}600(13.85)$
$$= 604{,}000 \text{ in-lb} \blacktriangleright$$

5-66 Wt. 1

The stiffness of a rectangular beam varies

(a) directly as the square of the depth and directly as the length
(b) directly as the cube of the depth and directly as the length
(c) directly as the cube of the depth and inversely as the length
(d) directly as the square of the depth and inversely as the length
(e) inversely as the square of the depth and directly as the length

Solution

$$\text{Moment rotational stiffness} = \frac{4EI}{L} = k\frac{bd^3}{L} \blacktriangleright$$

Answer is (c)

5-67 Wt. 1

The section modulus about the central axis of a 4 in × 10 in. beam on edge is

(a) 25 in.^{3}
(b) 40 in.^{3}
(c) 67 in.^{3}
(d) 80 in.^{3}
(e) 333 in.^{3}

Solution

$$\text{section modulus } S = \frac{I}{c}$$

where I is the moment of inertia and c is the distance from the neutral axis to the extreme fiber.

For a rectangular cross section,

$$I = \frac{bh^3}{12} \quad c = \frac{h}{2}$$

$$S = \frac{I}{c} = \frac{bh^3/12}{h/2} = \frac{bh^2}{6}$$

In this case

$$S = \frac{4(10)^2}{6} = 66.7 \text{ in.}^3 \blacktriangleright$$

Answer is (c)

5-68 Wt. 1

A cantilevered beam 20 in. long and of square cross section, 1 in. on a side, is loaded at its end, through the centroid of the cross section by a vertical force of magnitude 150 lb. The magnitude of the maximum bending stress is nearest to which value?

(a) 5,000 psi
(b) 15,000 psi
(c) 125 psi
(d) 17,000 psi
(e) 250 psi

Solution

The stress $S = Mc/I$ with $I = bh^3/12 = 1(1)^3/12 = \frac{1}{12}$ in.^{4} and $M = (150 \text{ lb})(20 \text{ in.}) = 3000$ in-lb.

$$S = \frac{(3000)(\frac{1}{2})}{\frac{1}{12}} = 18,000 \text{ psi} \blacktriangleright$$

Answer is (d)

5-69 Wt. 1

The maximum unit fiber stress at any vertical section in a beam is obtained by dividing the moment at that section by

(a) the section modulus
(b) the cross-sectional area
(c) one-half the distance to the point where the shear is zero
(d) the radius of gyration
(e) the moment of inertia

Solution

The basic formula for maximum fiber stress in a beam is

$$\text{Unit stress} = \frac{Mc}{I}$$

where M = moment at the section
c = distance from the neutral axis to the extreme fiber
I = moment of inertia

The section modulus S is defined as I/c, so the basic formula reduces to

$$\text{Unit stress} = \frac{M}{S} = \frac{\text{Moment}}{\text{Section modulus}} \blacktriangleright$$

Answer is (a)

Chapter 6

Fluid Mechanics

Fluid mechanics studies fluids at rest and in motion by the application of basic principles of mechanics. In general the discipline covers a very broad field since it endeavors to study both liquids and gases under all conditions. In this chapter we restrict our attention almost entirely to a study of the mechanical behavior of incompressible liquids such as water. Topics related to the internal energy changes and the compressibility and chemical behavior of gases may be found in Chapter 7 on Thermodynamics and Chapter 8 on Chemistry.

Paralleling the study of the mechanics of solid bodies, fluid mechanics is normally divided into fluid statics and fluid dynamics. Hydrostatics studies the variation of pressure in a stationary fluid body and leads to an understanding of such subjects as buoyancy and manometry; one also learns how to compute the forces exerted on submerged bodies because of this pressure variation. The fundamental principles underlying fluid dynamics are contained in three conservation laws: those expressing the conservation of mass, momentum, and energy. In connection with these principles the calculation of horsepower, head loss, and flow rate is considered. In almost all of the dynamics problems the flow is assumed to be steady and also one-dimensional, that is, the flow variables are only a function of the distance along the conveyance structure and do not vary across the flow cross section.

HYDROSTATICS

The fundamental equation of fluid statics indicates that the rate of change of the pressure p is directly proportional to the rate of change of the depth z, or

$$\frac{dp}{dz} = -\gamma = -\rho g \qquad (6\text{-}1)$$

where γ is the unit weight of the fluid (for water $\gamma_w = 62.4 \text{ lb/ft}^3$) and z is

286 / Fluid Mechanics

positive upward. Also ρ is the fluid density and g is the acceleration of gravity. For incompressible fluids γ is constant and equation (6-1) may be integrated to give

$$p_2 = p_1 + \gamma h \tag{6-2}$$

Here point 2 is located a distance h below point 1. In most problems in fluid mechanics one may work in either absolute or gage pressures if proper care is taken to be consistent. Gage pressure is equal to the difference between the absolute pressure and atmospheric pressure; when it is negative, it is often called a vacuum.

Manometers are devices which measure pressure differences by a direct application of the hydrostatic principle given in equation (6-2). Pressure differences in manometers can easily be computed by systematically applying the equation to each manometer limb.

Example 1

Mercury (Hg) is poured into a U-tube. Then 18-in. of oil of specific gravity $S_0 = 0.90$ is poured into one leg on top of the mercury (Fig. 6-1). What suction, in pounds per square inch, applied to the leg containing only mercury will bring the upper surfaces of the oil and mercury in the two legs to the same level?

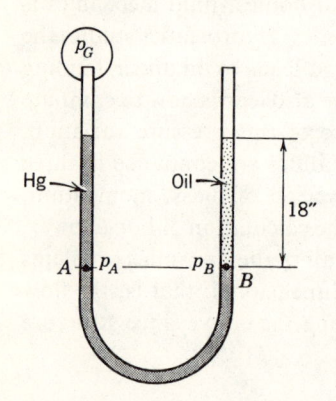

Fig. 6-1

Solution

For one fluid, equation (6-2) shows that the pressures at two points are equal when the elevation difference between the points is zero so that $p_A = p_B$. But equation (6-2) also shows that

$$p_A = p_G + \gamma_M h$$

and

$$p_B = \gamma_0 h$$

Hence
$$p_G = h(\gamma_0 - \gamma_M) = h\gamma_w(S_0 - S_M)$$
where the specific gravity of mercury $S_M = 13.55$.

$$p_G = \tfrac{18}{12}(62.4)(0.90 - 13.55) = -1184 \text{ psfg}$$

$$p_G = \frac{-1184}{144} = -8.22 \text{ psig} = 8.22 \text{ psi vacuum} \blacktriangleright$$

The buoyant force F_B exerted upward on a floating or submerged object is equal to the weight of the fluid displaced by that object, as can be shown by properly integrating equation (6-2) over the surface of the body. Restated,

$$F_B = \gamma_f(\text{volume}) \qquad (6\text{-}3)$$

where γ_f is the unit weight of the fluid, and the volume is the amount of fluid displaced.

Example 2

A piece of lead (specific gravity $S_L = 11.3$) is tied to 8 in.³ of cork whose specific gravity is $S_C = 0.25$. They float submerged in water (Fig. 6-2). What is the weight of the lead?

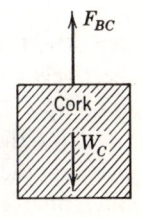

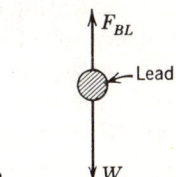

Fig. 6-2

Solution

The net upward force on the cork is $F_{BC} - W_C$, which is just balanced by the net downward force on the lead $W_L - F_{BL}$, or

$$F_{BC} - W_C = W_L - F_{BL}$$

Thus

$$(1 - S_C)\gamma_w \left(\frac{8}{12^3}\right) = (S_L - 1)\gamma_w V_L$$

where V_L is the unknown volume of the lead. Solving first for V_L and then for the weight W_L, we have

$$W_L = S_L \gamma_w V_L = \gamma_w \left(\frac{S_L}{S_L - 1}\right)(1 - S_C)\frac{8}{12^3}$$

$$W_L = 62.4 \left(\frac{11.3}{10.3}\right)(0.75)\frac{8}{12^3}$$

$$W_L = 0.238 \text{ lb lead} \blacktriangleright$$

Distributed fluid pressures cause hydrostatic forces to act on submerged surfaces. It is important to be able to compute the direction, magnitude, and location of these resultant forces. The force on any submerged surface is determinable if one knows how to compute the force acting on a submerged plane area and the centroid or center of gravity of areas and volumes.

The magnitude of the force F on a submerged plane surface is

$$F = \int_A p \, dA = p_C A = \gamma h_C A \tag{6-4}$$

where A is the surface area, γ is the fluid unit weight, and p_C and h_C are respectively the pressure and submerged depth of the centroid of the area. This compressive force acts normal to the plane area. Simple formulas can be derived for the location of this force, but they are easily misused. A more direct approach is to compute the first moment of the pressure distribution about some convenient axis, according to the principles of statics, and equate this to the product of F and the distance to the line of action of F.

An extension of these principles allows the computation of the magnitude and location of a hydrostatic force which acts on a submerged curved surface. The horizontal component of this force is computed from equation (6-4), where the area A is now the vertical projection of the curved surface. The vertical force component is basically equal to the weight of the fluid which lies vertically above the curved surface, but some care is needed in applying this principle. When all the fluid lies above the surface, the principle applies exactly. If some fluid also lies beneath a portion of the curved surface, then the *net* upward vertical force is a buoyant force equal to the weight of fluid displaced by the presence of the curved surface. The line of action of each component force and the magnitude and direction of the resultant force are then found by direct application of the principles of statics.

Example 3

A closed circular tank 2 ft in diameter and 6 ft deep, with its axis vertical, contains 4 ft of water. Air at a pressure of 5 psig is pumped into the cylinder. Find the normal force on the vertical wall of the tank and the distance to the center of pressure from the base of the tank.

Solution

First we plot a diagram, Fig. 6-3, of the gage pressure which is exerted on the wall. The hydrostatic pressure p_1 at the base of the tank is

$$p_1 = \gamma h = \frac{(62.4)(4)}{144} = 1.73 \text{ psi}$$

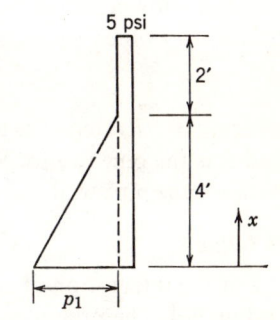

Fig. 6-3

The force per inch of circumference equals the area of the pressure diagram, or

$$F = \int p \, dA = 5(6)(12) + \tfrac{1}{2}(1.73)(4)(12)$$

$$= 360 + 41.5 = 401.5 \text{ lb/in.} \blacktriangleright$$

The distance to the center of pressure x_{cp}, by statics, is

$$x_{cp} = \frac{1}{F}\int px \, dA$$

$$= \frac{1}{401.5}[360(3) + 41.5(\tfrac{4}{3})] = 2.83 \text{ ft} = 34 \text{ in.} \blacktriangleright$$

CONTINUITY

The law of conservation of mass states that matter is neither created nor destroyed. Applied to a streamtube in incompressible flow, the law requires the flow to vary in a continuous way from cross section to cross section along the streamtube so that at any section

$$Q = \int_A V \, dA = \text{constant} \tag{6-5}$$

where Q is the volume rate of flow and V is the velocity at a point in the cross-sectional streamtube area A. When the velocity is assumed to be constant across the section, we obtain the familiar continuity equation

$$Q = A_1 V_1 = A_2 V_2 \tag{6-6}$$

This equation is widely used in combination with the energy or momentum equations to solve fluid flow problems.

Example 4

A fluid flowing steadily through a constant-diameter pipeline has a velocity profile $u(r)$ which varies parabolically across the pipe. Specifically,

$$u = m(1 - s^2)$$

where $s = r/R$. Here m is a constant, r is the distance from the pipe centerline, and R is the pipe radius. What fraction of the total flow in the pipe is flowing between the pipe wall and a distance of 10% of the pipe radius from the wall?

Solution

For the integration of equation (6-5) we choose the differential area shown in Fig. 6-4. Then the total volume rate of flow in the pipe is

$$Q_T = \int_A V \, dA = \int_0^R m(1 - s^2)(2\pi r \, dr)$$

$$= 2\pi m R^2 \int_0^1 (1 - s^2) s \, ds$$

$$= 2\pi m R^2 \left[\frac{s^2}{2} - \frac{s^4}{4} \right]_0^1$$

$$Q_T = \tfrac{1}{2} \pi m R^2$$

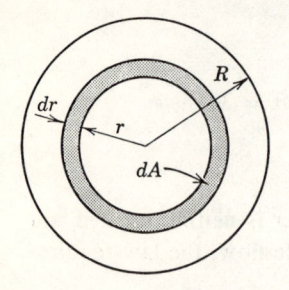

Fig. 6-4

The volume rate of flow near the wall is

$$Q = \int_{0.9R}^{R} m(1 - s^2)(2\pi r\, dr)$$

$$= 2\pi m R^2 \int_{0.9}^{1} (1 - s^2) s\, ds$$

$$= 2\pi m R^2 \left[\frac{s^2}{2} - \frac{s^4}{4}\right]_{0.9}^{1}$$

$$Q = 0.018 \pi m R^2$$

Thus the flow fraction is

$$\frac{Q}{Q_T} = \frac{0.018}{0.5} = 0.036 \blacktriangleright$$

ENERGY EQUATION

For one-dimensional, steady incompressible flow a general energy equation which expresses the changes in energy, *per pound* of flowing fluid, between points 1 and 2 is

$$\frac{V_1^2}{2g} + \frac{p_1}{\gamma} + z_1 = \frac{V_2^2}{2g} + \frac{p_2}{\gamma} + z_2 + h_L + E_m \qquad (6\text{-}7)$$

In this equation $V^2/2g$ is the kinetic energy or velocity head, p/γ is the pressure energy or pressure head, and z is the potential energy or elevation head. The head loss h_L represents the loss in energy between points 1 and 2 for any of a variety of causes. The last term E_m is the mechanical energy added to the fluid between the two points; this is accomplished by hydraulic machinery. For pumps, energy is added to the flow and E_m is positive. For turbines, E_m is negative since energy is then extracted from the fluid. When E_m and h_L are both zero, equation (6-7) is the classic Bernoulli equation.

The head loss term h_L represents a loss in energy that is due primarily to viscosity and the turbulence in the flow. It may be written as

$$h_L = \sum K \frac{V^2}{2g}$$

where V is a representative velocity for the point where the loss occurs, K is a loss coefficient related to the nature of the loss-producing element, and one sums the effects of all loss elements between the two points. Particularly important is the head loss caused by pipe friction in a pipe of length L and diameter d. In this case

$$h_L = f \frac{L}{d} \frac{V^2}{2g} \qquad (6\text{-}8)$$

292 / Fluid Mechanics

The equation is the Darcy-Weisbach equation. The friction factor $f = f(\mathbf{R})$, and the Reynolds number is $\mathbf{R} = Vd\rho/\mu$. For laminar flow $\mathbf{R} < 2100$ and $f = 64/\mathbf{R}$, but in turbulent flow, when $\mathbf{R} > 4000$, f is also a function of the relative roughness e/d for rough pipes. The height of a representative roughness projection is e. Between the laminar and turbulent flow zones lies an unpredictable transition zone. Most losses other than that due to pipe friction are termed minor losses, since their magnitude is usually small in comparison to the pipe friction loss in practical problems. A different K value is needed for each type of loss element.

The horsepower produced or expended in a given situation is directly related to the change in energy in the flow. The relation is

$$\text{Horsepower} = \frac{Q\gamma E}{550} \tag{6-9}$$

The weight rate of flow is $Q\gamma$, and the energy change per pound of fluid is E. For turbines or pumps E is the same as E_m.

Example 5

A Venturi meter with a throat diameter of 6 in. is placed in a 12-in.-diameter pipeline. It meters the flow of oil having a specific gravity $S_0 = 0.80$. A differential manometer containing a fluid of specific gravity $S_M = 3.20$ is connected between the pipe and the throat section and shows a deflection of 24 in. Above the manometer liquid the tubes are filled with oil (Fig. 6-5). Neglecting energy losses, what is the indicated flow through the meter?

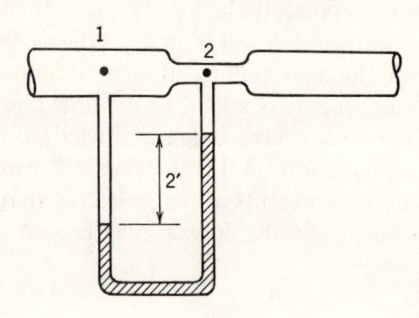

Fig. 6-5

Solution

Here we apply the Bernoulli and continuity equations in combination. The Bernoulli equation is

$$\frac{V_1^2}{2g} + \frac{p_1}{\gamma_0} + z_1 = \frac{V_2^2}{2g} + \frac{p_2}{\gamma_0} + z_2$$

Since $z_1 = z_2$,
$$\frac{p_1 - p_2}{\gamma_0} = \frac{V_2^2 - V_1^2}{2g}$$
Considering the manometer,
$$p_1 + S_0\gamma_w(2) = p_2 + S_M\gamma_w(2)$$
and by noting that $S_0\gamma_w = \gamma_0$, we obtain
$$\frac{p_1 - p_2}{\gamma_0} = 2\left(\frac{S_M}{S_0} - 1\right)$$
From equation (6-6), the continuity equation is $V_1A_1 = V_2A_2$, or
$$V_2 = \frac{A_1}{A_2}V_1 = \left(\frac{d_1}{d_2}\right)^2 V_1 = \left(\tfrac{12}{6}\right)^2 V_1 = 4V_1$$
Therefore
$$2\left(\frac{S_M}{S_0} - 1\right) = \frac{(4V_1)^2 - V_1^2}{2g} = \frac{15V_1^2}{2g}$$
$$V_1^2 = \frac{1}{15}\left[2(32.2)(2)\left(\frac{3.20}{0.80} - 1\right)\right] = 25.8$$
$$V_1 = 5.09 \text{ fps}$$
Finally,
$$Q = A_1V_1 = \frac{\pi}{4}(1)^2(5.09) = 4.0 \text{ cfs} \blacktriangleright$$

Example 6

A water main with a 24-in. I.D. carries a flow of 20 cfs. If the friction factor is 0.02 and the pump is 85% efficient, how much horsepower is required to pump the water through 10,000 ft of pipeline?

Solution

The velocity in the main is
$$V = \frac{Q}{A} = \frac{20}{(\pi/4)(2)^2} = 6.37 \text{ fps}$$
Using equation (6-8), the Darcy-Weisbach formula for head loss caused by friction, we obtain
$$h_L = f\frac{L}{d}\frac{V^2}{2g}$$
$$h_L = (0.02)\left(\frac{10,000}{2}\right)\frac{(6.37)^2}{2(32.2)}$$
$$h_L = 62.9 \text{ ft}$$

This head loss is equal to the amount of energy E that must be added per pound of fluid. The net power requirement is

$$\frac{Q\gamma h_L}{550} = \frac{(20)(62.4)(62.9)}{550} = 142.4 \text{ hp}$$

Since the pump efficiency is only 85%, the total power requirement is $142.4/0.85 = 167.5$ hp ▶

Pipelines of different sizes and lengths commonly occur in combination. When pipes are connected end-to-end in series, the head loss of the combination is the sum of the head losses in the individual pipes, while the flow rate is the same for each element. The situation is reversed for a parallel combination of pipes, that is, a case where two or more pipes are connected between the same two end points. Then the total flow rate is the sum of the individual flow rates, while the head lost is identical for each pipe.

MOMENTUM EQUATION

The steady-flow momentum conservation equation takes the form

$$\sum \mathbf{F}_S = \int_S \rho \mathbf{V} V_n \, dA \qquad (6\text{-}10)$$

This is a vector equation. The first term represents the net external surface force acting *on* a fluid volume which is enclosed by the surface S; this term could include contributions arising from the weight of the fluid or the pressure distribution acting on S. The right term is the net flux of momentum flowing out of the region enclosed by S. V_n is the fluid velocity component exiting normal or perpendicular to S. For uniform flow across an entrance section 1 and an exit section 2, we may write

$$\sum F_x = \rho Q(V_{2x} - V_{1x})$$
$$\sum F_y = \rho Q(V_{2y} - V_{1y}) \qquad (6\text{-}11)$$

for flow in two dimensions. The volume flow rate is Q. Note that V_x and V_y may be either positive or negative.

Example 7

The 18-in. to 12-in. reducing bend shown in Fig. 6-6 is in a horizontal pipeline conveying oil (of specific gravity $S_0 = 0.90$) at the rate of 10 cfs. The pressure in the line at the 18-in. section is 50 psia when the atmospheric pressure $p_a = 14.7$ psia. Compute the x and y components of thrust caused by the flowing fluid.

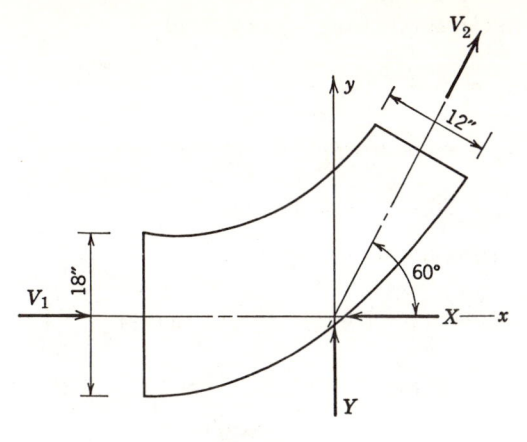

Fig. 6-6

Solution

The force components we seek are equal and opposite to the external forces X and Y shown in Fig. 6-6.

The reducing bend and the fluid within it are chosen as a control volume. Atmospheric pressure acts on the entire control volume surface and thus produces no net force. Employing equations (6-11) we find

$$\sum F_x = (p_1 - p_a)A_1 - (p_2 - p_a)A_2 \cos 60° - X$$

and

$$\rho Q(V_{2x} - V_{1x}) = \rho Q(V_2 \cos 60° - V_1)$$

Here

$$Q = 10 \text{ cfs}$$

$$V_1 = \frac{Q}{A_1} = \frac{10}{(\pi/4)(\tfrac{18}{12})^2} = 5.66 \text{ fps}$$

$$V_2 = \frac{Q}{A_2} = \frac{10}{(\pi/4)(\tfrac{12}{12})^2} = 12.73 \text{ fps}$$

Assuming no energy losses occur in the bend, we apply the Bernoulli equation between the entrance and exit to find p_2:

$$\frac{V_1^2}{2g} + \frac{p_1}{\rho g} + z_1 = \frac{V_2^2}{2g} + \frac{p_2}{\rho g} + z_2$$

Since there is no elevation change $z_1 = z_2$, and

$$p_2 = p_1 + \frac{\rho}{2}(V_1^2 - V_2^2)$$

$$p_2 = 50 + \tfrac{1}{2}(0.90)(1.94)[(5.66)^2 - (12.73)^2]\tfrac{1}{144}$$

$$p_2 = 50 - 0.79 = 49.2 \text{ psia}$$

The momentum equation now becomes

$$(50 - 14.7)(144)\frac{\pi}{4}\left(\frac{18}{12}\right)^2 - (49.2 - 14.7)(144)\frac{\pi}{4}\left(\frac{12}{12}\right)^2 \cos 60° - X$$

$$= (0.90)(1.94)(10)[12.73 \cos 60° - 5.66]$$

$$X = 7040 \text{ lb}$$

The momentum equation in the y direction is

$$\sum F_y = \rho Q(V_{2y} - V_{1y})$$

$$Y - (p_2 - p_a)A_2 \sin 60° = \rho Q(V_2 \sin 60° - 0)$$

$$Y - (49.2 - 14.7)(144)\frac{\pi}{4}\left(\frac{12}{12}\right)^2 \sin 60° = (0.90)(1.94)(10)[12.73 \sin 60°]$$

$$Y = 3570 \text{ lb}$$

The thrust components of the fluid are equal and opposite to X and Y, so the x component is 7040 lb to the right and the y component is 3570 lb downward ▶

OPEN CHANNEL FLOW RATE

The fundamental equation for the flow rate Q in uniform flow as a function of depth of flow and channel characteristics is the Manning equation

$$Q = \frac{1.49}{n} A R^{2/3} S^{1/2} \qquad (6\text{-}12)$$

In this equation n is a roughness coefficient which may vary from 0.01 for smooth uniform channels to 0.03 or higher for irregular natural river channels, and S is the channel bottom slope. The cross-sectional area of flow is A, and R is a shape parameter for the channel section. Called the hydraulic radius, $R = A/P$, where P is the wetted perimeter of the cross section. Using this equation to find Q when the other factors are known is straightforward, but solving for the flow depth y, when Q is known, usually requires trial-and-error computations.

Example 8

The trapezoidal channel shown in Fig. 6-7, with $S = 0.0009$ and $n = 0.025$, carries a discharge $Q = 300$ cfs. Compute the flow depth y and the average velocity V.

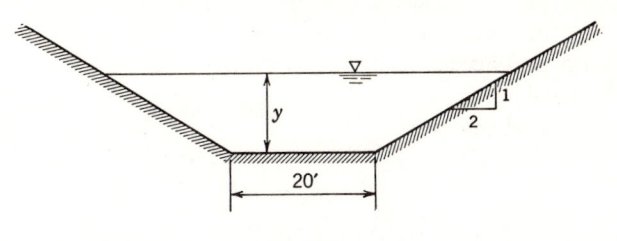

Fig. 6-7

Solution

In terms of y, the area A and hydraulic radius R are

$$A = y(20 + 2y)$$

$$R = \frac{A}{P} = \frac{y(20 + 2y)}{20 + 2y\sqrt{5}}$$

The Manning equation then gives

$$300 = \frac{1.49}{0.025} [y(20 + 2y)] \left| \frac{y(20 + 2y)}{20 + 2y\sqrt{5}} \right|^{2/3} (0.0009)^{1/2}$$

or

$$7680 + 1720y = [y(10 + y)]^{5/2}$$

This equation must be solved by trial and error.

Trial y	$7680 + 1720y$	$[y(10 + y)]^{5/2}$
3.00	12,840	9,500
3.30	13,356	12,750
3.40	13,528	14,050
3.36	13,460 ~	13,500

The depth of flow is thus $y = 3.36$ ft ▶

For this depth the area is

$$A = 3.36[20 + 2(3.36)] = 89.7 \text{ ft}^2$$

and the average velocity is

$$V = \frac{Q}{A} = \frac{300}{89.7} = 3.34 \text{ fps} \blacktriangleright$$

298 / Fluid Mechanics

6-1 Wt. 1

The anchor block at a bend in a pipeline must be designed primarily to resist forces caused by

(*a*) friction and acceleration
(*b*) friction and pressure
(*c*) pressure and acceleration due to gravity
(*d*) static head
(*e*) pressure and velocity

Solution

The anchor block at a bend in a pipeline must be designed primarily to resist forces caused by pressure and velocity ▶

Answer is (*e*)

6-2 Wt. 1

Pressure applied anywhere on a confined liquid is transmitted undiminished in every direction. The force thus exerted by the confined fluid acts at right angles to every portion of the surface of the container and is equal upon equal areas. This principle was formulated by

(*a*) Archimedes
(*b*) Pascal
(*c*) Hooke
(*d*) Bernoulli
(*e*) Charles

Solution

The principle was formulated by Pascal ▶

Answer is (*b*)

6-3 Wt. 1

The line showing the pressure head plus the potential head plus the velocity head at any section of a pipe is called the

(*a*) total head line
(*b*) energy gradient
(*c*) energy head line
(*d*) hydraulic gradient
(*e*) combined head line

Solution

The line showing the sum of the pressure, potential, and velocity heads is the total head line; it graphically shows the sum of the three terms in the Bernoulli equation ▶

Answer is (a)

6-4 Wt. 1

Cavitation results from

(a) insufficient carbon content in steel
(b) low soil permeability
(c) turbulent water flow
(d) inadequate concrete curing
(e) excessive pore water pressure in soil

Solution

Cavitation is a high-speed water phenomenon that occurs when the fluid pressure approaches vapor pressure. This can only occur in turbulent flow ▶

Answer is (c)

6-5 Wt. 1

Capillarity results from

(a) excess pore water pressure
(b) surface tension
(c) seismic forces
(d) inadequate compaction
(e) second-degree indeterminacy

Solution

Capillarity results from surface tension ▶

Answer is (b)

6-6 Wt. 1

The hydraulic jump is utilized for

(a) energy dissipation
(b) pressure regulation
(c) lifting of water
(d) transport of sediment
(e) evaporation rate increase

Solution

In the hydraulic jump energy is dissipated ▶

Answer is (a)

6-7 Wt. 1

In a power plant the pipe which delivers water from the forebay to the scroll case of a turbine is called a

(*a*) siphon
(*b*) surge chamber
(*c*) culvert
(*d*) wasteway
(*e*) penstock

Solution

The pipeline used to deliver water from the forebay to the turbine is called a penstock ▶

Answer is (*e*)

6-8 Wt. 1

The primary function of a weir is

(*a*) wildlife conservation
(*b*) channel diversion
(*c*) measurement of discharge
(*d*) prevention of scour
(*e*) energy dissipation

Solution

Weirs are used primarily to measure discharge in channels ▶

Answer is (*c*)

6-9 Wt. 1

If the pressure in a plenum was noted as 6 in. of water column, this would be equivalent to

(*a*)   6   psf
(*b*) 392   psf
(*c*)   8.8 psf
(*d*) 31.2 psf
(*e*) 39.2 psf

Solution

$$p = \gamma z = 62.4 \text{ lb/ft}^3 \times 0.5 \text{ ft} = 31.2 \text{ psf} \blacktriangleright$$

Answer is (*d*)

6-10 Wt. 3

The value of the coefficient of viscosity of air at 19.2°C is 1.828×10^{-4} g/cm-sec. Calculate the equivalent value expressed in pounds per foot-second.

Solution

$$1.828 \times 10^{-4} \times \frac{1 \text{ lb}}{454 \text{ g}} \times \frac{2.54 \times 12 \text{ cm}}{1 \text{ ft}} = 1.227 \times 10^{-5} \text{ lb/ft-sec} \blacktriangleright$$

6-11 Wt. 1

A fluid flows at a constant velocity in a pipe. The fluid completely fills the pipe, and the Reynolds number is such that the flow is just subcritical and laminar. If all other parameters remain unchanged and the viscosity of the fluid is decreased a significant amount, one would generally expect the flow to

(a) not change
(b) become turbulent
(c) become more laminar
(d) increase
(e) temporarily increase

Solution

Considering the Reynolds number,

$$\mathbf{R} = \frac{VD\rho}{\mu}$$

where V = average velocity in the pipe
D = diameter of the pipe
ρ = density of the fluid flowing
μ = viscosity of the fluid flowing

we see that a decrease in μ will increase $\mathbf{R}$ so that it will be greater than the Reynolds number for the transition to turbulent flow, and the flow will become turbulent $\blacktriangleright$

Answer is (b)

6-12 Wt. 1

The transition between laminar and turbulent flow usually occurs at a Reynolds number of approximately

(a) 350
(b) 900
(c) 1800
(d) 2100
(e) 3850

302 / Fluid Mechanics

Solution

The lower critical Reynolds number, below which laminar flow will always occur, is approximately 2100 ▶

Answer is (d)

6-13 Wt. 5

Given a cube with sides of length L and of density ρ_0 floating in two fluids of different densities, as shown in the figure. The heavier fluid has a density ρ_1 and depth h_1. The lighter fluid has a density ρ_2 and forms a layer of thickness h_2.

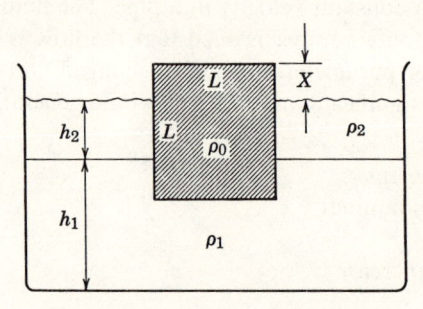

Fig. 6-8

Find the distance X that the cube projects above the surface of the lighter fluid. (Neglect surface tension.)

Solution

Archimedes' principle tells us that the buoyant force is equal to the weight of the liquid displaced. The weight of the displaced liquid is

$$\text{Volume} \times \text{density} = L^2 h_2 \rho_2 g + L^2(L - h_2 - X)\rho_1 g$$

for the two liquid layers. The weight of the block ($L^3 \rho_0 g$) is equal to this buoyant force.

$$L^3 \rho_0 g = L^2 g[h_2 \rho_2 + \rho_1(L - h_2 - X)]$$

$$L \rho_0 = h_2 \rho_2 + \rho_1 L - \rho_1 h_2 - X \rho_1$$

$$X \rho_1 = h_2 \rho_2 + \rho_1 L - \rho_1 h_2 - L \rho_0$$

$$X = \frac{h_2 \rho_2}{\rho_1} + L - h_2 - L \frac{\rho_0}{\rho_1}$$

$$X = h_2 \left(\frac{\rho_2}{\rho_1} - 1\right) - L\left(\frac{\rho_0}{\rho_1} - 1\right) \blacktriangleright$$

6-14 Wt. 1

An "overflow" can brimful of water is suspended from the hook of a spring balance. After a small block of wood is placed in the water, the balance reading is

(a) increased by the weight of the wood block
(b) decreased by the weight of the wood block
(c) decreased by the weight of the water displaced
(d) increased by the weight of the wood above the water surface
(e) unaffected

Solution

A floating body when placed in a liquid sinks until it displaces its own weight of liquid.

Assuming that the wood floats, the weight of water that will overflow is equal to the weight of the wood block. The balance reading will be unchanged ▶

$$\text{Answer is } (e)$$

6-15 Wt. 1

If a body weighs 100 lb in air and 25 lb in fresh water, its volume in cubic feet is

(a) 0.75
(b) 1.1
(c) 1.2
(d) 1.3
(e) 1.5

Solution

$$\text{Buoyant force} = (\text{weight in air}) - (\text{weight in water})$$

$$= 100 - 25 = 75 \text{ lb}$$

$$\text{Buoyant force} = \text{weight of the displaced water}$$

Knowing the weight of the displaced water we can calculate the volume of the body:

$$\text{Volume} = \frac{\text{weight of displaced water}}{\text{unit weight of water}} = \frac{75 \text{ lb}}{62.4 \text{ lb/ft}^3} = 1.2 \text{ ft}^3 \blacktriangleright$$

$$\text{Answer is } (c)$$

304 / Fluid Mechanics

6-16 Wt. 1

If a body weighs 100 lb in air and 25 lb in fresh water, its specific gravity (to the nearest tenth) is

(a) 0.75
(b) 1.1
(c) 1.2
(d) 1.3
(e) 1.5

Solution

$$\text{Buoyant force} = (\text{weight in air}) - (\text{weight in water})$$

$$= 100 - 25 = 75 \text{ lb}$$

$$\text{Buoyant force} = \text{weight of an equal volume of water}$$

By definition,

$$\text{Specific gravity } S = \frac{\text{weight in air}}{\text{weight of equal volume of water}} = \frac{100}{75} = 1.3 \blacktriangleright$$

Answer is (d)

6-17 Wt. 4

It is said that Archimedes discovered his principle while seeking to detect a suspected fraud in the construction of a crown. The crown was thought to have been made from an alloy of gold and silver instead of from pure gold. If the crown weighed 1000 g in air and 940 g in pure water, how much gold and how much silver did it contain? Assume that the volume of the alloy was the combined volumes of the components (density of gold = 19.3 g/cm^3, density of silver = 10.5 g/cm^3).

Solution

Archimedes' principle: Any object immersed in a fluid will suffer an apparent loss of weight equal to the weight of the fluid displaced.

The weight of displaced water = 1000 − 940 = 60 g. Since water weighs 1 g/cm^3, the volume of displaced water = volume of crown = 60 cm^3.

Let

$$x = \text{weight of gold in the crown}$$

$$1000 - x = \text{weight of silver in the crown}$$

Then the volume of gold plus the volume of silver equals the volume of the crown, or

$$\frac{x}{19.3} + \frac{1000 - x}{10.5} = 60 \qquad 0.0518x + 95.24 - 0.0952x = 60$$

$$0.0434x = 35.24$$

$$\text{Weight of gold} = x = \frac{35.24}{0.0434} = 812 \text{ g} \blacktriangleright$$

$$\text{Weight of silver} = 1000 - x = 1000 - 812 = 188 \text{ g} \blacktriangleright$$

6-18 Wt. 5

A uniform solid rod 24 in. long is supported at one end by a string 6 in. above the water. If the specific gravity of the rod is $\frac{5}{9}$, find the length of the rod that is immersed in water. Assume the cross section of the rod is small.

Solution

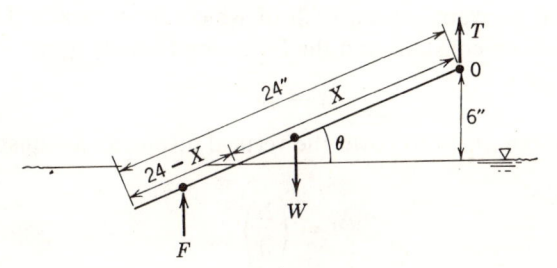

Fig. 6-9

There are three forces acting on the rod:

(1) Its weight $W = \gamma V = \frac{5}{9}\gamma_{H_2O}(24)(A)$, where A = cross-sectional area of the rod.
(2) The buoyant force $F = \gamma_{H_2O}(24 - X)(A)$.
(3) The tensile force T.

The rod will be in equilibrium if $\sum M_0 = 0$

$$W(12\cos\theta) - F\left(24 - \frac{24 - X}{2}\right)\cos\theta = 0$$

306 / Fluid Mechanics

Substituting the values of W and F in this equation, we have

$$12(\tfrac{5}{9}\gamma_{H_2O})(24A) - \gamma_{H_2O}(24 - X)(A)\left(24 - \frac{24 - X}{2}\right) = 0$$

$$\frac{12(5)(24)}{9} - (24 - X)\left(\frac{24 + X}{2}\right) = 0$$

$$160 - \frac{24^2 - X^2}{2} = 0 \qquad X^2 = 256 \qquad X = 16 \text{ in.}$$

Therefore the immersed portion of the rod is 24 in. − 16 in. = 8 in. ▶

6-19 Wt. 4

A brass weight of density 8.4 g/cm³ is dropped from the water surface in a tank. The water in the tank is 8.0 m deep. Neglecting water viscosity, find the time it takes for the weight to reach the bottom of the tank.

Solution

Since viscosity directly or indirectly causes all fluid drag, both friction drag (shear) and pressure drag (fore and aft), neglecting viscosity implies neglecting drag forces. Thus the only forces acting on the brass weight are gravity and buoyant forces, both of which are constant. Therefore, the acceleration will be constant and the following formula applies:

$$s = \tfrac{1}{2}at^2 \qquad (1)$$

or, solving for the time t to cover the vertical distance s at constant acceleration a,

$$t = \left(\frac{2s}{a}\right)^{1/2} \qquad (2)$$

Using $\sum F = ma$, we have

$$a = \frac{\sum F}{m} = \frac{\gamma_{brass}\text{Vol} - \gamma_{water}\text{Vol}}{\rho_{brass}\text{Vol}} \qquad (3)$$

where γ = specific weight
ρ = mass density
Vol = volume of the brass weight

Since $\gamma = \rho g$, equation (3) becomes

$$a = \frac{g(\rho_{brass} - \rho_{water})\text{Vol}}{\rho_{brass}\text{Vol}} = g\left(1 - \frac{\rho_{water}}{\rho_{brass}}\right) = 9.8 \text{ m/sec}^2\left(1 - \frac{1}{8.4}\right)$$

$$a = 9.8(0.881) = 8.64 \text{ m/sec}^2$$

Substituting back into equation (2),

$$t = \left(\frac{2 \times 8}{8.64}\right)^{1/2} = (1.85)^{1/2} = 1.36 \text{ sec} \blacktriangleright$$

6-20 Wt. 6

A rectangular barge 25 ft wide × 46 ft long × 8 ft deep floats in a canal lock which is 32 ft wide × 60 ft long × 12 ft deep. With no load on the barge other than its own weight, the bottom of the barge is 3 ft beneath the water surface, and the depth of the water in the lock is 7 ft. What is the new water depth in the lock if a load of steel which weighs 75 tons is added to the barge?

Solution

Barge displacement with no load = 25 × 46 × 3 = 3450 ft³
Weight of steel = weight of additional water displaced

$$\text{Volume of additional water displaced} = \frac{75 \times 2000}{62.4} = 2404 \text{ ft}^3$$

Therefore

Barge displacement with load = 3450 + 2404 = 5854 ft³

Volume of water in lock = (32 × 60 × 7) − 3450 =
13,440 − 3450 = 9990 ft³

New apparent volume of water in lock = volume of water in lock + barge displacement, loaded

= 9990 + 5854 = 15,844 ft³

Therefore

$$\text{New water depth in lock} = \frac{15,844}{32 \times 60} = 8.25 \text{ ft} \blacktriangleright$$

6-21 Wt. 1

A barge loaded with rocks floats in a canal lock with both the upstream and the downstream gates closed. If the rocks are dumped into the canal lock water with both gates still in the closed position, the water level in the lock will theoretically

(*a*) rise
(*b*) rise and then return to original level
(*c*) fall
(*d*) fall and then return to original level
(*e*) remain the same

308 / Fluid Mechanics

Solution

With the rocks in the barge, the barge displaces additional water equal to the weight of the rocks. When the rocks are thrown into the canal, they displace their own volume of water. Thus with rocks of specific gravity greater than 1, the water level in the lock will fall ▶

Answer is (c)

6-22 Wt. 1

A cylindrical water tank has a spherical dome with the dimensions shown in the figure. When the tank is full, the total force exerted by the water on the bottom of the tank is approximately

(a) 500 kips
(b) 520 kips
(c) 550 kips
(d) 590 kips
(e) 640 kips

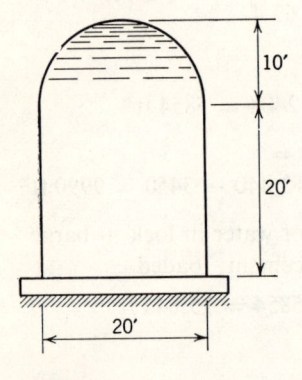

Fig. 6-10

Solution

The pressure on the bottom of the tank is due to the total head of 30 ft of water. The force is equal to the pressure times the area.

$$30 \text{ ft of water} = 30 \times 62.4 = 1872 \text{ psf}$$

$$F = PA = 1872\left(\frac{\pi}{4}\right)20^2 = 1872 \times 314.16 = 589{,}000 \text{ lb} = 589 \text{ kips} \blacktriangleright$$

Answer is (d)

6-23 Wt. 5

A hollow steel cone, with internal dimensions as shown, has a small hole at the apex. The cone is filled with water (62.4 lb/ft³). What is the minimum weight of the cone V which will prevent the water from uplifting the cone and flowing out?

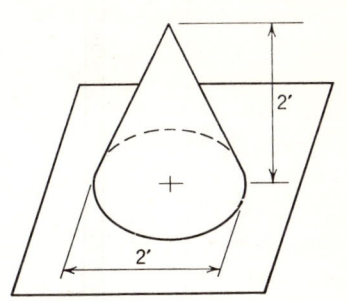

Fig. 6-11

Solution

Unit pressure at base of cone $= \gamma h = 62.4 \times 2 = 124.8$ psf

$R =$ base uplift force $= PA = 124.8 \times \pi 1^2 = 392$ lb

But since a liquid exerts a pressure normal to all surfaces with which it is in contact, there is another vertical force to be considered. The vessel in turn exerts an equal and opposite reaction on the liquid.

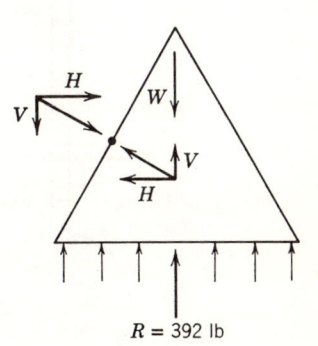

Fig. 6-12 $R = 392$ lb

$$W = V_{\text{cone}} \times \gamma_{\text{H}_2\text{O}} = \tfrac{1}{3}(\text{area of base})(h)(\gamma)$$
$$= \tfrac{1}{3}\pi 1^2 (2)(62.4) = 131 \text{ lb}$$

$\Sigma F_y = 0$

$R - V - W = 0 \qquad V = 392 - 131 = 261$ lb

For equilibrium, that is, to prevent uplift, the cone must weigh at least 261 lb ▶

6-24 Wt. 4

What is the difference in pressure between points A and B? The specific weight of fluid 1 is γ_1 and of fluid 2 is γ_2.

310 / Fluid Mechanics

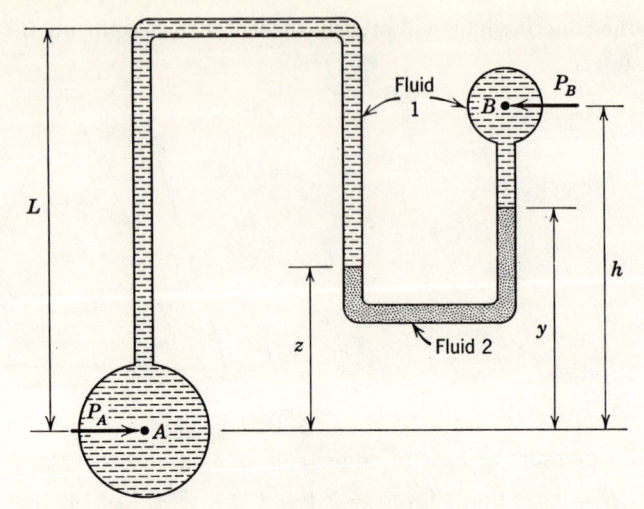

Fig. 6-13

Solution

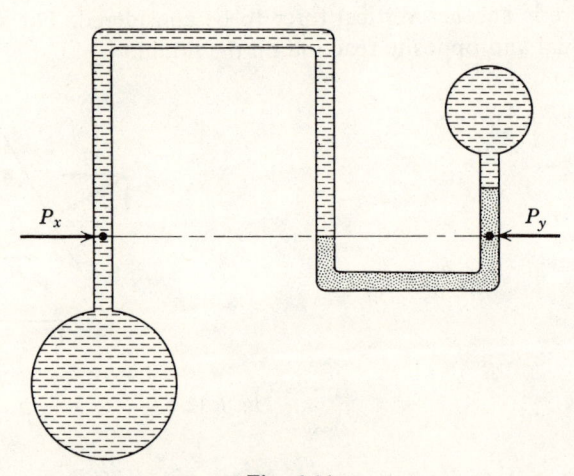

Fig. 6-14

$$\Delta P = P_A - P_B$$
$$P_B = P_y - (y - z)\gamma_2 - (h - y)\gamma_1$$
$$P_A = P_x + z\gamma_1 \quad \text{or} \quad P_A - z\gamma_1 = P_x$$

Since $P_x = P_y$,

$$P_A - z\gamma_1 = P_B + (y - z)\gamma_2 + (h - y)\gamma_1$$
$$\Delta P = P_A - P_B$$
$$= (h - y + z)\gamma_1 + (y - z)\gamma_2 \blacktriangleright$$

6-25 Wt. 1

The moment tending to overturn the dam about the toe will increase in proportion to

(a) $\sqrt{h}$
(b) h
(c) $h^{3/2}$
(d) h^2
(e) h^3

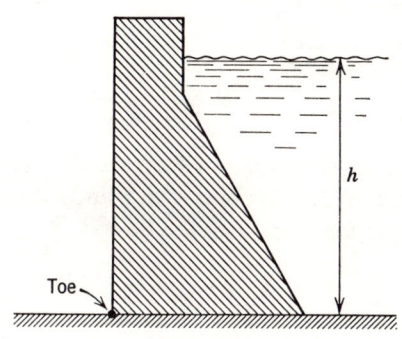

Fig. 6-15

Solution

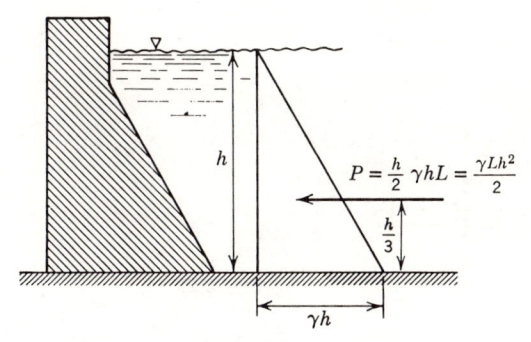

Fig. 6-16

$$L = \text{length of dam}$$

$$\text{Moment} = \frac{\gamma L h^2}{2} \times \frac{h}{3} = \frac{\gamma L h^3}{6}$$

The moment will increase in proportion to h^3 ▶

$$\text{Answer is } (e)$$

6-26 Wt. 6

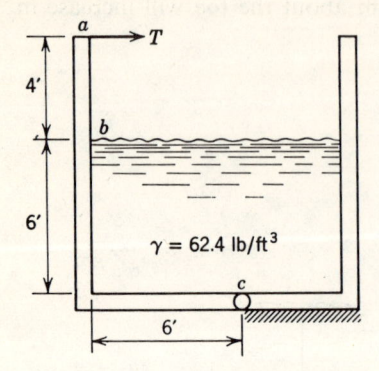

Fig. 6-17

The figure shows a cross-sectional view of a 10-ft-long rectangular water tank. The wall of the tank, abc, is hinged at c and supported by a horizontal tie rod at a. Determine the force T in the tie rod.

Solution

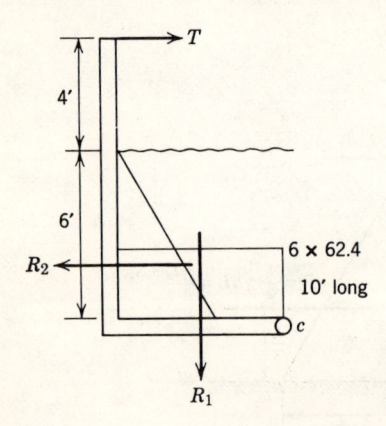

Fig. 6-18

$$R_1 = 10(6)(62.4)(6) = 22{,}460 \text{ lb}$$
$$R_2 = 10(\tfrac{1}{2})(6)(62.4)(6) = 11{,}230 \text{ lb}$$

$\sum M_c = 0$

$$2R_2 + 3R_1 - 10T = 0$$

$$T = \frac{2(11{,}230) + 3(22{,}460)}{10} = \frac{89{,}840}{10}$$

$$T = 8{,}980 \text{ lb} \blacktriangleright$$

6-27 Wt. 6

The gate AB shown in the figure rotates about an axis through B. If the width of gate is 4 ft, what torque applied to the shaft through B is required to keep the gate closed? (The unit weight of the water is 62.4 lb/ft^3.)

Fig. 6-19

Solution

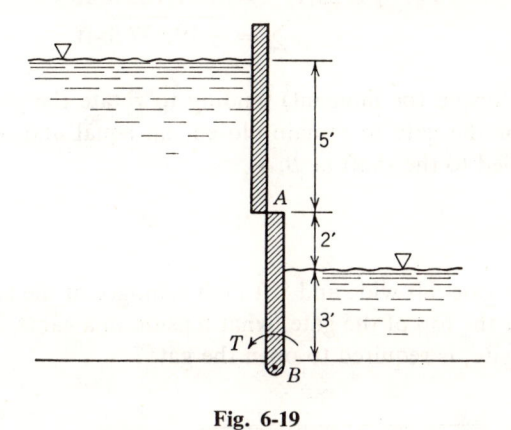

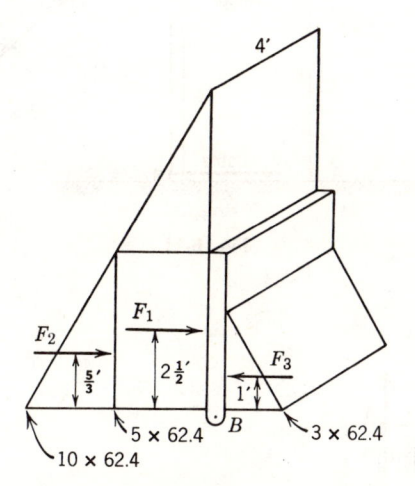

Fig. 6-20

$$F_1 = 5(62.4)(5)(4) = 6240 \text{ lb}$$
$$F_2 = \tfrac{1}{2}(5)(62.4)(5)(4) = 3120 \text{ lb}$$
$$F_3 = \tfrac{1}{2}(3)(62.4)(3)(4) = 1123 \text{ lb}$$

314 / Fluid Mechanics

$$\Sigma M_B = 0$$

$$F_1: \ -6240(2.5) = -15{,}600 \text{ lb-ft}$$
$$F_2: \ -3120(\tfrac{5}{3}) = - \ 5{,}200 \text{ lb-ft}$$
$$F_3: \ +1123(1) \ \ = + \ 1{,}123 \text{ lb-ft}$$
$$\Sigma = -19{,}677 \text{ lb-ft}$$

Therefore, the torque (or moment) tending to rotate the gate clockwise is 19,677 lb-ft. For the gate to remain closed, an equal and opposite torque T must be applied to the shaft at B. ▶

6-28 Wt. 6

A rectangular gate 5 ft wide and 8 ft high is hinged at the top. If the water level is 2 ft over the top of the gate, what tension in a cable, attached at the bottom of the gate, is required to open the gate?

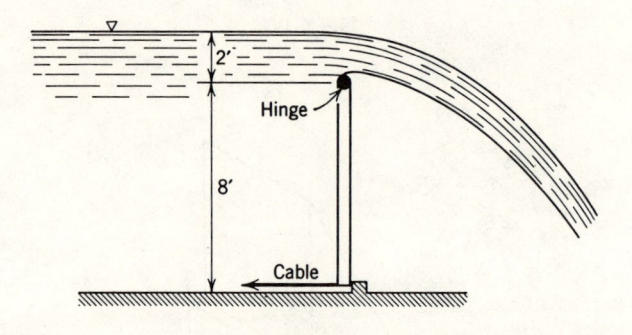

Fig. 6-21

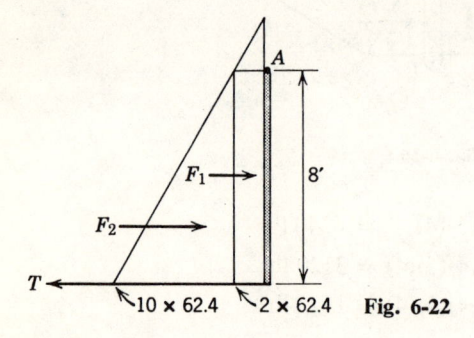

Solution

Fig. 6-22

We use the pressure distribution shown in the figure as a reasonable approximation to the true distribution. Actually the pressure is locally zero at A.

$$F_1 = 2(62.4)(8)(5) = 4992 \text{ lb}$$
$$F_2 = \tfrac{1}{2}(8)(62.4)(8)(5) = 9984 \text{ lb}$$

$\sum M_A = 0$

$$4F_1 + \tfrac{2}{3}(8)(F_2) - 8T = 0$$
$$4(4992) + \tfrac{16}{3}(9984) - 8T = 0$$
$$8T = 19{,}968 + 53{,}248 = 73{,}216 \text{ lb}$$

For equilibrium, the tension in the cable $T = 9137$ lb. Since the assumed pressure distribution very slightly overestimates the true pressure distribution, $T = 9137$ lb will open the gate ▶

6-29 Wt. 6

A dam has a flashboard AB which is pivoted at C. If it is required that the flashboard tip over when the height of the water exceeds 1 ft above B, what is the height h of the pivot C above A?

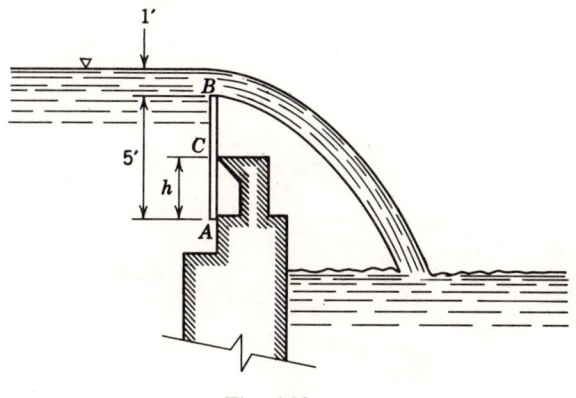

Fig. 6-23

Solution

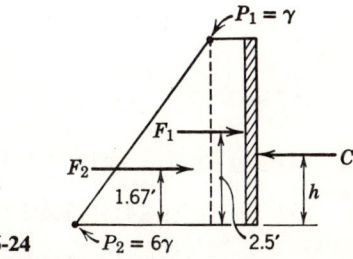

Fig. 6-24

316 / Fluid Mechanics

Use the same pressure distribution approximation as in the preceding problem.

$$F_1 = 5\gamma$$
$$F_2 = 5\gamma(\tfrac{5}{2}) = 12.5\gamma$$
$$C = F_1 + F_2 = 17.5\gamma$$

$\sum M_A = 0$

$$+5\gamma(2.5) + 12.5\gamma(1.67) - 17.5\,\gamma h = 0$$
$$12.5\,\gamma + 20.8\gamma - 17.5\gamma h = 0$$
$$h = \frac{33.3\gamma}{17.5\gamma} = 1.9 \text{ ft} \blacktriangleright$$

6-30 Wt. 6

Calculate the tension in the cord. The gate is 10 ft wide.

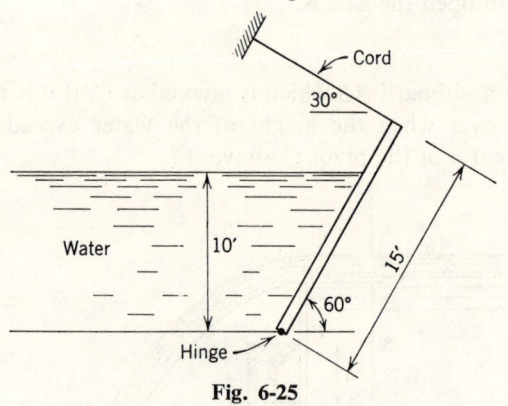

Fig. 6-25

Solution

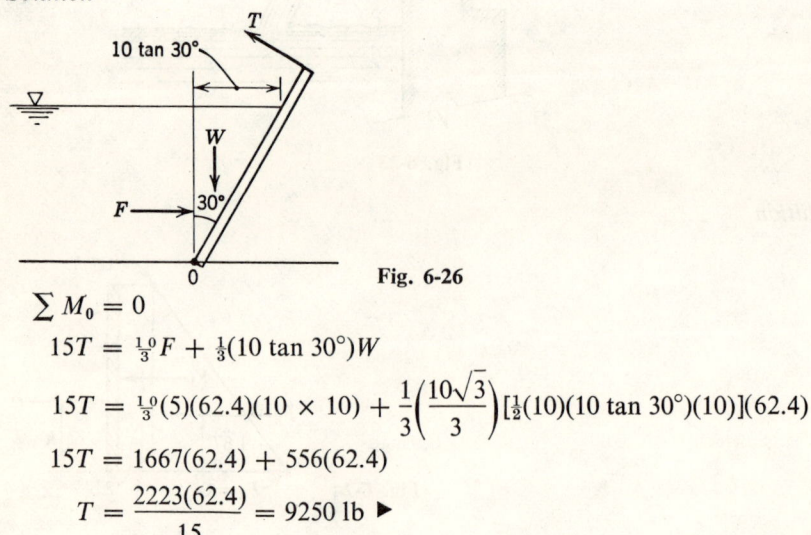

Fig. 6-26

$\sum M_0 = 0$

$$15T = \tfrac{10}{3}F + \tfrac{1}{3}(10\tan 30°)W$$

$$15T = \tfrac{10}{3}(5)(62.4)(10 \times 10) + \frac{1}{3}\left(\frac{10\sqrt{3}}{3}\right)[\tfrac{1}{2}(10)(10\tan 30°)(10)](62.4)$$

$$15T = 1667(62.4) + 556(62.4)$$

$$T = \frac{2223(62.4)}{15} = 9250 \text{ lb} \blacktriangleright$$

6-31 Wt. 5

What value of b is required to prevent overturning of the masonry dam? Masonry weighs 150 lb/ft^3 and the coefficient of friction between the bottom of the dam and the stream bed is 0.4.

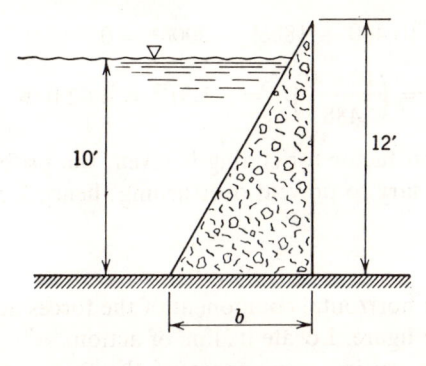

Fig. 6-27

Solution

The problem can be readily solved by calculating the horizontal and vertical components of the hydrostatic force.

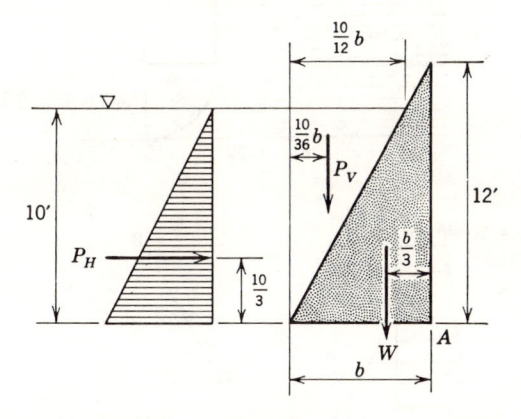

Fig. 6-28

$$P_H = \frac{\gamma h}{2} h = \frac{(62.4)(10)}{2}(10) = 3120 \text{ lb}$$

$$P_V = \frac{\gamma h}{2} h = \frac{(62.4)(10)}{2}(\tfrac{10}{12}b) = 260b \text{ lb}$$

$\Sigma M_A = 0$

$$-\tfrac{10}{3}P_H + \tfrac{26}{36}bP_V + \tfrac{b}{3}W = 0$$

$$-\tfrac{10}{3}(3120) + \tfrac{26}{36}b(260b) + \tfrac{b}{3}(900b) = 0$$

$$-10,400 + 188b^2 + 300b^2 = 0$$

$$b = \left(\frac{10,400}{488}\right)^{\frac{1}{2}} = (21.3)^{\frac{1}{2}} = 4.62 \text{ ft} \blacktriangleright$$

Although a friction factor for sliding is given, the problem asks only for the base width necessary to prevent overturning; hence a check on sliding is not needed.

6-32 Wt. 6

(*a*) Determine the horizontal component of the forces acting on the radial gate as shown in the figure. Locate its line of action.

(*b*) Determine the vertical component of the forces acting on the gate and its line of action.

(*c*) What force *F* is required to open the gate, neglecting its own weight?

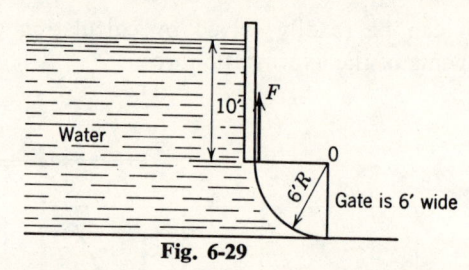

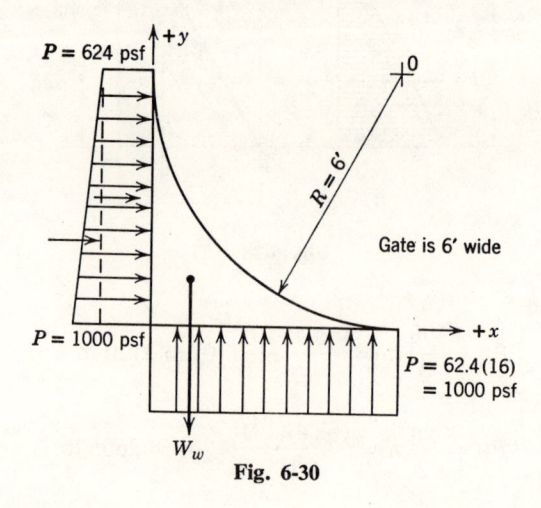

Fig. 6-29

Solution

Fig. 6-30

(*a*) Horizontal force

$$F_h = 624(6)(6) + \tfrac{376}{2}(6)(6) = 22{,}460 + 6770 = 29{,}230 \text{ lb} \blacktriangleright$$

Line of action $\bar{y} = \dfrac{\Sigma yF}{\Sigma F} = \dfrac{3(22{,}460) + 2(6770)}{29{,}230} = 2.77$ ft above base ▶

(*b*) Vertical force

$$F_v = 1000(36) - (62.4)\overbrace{\left(36 - \dfrac{36\pi}{4}\right)}^{W_w}(6) = 36{,}000 - 2900 = 33{,}100 \text{ lb} \blacktriangleright$$

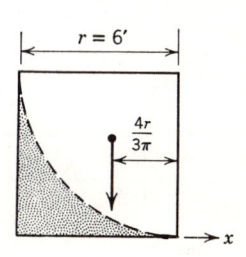

Fig. 6-31

Centroid of water area

$$\bar{x} = \dfrac{\Sigma Ax}{A} = \dfrac{6(6)(3) - \dfrac{\pi}{4}(6)^2\left[6 - \dfrac{4(6)}{3\pi}\right]}{6(6) - \dfrac{\pi}{4}(6)^2}$$

$$\bar{x} = \dfrac{108 - 97.7}{7.75} = \dfrac{10.3}{7.75} = 1.33 \text{ ft}$$

Line of action $\bar{x}$ for F_v:

$$\bar{x} = \dfrac{36{,}000(3) - 2900(1.33)}{33{,}100} = \dfrac{108{,}000 - 3860}{33{,}100}$$

$$\bar{x} = \dfrac{104{,}140}{33{,}100} = 3.15 \text{ ft} \blacktriangleright$$

(*c*) The force to open the gate is found by using $\Sigma M_0 = 0$

$$6F = -33{,}100(2.85) + 29{,}230(3.23)$$
$$= -94{,}500 + 94{,}500$$

Therefore $F = 0$ ▶

This is to be expected since the pressure forces acting directly on the cylinder have lines of action that pass through the center of rotation of the gate.

320 / Fluid Mechanics

6-33 Wt. 6

Find the resultant force and its point of application for the hydrostatic pressure acting upon the triangular gate shown. Use the water surface as the reference plane in locating the resultant force.

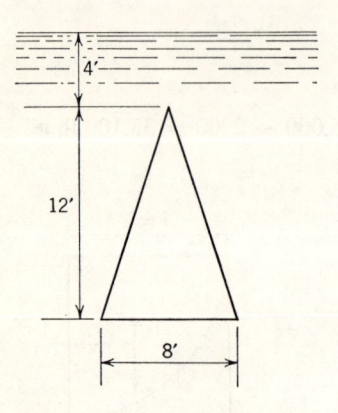

Fig. 6-32

Solution

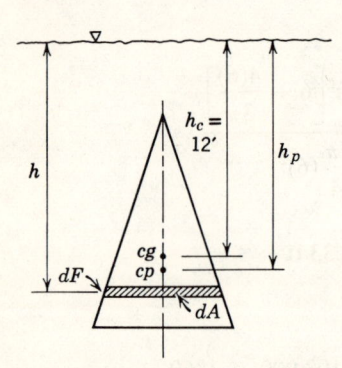

Fig. 6-33

Integrate the pressure distribution over the gate to find the force:

dF on the area $dA = \gamma h\, dA$

$$F = \int dF = \gamma \int h\, dA$$

But $\int h\, dA$ is the moment of the area or $h_c A$; hence

$$F = \gamma h_c A = 62.4(12)(\tfrac{8}{2})(12) = 36{,}000 \text{ lb} \blacktriangleright$$

The point of application of the force will next be found. Since the gate is symmetrical, the force is located on its vertical centerline. With the water surface as the reference plane, h_p is equal to the moment of the force divided by the force: $h_p = M/F$. $dM = h\,dF$ with $dF = \gamma h\,dA$, so $dM = \gamma h^2\,dA$ and $M = \gamma \int h^2\,dA$. Since $\int h^2\,dA$ is the moment of inertia of the area with respect to the water surface, $M = \gamma I_{\text{water surface}}$

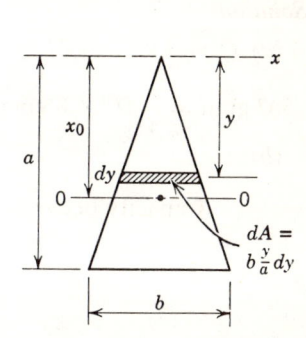

Fig. 6-34

We must now determine the moment of inertia of the triangle. Using the transfer equation: $I_{\text{water surface}} = I_0 + Ah_c^2$

$$I_x = \int y^2\,dA = \frac{b}{a}\int_0^a y^3\,dy = \frac{b}{a}\frac{y^4}{4}\Big]_0^a = \frac{ba^4}{4a} = \frac{ba^3}{4}$$

$$I_0 = I_x - Ax_0^2$$

$$I_0 = \frac{8(12)^3}{4} - \frac{8}{2}(12)(8)^2 = 3456 - 3072 = 384 \text{ ft}^4$$

$$I_{\text{water surface}} = 384 + \tfrac{8}{2}(12)(12)^2$$

$$= 384 + 6912 = 7296 \text{ ft}^4$$

$$h_p = \frac{M}{F} = \gamma\frac{I_{\text{water surface}}}{F}$$

$$= \frac{(62.4)(7296)}{36{,}000} = 12.63 \text{ ft} \blacktriangleright$$

6-34 Wt. 5

A vertical cylindrical water softener is to operate under the following conditions:

1. Water flow of 307 gpm.
2. Maximum flow rate of 8.0 gpm/ft^2 of area.
3. Supply water with a hardness of 12.0 grains/gal.
4. Softener contains 95.0 ft^3 of exchange resin.
5. Exchange value of resin is 24,000 grains/ft^3.

322 / Fluid Mechanics

Determine the

(a) diameter of softener to the nearest foot
(b) number of gallons of water softened between regenerations
(c) length of time the softener will operate before requiring regeneration

Solution

(a) $Q = AV$

$$307 \text{ gpm} = \frac{\pi}{4} D^2 \times 8 \text{ gpm/ft}^2 \qquad D = \left[\left(\frac{307}{8}\right)\left(\frac{4}{\pi}\right)\right]^{1/2} = \sqrt{48.9} = 7 \text{ ft} \blacktriangleright$$

(b)

$$\text{Capacity between regenerations} = \frac{\text{total exchange capacity}}{\text{supply water hardness}}$$

$$= \frac{24{,}000 \text{ grains/ft}^3 \times 95 \text{ ft}^3}{12 \text{ grains/gal}}$$

$$= 190{,}000 \text{ gal} \blacktriangleright$$

(c)

$$\text{Time between regenerations} = \frac{190{,}000 \text{ gallons}}{307 \text{ gpm}} \times \frac{1}{60 \text{ min/hr}} = 10.3 \text{ hrs} \blacktriangleright$$

6-35 Wt. 1

The theoretical head required to push water through a 1-in.-round orifice with a velocity of 55 fps is

(a) 40 ft
(b) 47 ft
(c) 52 ft
(d) 55 ft
(e) 94 ft

Solution

$$v = \sqrt{2gh} \qquad h = \frac{v^2}{2g} = \frac{(55)^2}{64.4} = 47 \text{ ft} \blacktriangleright$$

Answer is (b)

6-36 Wt. 3

Water is flowing through a pipe. The following data are known:

$$D = 2 \text{ in. I.D.} \qquad h_f = 20 \text{ ft}$$
$$p = 70 \text{ psig} \qquad R = 1590$$
$$n = 0.015 \qquad V = 25 \text{ fps}$$

What is the rate of flow in gallons per minute?

Solution

$$Q = VA = V\frac{\pi}{4}D^2 = (25 \text{ fps})\left(\frac{\pi}{4}\right)\left(\frac{2}{12}\text{ ft}\right)^2 = 0.545 \text{ cfs}$$

$$= 0.545 \times 60 \text{ sec/min} \times 7.48 \text{ gal/ft}^3 = 245 \text{ gpm} \blacktriangleright$$

6-37 Wt. 5

Water flows through two orifices in the side of a large water tank. The water surface in the tank is held constant. The upper orifice is 16 ft above the ground surface, and this stream strikes the ground 8 ft from the base of the tank. The stream from the lower orifice strikes the ground 10 ft from the base of the tank. (Assume $g = 32$ fps².)

(a) Find the height h_w of the water surface above the ground.
(b) Find the height d of the lower orifice above the ground.

Solution

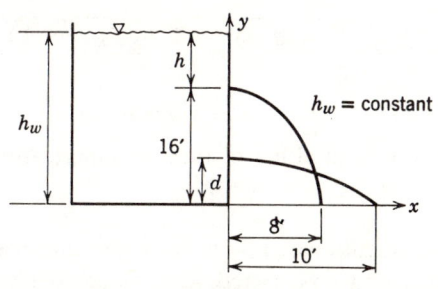

Fig. 6-35

(a) For the upper stream, $X = vt$ and $Y = \frac{1}{2}gt^2$ on the trajectory of the stream. Eliminating t, $X^2 = (2v^2/g)Y$, where v is the velocity at the vena contracta of the stream.

$$v^2 = \frac{X^2 g}{2Y} = \frac{(8^2)(32)}{(2)(16)} = 64 \qquad v = 8 \text{ fps}$$

Neglecting friction, the Bernoulli equation may be written as

$$\frac{V_1^2}{2g} + \frac{P_1}{\gamma} + h = \frac{V_2^2}{2g} + \frac{P_2}{\gamma}$$

where subscript 2 refers to the vena contracta and subscript 1 refers to the surface of the water in the tank.

$$V_2 = v = \sqrt{2g\left(h + \frac{P_1 - P_2}{\gamma} + \frac{V_1^2}{2g}\right)^{1/2}}$$

324 / Fluid Mechanics

where $P_1 = P_2 =$ atmospheric pressure and $V_1 = 0$. Therefore
$$V_2 = v = \sqrt{2gh}$$
Using the velocity we calculated above for the vena contracta, we find h as
$$8 \text{ fps} = \sqrt{2(32)(h)} \qquad h = 1 \text{ ft}$$
$$h_w = 16 + h = 17 \text{ ft} \blacktriangleright$$

(b) The velocity at the vena contracta of the lower orifice will be
$$v = \sqrt{2(32)(17 - d)}$$
Using the equation $X^2 = (2v^2/g)Y$, where now $Y = d$,
$$10^2 = \frac{2[2(32)(17 - d)]}{32}(d) \qquad \text{or} \qquad 100 = 4(17 - d)(d)$$
Solving for d,
$$4d^2 - 68d + 100 = 0$$
$$d^2 - 17d + 25 = 0$$
$$d = \frac{17 \pm \sqrt{17^2 - 4(25)}}{2} = \frac{17 \pm \sqrt{189}}{2} = \frac{17 \pm 13.8}{2}$$
$$d = 15.4 \text{ ft} \qquad \text{or} \qquad d = 1.6 \text{ ft} \blacktriangleright$$

The lower orifice may be at either of these distances above the ground.

6-38 Wt. 4

In a Venturi water meter, water flows through a constriction. The pressures at the unconstricted and constricted sections are determined by means of gages, as shown. Neglecting pipe friction, show that Q, the flow rate, is given by
$$Q = A_1 \left[\frac{2gh}{(A_1/A_2)^2 - 1} \right]^{1/2}$$
when A_1 and A_2 are the cross-sectional areas at points 1 and 2 and h indicates the difference in pressure between these two points expressed as a piezometric head difference (length).

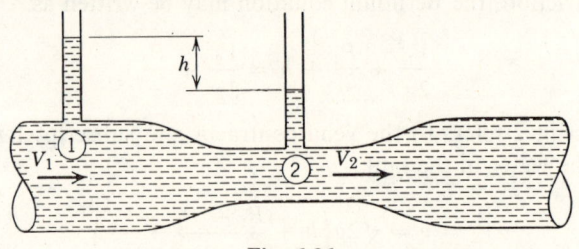

Fig. 6-36

Solution

The continuity equation is
$$Q = A_1 V_1 = A_2 V_2 \tag{1}$$
The Bernoulli equation is
$$\frac{P_1}{\gamma} + \frac{V_1^2}{2g} + z_1 = \frac{P_2}{\gamma} + \frac{V_2^2}{2g} + z_2 \tag{2}$$
In our case, $z_1 - z_2 = 0$ and $\dfrac{P_1}{\gamma} - \dfrac{P_2}{\gamma} = h$, so the equation reduces to
$$\frac{V_1^2}{2g} + h = \frac{V_2^2}{2g} \tag{3}$$
Squaring (1),
$$Q^2 = A_1^2 V_1^2 = A_2^2 V_2^2 \qquad V_1^2 = \frac{Q^2}{A_1^2} \qquad V_2^2 = \frac{Q^2}{A_2^2}$$
Substituting these values into (3),
$$\frac{Q^2}{A_1^2 2g} + h = \frac{Q^2}{A_2^2 2g}$$
Factoring,
$$Q^2 \left(\frac{1}{A_1^2} - \frac{1}{A_2^2} \right) = -2gh$$
$$Q^2 \left[\left(\frac{A_1}{A_2}\right)^2 - 1 \right] = 2gh A_1^2 \qquad Q = A_1 \left[\frac{2gh}{(A_1/A_2)^2 - 1} \right]^{1/2} \blacktriangleright$$

6-39 Wt. 6

Compute the rate of flow of water in gallons per minute for the system shown. Gage A indicates a pressure of 25 psi, and Gage B indicates a pressure of 15 psi. Assume that losses are negligible in the transition from the 6-in. pipe to the 2-in. pipe.

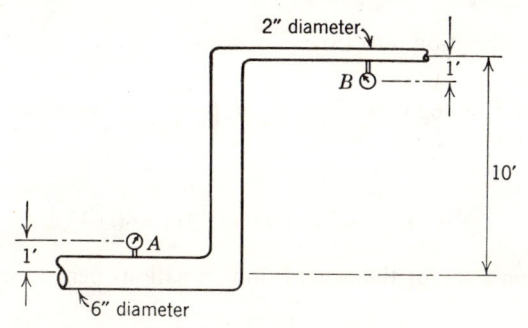

Fig. 6-37

326 / Fluid Mechanics

Solution

The Bernoulli equation is

$$\frac{P_1}{\gamma} + \frac{V_1^2}{2g} + z_1 = \frac{P_2}{\gamma} + \frac{V_2^2}{2g} + z_2 \qquad (1)$$

Also, 1 ft of water = 0.433 psi. Each of the terms can be evaluated with reference to a datum plane passing through the 6-in.-diameter pipe.

$P_6 = 25$ psig + 1 ft water = $25 + 0.433 = 25.4$ psig $\qquad z_6 = 0$

$P_2 = 15$ psig − 1 ft water = $15 − 0.433 = 14.6$ psig $\qquad z_2 = 10$ ft

$$\frac{25.4}{0.433} + \frac{V_6^2}{(2)(32.2)} + 0 = \frac{14.6}{0.433} + \frac{V_2^2}{(2)(32.2)} + 10 \qquad (2)$$

$$58.6 + 0.0155 V_6^2 = 33.7 + 0.0155 V_2^2 + 10$$

$$V_2^2 - V_6^2 = \frac{58.6 - 43.7}{0.0155} = 962 \text{ ft}^2/\text{sec}^2 \qquad (3)$$

The continuity equation shows

$$Q = A_6 V_6 = A_2 V_2 \qquad (4)$$

Using

$$A_2 = \frac{\pi}{4}\left(\frac{2}{12}\right)^2 = \frac{\pi}{144} = 0.0218 \text{ ft}^2$$

and

$$A_6 = \frac{\pi}{4}\left(\frac{6}{12}\right)^2 = \frac{9\pi}{144} = 0.1965 \text{ ft}^2$$

$$V_6 = \frac{A_2}{A_6} V_2 = \frac{0.0218}{0.1965} V_2 = 0.111 V_2 \qquad V_6^2 = 0.0123 V_2^2$$

Substituting this relation into (3),

$$V_2^2 - 0.0123 V_2^2 = 962 \text{ ft}^2/\text{sec}^2 \qquad V_2 = \left(\frac{962}{0.9877}\right)^{1/2} = (974)^{1/2} = 31.21 \text{ fps}$$

Now using (4),

$$Q = A_2 V_2 = (0.0218)(31.21) = 0.68 \text{ cfs}$$

But the problem asks for the rate of flow in gallons per minute. Here

$$Q = 0.68 \text{ cfs} \times 7.48 \text{ gal/ft}^3 \times 60 \text{ sec/min} = 306 \text{ gpm} \blacktriangleright$$

6-40 Wt. 5

In the figure below, the pipe is of uniform diameter. The gage pressure at A is 20 psi and at B is 30 psi. In which direction is the flow, and what is the head loss if the liquid has a specific weight of 30 lb/ft³?

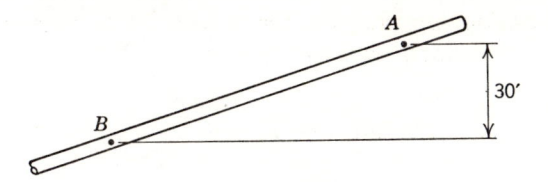

Fig. 6-38

Solution

The assumed direction of flow is from B to A. The Bernoulli equation is

$$\frac{P_B}{\gamma} + \frac{V_B^2}{2g} + z_B = \frac{P_A}{\gamma} + \frac{V_A^2}{2g} + z_A + h_L$$

Since the diameter is uniform, the continuity equation $Q = AV$ tells us that $V_B = V_A$. Therefore the Bernoulli equation reduces to

$$\frac{P_B}{\gamma} + z_B = \frac{P_A}{\gamma} + z_A + h_L$$

$$\frac{30(144)}{30} + 0 = \frac{20(144)}{30} + 30 \text{ ft} + h_L$$

$$144 = 96 + 30 + h_L \qquad \text{head loss } h_L = 18 \text{ ft} \blacktriangleright$$

Since h_L is positive, the assumed direction of flow (from B to A) is correct ▶

6-41 Wt. 5

A 6-in. I.D. pipe discharges water at an elevation 20 ft below the surface of a reservoir. If the total head loss to the point of discharge is 10 ft, compute the quantity of discharge in cubic feet per second.

Solution

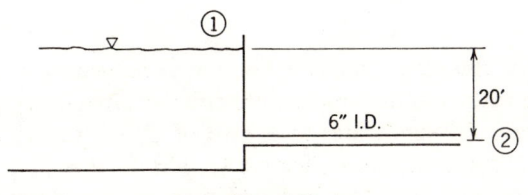

Fig. 6-39

The Bernoulli equation is

$$\frac{P_1}{\gamma} + \frac{V_1^2}{2g} + h_1 = \frac{P_2}{\gamma} + \frac{V_2^2}{2g} + h_2 + h_L$$

P_1 and V_1 are zero at the free surface of the water. P_2 is zero since the water discharges at atmospheric pressure. Using point 2 as the reference elevation, $h_2 = 0$. Thus the Bernoulli equation reduces to

$$h_1 = \frac{V_2^2}{2g} + h_L$$

$$20 = \frac{V_2^2}{2g} + 10 \qquad V^2 = 20g$$

$$V = \sqrt{20(32.2)} = 25.4 \text{ fps}$$

$$Q = AV = \frac{\pi}{4}(0.5)^2(25.4) = 5 \text{ cfs} \blacktriangleright$$

6-42 Wt. 1

Friction head in a pipe carrying water varies

(a) inversely with gravity squared
(b) directly with diameter
(c) inversely with diameter
(d) directly with velocity
(e) inversely with the coefficient of friction f

Solution

The formula for friction losses in a pipe is

$$h_L = f \frac{L}{D} \frac{V^2}{2g}$$

From the formula we see that h_L varies inversely with diameter $\blacktriangleright$

Answer is (c)

6-43 Wt. 6

Shown below is the plan view of a horizontal branching and converging water main. The main and its branches are circular pipes in which the flow is laminar. The main delivers water at a rate of Q_0 cfs to the branch lines. These lines have diameters and effective lengths of d_1, d_2, L_1, and L_2 ft, respectively. If the only head loss is due to pipe friction, determine the flow rates Q_1 and Q_2 in the branch lines for the following conditions:

$$Q_0 = 9 \text{ cfs}$$
$$d_2 = 2d_1$$
$$L_2 = 2L_1$$

Hint: The friction factor for laminar flow in a circular pipe is inversely proportional to the Reynolds number.

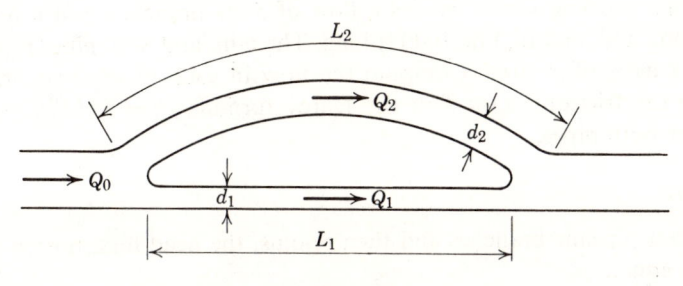

Fig. 6-40

Solution

Head loss varies directly with velocity head and pipe length and inversely with pipe diameter. With a coefficient of proportionality f (the friction factor) the equation becomes

$$h_L = f \frac{L}{d} \frac{V^2}{2g}$$

The problem also indicates that $f \propto 1/$Reynolds number. Since the Reynolds number for a given fluid is directly proportional to Vd,

$$h_L \propto \frac{1}{Vd} \frac{L}{d} \frac{V^2}{2g} \propto \frac{LV}{2d^2 g} \propto \frac{LV}{d^2}$$

The head loss in each branch must be identical, so

$$\frac{L_1 V_1}{d_1^{\,2}} = \frac{L_2 V_2}{d_2^{\,2}}$$

Substituting $d_2 = 2d_1$ and $L_2 = 2L_1$

$$\frac{L_1 V_1}{d_1^{\,2}} = \frac{2L_1 V_2}{(2d_1)^2} \qquad \frac{V_1}{V_2} = \frac{2L_1 d_1^{\,2}}{4L_1 d_1^{\,2}} = \frac{1}{2}$$

$$Q = AV \qquad Q = Q_1 + Q_2$$

330 / Fluid Mechanics

Therefore,

$$Q = Q_1 + Q_2 = \frac{\pi}{4}d_1^2 V_1 + \frac{\pi}{4}(2d_1)^2(2V_1) = 9 \qquad Q = \frac{\pi}{4}d_1^2 V_1 + \frac{8\pi}{4}d_1^2 V_1 = 9$$

$$Q_1 + 8Q_1 = 9 \qquad Q_1 = 1 \text{ cfs} \blacktriangleright$$

$$Q_2 = Q - Q_1 = 9 - 1 = 8 \text{ cfs} \blacktriangleright$$

6-44 Wt. 6

A 10-in. pipeline which carries a flow of 5 cfs branches into a 6-in. line 500 ft long and an 8-in. line 1000 ft long. The 6-in. and 8-in. pipes rejoin and continue as a 10-in. line. Compute the flow in each of the two branches. Assume the friction factor f in the Darcy formula ($h = f(L/d)(V^2/2g)$) is 0.022 for both pipes.

Solution

Where a pipeline branches and then rejoins, the head loss in each branch must be equal.

$$h_6 = \left[f\frac{L}{d}\frac{V^2}{2g}\right]_6 \qquad h_8 = \left[f\frac{L}{d}\frac{V^2}{2g}\right]_8$$

Since f and $2g$ are constants and $h_6 = h_8$,

$$\left[\frac{L}{d}V^2\right]_6 = \left[\frac{L}{d}V^2\right]_8 \qquad \text{or} \qquad \frac{500}{0.5}V_6^2 = \frac{1000}{0.67}V_8^2$$

$$V_6^2 = \frac{(1000)(0.5)}{(500)(0.67)}V_8^2 = 1.5V_8^2 \qquad V_6 = (1.5V_8^2)^{1/2} = 1.22V_8$$

For continuity, $Q_{10} = Q_6 + Q_8 = 5$ cfs and also $Q = AV$.

$$5 = \frac{\pi}{4}0.5^2 V_6 + \frac{\pi}{4}0.67^2 V_8 \qquad 5 = 0.196V_6 + 0.349V_8$$

Using $V_6 = 1.22V_8$,

$$5 = 0.196(1.22V_8) + 0.349V_8 = 0.589V_8$$

$$V_8 = \frac{5}{0.589} = 8.50 \text{ fps}$$

$$Q_8 = A_8 V_8 = (0.349)(8.50) = 2.97 \text{ cfs}$$
$$Q_6 = A_6(1.22V_8) = (0.196)(1.22)(8.50) = 2.03 \text{ cfs}$$

Flow in 8-in. line = 2.97 cfs $\blacktriangleright$

Flow in 6-in. line = 2.03 cfs $\blacktriangleright$

6-45 Wt. 1

A pump requires 100 hp to pump water (with a specific gravity of 1.0) at a certain capacity to a given elevation. What horsepower is required if the capacity and elevation conditions are the same but the fluid pumped has a specific gravity of 0.8?

(a) 60 hp
(b) 80 hp
(c) 100 hp
(d) 125 hp
(e) 130 hp

Solution

$$\text{Power} = \text{work per unit time} = \frac{\text{weight} \times \text{height}}{\text{time}}$$

If the specific gravity of the fluid is reduced from 1.0 to 0.8, then the weight is reduced by this ratio. Consequently, the horsepower will be reduced similarly to 80 hp ▶

Answer is (b)

6-46 Wt. 3

The head loss through a 24-in. butterfly valve is approximately 0.3 of a velocity head. If water is being pumped through the valve at the rate of 30 cfs for 6 months out of the year, what is the annual energy charge for pumping through the valve if the cost of energy is $0.05 per kilowatt hour?

(a) $185
(b) $200
(c) $235
(d) $270
(e) $300

Solution

$$Q = AV \qquad V = \frac{Q}{A} = \frac{30}{(\pi/4)2^2} = 9.53 \text{ fps}$$

$$h_L = 0.3 \frac{V^2}{2g} = 0.3 \frac{9.53^2}{(2)(32.2)}$$

$$= 0.424 \text{ ft}$$

$$\text{Lost hp} = \frac{Q\gamma h_L}{550} = \frac{(30)(62.4)(0.424)}{550} = 1.44 \text{ hp}$$

Since 1 hp = 0.746 kw, the cost per year is

$$(1.44)(0.746)(\tfrac{365}{2})(24)(0.05) = \$235 \blacktriangleright$$

Answer is (c)

6-47 Wt. 6

Water falling from a height of 120 ft at the rate of 1000 cfm drives a turbine which is directly connected to an electric generator. The generator rotates at 120 rpm. If the total resisting torque due to friction and other losses is 250 lb-ft and the water leaves the turbine blades with a velocity of 15 fps, find the horsepower developed by the generator. (Assume: Unit weight of water $= 62.4$ lb/ft^3 and $g = 32.2$ ft/sec^2.)

Solution

The exit velocity head is $V^2/2g = 15^2/(2)(32.2) = 3.5$ ft, so the net head is $120 - 3.5 = 116.5$ ft.

$$\text{Power} = Q\gamma h = (1000)(62.4)(116.5) = (7.27)10^6 \text{ ft-lb/min}$$

From this we must deduct the friction and other losses to obtain the generator output.

$$\text{Lost power} = 2\pi NT = (2\pi)(120)(250) = 188{,}500 \text{ ft-lb/min}$$

$$\text{Generator horsepower} = \frac{7{,}270{,}000 - 188{,}500}{33{,}000} = 214.2 \text{ hp} \blacktriangleright$$

6-48 Wt. 6

Water is supplied by pumping for a certain industrial application. Owing to fluctuations in the water demand, two identical pumps are connected in parallel as indicated in the plot of their performance. Determine the following:

(a) The range of pumping (i.e., minimum and maximum pump capacity for each pump).
(b) Horsepower required when one pump is pumping.
(c) Horsepower required when both pumps are pumping.
(d) What standard commercial motor would you select, assuming that it is permissible to overload the motor a maximum of 10%?

Solution

The maximum pump capacity is given by the intersection of the pump head and pump discharge curves:

(a) One pump: Range of capacity 0–210 gpm $\blacktriangleright$
     Two pumps: Range of capacity 0–360 gpm $\blacktriangleright$
                                   or 0–180 gpm per pump

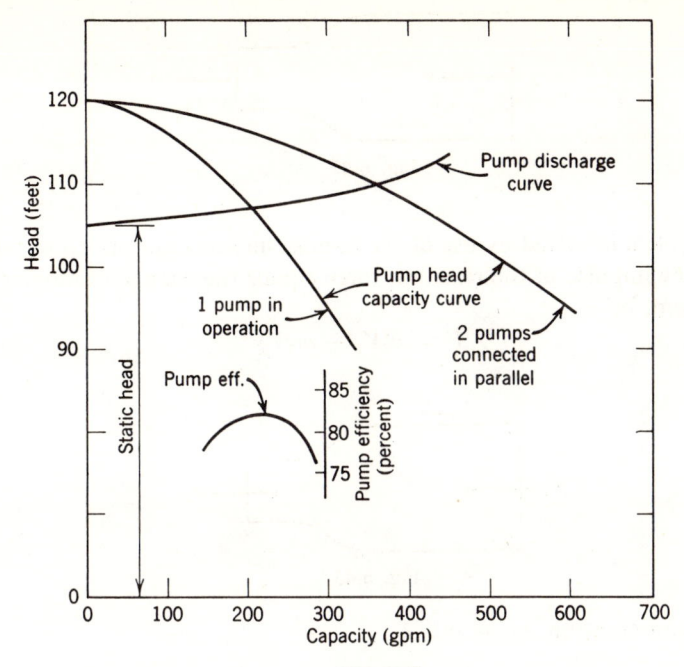

Fig. 6-41

(b)

$$\text{Work rate} = \frac{107 \text{ ft} \times 210 \text{ gpm} \times 8.33 \text{ lb/gal}}{0.82 \text{ eff} \times 33,000 \text{ ft-lb/min}}$$

$$= 6.93 \text{ hp (one pump pumping)} \blacktriangleright$$

(c)

$$\text{Work rate} = \frac{110 \text{ ft} \times 360 \text{ gpm} \times 8.33 \text{ lb/gal}}{0.81 \text{ eff} \times 33,000 \text{ ft-lb/min}}$$

$$= 12.35 \text{ hp total for both pumps or}$$
$$6.175 \text{ hp for each pump when}$$
$$\text{both pumps are pumping} \blacktriangleright$$

(d)

$$\text{Required motor size} = \frac{6.93}{1.10} = 6.3 \text{ hp}$$

Use two $7\frac{1}{2}$-hp motors, one for each pump. Use a synchronous motor of 1750 rpm ▶

6-49 Wt. 5

Given a nozzle, as shown, with a large cross-sectional area of A_1 acted on by pressure P_1 and a small area of A_2 acted on by P_2. With water flowing into the nozzle at velocity V_1 and out at V_2, calculate the external force required to hold the nozzle stationary.

334 / Fluid Mechanics

Fig. 6-42

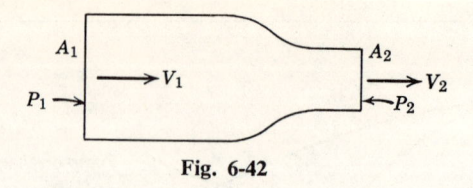

Solution

This problem is solved by use of the impulse-momentum principle, which says that the impulse of the resultant force equals the change of momentum of the system, or

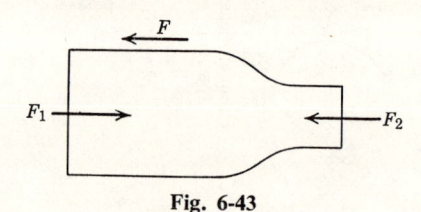

$$\sum F = m_2 V_2 - m_1 V_1$$

Fig. 6-43

For mass conservation, $m_1 = m_2$.

$$\text{Mass} = \text{area} \times \text{velocity} \times \text{density} = A_1 V_1 \frac{\gamma}{g} = A_2 V_2 \frac{\gamma}{g}$$

$$F_1 - F_2 - F = A_1 V_1 \frac{\gamma}{g} (V_2 - V_1)$$

Since $F_1 = A_1 P_1$ and $F_2 = A_2 P_2$, the external force F is

$$F = A_1 P_1 - A_2 P_2 - A_1 V_1 \frac{\gamma}{g} (V_2 - V_1) \blacktriangleright$$

6-50 Wt. 1

An open channel moves a given discharge most efficiently when the water is flowing

(a) at critical depth
(b) through a sharp crested weir
(c) so that the Reynolds number is 4200
(d) so that the depth equals one-half the width
(e) at optimum energy gradient

Solution

An open channel moves a given discharge most efficiently when the water is flowing at critical depth ▶

$$\text{Answer is } (a)$$

6-51 Wt. 1

An equation from hydraulics gives the critical velocity V_c of flow through a pipe of diameter D as

$$V_c = \frac{Kg\mu}{\gamma_w D}$$

where K = a constant
g = acceleration due to gravity
μ = coefficient of viscosity
γ_w = unit weight of water
D = pipe diameter

Select the one true statement regarding V_c:

(a) V_c varies directly as the temperature of the fluid.
(b) V_c varies inversely as the temperature of the fluid.
(c) V_c varies directly as the square root of the temperature.
(d) V_c varies inversely as the square root of the temperature.
(e) V_c is independent of the temperature of the fluid.

Solution

Here one must recognize the functional dependence of the factors on temperature. The viscosity μ is approximately inversely proportional to temperature. The unit weight of water γ_w slightly decreases with an increase in temperature (4% for 180°F change). K, g, and D are uninfluenced by temperature. Since viscosity is the only variable to be significantly affected by temperature, V_c varies inversely as the temperature of the fluid ▶

Answer is (b)

6-52 Wt. 3

A 6-ft-diameter pipe has a depth of flow of 5.6 ft. What is the hydraulic radius of the pipe for this depth of flow?

Solution

$$\text{Hydraulic radius} = \frac{\text{area}}{\text{wetted perimeter}}$$

To work this problem without tables, the easiest way is to find θ. Referring to Fig. 6-44,

$$\theta = \cos^{-1}\left(\frac{2.6}{3.0}\right) = 30°$$

336 / Fluid Mechanics

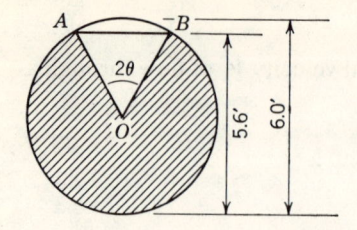

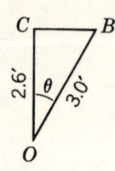

Fig. 6-44

Therefore $BC = 1.5$ ft

$$\text{Area of } \triangle OAB = 2(2.6)(1.5)\tfrac{1}{2} = 3.9 \text{ ft}^2$$

$$\text{Shaded area} = \pi R^2 (\tfrac{300}{360}) = \pi(3)^2 (\tfrac{5}{6})$$

$$= 23.6 \text{ ft}^2$$

$$\text{Area of wetted section} = 3.9 + 23.6 = 27.5 \text{ ft}^2$$

$$\text{Wetted perimeter} = \pi D (\tfrac{300}{360}) = \pi(6.0)\tfrac{5}{6}$$

$$= 15.71 \text{ ft}$$

$$\text{Hydraulic radius} = \frac{27.5}{15.71} = 1.75 \text{ ft} \blacktriangleright$$

6-53 Wt. 6

A rectangular canal 20 ft wide is designed to flow 10 ft deep at a maximum capacity of 1000 cfs. The canal has an n value of 0.014. The operating efficiency and ability of the turnouts will be impaired if the canal is not maintained near maximum depth, even under reduced capacity. This can be accomplished by placing check gates (or weirs) in the canal. What is the spacing between checks if they are placed so as not to allow the water to be lower than 1 ft below the design water surface at 20% capacity?

Solution

The Manning equation for open channel flow is

$$Q = \frac{1.49}{n} A R^{2/3} S^{1/2}$$

Under full flow conditions,

$$Q = 1000 \text{ cfs} \qquad A = (20)(10) = 200 \text{ ft}^2 \qquad n = 0.014$$

$$\text{Hydraulic radius } R = \frac{\text{area}}{\text{wetted perimeter}} = \frac{200}{20 + 2(10)} = 5 \text{ ft}$$

Substituting into the Manning equation,

$$1000 = \left(\frac{1.49}{0.014}\right)(200)(5^{2/3})S^{1/2}$$

$$S = \left[\frac{(1000)(0.014)}{(1.49)(200)(5^{2/3})}\right]^2 = \left[\frac{14}{(1.49)(200)(2.92)}\right]^2 = (0.0161)^2 = 0.00026$$

[Calculation of $5^{2/3}$: Set C scale index opposite 5 on LL3. Opposite 0.667 on C scale read answer (2.92) on LL3.]

Under the reduced flow condition,

$$A = 20\left(\frac{9+10}{2}\right) = 190 \text{ ft}^2 \qquad R = \frac{190}{20 + 2(9.5)} = 4.87 \text{ ft}$$

$$S = \left(\frac{Qn}{1.49AR^{2/3}}\right)^2 = \left[\frac{(0.2)(1000)(0.014)}{(1.49)(190)(4.87^{2/3})}\right]^2 = \left[\frac{2.8}{(1.49)(190)(2.875)}\right]^2$$

$$= 0.000012$$

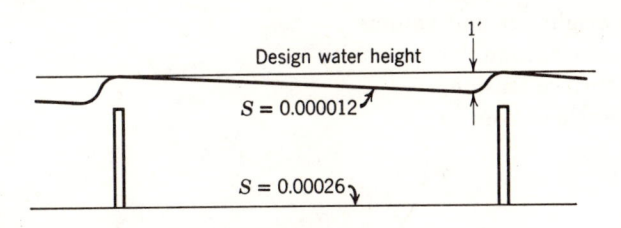

Fig. 6-45

The relation between the bottom slopes and the check spacing is shown below:

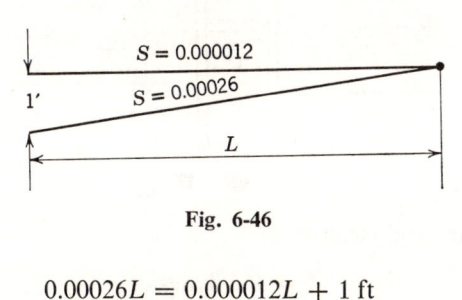

Fig. 6-46

$$0.00026L = 0.000012L + 1 \text{ ft}$$

$$0.000248L = 1 \text{ ft}$$

$$L = \frac{1}{0.000248} = 4030 \text{ ft} \blacktriangleright$$

338 / Fluid Mechanics

6-54 Wt. 1

The tendency of a free liquid surface to contract is called

(a) elasticity
(b) adhesion
(c) cohesion
(d) capillarity
(e) surface tension

Solution

This phenomenon is called surface tension ▶

Answer is (e)

6-55 Wt. 4

The general equation for the rise h in a capillary tube is

$$h = \frac{2T \cos \theta}{rdg}$$

where r = radius of tube
d = weight per unit volume
g = acceleration due to gravity
T = surface tension
θ = angle of contact

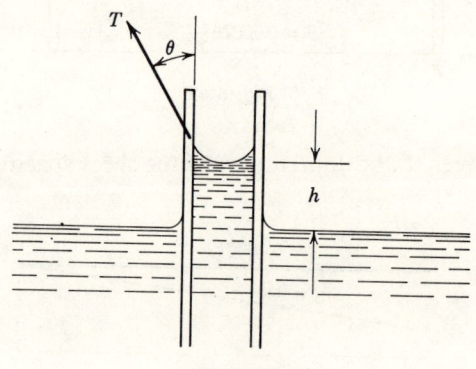

Fig. 6-47

Given for mercury and clean glass

$$T = 465 \text{ dynes/cm}$$
$$\theta = 128°$$
$$d = 13.6 \text{ g/cm}^3$$

Find h for a tube 2 mm in diameter.

Solution

$$g = 980 \text{ cm/sec}^2$$
$$\cos 128° = -\sin 38° = -0.6157$$
$$\text{Tube radius} = 1 \text{ mm} = 0.1 \text{ cm}$$
$$h = \frac{2(465)(-0.6157)}{(0.1)(13.6)(980)} = -0.43 \text{ cm} \blacktriangleright$$

The negative value tells us that the mercury does not rise in the capillary tube. Instead, the mercury falls in the tube. The illustration given in the problem is incorrect for this case. It should look like the figure below:

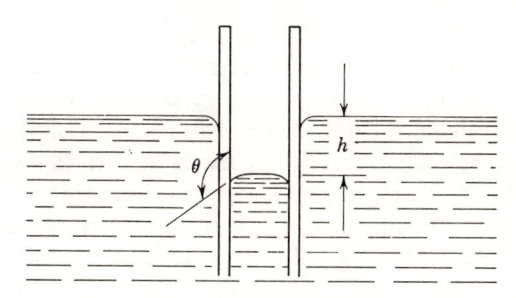

Fig. 6-48

Chapter 7

Thermodynamics

The science of thermodynamics is concerned with the relations and inter-relations of thermal and mechanical energy transfer. Engineering thermodynamics applies this knowledge to the analysis and design of a myriad of engineering devices, including engines of all kinds. Being both broadly based and practically useful, thermodynamics is founded on a small number of fundamental laws. We shall first review some of these basic principles and then look into the application of thermodynamics to some engineering problems of general interest.

We begin by reviewing the concepts of absolute temperature and absolute pressure which are used in gas law calculations both in thermodynamics and in chemistry. The first and second laws of thermodynamics are then examined. Some properties of gases and that important working fluid, steam, are reviewed. This is followed by an examination of some elements of compression processes and cycles. The chapter concludes with a brief review of some heat transfer principles.

ABSOLUTE TEMPERATURE AND PRESSURE

The temperature scales in common use around the world are relative temperature scales. Two such scales, called the Fahrenheit and Celsius or Centigrade scales, have in the past been constructed by assuming a linear temperature variation between two arbitrary temperature points, namely, the freezing and boiling points for water. The Centigrade scale was divided into 100 parts between 0°C (freezing) and 100°C (boiling); the Fahrenheit scale had 180 divisions between 32°F and 212°F.

An absolute temperature scale can be associated with each of the two foregoing relative temperature scales. The lower end of each absolute scale is called absolute zero and is the lowest temperature that can exist. One of these absolute scales, the Kelvin scale, assigns the value 273°K to the freezing point of water; the other, the Rankine scale, assigns the value 492°R to this point. Also $1K° = 1C°$ and $1R° = 1F°$. The conversion from one temperature scale to another is easily accomplished by deriving the relation from Fig. 7-1.

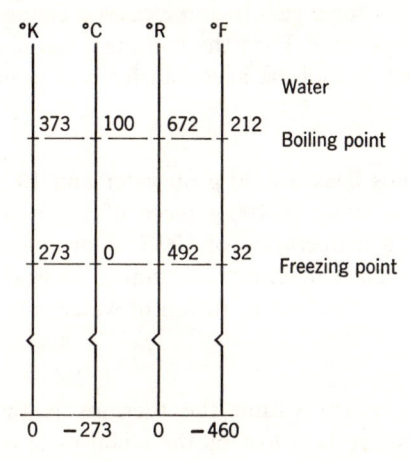

```
         °K    °C    °R    °F

                                  Water
        373   100   672   212
        ---   ---   ---   ---     Boiling point


        273    0    492    32
        ---   ---   ---   ---     Freezing point




         0   -273    0   -460
```

Fig. 7-1

Many pressure gages register differential pressures, usually the difference in pressure between the point of interest and the surrounding atmosphere; this is a gage pressure. In many problems in thermodynamics, however, it is important to use absolute pressures which are measured relative to a perfect vacuum. Assuming that atmospheric pressure is known, we can easily convert values from one system to the other since the absolute pressure is equal to the sum of the gage pressure and the local atmospheric pressure. At sea level, atmospheric pressure is normally 14.7 psia.

THERMAL EQUILIBRIUM

A body is in thermal equilibrium when it is not exchanging thermal energy in the form of heat with another body. Directly related to this is the zeroth law of thermodynamics: two bodies each in thermal equilibrium with a third body will also be in thermal equilibrium with each other. On the other hand, when two bodies are placed together when they are not in thermal equilibrium, heat transfer will take place from one body to the other so that equilibrium will eventually be established. The amount of heat transfer Q is

$$Q = mc(T_2 - T_1) \qquad (7\text{-}1)$$

where m is the mass of the body, T_2 and T_1 are, respectively, the final and initial temperatures of the body, and c is the specific heat of the body (the

342 / Thermodynamics

amount of heat transfer per unit mass per degree). Care must be taken to use consistent units in the equation. For nongases c is relatively independent of the nature of the heat transfer process. For gases this is not true; the specific heats c_p and c_v are normally given for a constant-pressure and a constant-volume process for a gas. In some cases a change of state occurs during the heat transfer process. Then the constant characteristic of the state change is the latent heat L, and the heat transfer is $Q = mL$.

Example 1

In a thermos flask are 90 g of water and 10 g of ice in equilibrium at a temperature of 0°C. A 100-g piece of metal with a specific heat of 0.40 cal/g-°C and a temperature of 100°C is dropped into the flask. What is the final equilibrium temperature, assuming no heat loss or gain to or from the surroundings? (The heat of fusion of water is 80 cal/g.)

Solution

Since the system within the thermos is thermally insulated from its surroundings, the heat lost by the metal must equal the heat gained by the ice and water. Also, the ice and water are initially in equilibrium so that all the ice will melt before the water temperature is raised. Stated mathematically with T_e as the unknown equilibrium temperature, equation (7-1) gives

$$(100 \text{ g})(0.40 \text{ cal/g-°C})(100°C - T_e)$$
$$= (10 \text{ g})(80 \text{ cal/g}) + (90 \text{ g} + 10 \text{ g})(1.0 \text{ cal/g-°C})T_e$$

or

$$4000 - 40T_e = 800 + 100T_e$$
$$140T_e = 3200$$
$$T_e = 22.8°C \blacktriangleright$$

FIRST LAW OF THERMODYNAMICS

This law is a statement that energy is conserved. It is a very general law. For different classes of problems the law may be written in varying mathematical forms for ease of application. In all cases, however, the equations state that during a given process the net amount of heat Q transferred into a system is equal to the net work output W of the system plus the change in the system internal energy ΔE, that is,

$$Q = W + \Delta E \tag{7-2}$$

Other definitions or sign conventions may also be selected here; what is important is that the net energy change of the system is the difference between the amount of energy entering and leaving the system. In this equation the internal energy is a property of the system, while heat and work generally are not properties. One exception, however, is the adiabatic process ($Q = 0$) where the net work is identified with the change in internal energy.

Most constant-mass or nonflow processes can be analyzed directly by equation (7-2), but for flow processes the terms are usually modified to make more explicit the different contributing energy terms. The foregoing work term for a flow process represents both shaft work and the flow work pv, where p is the fluid pressure and v is the fluid specific volume (reciprocal of the density ρ); let W now represent shaft work only. We also split the internal energy per unit mass into kinetic, potential, and specific internal energy. We also introduce the enthalpy $h = u + pv$. The specific internal energy is u. The steady flow energy equation, written on a unit time basis, is then

$$\left(\frac{V^2}{2g} + z + h\right)_1 \dot{m} + Q = \left(\frac{V^2}{2g} + z + h\right)_2 \dot{m} + W \qquad (7\text{-}3)$$

The mass rate of flow $\dot{m}$ is in steady flow the same for the entering and exiting stream. The subscript 1 denotes terms related to the entering fluid stream and the subscript 2 refers to the exiting stream. For additional fluid streams, additional groups of terms must be added. For only one stream entering and exiting, equation (7-3) may be divided by $\dot{m}$ to express the first law on a unit mass basis.

SECOND LAW OF THERMODYNAMICS

The second law of thermodynamics, unlike the first law, is not a conservation law. One statement of the second law, following Kelvin and Planck, says that a system cannot operate cyclically and produce a net work output while exchanging heat only at one fixed temperature. Systems whose operations violate the second law are impossible. Related to the second law are the concepts of reversibility, entropy, and thermal efficiency.

A reversible process is one where *both* the system *and* its surroundings can be restored exactly to a prior state; all other processes are irreversible. Strictly speaking, no real process is reversible, but it is a useful concept and many actual processes closely approximate this condition. One system property, the entropy s, is defined in terms of a reversible heat transfer process as

$$s_B - s_A = \Delta s_{AB} = \int_A^B ds = \int_A^B \frac{dQ_\text{rev}}{T} \qquad (7\text{-}4)$$

between any system states A and B. This definition of entropy in terms of the reversible heat transfer dQ_rev applies regardless of whether the actual process

344 / Thermodynamics

is reversible or irreversible. From equation (7-4) we note that a reversible, adiabatic ($dQ = 0$) process is isentropic ($s =$ constant). In general, any process satisfying two conditions, those of being reversible and adiabatic or isentropic, also satisfies the third condition. Finally, in all nonisentropic processes the overall entropy of the process must increase, according to the second law.

The efficiency η of a cyclical process is the ratio of the net work output to the total heat input. If Q_1 is the heat input and Q_2 is the heat output of a cycle, then the first law shows the thermal efficiency of the cycle is

$$\eta = 1 - \frac{Q_2}{Q_1} \tag{7-5}$$

The maximum thermal efficiency occurs for a reversible engine; the second law then shows the thermal efficiency to be $\eta = 1 - T_2/T_1$, where T_1 and T_2 are, respectively, the (absolute) temperature of the heat received and rejected.

Example 2

A mass flow of 200 lb of air per minute is passing through a steady flow machine. Entrance conditions are: pressure = 400 psia, specific volume = 1.387 ft³/lb, temperature = 1040°F, velocity = 100 fps. Exit conditions are: pressure = 20 psia, specific volume = 9.25 ft³/lb, temperature = 40°F, velocity = 110 fps. The transferred heat given up by the air is 40 Btu/lb. The entrance and exit connections are at the same elevation.

For air $c_v = 0.1715$, $c_p = 0.24$, $k = 1.4$, $R = 53.3$. Find the shaft horsepower. Is the work done on or by the air?

Solution

We apply the steady flow energy equation, equation (7-3), to this process. Since the equation is written on a unit time basis, Q and W represent heat transfer and work per unit time. Since the entrance and exit elevations are identical, equation (7-3) is

$$W = Q + \dot{m}\left(h_1 - h_2 + \frac{V_1^2 - V_2^2}{2g}\right)$$

The entrance and exit temperatures are $T_1 = 1040 + 460 = 1500°R$, $T_2 = 40 + 460 = 500°R$. It can be shown for perfect gases that $h = c_p T$, so that

$$h_1 - h_2 = c_p(T_1 - T_2)$$
$$h_1 - h_2 = 0.24(1500 - 500) = 240 \text{ Btu/lbm}$$

In computing the velocity terms, we must convert from mechanical to thermal units:

$$\frac{V_1^2 - V_2^2}{2g} = \frac{(100)^2 - (110)^2}{2(32.2)(778)} = -0.042 \frac{\text{Btu}}{\text{lbm}}$$

Hence

$$W = \left(-40 \frac{\text{Btu}}{\text{lbm}}\right)\left(200 \frac{\text{lbm}}{\text{min}}\right) - \left(200 \frac{\text{lbm}}{\text{min}}\right)\left(240 - 0.042\right)\frac{\text{Btu}}{\text{lbm}}$$

$$W = 40{,}000 \frac{\text{Btu}}{\text{min}}$$

$$W = \frac{\left(40{,}000 \frac{\text{Btu}}{\text{min}}\right)\left(\frac{1 \text{ min}}{60 \text{ sec}}\right)\left(778 \frac{\text{ft-lb}}{\text{Btu}}\right)}{\left(550 \frac{\text{ft-lb}}{\text{sec-hp}}\right)}$$

$W = 942$ hp done by the air ▶

STEAM TABLES

In thermodynamic analyses it is regularly necessary to know the properties of fluids which do not obey a simple equation of state. This is the case for water in the liquid and vapor (steam) states, since it is so commonly used as the working fluid in machines. Steam tables provide these data in tabular form;* much of the same data is alternately displayed in thermodynamic charts such as the Mollier diagram, which is an enthalpy-entropy chart for steam.

For liquid-vapor mixtures of water, called unsaturated steam, the value of any extensive property of the mixture, such as the specific volume v, enthalpy h, entropy s, or internal energy u, is the sum of individual property values for the liquid and vapor phases. As an example, consider specific volume. Let the subscript f indicate the saturated liquid state and the subscript g denote the saturated vapor state. The difference in specific volume between the two saturated states is v_{fg}, that is,

$$v_{fg} = v_g - v_f \tag{7-6}$$

For a mass fraction x (called quality) of vapor in a mixture, the mixture specific volume v_x is found by summing the liquid and vapor fractional

* J. H. Keenen and F. G. Keyes, *Thermodynamic Properties of Steam*, New York, Wiley, 1937. Abridged versions of these tables appear in many thermodynamics textbooks.

components; thus
$$v_x = (1-x)v_f + xv_g$$
or (7-7)
$$v_x = v_f + xv_{fg}$$
In many problems the steam quality x is not given directly but instead is determinable from two stated properties. Once x is found, the other fluid properties can also be computed in the same manner as v_x.

Example 3

One pound of a mixture of steam and water at 160 psia is contained in a rigid vessel. Heat is added to the vessel until the contents are at 560 psia and 600°F. Determine the quantity of heat, in Btu's, added to the tank contents.

Solution

Using the tables for steam at 560 psia and 600°F, we find the final enthalpy to be $h_2 = 1293.4$ Btu/lb, and the vapor is in a superheated state. The associated specific volume is $v_2 = 1.0224$ ft³/lb. Since the containing vessel is rigid, this is also the specific volume at 160 psia when the process began. At 160 psia, $v_f = 0.01815$ ft³/lb and $v_g = 2.834$ ft³/lb. Hence

$$v_1 = (1-x)v_f + xv_g$$
$$1.0237 = (1-x)(0.0182) + x(2.834)$$

The steam quality is
$$x = 0.357$$

Then the original enthalpy was
$$h_1 = (1-x)h_f + xh_g$$
$$h_1 = (1-0.357)(335.93) + (0.357)(1195.1)$$
$$h_1 = 642.7 \text{ Btu/lb}$$

Since no work can be done on a fluid in a rigid container, the heat added, according to the first law, is

$$Q = E_2 - E_1 = u_2 - u_1$$

Internal energy can be found from enthalpy $h = u + pv$, and the result is

$$Q = h_2 - h_1 - (p_2 - p_1)v = 12\,93.4 - 642.7 - (560-160)(144)(1.0224)/778$$
$$Q = 575 \text{ Btu} \blacktriangleright$$

GAS LAW RELATIONS

The behavior of many common gases is reasonably well described by a set of simple equations. These equations are called gas laws, and the gases which are assumed to obey these laws exactly are called perfect (or ideal) gases.

The equation of state for a mass m of a gas is

$$pV = mRT \qquad (7\text{-}8)$$

Here V is the volume occupied by the gas and p and T are the absolute pressure and temperature. R is a gas constant related to the universal gas constant $\bar{R}$ by the relation $R = \bar{R}/M$, M being the molecular weight of the gas. Numerically, $\bar{R} = 1544$ ft-lbf/lbm-mole-°R, and for air $R = 53.35$ ft-lbf/lbm-°R. From the state equation we obtain Charles' law $p/T = $ constant for a constant-volume process and Boyle's law $pV = $ constant for a constant-temperature process. For a given mass of gas they may be combined to give $pV/T = $ constant, which is sometimes called the universal gas law. (Additional gas law problems are presented in Chapter 8 on Chemistry.)

Some other simple relations are useful in describing further the behavior of perfect gases. For a perfect gas the specific heats c_p and c_v are each constant, and it can be further shown that $c_p - c_v = R$. Also, the internal energy and enthalpy are each directly proportional to the absolute temperature; specifically $u = c_v T$ and $h = c_p T$. By evaluating equation (7-4) for different paths, several different expressions for the entropy change of a perfect gas between two points can be derived. One relation is

$$s_2 - s_1 = c_p \ln\left(\frac{v_2}{v_1}\right) + c_v \ln\left(\frac{p_2}{p_1}\right) \qquad (7\text{-}9)$$

Equation (7-9) is useful in learning about compression processes. Using the specific heat ratio $k = c_p/c_v$, equation (7-9) can be rearranged to show that an isentropic or reversible adiabatic compression process satisfies the relation

$$pV^k = \text{constant} \qquad (7\text{-}10)$$

By using the equation of state, the alternative relations $TV^{k-1} = $ constant and $Tp^{(1-k)/k} = $ constant can be derived. The isentropic compression process is actually a special case of the more general polytropic process described by $pV^n = $ constant, where n is not equal to k. The polytropic process is generally not isentropic.

Example 4

The air pressure in an automobile tire was checked at a service station and found to be 30 psig when the temperature was 65°F. Later the same tire was checked again, and the pressure gage read 35 psi. Assuming that the atmospheric pressure of 14.7 psi did not change, what was the new temperature of the air in the tire?

348 / Thermodynamics

Solution

If we assume that the volume of the tire remained constant, a special case of the universal gas law, called Charles' law, may be applied. The law shows $p/T = $ constant for a constant-volume process. Here

$$p_1 = 30 + 14.7 = 44.7 \text{ psia}$$
$$p_2 = 35 + 14.7 = 49.7 \text{ psia}$$
$$T_1 = 65 + 460 = 525°R$$

and T_2 is to be found. Thus

$$\frac{p_1}{T_1} = \frac{p_2}{T_2}$$

$$T_2 = T_1\left(\frac{p_2}{p_1}\right) = 525\left(\frac{49.7}{44.7}\right) = 584°R$$

$$T_2 = 584 - 460 = 124°F \blacktriangleright$$

Example 5

Five pounds of a gas initially at 0.0 psig and 60°F are compressed. Later it is found that the increase in pressure was 1900% and the decrease in volume was 90%. The barometric pressure was 24.44 in. of mercury during the compression process. Find

(a) the value of n in $pV^n = $ constant
(b) the final pressure and volume if the initial volume was 10 ft³.

Solution

(a) From $pV^n = $ constant,

$$\frac{p_2}{p_1} = \left(\frac{V_1}{V_2}\right)^n$$

Here $p_2 = 19p_1$ and $V_2 = 0.10V_1$, so that

$$19 = \left(\frac{1}{0.10}\right)^n = (10)^n$$

Taking logarithms,

$$\log_{10}(19) = n \log_{10}(10) = n$$

Hence
$$n = 1.279 \blacktriangleright$$

(b) $\qquad V_2 = 0.10 V_1 = 0.10(10) = 1.0 \text{ ft}^3 \blacktriangleright$

The initial pressure $p_1 = 0.0$ psig = 24.44 in. of mercury on the absolute scale.

$$24.44 \text{ in. of mercury} = \frac{24.44}{12} \frac{(62.4)(13.55)}{144} = 12.0 \text{ psia}$$

The final pressure is therefore $p_2 = 19 p_1 = 19(12.0) = 228$ psia, or

$$p_2 = 228 - 12 = 216 \text{ psig} \blacktriangleright$$

CYCLES

Cyclical processes play a significant role in engineering thermodynamics. The characteristics of several basic cycles are briefly reviewed here. Each cycle is idealized in terms of reversible processes.

The Carnot cycle is a reversible four-process cycle consisting of alternating isothermal heat transfer processes and adiabatic compression or expansion processes. Since the second law shows that a reversible cycle is the most efficient of all cycles operating between two given temperature levels, the simple Carnot cycle is often used as a standard against which the performance characteristics of all other cycles may be measured.

The Otto, Diesel, and Brayton cycles are four-process gas cycles. In all three cycles every second process is an isentropic expansion or compression process. The Otto cycle is used for reciprocating engines; the other two processes in this cycle are constant-volume heat transfer processes. The Diesel cycle is thermodynamically like the Otto cycle except that the heat supply process occurs at constant pressure.* In the Brayton cycle both heat transfer processes occur at constant pressure; it is the standard cycle for gas turbines.

The idealized cycle which describes the operation of steam power plants is the Rankine cycle, which, of course, is a vapor cycle. The cycle consists, in turn, of an isentropic expansion process for the steam, a constant-pressure heat transfer process, an isentropic compression of the liquid, and a constant-pressure heat transfer process to the liquid. The first process could occur at any of several states (see Fig. 7-2).

State diagrams for these basic cycles are given next. A variable listed beside a process line indicates that the variable is constant during the process. The extra line on the Rankine cycle plots divides the vapor and liquid phases.

---
* In operating diesel engines the actual heat supply process is not exactly constant pressure. This is due to the inability to control the fuel injection at the rate required to give constant pressure.

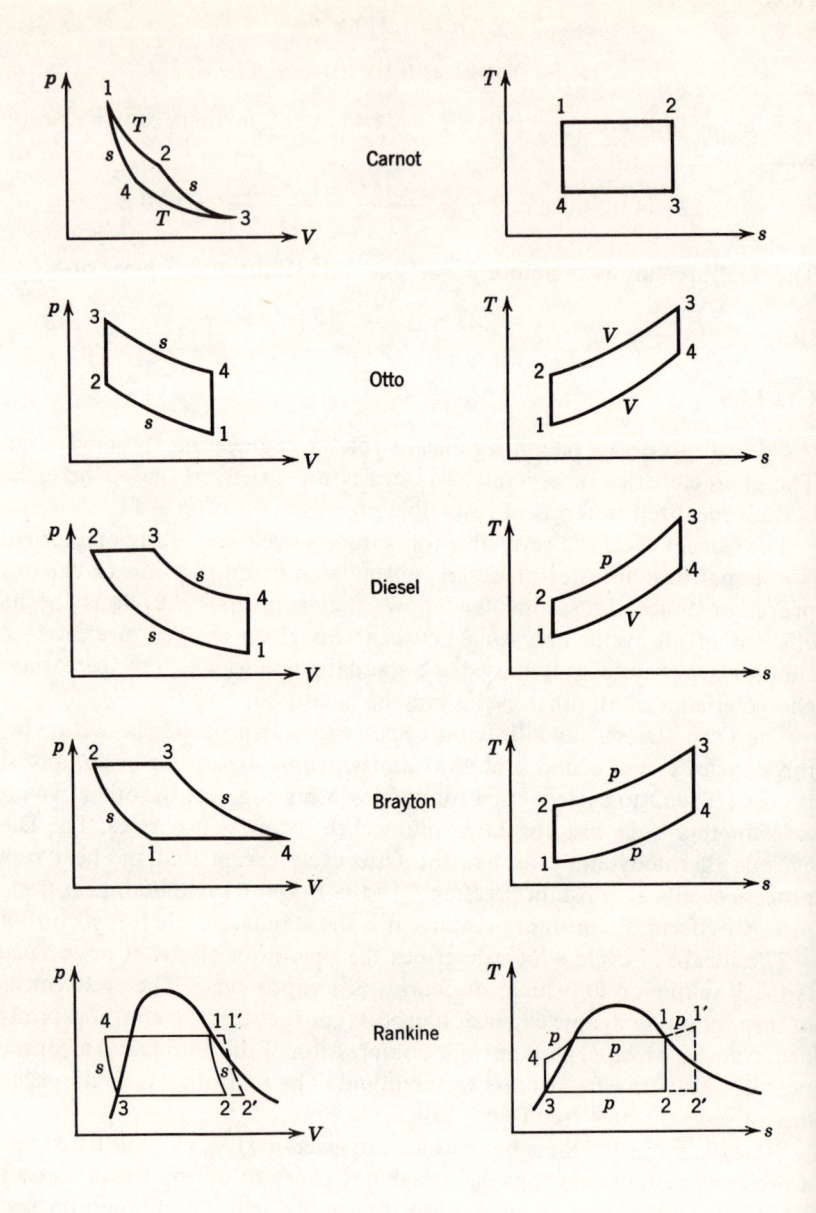

HEAT TRANSFER

Engineers often want to know not only the amount but also the rate of heat transfer during a process. For this reason the various heat transmission modes are outlined here. All heat transfer occurs by some combination of the three mechanisms of conduction, convection, and radiation.

Conductive heat transfer occurs in the absence of mass transfer. The rate of heat transfer q, in Btu/hour, is given in this case by Fourier's law

$$q = -kA\frac{dT}{dx} \qquad (7\text{-}11)$$

where k is the thermal conductivity in Btu/hr-ft-°F, A is the cross-sectional area in square feet, and dT/dx is the temperature gradient in the direction of heat flow.

Convective heat transfer involves the mass transfer of fluid from regions of high temperature to regions of lower temperature. The process is described by Newton's cooling law as

$$q = hA\,\Delta T \qquad (7\text{-}12)$$

Here h is called the convective heat transfer coefficient or film coefficient and is measured in Btu/hr-ft²-°F. ΔT is the temperature difference causing the process.

Radiation depends on the transmission of electromagnetic waves to achieve heat transfer. The rate of emission of heat is governed by the Stefan-Boltzmann law for black-body radiation, which may be written

$$q = A\sigma(T_1^4 - T_2^4) \qquad (7\text{-}13)$$

This expression applies for a small body having an area A which emits radiation at a temperature T_1 (absolute) to some point which is at temperature T_2. The Stefan-Boltzmann constant is σ. For nonblack bodies this constant is replaced by $e\sigma$, e being the dimensionless emissivity of the particular body.

Example 6

A masonry wall has a 4-in.-thick facing wall bonded to an 8-in.-thick concrete backing. On a day when the room temperature is 68°F and the outside temperature is 12°F, the inner surface temperature of the concrete is 57°F and the outer surface temperature of the brick is 19°F. See Fig. 7-3.

The thermal conductivity of the brick is 0.36 and the value for concrete is

352 / Thermodynamics

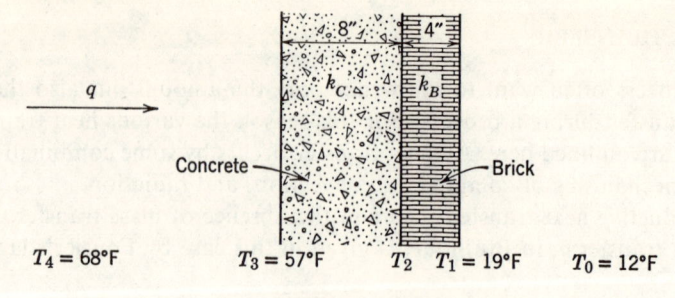

Fig. 7-3

$T_4 = 68°F$ $T_3 = 57°F$ T_2 $T_1 = 19°F$ $T_0 = 12°F$

Concrete — 8" k_C | Brick — 4" k_B

0.68 Btu/hr-ft-°F. Determine

(a) the overall heat transfer coefficient
(b) the convective heat transfer coefficients of the vertical concrete and brick walls.

Solution

Here we assume a steady-state heat flow. A combination of conduction and convection is occurring. By analogy to equations (7-11) and (7-12), an overall heat transfer coefficient U may be defined by the equation

$$q = UA\,\Delta T = UA(T_4 - T_0)$$

From equations (7-11) and (7-12) we may write several expressions for the ratio q/A:

$$\frac{q}{A} = h_C(T_4 - T_3) = \frac{k_C}{L_C}(T_3 - T_2) = \frac{k_B}{L_B}(T_2 - T_1) = h_B(T_1 - T_0)$$

Using the two conduction expressions, we can solve for the unknown temperature T_2 and then determine q/A. Then U can be computed. Thus

$$\frac{q}{A} = \frac{0.68}{\frac{8}{12}}(57 - T_2) = \frac{0.36}{\frac{4}{12}}(T_2 - 19)$$

$$1.02(57 - T_2) = 1.08(T_2 - 19)$$

$$T_2 = 37.4°F$$

and

$$\frac{q}{A} = 1.02(57 - T_2) = 1.02(57 - 37.4)$$

$$\frac{q}{A} = 20 \text{ Btu/hr-ft}^2$$

Therefore

$$U = \frac{q}{A}\frac{1}{\Delta T} = 20\frac{1}{(68-12)}$$

$$U = 0.357 \text{ Btu/hr-ft}^2\text{-°F} \blacktriangleright$$

The convective heat transfer coefficients can now be found.

$$\frac{q}{A} = h_C(T_4 - T_3) = h_B(T_1 - T_0)$$

$$20 = h_C(68 - 57) = h_B(19 - 12)$$

For the concrete,

$$h_C = \tfrac{20}{11} = 1.82 \text{ Btu/hr-ft}^2\text{-°F} \blacktriangleright$$

For the brick,

$$h_B = \tfrac{20}{7} = 2.86 \text{ Btu/hr-ft}^2\text{-°F} \blacktriangleright$$

7-1 Wt. 1

The temperature 45°C is equal to

(a) 45°F
(b) 57°F
(c) 113°F
(d) 81°F
(e) 25°F

Solution

The melting point of ice and the boiling point of water are respectively 0°C = 32°F and 100°C = 212°F. By direct proportion,

$$45°\text{C} = 32°\text{F} + 45\left(\frac{212-32}{100}\right)$$

$$= 113°\text{F} \blacktriangleright$$

This solution is equivalent to using the formula

$$°\text{C} = \tfrac{5}{9}(°\text{F} - 32)$$

Answer is (c)

7-2 Wt. 1

How much heat is required to raise 10 g of water from 0°C to 1°C?

(a) 10 Btu
(b) 1 Btu
(c) 1 cal
(d) 5 joules
(e) 10 calories

354 / Thermodynamics

Solution

One cal will raise the temperature of 1 g of water 1°C. Hence 10 g of water will require 10 cal of heat ▶

Answer is (e)

7-3 Wt. 1

An adiabatic process is one in which

(a) the pressure is constant
(b) internal energy is constant
(c) no work is done
(d) no heat is transferred
(e) friction is not considered

Solution

By definition, an adiabatic process is one in which no heat is transferred into or out of the system ▶

Answer is (d)

7-4 Wt. 3

A room experiences a heat gain of 100,000 Btu/hr and must be maintained at 80°F. How many cubic feet per minute of 64°F air is required to maintain the desired temperature?

Solution

The ratio of the quantity of heat ΔQ supplied to a body to the change in temperature ΔT experienced by that body is called the heat capacity C. The heat capacity of 1 ft^3 of air is known to be

$$C = \frac{\Delta Q}{\Delta T} = 0.018 \text{ Btu/°F}$$

Here $\Delta T = 80 - 64 = 16°F$, and the amount of heat to be removed is $100,000/60 = 1667$ Btu/min. Hence the required volume of air is

$$V = \frac{1667}{16(0.018)} = 5800 \text{ cfm} \blacktriangleright$$

7-5 Wt. 3

How much boiling water is required to melt 1000 g of ice at 0°C and produce a mixture at 20°C?

Solution

The amount of heat lost by the boiling water must equal the heat gained by the ice and cold water.

Let L = heat of fusion of ice = 80 cal/g
C = heat capacity of water = 1 cal/g-°C
M_w = mass of boiling water at 100°C
M_i = mass of ice = 1000 g
ΔT_w = decrease in hot water temperature
ΔT_i = increase in temperature for ice

Then
$$M_w C \Delta T_w = L M_i + M_i C \Delta T_i$$
$$M_w(1)(100 - 20) = 80(1000) + (1000)(1)(20 - 0)$$
$$80 M_w = 100{,}000$$
$$M_w = 1250 \text{ g of boiling water} \blacktriangleright$$

7-6 Wt. 3

200 g of a metal having a temperature of 100°C are plunged into 40 g of water at 20°C. The temperature of the water and metal becomes 48°C. If the latent heat of ice at 0°C is 80 cal/g, compute

(a) the specific heat of the metal
(b) the number of grams of ice at 0°C which can be melted by 200 g of this metal at 100°C

Solution

Assuming no heat loss to the surroundings, we equate the amount of heat lost by the metal and the amount of heat gained by the water. We employ the formula Heat = $MC \Delta T$ and we use the subscripts m for metal and w for water. Then

(a)
$$M_m C_m \Delta T_m = M_w C_w \Delta T_w$$
$$200 C_m (100 - 48) = 40(1)(48 - 20)$$

The specific heat of the metal is therefore
$$C_m = 0.108 \text{ cal/g-°C} \blacktriangleright$$

(b) Again we equate the heat lost by the metal and the heat gained by the ice. Let L = latent heat of ice. Then
$$M_m C_m \Delta T_m = L M_i$$
$$200(0.108)(100 - 0) = 80 M_i$$

Hence
$$M_i = 27 \text{ g} \blacktriangleright$$

7-7 Wt. 3

The melting rate of snow will normally be accelerated when warm rains fall. Using the following data, determine the percentage of a 50-in. snow pack that will be melted by the rain water.

Snow	Rain
Depth 50 in.	Amount 2 in.
Water content 40%	Temperature 68°F
Temperature 32°F (melting point)	
Heat of fusion 144.0 Btu/lb of water	

Assume the weight of melted snow and rain water to be 62.4 lb/ft³.

Solution

To solve the problem, we first calculate the amount of heat available in the 68°F rain. Then we calculate the amount of snow at 32°F that will be converted to water at 32°F in absorbing this amount of heat.

The quantity of heat available above 32°F, per square foot of surface, is

$$\tfrac{2}{12}(62.4)(68 - 32) = 374.4 \text{ Btu/ft}^2$$

This amount of heat will melt an amount of snow equivalent to 374.4/144.0 = 2.6 lb of water, or

$$\left(\frac{2.6}{62.4}\right)\left(\frac{1}{0.4}\right) = 0.104 \text{ ft}^3 \text{ of snow/ft}^2 \text{ of surface}$$

This equals 0.104 × 12 = 1.25 in. of snow. The percentage of the 50-in. snow pack that is melted is

$$\frac{1.25}{50}(100) = 2.5\% \blacktriangleright$$

7-8 Wt. 1

The first law of thermodynamics may be referred to as the principle of the conservation of

(a) mass
(b) momentum
(c) heat
(d) energy
(e) enthalpy

Solution

The first law of thermodynamics is a statement of the principle of energy conservation which relates the net heat Q and net work W to the internal energy change ΔE of a closed system: $\Delta E = Q - W$ ▶

Answer is (d)

7-9 Wt. 3

A perfect gas is contained in a piston-and-cylinder machine. Within the machine the pressure of the gas is always directly proportional to its volume. Initially the gas is at a pressure of 15 psia and a volume of 1 ft³. Heat is transferred reversibly to the gas until its pressure is 150 psia. If the movement of the piston is frictionless, determine the work done by the gas. Express your answer in Btu.

Solution

The gas pressure p is directly proportional to its volume V, or $p = KV$. The initial pressure p_0 is $p_0 = 15(144)$ psf absolute when the initial volume $V_0 = 1$ ft³. Hence

$$15(144) = K(1) \qquad K = 2160 \text{ lb/ft}^5$$

When the final pressure $p_1 = 150(144)$ psf absolute, the volume is

$$V_1 = \frac{p_1}{K} = \frac{150(144)}{2160} = 10 \text{ ft}^3$$

The work done is

$$W = \int_{V_0}^{V_1} p\, dV = \int_1^{10} KV\, dV = 2160 \left[\frac{V^2}{2}\right]_1^{10}$$

$$W = 2160(\tfrac{1}{2})(10^2 - 1) = 106{,}920 \text{ ft-lb}$$

The work done by the gas, in Btu's, is

$$W = \frac{106{,}920}{778} = 137.5 \text{ Btu} \blacktriangleright$$

7-10 Wt. 4

Steam enters a turbine at a velocity of 100 fps and an enthalpy of 1410 Btu/lb; it leaves the turbine at 390 fps and an enthalpy of 990 Btu/lb. Heat is lost to the surroundings at a rate of 22 Btu/lb of steam. If the steam flow rate is 75,000 lb/hr, determine the

(a) work output in Btu per pound of steam
(b) power output in kilowatts

Note: 1 kw = 3413 Btu/hr

358 / Thermodynamics

Solution

The turbine work output can be found by using the steady flow energy equation, written on a unit weight basis and neglecting changes in potential energy.

Notation: W = work done
h = enthalpy
V = velocity
q = heat added (heat loss is negative)
Subscripts: i = inlet, e = exit

Energy equation:
$$W = (h_i - h_e) + \frac{V_i^2 - V_e^2}{2g} + Q$$

$$= (1410 - 990) + \frac{(100)^2 - (390)^2}{2(32.2)(778)} - 22$$

$$= 420 - 2.84 - 22$$

$$W = 395.2 \text{ Btu/lb} \blacktriangleright$$

Power output P equals work times mass rate of flow.

$$P = \frac{(395.2)(75,000)}{3413} = 8680 \text{ kw} \blacktriangleright$$

7-11 Wt. 5

Fluid enters a turbine with a velocity of 5 fps and an enthalpy of 900 Btu/lb and leaves with an enthalpy of 850 Btu/lb and a velocity of 300 fps. Heat losses are 30 Btu/min, and the flow rate is 1 lb/sec. The inlet to the turbine is 10 ft higher than the outlet. What is the maximum theoretical horsepower that can be developed by the turbine?

Note: 1 Btu = 778 ft-lb
1 hp = 550 ft-lb/sec

Solution

The steady flow energy equation, per unit mass, is

$$\frac{V_1^2}{2g} + z_1 + h_1 + \frac{dQ}{dm} = \frac{V_2^2}{2g} + z_2 + h_2 + \frac{dW}{dm}$$

Here z = elevation above a chosen datum, and the subscripts 1 and 2 represent the turbine entrance and exit, respectively.

$$\frac{dQ}{dm} = \frac{dQ/dt}{dm/dt} = \frac{-30/60}{1/1} = -0.5 \frac{\text{Btu}}{\text{lb}}$$

Then

$$\frac{(5)^2}{2(32.2)(778)} - \frac{(300)^2}{2(32.2)(778)} + \frac{10}{778} + 900 - 850 - 0.5 = \frac{dW}{dm}$$

$$\frac{dW}{dm} = 47.7 \frac{\text{Btu}}{\text{lb}}$$

$$\frac{dW}{dt} = \left(\frac{dW}{dm}\right)\left(\frac{dm}{dt}\right) = \left(47.4 \frac{\text{Btu}}{\text{lb}}\right)\left(1 \frac{\text{lb}}{\text{sec}}\right) = 47.7 \frac{\text{Btu}}{\text{sec}}$$

Thus the maximum horsepower is

$$\frac{(47.7)(778)}{550} = 67.5 \text{ hp} \blacktriangleright$$

7-12 Wt. 1

Entropy

(a) remains constant during an irreversible process
(b) is independent of temperature
(c) is a maximum at absolute zero
(d) is a measure of unavailable energy
(e) is the reciprocal of enthalpy

Solution

Entropy is a measure of unavailable energy ▶

Answer is (d)

7-13 Wt. 3

Given the temperature-entropy diagram shown below. Find the thermal efficiency of the process.

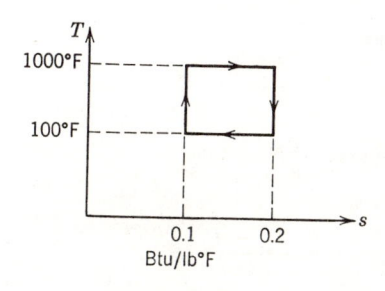

Fig. 7-4

360 / Thermodynamics

Solution

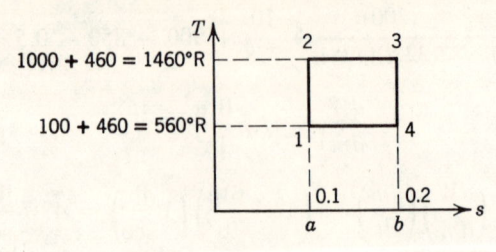

Fig. 7-5

Note that temperatures in this problem must be expressed in an absolute system, in this case degrees Rankine.

Thermal efficiency is defined as

$$\eta = \frac{W}{Q} = \frac{\text{net work}}{\text{heat added}} = \frac{\text{Area 1-2-3-4}}{\text{Area }a\text{-2-3-}b} = \frac{(0.2 - 0.1)(1460 - 560)}{(0.2 - 0.1)(1460)}$$

$$\eta = \tfrac{900}{1460} = 0.616 = 61.6\% \;\blacktriangleright$$

7-14 Wt. 3

A Carnot engine uses steam (a vapor) as the thermodynamic medium. 1000 Btu/min is supplied by a source at 500°F. The temperature of the refrigerator is 120°F. What is the efficiency of the engine? What horsepower does it develop?

Solution

Converting to degrees Rankine,

$$T_1 = 500 + 460 = 960°\text{R} \qquad T_2 = 120 + 460 = 580°\text{R}$$

The Carnot cycle could be shown on a temperature-entropy diagram.

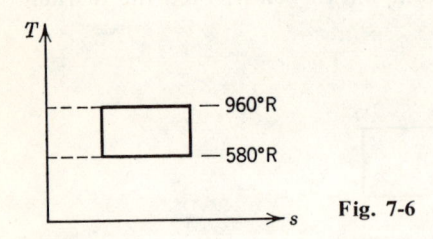

Fig. 7-6

The thermal efficiency η is

$$\eta = \frac{T_1 - T_2}{T_1} = \frac{960 - 580}{960}$$

$$\eta = 0.396 = 39.6\%$$

The net work output is then

$$\eta Q_1 = (0.396)\left(1000 \ \frac{\text{Btu}}{\text{min}}\right) = 396 \ \frac{\text{Btu}}{\text{min}}$$

Since 1 hp = 550 ft-lb/sec and 1 Btu = 778 ft-lb, this is equivalent to a horsepower output

$$\text{hp} = \frac{\left(396 \ \frac{\text{Btu}}{\text{min}}\right)\left(778 \ \frac{\text{ft-lb}}{\text{sec}}\right)}{\left(60 \ \frac{\text{sec}}{\text{min}}\right)\left(550 \ \frac{\text{ft-lb}}{\text{sec-hp}}\right)} = 9.32 \ \text{hp} \ \blacktriangleright$$

7-15 Wt. 5

Each different, physically homogeneous part of a system is called a phase. In the thermodynamic sense, the chemical composition of the phase alone is insufficient to describe the phase completely.

(a) Give at least three examples of the additional parameters necessary to specify a gaseous phase completely.

(b) What is the minimum number of these additional parameters required to specify a phase completely?

Solution

(a) Any intensive property is a suitable parameter. Examples are temperature, pressure, density, molar energy, and surface energy. Any ratio relating two extensive properties is also suitable ▶

(b) For a single component system, we require two parameters to define (i.e., specify) a single phase ▶

7-16 Wt. 3

At the place where Piccard started his ascent in the stratosphere balloon, the temperature was 17°C and the pressure 640 mm of Hg. At the highest altitude reached, the temperature was −48°C and the pressure 310 mm of Hg. If none of the gas was vented, to what fractional part of its total capacity was the balloon filled before ascending so that it would be fully expanded at the highest altitude reached?

Solution

If the air is assumed to be a perfect gas, then Boyle's law $p_1 V_1 = p_2 V_2$ and Charles' law $p_1/T_1 = p_2/T_2$ each apply. When combined, they form the universal gas law:

$$\frac{p_1 V_1}{T_1} = \frac{p_2 V_2}{T_2}$$

In this equation absolute temperatures and pressures must be used. Initially,

$$T_1 = 17°C = 17 + 273 = 290°K$$

$$p_1 = 640 \text{ mm of Hg}$$

$$V_1 \text{ is unknown}$$

At the highest altitude,

$$T_2 = -48°C = -48 + 273 = 225°K$$

$$p_2 = 310 \text{ mm of Hg}$$

$$V_2 = \text{Total capacity of balloon}$$

Hence

$$\frac{V_1}{V_2} = \frac{p_2}{p_1}\frac{T_1}{T_2} = \left(\frac{310}{640}\right)\left(\frac{290}{225}\right) = 0.625$$

and the balloon therefore was filled to 0.625 or five-eighths of its capacity before it was launched ▶

7-17 Wt. 3

An automobile tire registered a gage pressure of 28 psi when the temperature was 70°F. After driving for a while, the gage pressure was found to be 31 psi. Assuming the volume to be constant, what was the temperature of the tire at the time of the second reading? Barometric pressure was constant at 14.3 psi.

Solution

Assuming air to behave as a perfect gas while undergoing a constant-volume process, we may apply Charles' law $p_1/T_1 = p_2/T_2$. In this equation we must use absolute pressures and temperatures.

$$p_1 = 28 + 14.3 = 42.3 \text{ psia}$$

$$p_2 = 31 + 14.3 = 45.3 \text{ psia}$$

$$T_1 = 70 + 460 = 530°R$$

Thus

$$T_2 = T_1 \frac{p_2}{p_1} = 530\left(\frac{45.3}{42.3}\right) = 568°R$$

$$T_2 = 568 - 460 = 108°F \blacktriangleright$$

7-18 Wt. 6

Nitrogen is pumped into a 10-ft³ tank until the pressure gage reads 185.3 psi and the temperature of the gas is 200°F. The tank is cooled until the temperature of the nitrogen is 80°F. Assume atmospheric pressure of 14.7 psia.

(a) Determine the amount of heat removed from the nitrogen gas.
(b) Compute the final pressure indicated by the gage.

TABLE OF GAS-CONSTANT VALUES

Gas	Chemical Formula	Molecular Weight	R ft-lbf lbm-°R	c_p Btu lbm-°R	c_v Btu lbm-°R	k $\dfrac{c_p}{c_v}$
Nitrogen	N_2	28.0	55.1	0.248	0.177	1.40

Solution

We first determine the mass of the nitrogen in the tank by using the equation of state for a perfect gas $pV = mRT$, where p and T must be in absolute units.

$$m = \frac{pV}{RT} = \frac{(185.3 + 14.7)(12^2)(10)}{(55.1)(200 + 460)} = 7.92 \text{ lbm}$$

We can find the heat removed from the equation $Q = mc(T_2 - T_1)$. For this constant-volume process, the specific heat $c = c_v$, the specific heat at constant volume. Hence

$$Q = mc_v(T_2 - T_1) = (7.92)(0.177)(80 - 200) = -168.5 \text{ Btu} \blacktriangleright$$

The minus sign indicates heat removed. Using the gas law at constant volume (Charles' law) $V_1 = V_2$,

$$\frac{p_2}{p_1} = \frac{T_2}{T_1}$$

or

$$p_2 = (185.3 + 14.7)\left(\frac{80 + 460}{200 + 460}\right) = 200\left(\frac{540}{660}\right) = 163.5 \text{ psia}$$

$$p_2 = 163.5 - 14.7 = 148.8 \text{ psig} \blacktriangleright$$

7-19 Wt. 6

Two moles of oxygen at 50 psia and 40°F are in a container that is connected by a valve to a second container filled with 5 moles of nitrogen at 30 psia and 140°F. The valve is opened, and adiabatic mixing occurs. Determine the equilibrium temperature and pressure of the mixture.

TABLE OF GAS-CONSTANT VALUES

Gas	Chemical Formula	Molecular Weight	R ft-lbf lbm-°R	c_p Btu lbm-°R	c_v Btu lbm-°R	k $\dfrac{c_p}{c_v}$
Nitrogen	N_2	28.0	55.1	0.248	0.177	1.40
Oxygen	O_2	32.0	48.3	0.219	0.157	1.39

Solution

First we determine the initial volume of each gas using the state equation $pV = mRT$.

$$2 \text{ lb-moles of oxygen} = 2(32) = 64 \text{ lbm oxygen}$$
$$5 \text{ lb-moles of nitrogen} = 5(28) = 140 \text{ lbm nitrogen}$$

$$N_2: \quad V = \frac{mRT}{p} = \frac{(140)(55.1)(140+460)}{(30)(12^2)} = 1070 \text{ ft}^3$$

$$O_2: \quad V = \frac{mRT}{p} = \frac{(64)(48.3)(40+460)}{(50)(12^2)} = 215 \text{ ft}^3$$

An adiabatic process is one in which no net heat transfer to the surroundings occurs, so the heat lost by the nitrogen in mixing is gained by the oxygen. The mixing is a constant-volume process so that c_v, the specific heat at constant volume, is the proper specific heat value to use. Since the specific heat of a substance is that amount of heat which will change the temperature of 1 lb of the substance by 1°, the total heat transfer Q in this process is $Q = mc_v(T_2 - T_1)$. Let X be the final, or equilibrium, temperature of the mixture. Then

Heat lost by the nitrogen = $140(0.177)(140 - X) = 3470 - 24.8X$

Heat gained by the oxygen = $64(0.157)(X - 40) = 10.0X - 400$

$$3470 - 24.8X = 10.0X - 400$$
$$34.8X = 3870$$

The equilibrium temperature is

$$X = 111°F \blacktriangleright$$

For a mixture of gases the pressure P of the mixture is equal to the sum of the partial pressures of the constituent gases (Dalton's law). Since T and V are now common to both gases and $p = mRT/V$,

$$P = p_{N_2} + p_{O_2} = [140(55.1) + 64(48.3)]\left(\frac{111 + 460}{1070 + 215}\right)$$

$$= 4800 \text{ psf abs} = \frac{4800}{144} = 33.4 \text{ psia} \blacktriangleright$$

7-20 Wt. 1

The condensation temperature of air which is the saturation temperature corresponding to the vapor pressure is called the _____ _____ .

Solution

The answer is "dew point" ▶

7-21 Wt. 3

Given a closed cylinder (Fig. 7-7) containing 3.00 ft³ of dry air at a temperature of 60°F and a gage pressure of 10 psi. A piston very suddenly reduces the enclosed volume to 2.00 ft³. What is

(a) the new temperature of the air?
(b) the new gage pressure of the air?

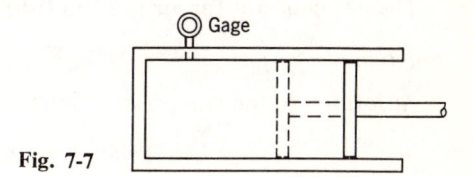

Fig. 7-7

Solution

We assume the air behaves as an ideal gas. From the description of the piston movement we may also reasonably assume the volume change to be adiabatic.

$$p_1 = 10 \text{ psig} = 10 + 14.7 = 24.7 \text{ psia} \qquad p_2 = ?$$
$$T_1 = 60°F = 60 + 460 = 520°R \qquad T_2 = ?$$
$$V_1 = 3.00 \text{ ft}^3 \qquad V_2 = 2.00 \text{ ft}^3$$

366 / Thermodynamics

The specific heat ratio for air is $k = 1.4$. From the above assumptions it can be shown that both TV^{k-1} and pV^k are constant during the compression process.

(a)
$$\frac{T_1}{T_2} = \left(\frac{V_2}{V_1}\right)^{k-1}$$

$$T_2 = T_1\left(\frac{V_1}{V_2}\right)^{k-1} = 520\left(\frac{3.00}{2.00}\right)^{1.4-1}$$

$$T_2 = 520(1.5)^{0.4} = 520(1.176)$$

$$T_2 = 612°R = 612 - 460 = 152°F \blacktriangleright$$

(b)
$$\frac{p_1}{p_2} = \left(\frac{V_2}{V_1}\right)^k$$

$$p_2 = p_1\left(\frac{V_1}{V_2}\right)^k = 24.7\left(\frac{3.00}{2.00}\right)^{1.4} = 24.7(1.765)$$

$$p_2 = 43.6 \text{ psia} = 43.6 - 14.7 = 28.9 \text{ psig} \blacktriangleright$$

7-22 Wt. 4

Air is compressed polytropically according to the relation $pV^{1.4} = C$ (constant). If 0.6 ft³ of air at atmospheric pressure (14.7 psi) and 40°F are compressed to a gage pressure of 58.8 psi, what is the volume and temperature of the air?

The gas constant for air is 53.35 ft-lbf/lbm-°R.

Solution

According to the compression relation, we have in this situation

$$(14.7)(0.6)^{1.4} = (58.8 + 14.7)V^{1.4}$$

$$(14.7)(0.49) = 73.5 V^{1.4}$$

$$V^{1.4} = 0.098 \qquad V = 0.191 \text{ ft}^3 \blacktriangleright$$

Using the gas law,

$$\frac{p_1 V_1}{T_1} = \frac{p_2 V_2}{T_2}$$

$$\frac{(14.7)(0.6)}{40 + 460} = \frac{(73.5)(0.191)}{T_2}$$

$$T_2 = 795°R = 795 - 460 = 335°F \blacktriangleright$$

7-23 Wt. 4

(a) The volume in the cylinder of a one-cylinder air compressor is 0.57 ft³ at the beginning of the compression stroke with air at atmospheric pressure. The piston compresses the air polytropically to 69.8 psig according to the law $pV^{1.35}$ = constant. What is the volume under compression?

(b) If an ideal gas is compressed from a lower pressure to a higher pressure at a constant temperature, which of the following is true?

(1) The work required will be zero.
(2) The volume remains constant.
(3) The volume will vary inversely as the absolute pressure.
(4) Heat is being absorbed.
(5) None of these.

Solution

(a) Assuming atmospheric pressure to be 14.7 psia, we have

$$p_1 V_1^{1.35} = p_2 V_2^{1.35}$$

or

$$V_2 = V_1 \left(\frac{p_1}{p_2}\right)^{1/1.35}$$

$$V_2 = (0.57)\left(\frac{14.7}{14.7 + 69.8}\right)^{1/1.35}$$

$$V_2 = (0.57)(0.174)^{1/1.35} = (0.57)(0.274)$$

$$V_2 = 0.156 \text{ ft}^3 \blacktriangleright$$

(b) For an isothermal compression the pressure-volume relation is governed by the universal gas law, which gives pV = constant. Thus the answer is

(3) The volume will vary inversely as the absolute pressure ▶

7-24 Wt. 6

Air is taken from the atmosphere at 14.7 psia and 70°F and delivered to a tank in which the pressure is 100 psia and 300°F. How much heat is removed from the air during its compression and delivery from the compressor? Consider 1 lb of air.

Solution

For a polytropic process it can be shown that

$$\frac{T_2}{T_1} = \left(\frac{p_2}{p_1}\right)^{(n-1)/n}$$

Here

$$p_1 = 14.7 \text{ psia} \qquad p_2 = 100 \text{ psia}$$
$$T_1 = 70 + 460 = 530°\text{R} \qquad T_2 = 300 + 460 = 760°\text{R}$$

$$\frac{760}{530} = \left(\frac{100}{14.7}\right)^{1-1/n}$$

$$1.433 = (6.80)^{1-1/n}$$

Taking the logarithm of each term in the equation,

$$\ln(1.433) = \left(1 - \frac{1}{n}\right)\ln(6.80)$$

$$0.360 = \left(1 - \frac{1}{n}\right)(1.92) \qquad n = 1.23$$

The work done is

$$W = \frac{R(T_2 - T_1)}{1 - n} = \frac{(53.3)(760 - 530)}{(1 - 1.23)(778)} = -68.4 \frac{\text{Btu}}{\text{lbm}}$$

The heat transfer Q can now be found by using the first law of thermodynamics. For air, $c_p = 0.24$ Btu/lb-°F.

$$Q = W + (h_2 - h_1)$$
$$Q = W + c_p(T_2 - T_1)$$
$$Q = -68.4 + 0.24(760 - 530)$$
$$Q = -13.2 \text{ Btu/lb} \blacktriangleright$$

The minus sign indicates that the heat is transferred from, not to, the air.

7-25 Wt. 6

Air enters an engine in which the expansion ratio is 5 at a temperature of 70°F. The expansion is polytropic in accordance with $pV^{1.36} = C$. What is the final temperature, and how much heat per pound was added to the air during expansion?

Note: $c_n = c_v \dfrac{k - n}{1 - n}$. For air, $c_v = 0.1715$ and $k = 1.4$.

Solution

For polytropic compression of a perfect gas

$$\frac{T_2}{T_1} = \left(\frac{V_1}{V_2}\right)^{n-1}$$

or

$$T_2 = (70 + 460)(\tfrac{1}{5})^{1.36-1} = 297°R$$
$$T_2 = 297 - 460 = -163°F \blacktriangleright$$
$$c_n = c_v \frac{k-n}{1-n} = (0.1715)\frac{1.4 - 1.36}{1 - 1.36} = -0.0191 \frac{\text{Btu}}{\text{lbm-}°R}$$

The heat added is therefore

$$Q = c_n(T_2 - T_1) = -0.0191(297 - 530) = 4.45 \text{ Btu/lbm} \blacktriangleright$$

7-26 Wt. 6

Air having an initial pressure $p_i = 14$ psia, temperature $T_i = 80°F$, and volume $V_i = 28.6$ ft³ is compressed isentropically in a nonflow process to a final pressure $p_2 = 120$ psia. Calculate

(a) V_2 (cubic feet)
(b) T_2 (degrees Fahrenheit)
(c) ΔU, the change in internal energy (Btu)
(d) work W (Btu)
(e) change in entropy ΔS
(f) heat transmitted Q (Btu)

Solution

(a) For an isentropic process $pV^k = $ constant. For air, $k = 1.4$. Hence

$$\frac{p_2}{p_i} = \left(\frac{V_i}{V_2}\right)^k$$

$$V_2 = V_i\left(\frac{p_i}{p_2}\right)^{1/k}$$

$$V_2 = 28.6(\tfrac{14}{120})^{1/1.4} = 6.15 \text{ ft}^3 \blacktriangleright$$

(b) Also for an isentropic process (with $T_i = 80 + 460 = 540°R$)

$$\frac{T_2}{T_i} = \left(\frac{p_2}{p_i}\right)^{(k-1)/k}$$

$$T_2 = 540(\tfrac{120}{14})^{(1.4-1)/1.4}$$

$$T_2 = 540(8.59)^{0.286}$$

$$T_2 = 998°R = 998 - 460 = 538°F \blacktriangleright$$

370 / Thermodynamics

(c) To determine the change in internal energy ΔU, we must find the mass of the air in this process. Using the equation of state $pV = mRT$, we have

$$m = \frac{p_i V_i}{RT_i} = \frac{\left(14 \frac{\text{lbf}}{\text{in}^2}\right)\left(144 \frac{\text{in}^2}{\text{ft}^2}\right)(28.6 \text{ ft}^3)}{\left(53.35 \frac{\text{ft-lbf}}{\text{lbm-}°\text{R}}\right)(540°\text{R})}$$

$$m = 2.00 \text{ lbm}$$

Now $\Delta U = mc_v(T_2 - T_1)$. Using $c_v = 0.1715$ Btu/lbm-°R for air,

$$\Delta U = (2.00)(0.1715)(998 - 540) = 157 \text{ Btu} \blacktriangleright$$

(d, e, f) Since an isentropic process is also a reversible adiabatic process, the heat transmitted is

$$(f) \quad Q = 0 \blacktriangleright$$

The change in entropy is then

$$(e) \quad \Delta S = 0 \blacktriangleright$$

and from the first law, the work W is equal to the negative of the change in internal energy, since $Q = \Delta U + W$.

$$(d) \quad \Delta W = -157 \text{ Btu} \blacktriangleright$$

7-27 Wt. 6

100 gal/min of kerosene is to be heated from 85°F to 195°F (zero vaporization) in an exchanger using 30 psia steam of 96% quality. The heat losses to the surrounding air have been estimated to be 3% of the heat transferred from the condensing steam to the kerosene. If the steam condensate leaves at its saturation point, how many pounds of steam per hour will be used in the exchanger?

Data:
For kerosene, specific gravity = 0.82

$$\text{specific heat} = 0.52 \text{ Btu/lbm-°F}$$

For saturated water at 30 psia, $h_f = 218.8$ Btu/lbm

$$v_f = 0.0170 \text{ ft}^3/\text{lbm}$$

For saturated steam, $h_g = 1164.0$ Btu/lbm

$$v_g = 13.76 \text{ ft}^3/\text{lbm}$$

Solution

[Note: 1 gal of kerosene weighs $\left(62.4 \dfrac{\text{lbm}}{\text{ft}^3}\right)\left(\dfrac{1 \text{ ft}^3}{7.48 \text{ gal}}\right)(0.82) = 6.83$ lbm/gal.]

The heat from the condensing steam Q_{st} will be equal to the heat gained by the kerosene Q_k plus 3% of Q_k due to the heat losses. The amount of heat gained by the kerosene is

$$Q_k = \dot{m}c_p \Delta T = \left(100 \dfrac{\text{gal}}{\text{min}} \times 60 \dfrac{\text{min}}{\text{hr}} \times 6.83 \dfrac{\text{lbm}}{\text{gal}}\right)\left(0.52 \dfrac{\text{Btu}}{\text{lbm-°F}}\right)(195 - 85)°\text{F}$$

$$= 23.45 \times 10^5 \text{ Btu/hr}$$

Therefore
$$Q_{st} = 1.03 Q_k = 24.15 \times 10^5 \text{ Btu/hr}$$

We must now calculate the heat transfer Q'_{st} per lbm of steam, which is equal to the difference in enthalpies before and after the process. At 30 psia and a steam quality $x = 96\% = 0.96$, the initial enthalpy is

$$h_x = xh_g + (1 - x)h_f$$
$$= 0.96(1164.0) + (1 - 0.96)(218.8) = 1119 + 8.76$$
$$= 1127.8 \text{ Btu/lbm}$$

Thus
$$Q'_{st} = (h_x - h_f) = (1127.8 - 218.8) = 909.0 \text{ Btu/lbm}$$

The required amount of steam in pounds per hour is then

$$\dfrac{Q_{st}}{Q'_{st}} = \dfrac{24.15 \times 10^5 \text{ Btu/hr}}{909.0 \text{ Btu/lbm}} = 2656.8 \dfrac{\text{lbm}}{\text{hr}} \blacktriangleright$$

7-28 Wt. 6

Steam enters the blades of a turbine at 300 psia and 1000°F. The discharge is at 3 psia.

(a) What is the work done in Btu per pound of steam for an isentropic expansion through the turbine?

(b) If the efficiency of the turbine is 90%, what is the quality of the exhaust?

Solution

The solution of this problem requires the use of either a Mollier diagram or steam tables.

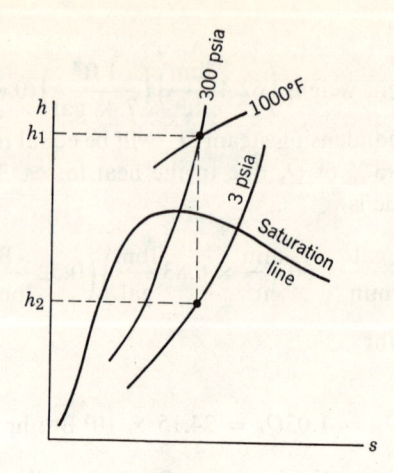

Fig. 7-8

Using a Mollier Diagram (Fig. 7-8),

$$h_1 = 1524 \text{ Btu/lbm} \quad \text{and} \quad h_2 = 1067 \text{ Btu/lbm}$$

for isentropic expansion. The work W is

$$W = h_1 - h_2 = 457 \text{ Btu/lbm} \blacktriangleright$$

If the turbine efficiency is 90%, the actual work done W' is

$$W' = 0.9W = 0.9(457) = 411 \text{ Btu/lbm}$$

Also, $W' = h_1 - h_2' = 1524 - h_2'$, so that

$$h_2' = 1524 - 411 = 1113 \text{ Btu/lbm}$$

On a diagram at the intersection of the lines $h_2' = 1113$ Btu/lbm and $p = 3$ psia, we read the steam quality x as

$$x = 99\% \blacktriangleright$$

7-29 Wt. 6

A small steam turbine is supplied with steam at 1000 psia and 100% quality, and exhausts at 14.7 psia. The turbine uses 40 lb of steam per hour for each horsepower delivered at the turbine shaft. Heat losses from the turbine to its surroundings are negligible. What is the entropy per pound of the exhaust steam?

Solution

From steam tables it is found that the initial enthalpy of the steam is $h_1 = 1191.8$ Btu/lbm. Since heat losses Q are negligible and the work done per unit mass w is

$$\frac{W}{\dot{m}} = w = \frac{1 \text{ hp}}{40 \text{ lbm steam/hr}} = (1 \text{ hp}) \frac{2545 \text{ Btu/hr-hp}}{40 \text{ lbm/hr}} = 63.6 \frac{\text{Btu}}{\text{lbm}}$$

the first law yields

$$0 = Q = W + \dot{m}(h_2 - h_1)$$
$$h_2 = h_1 - w = 1191.8 - 63.6 = 1128.2 \text{ Btu/lbm}$$

The enthalpy h_2 and the exhaust pressure $p_2 = 14.7$ psia determine the final thermodynamic state of the steam. The steam tables show that the steam quality at this state is less than 100%. Consequently we must interpolate in the tables to find the final entropy.

$$h_g = 1150.4 \qquad s_g = 1.7566$$
$$h_f = 180.1 \qquad s_f = 0.3120$$
$$\overline{h_{fg} = 970.3} \qquad \overline{s_{fg} = 1.4446}$$

Interpolating,

$$\frac{s_2 - s_g}{s_{fg}} = \frac{h_2 - h_g}{h_{fg}}$$

$$s_2 = 1.7566 + \frac{1.4446}{970.3}(1128.2 - 1150.4)$$

$$= 1.7566 - 0.0331 = 1.7235 \text{ Btu/lbm-°F} \blacktriangleright$$

7-30 Wt. 6

Three lb of steam expand isentropically (nonflow) from $p_1 = 300$ psia and $T_1 = 700°$F to $T_2 = 200°$F. Determine

(a) the quality of the steam x
(b) the work done in this process

Data: $u = h - pv/J$.

Initial conditions:

$p_1 = 300$ psia
$T_1 = 700°$F
$v_1 = $ specific volume of steam $= 2.227$ ft³/lbm
$h_1 = $ specific enthalpy of steam $= 1368.3$ Btu/lbm
$s_1 = $ specific entropy of steam $= 1.6751$ Btu/lbm-°R

374 / Thermodynamics

Final conditions:

$T_2 = 200°F$
h_f = enthalpy of water = 167.99 Btu/lbm
h_{fg} = change of enthalpy during evaporation of water = 977.9 Btu/lbm
s_f = entropy of saturated water = 0.2938 Btu/lbm-°R
s_{fg} = change of entropy during evaporation of water
    = 1.4824 Btu/lbm-°R
v_g = specific volume of steam = 33.64 ft³/lbm
p_2 = final pressure = 11.526 psia

Solution

In an isentropic (reversible, adiabatic) process, no heat is transferred ($Q = 0$), and there is no change in entropy ($\Delta s = 0$).

(a) Since the steam is initially superheated,

$$s_1 = s_2 = s_f + x s_{fg} = 0.2938 + x(1.4824) = 1.6751$$

The final quality of the steam is then

$$x = \frac{1.6751 - 0.2938}{1.4824} = 0.932 = 93.2\% \blacktriangleright$$

(b) The first law states, per pound, $Q = \Delta u + W$. Since $Q = 0$,

$$W = -\Delta u = u_1 - u_2$$

If h_2 and v_2 are found, then both specific energies can be calculated by the relation $u = h - pv/J$.

$$h_2 = h_f + x h_{fg} = 167.99 + 0.932(977.9) = 1079 \text{ Btu/lbm}$$
$$v_2 = x v_g + (1 - x) v_f$$

In this case v_f is unknown, but since $1 - x = 0.068$ and v_f is normally much smaller than v_g, it is a good approximation to neglect the term $(1 - x)v_f$ and compute v_2 as

$$v_2 = x v_g = 0.932(33.64) = 31.4 \text{ ft}^3$$

Now we can solve for u_1 and u_2:

$$u_1 = h_1 - \frac{p_1 v_1}{J} = 1368.3 - \frac{300(144)(2.227)}{778} = 1244.4 \text{ Btu/lbm}$$

$$u_2 = h_2 - \frac{p_2 v_2}{J} = 1079 - \frac{11.526(144)(31.4)}{778} = 1012 \text{ Btu/lbm}$$

Hence
$$W = u_1 - u_2 = 1244.4 - 1012 = 232.4 \text{ Btu/lbm}$$
or, for 3 lb,
$$W = 3(232.4) = 697.2 \text{ Btu} \blacktriangleright$$

7-31 Wt. 5

One lb of air completes a reversible cycle consisting of the following two processes: (1) From a volume of 2 ft³ and a temperature of 40°F, the air is compressed adiabatically to half the original volume. (2) Heat is then added at constant pressure until the original volume is reached. Calculate the heat transfer and work for each process.

Solution

Plots of the p-V and T-s planes for this process are shown in Fig. 7-9.

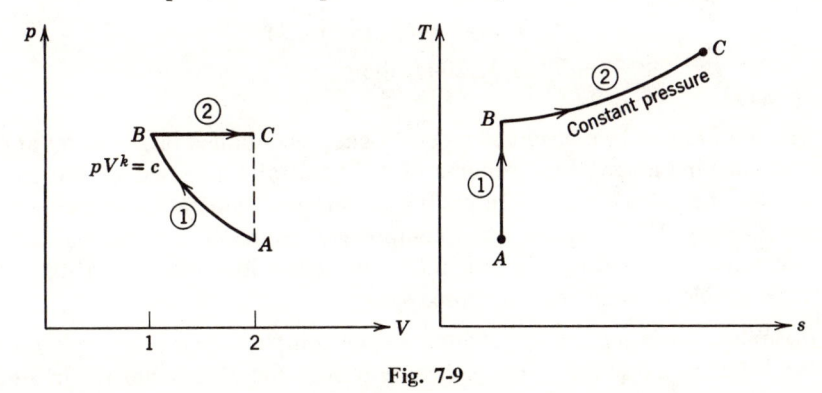

Fig. 7-9

Part 1: For a reversible adiabatic process the heat transfer $Q = 0$ ▶
The temperature change is given by

$$\frac{T_B}{T_A} = \left(\frac{V_A}{V_B}\right)^{k-1} = (2.0)^{0.4} = 1.32$$

with $k = 1.4$ for air. $T_A = 40 + 460 = 500°R$ so that $T_B = 1.32(500) = 660°R$. The work done is

$$W = mc_v(T_B - T_A) = m\frac{R}{k-1}(T_B - T_A)$$

$$W = (1 \text{ lbm})\left(\frac{53.35 \frac{\text{ft-lbf}}{\text{lbm-°R}}}{1.4 - 1}\right)(660 - 500)°R$$

$$W = 21{,}300 \text{ ft-lb} \blacktriangleright$$

Part 2: At constant pressure we may use the gas law to obtain

$$\frac{V_B}{V_C} = \frac{T_B}{T_C}$$

$$T_C = T_B \frac{V_C}{V_B} = 660(\tfrac{2}{1}) = 1320°R$$

The heat transfer Q, using $c_p = 0.24$ Btu/lbm-°R for air, is

$$Q = mc_p(T_C - T_B) = (1)(0.24)(1320 - 660)$$
$$Q = 158.5 \text{ Btu} \blacktriangleright$$

Using the equation of state,

$$p_B = p_C = \frac{mRT_B}{V_B} = \frac{(1)(53.35)(660)}{(1)} = 35,200 \text{ psf}$$

The work done is

$$W = p(V_C - V_B) = p_B(1)$$
$$W = 35,200 \text{ ft-lb} \blacktriangleright$$

7-32 Wt. 6

One lb of air, to be considered a perfect gas, is contained in a cylinder and piston machine at an initial volume of 1 ft³ and at a pressure of 100 psia (State 1). The gas is expanded reversibly at constant temperature to a volume of 2 ft³ (State 2). The gas is then compressed reversibly and adiabatically to a volume of 1 ft³ (State 3). It is then returned to its initial state (State 1) by a reversible constant-volume process.

(a) What is the net entropy change for the complete cycle?

(b) What is the net amount of heat transfer? State your answer in Btu and indicate whether it is into or out of the system.

(c) What is the net amount of work? State your answer in Btu and indicate whether work has been done on, or by, the air.

Solution

A T-s diagram of the cycle is shown in Fig. 7-10.

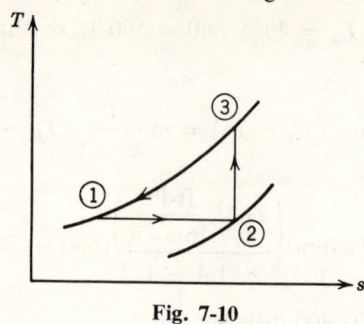

Fig. 7-10

(a) For this reversible cyclical process, the change in entropy Δs is zero, as is shown in the diagram ▶

The air mass $m = 1$ lbm. At State 1, $p_1 = 100$ psia and $V_1 = 1$ ft³. Also, $V_2 = 2$ ft³, $V_3 = 1$ ft³. Using the equation of state,

$$T_1 = \frac{p_1 V_1}{mR} = \frac{(100)(144)(1)}{(1)(53.35)} = 270°R$$

For a constant-temperature process the gas law gives

$$p_2 = p_1\left(\frac{V_1}{V_2}\right) = 100(\tfrac{1}{2}) = 50 \text{ psia}$$

Then T_3 can be found from the relation

$$\frac{T_3}{T_2} = \left(\frac{V_2}{V_3}\right)^{k-1} = \left(\frac{2}{1}\right)^{1.4-1} = 1.32$$

and

$$T_3 = 1.32 T_2 = 1.32(270) = 356°R$$

The net heat transfer is

$$\oint_{\text{cycle}} dQ = {}_1Q_2 + {}_2Q_3 + {}_3Q_1$$

$$= T_1 \frac{R}{J} \ln\left(\frac{V_2}{V_1}\right) + 0 + c_v(T_1 - T_3)$$

$$= 270\left(\frac{53.35}{778}\right) \ln(2) + (0.171)(270 - 356)$$

$$= 12.8 - 14.7 = -1.9 \text{ Btu}$$

(Here we have used $c_v = 0.171$ Btu/lbm-°R for air.)

(b) The net heat transfer = 1.9 Btu rejected by the gas ▶

Since the net internal energy change over the cycle is zero, the net heat transfer and the net work done on the gas are numerically equal. Hence

(c) The net work = 1.9 Btu done on the gas ▶

7-33 Wt. 5

A gas having a molecular weight of 36 and initially at 50 psia and 1200°F is expanded adiabatically in a turbine to a pressure of 15 psia. The specific heat ratio k of the gas is 1.25, and the isentropic efficiency of the turbine is 0.70. The gas is expanded at a steady rate of 1 lb/sec, and the gas velocity is low at both the inlet to and the exhaust from the turbine. What is the horsepower developed by the turbine?

Solution

The isentropic efficiency $\eta = W_A/W_s = 0.7$, where W_A is the actual work done and W_s is the work that would be done if the process were isentropic. We shall determine W_s and thus find W_A. For an adiabatic expansion we have

$$\frac{T_2}{T_1} = \left(\frac{p_2}{p_1}\right)^{(k-1)/k}$$

$$T_2 = (1200 + 460)(\tfrac{15}{50})^{(1.25-1)/1.25}$$

$$= (1660)(0.3)^{0.2} = 1660(0.786) = 1305°R$$

Neglecting changes in kinetic and potential energy for an ideal gas which undergoes a steady flow, adiabatic process, the isentropic work W_s is

$$W_s = \dot{m}c_p(T_1 - T_2) = \dot{m}\frac{kR}{k-1}(T_1 - T_2)$$

where $R = \bar{R}/m$ with $\bar{R}$ = universal gas constant and m = molecular weight

$$R = \frac{1544}{36} = 43.0 \, \frac{\text{ft-lbf}}{\text{lbm-°R}}$$

$$W_s = (1)\frac{(1.25)(43.0)}{1.25 - 1}(1660 - 1305)$$

$$W_s = 76{,}400 \text{ ft-lbf}$$

The actual work done is

$$W_A = \eta W_s = \frac{0.7(76{,}400)}{550 \, \frac{\text{ft-lbf}}{\text{sec-hp}}}$$

$$W_A = 97.1 \text{ hp} \blacktriangleright$$

7-34 Wt. 6

A turbine, which is part of a Rankine cycle, receives steam at a pressure of 180 psia and 400°F from the boiler. The turbine exhausts at a pressure of 5 psia. Neglecting pump work, calculate

(a) the heat efficiency of the cycle
(b) the pounds of steam required per horsepower-hour

Assume that the turbine operates adiabatically and reversibly.

Solution

Since the turbine process is both adiabatic and reversible, it is also isentropic so that $s_1 = s_2$.

From the tables for superheated steam at 180 psia and 400°F we find $h_1 = 1214.0$ Btu/lbm and $s_1 = 1.5745$ Btu/lbm-°R. At 5 psia, the tables give

$$h_f = 130.13 \text{ Btu/lbm} \qquad h_{fg} = 1001.0 \text{ Btu/lbm}$$
$$s_f = 0.2347 \text{ Btu/lbm-°R} \qquad s_{fg} = 1.6094 \text{ Btu/lbm-°R}$$

The steam quality x is determinable from the equation

$$s_1 = s_2 = s_f + x s_{fg}$$
$$1.5745 = 0.2347 + x(1.6094) \qquad \text{or} \qquad x = 0.833 = 83.3\%$$

Now we can find the final enthalpy h_2:

$$h_2 = h_f + x h_{fg}$$
$$h_2 = 130.13 + (0.833)(1001.0)$$
$$h_2 = 963.96 \text{ Btu/lbm}$$

(a) The heat efficiency $\eta = \dfrac{\text{net work output}}{\text{total heat input}}$

$$\eta = \frac{h_1 - h_2}{h_1 - h_{f2}} = \frac{1214.0 - 963.96}{1214.0 - 130.13} = 0.23 = 23\% \blacktriangleright$$

(b) The steam rate $w = \dfrac{\text{work/hp-hr}}{\text{work/lbm steam}}$

$$w = \frac{2545}{h_1 - h_2} = \frac{2545}{1214.0 - 963.96} = 10.2 \text{ lbm steam/hp-hr} \blacktriangleright$$

7-35 Wt. 4

Explain, using pertinent thermodynamic relationships, why it is desirable for gas-compressor inlet temperatures to be low.

Solution

For a flow process, the first law of thermodynamics may be written on a unit mass basis as

$$h_1 + Q + \frac{V_1^2}{2g} = h_2 + W + \frac{V_2^2}{2g}$$

In a compression process, heat transfer is so small that it may be neglected. Also, inlet and outlet sizes are chosen so that the change in kinetic energy is relatively small. For these conditions the required work is

$$W = h_1 - h_2 = -\int_1^2 dh$$

For any thermodynamic process

$$dQ_{rev} = T\,ds = dh - v\,dp$$

For zero heat transfer we then have $dh = v\,dp$ and

$$W = -\int_1^2 v\,dp$$

For a given pressure rise, a smaller average value for v will result in a smaller work requirement, as is shown in Fig. 7-11.

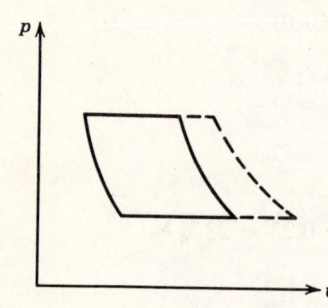

Fig. 7-11

In this figure we see clearly that a larger v will increase the net area under the cycle curve and cause a larger unit work requirement. If the inlet gas pressure is fixed, then a low inlet temperature results in a low specific volume, which in turn results in a theoretically low required energy input to the compressor.

7-36 Wt. 6

A nozzle is designed to expand air from 100 psia and 90°F to 20 psia. Assume an isentropic expansion and a zero initial velocity. The air flow rate is 3 lb/sec. Calculate

(a) the exit velocity (feet per second)
(b) the proper exit cross-sectional area (square inches)

Note: The universal gas constant $\bar{R} = 1544$ ft-lbf/lb-mole-°R. For air, $c_p = 0.24$ Btu/lbm-°R, $c_v = 0.171$ Btu/lbm-°R, and the molecular weight of air is $m = 29$. For an isentropic expansion,

$$\frac{T_2}{T_1} = \left(\frac{p_2}{p_1}\right)^{(k-1)/k} = \left(\frac{v_1}{v_2}\right)^{k-1}$$

Solution

(a) Applying the first law to the nozzle flow,

$$\frac{V_1^2}{2gJ} + h_1 = \frac{V_2^2}{2gJ} + h_2$$

Noting that the initial velocity $V_1 = 0$,

$$V_2 = [2gJ(h_1 - h_2)]^{1/2}$$

which for an ideal gas can be expressed as

$$V_2 = \left[2gJc_p T_1\left(1 - \frac{T_2}{T_1}\right)\right]^{1/2}$$

Since the expansion is isentropic and the initial and final pressures are known, we use the isentropic relation to obtain (using $k = c_p/c_v = 1.4$)

$$V_2 = \left\{2gJc_p T_1\left[1 - \left(\frac{p_2}{p_1}\right)^{(k-1)/k}\right]\right\}^{1/2}$$

$$V_2 = \{2(32.2)(778)(0.24)(90 + 460)[1 - (\tfrac{20}{100})^{(1.4-1)/1.4}]\}^{1/2}$$

$$V_2 = \{6.62(10^6)[1 - 0.631]\}^{1/2} = 1560 \text{ fps} \blacktriangleright$$

(b) Since $\dfrac{T_2}{T_1} = \left(\dfrac{p_1}{p_2}\right)^{(k-1)/k} = 0.631$,

$$T_2 = 0.631(90 + 460) = 347°R$$

By the equation of state, the specific volume v_2 is

$$v_2 = \frac{RT_2}{p_2} = \frac{\bar{R}T_2}{mp_2} = \frac{(1544)(347)}{(29)(20)(144)}$$

$$v_2 = 6.42 \text{ ft}^3/\text{lbm}$$

The weight rate of flow is 3 lbm/sec, so by conservation of mass

$$w = \frac{A_2 V_2}{v_2}$$

or

$$A_2 = \frac{wv_2}{V_2} = \frac{3(6.42)}{1560} = 0.0123 \text{ ft}^2$$

$$A_2 = (0.0123)(144) = 1.78 \text{ in}^2 \blacktriangleright$$

7-37 Wt. 6

(a) Draw a schematic diagram of a typical refrigeration system showing the major components, as well as the state of the refrigerant between the components (that is, liquid or gas).

(b) If the refrigeration system has a 10-ton capacity while operating with a 40°F suction temperature and a condensing temperature of 102°F, how many gallons per minute of cooling water are required for condensing if a 15°F temperature rise is allowed?

Assume the theoretical heat added to the vapor during compression is 29.6 Btu/min per ton of refrigeration and a ton of refrigeration equals 12,000 Btu/hr.

Solution

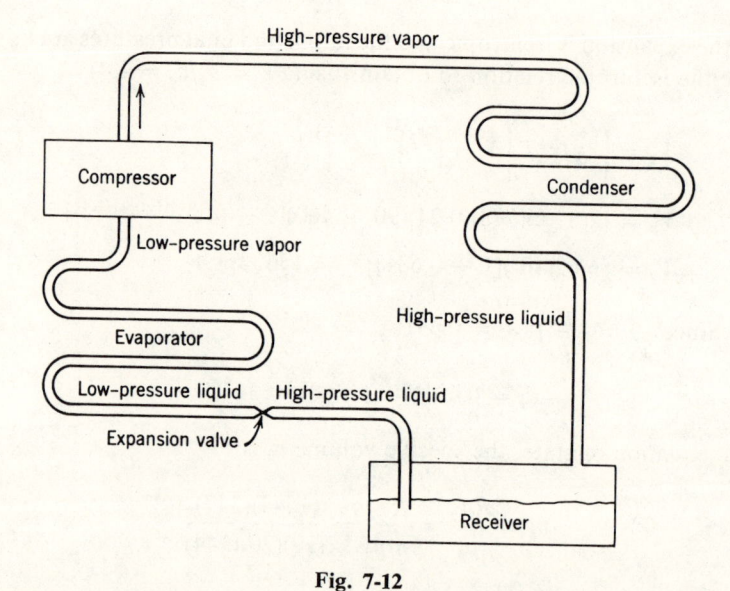

Fig. 7-12

The heat rejected by the system must equal the heat entering the system plus the work done on the system. The heat in is

$$\left(\frac{12{,}000 \text{ Btu/hr}}{1 \text{ ton}}\right)\left(\frac{1 \text{ hr}}{60 \text{ min}}\right)(10 \text{ ton}) = 2000 \frac{\text{Btu}}{\text{min}}$$

The heat rejected is then $2000 + 29.6(10) = 2296$ Btu/min. This heat must be accepted by the cooling water with a coolant temperature rise of 15°F. Since 1 Btu will raise 1 lb of water 1°F, the flow rate of cooling water must be

2296/15 = 153 lb/min of water. One gal of water weighs

$$\left(62.4 \frac{lb}{ft^3}\right)\left(231 \frac{in^3}{gal}\right)\left(\frac{1}{1728} \frac{ft^3}{in^3}\right) = 8.33 \frac{lb}{gal}$$

Hence 153/8.33 = 18.4 gal/min of cooling water are required ▶

7-38 Wt. 3

A counterflow heat exchanger is operating with the hot liquid entering at 400°F and leaving at 327°F. The cool liquid enters at 100°F and leaves at 283°F. What is the logarithmic mean temperature difference between the hot and cold liquids?

Solution

A diagram of the process is shown in Fig. 7-13.

T_h in 400°
T_c out 283°
327° T_h out
100° T_c in

Counterflow

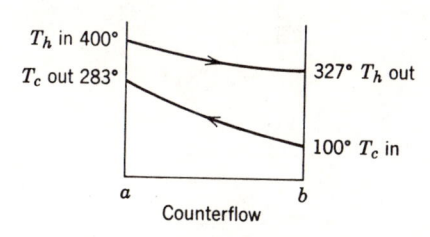

Fig. 7-13

The logarithmic mean temperature difference ΔT_m, as used in heat exchanger calculations, is defined as

$$\Delta T_m = \frac{\Delta T_a - \Delta T_b}{\ln(\Delta T_a/\Delta T_b)}$$

Here

$$\Delta T_a = 400 - 283 = 117°$$
$$\Delta T_b = 327 - 100 = 227°$$

Hence

$$\Delta T_m = \frac{117 - 227}{\ln(117/227)} = \frac{-110}{-0.663} = 166°F \blacktriangleright$$

7-39 Wt. 4

A windowless wall of a house 8 ft high by 15 ft long is of wood frame construction with a ¾-in stucco exterior, 3 in. of insulation, and a ½-in. plaster-board interior. The outside temperature is 100°F and the inside temperature is 75°F.

384 / Thermodynamics

Compute the Btu per hour of heat flow into the room. (You may ignore the effects of the studs and both the inside and outside surface conductances.) The conductivities of the stucco and the insulation are 12.00 and 0.27, respectively. The conductance of the plasterboard is 2.82.

Solution

A cross section of the wall is shown in Fig. 7-14.

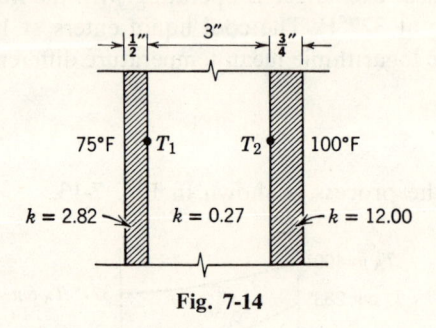

Fig. 7-14

The basic heat conduction equation is

$$Q = k \frac{A}{L}(T_a - T_b) \frac{\text{Btu}}{\text{hr}}$$

where k = conductivity in Btu-in./ft^2-°F-hr
A = wall area in square feet
L = wall thickness in inches
$T_a - T_b$ = temperature difference in degrees Fahrenheit

For steady heat flow through the wall, the heat flow Q is identical in each layer of material. Thus

$$(100 - T_2) = \frac{0.75}{12} \frac{Q}{A}$$

$$(T_2 - T_1) = \frac{3}{0.27} \frac{Q}{A}$$

$$(T_1 - 75) = \frac{0.5}{2.82} \frac{Q}{A}$$

Adding these three equations gives

$$(100 - 75) = \frac{Q}{A}\left[\frac{0.75}{12} + \frac{3}{0.27} + \frac{0.5}{2.82}\right]$$

Using $A = 8(15) = 120$ ft², we have

$$(25)(120) = Q(0.06 + 11.11 + 0.18)$$

$$Q = \frac{(25)(120)}{11.35} = 264 \frac{\text{Btu}}{\text{hr}} \blacktriangleright$$

7-40 Wt. 4

One end of a copper bar of cross-sectional area 4 cm² and length 80 cm is kept in steam at 1 atm pressure; the other end is in contact with melting ice. How many grams of ice will be melted in 10 min if the sides of the copper rod are insulated? Assume the thermal conductivity of copper $k = 1$ cal/sec/cm/°C and the latent heat of ice is 80 cal/g.

Solution

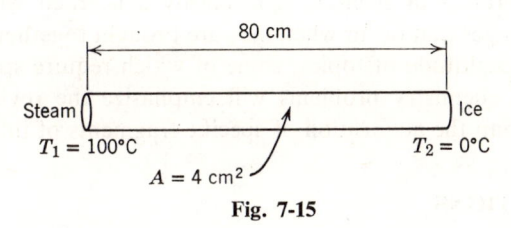

Fig. 7-15

The amount of heat Q transferred from the hot end to the cold end of the rod is

$$Q = k\frac{A}{L}(T_1 - T_2)$$

$$Q = (1)(\tfrac{4}{80})(100 - 0) = 5 \frac{\text{cal}}{\text{sec}}$$

In 10 min the total heat transfer is

$$\left(5 \frac{\text{cal}}{\text{sec}}\right)\left(60 \frac{\text{sec}}{\text{min}}\right)(10 \text{ min}) = 3000 \text{ cal}$$

The amount of ice melted is therefore

$$\frac{3000 \text{ cal}}{80 \text{ cal/g}} = 37.5 \text{ g} \blacktriangleright$$

Chapter 8

Chemistry

Chemistry, as a branch of science, is primarily concerned with kinds of matter and the changes that occur when they are brought together. Chemistry problems cover a multitude of topics, some of which require specific recall. This collection of chemistry problems will emphasize the review of basic principles rather than the acquisition of specific fragments of information.

GAS LAW RELATIONS

For ease of comparison, gas volumes often are given at a temperature of 0°C (32°F) and 760 mm of mercury (1 atm or 14.7 psia). These values are referred to as *standard temperature and pressure* (STP) or *standard conditions*.

Boyle's law and Charles' law are often combined to form the universal gas law, which is

$$\frac{P_1 V_1}{T_1} = \frac{P_2 V_2}{T_2}$$

Here the subscripts represent two different states. Absolute values of temperature and pressure are required in this relation.

At the same temperature and pressure, equal volumes of different gases contain equal numbers of molecules. One g-mole (gram molecular weight) of any gas contains 6.02×10^{23} molecules, and at standard conditions it occupies a volume of 22.4 liters. One lb-mole of any gas contains 453.6 times as much and occupies 359 ft^3 at standard conditions.

If different gases are mixed together, each component of the mixture acts as if it alone were present in the container. The pressure exerted by each component in the mixture is proportional to its mole concentration in the mixture. The total pressure of the mixture of gases is equal to the sum of the pressures of the components. Called Dalton's law, this can be written

$$P_T = P_1 + P_2 + P_3 + \cdots + P_n$$

The partial pressure ratio (P_1/P_T) is equal to the mole fraction (n_1/n_T) of that component, or $P_1/P_T = n_1/n_T$, where n_1 = moles of gas component 1 and n_T = total moles of gas. For additional discussion on gas laws, refer to Chapter 7.

Example 1

A candle is burned under an inverted beaker until the flame dies out. A sample of the mixture of perfect gases in the beaker after the flame has burnt out contains 8.30×10^{20} molecules of nitrogen, 0.70×10^{20} molecules of oxygen, and 0.50×10^{20} molecules of CO_2. If the total pressure of the mixture is 760 mm of mercury, how many moles of gas are present, and what is the partial pressure of each gas?

Solution

Law of partial pressures: The pressure exerted by each component in a gaseous mixture is proportional to its concentration in the mixture, and the total pressure of the gas is equal to the sum of those of its components. Therefore

$$P = P_{N_2} + P_{O_2} + P_{CO_2} = 760 \text{ mm of mercury}$$

Partial pressure is related to mole fraction:

$$P_{N_2} = P(n_{N_2})$$

where n_{N_2} = mole fraction of N_2 = (moles of N_2)/(total moles of gas). The number of moles of each gas is equal to the number of molecules divided by Avogadro's number (6.02×10^{23}):

Gas	No. molecules	Divided by	Moles of gas
N_2	8.3×10^{20}	6.02×10^{23}	13.78×10^{-4}
O_2	0.7×10^{20}	6.02×10^{23}	1.16×10^{-4}
CO_2	0.5×10^{20}	6.02×10^{23}	0.83×10^{-4}
		Total moles gas =	15.77×10^{-4} ▶

$$P_{N_2} = P \frac{\text{moles of } N_2}{\text{total moles of gas}} = (760) \frac{13.78 \times 10^{-4}}{15.77 \times 10^{-4}} = 664 \text{ mm of mercury} \blacktriangleright$$

$$P_{O_2} = (760) \frac{1.16 \times 10^{-4}}{15.77 \times 10^{-4}} = 56 \text{ mm of mercury} \blacktriangleright$$

$$P_{CO_2} = (760) \frac{0.83 \times 10^{-4}}{15.77 \times 10^{-4}} = 40 \text{ mm of mercury} \blacktriangleright$$

CHEMICAL BALANCE CALCULATIONS

Frequently a problem describes a chemical reaction and then the quantities of the various components are to be calculated. The starting point is a balanced chemical equation. Using the atomic weights of the elements, the molecular weights of the components of the chemical equation are determined. The coefficients in front of the formulas provide the relative numbers of each kind of molecule. The relative weights of the various components are thus known. The values may be set in a ratio to determine the quantities of each component in a specified situation.

Example 2

One of the principal scale-forming constituents of water is calcium bicarbonate, $Ca(HCO_3)_2$. This substance may be removed by treating the water with lime, $Ca(OH)_2$, in accordance with the following reaction:

$$Ca(HCO_3)_2 + Ca(OH)_2 \rightarrow 2CaCO_3 + 2H_2O$$

Atomic weights: $Ca = 40$ $H = 1$ $C = 12$ $O = 16$

Determine the pounds of lime required to remove 1 lb of calcium bicarbonate.

Solution

Molecular weights

$$Ca(HCO_3)_2 = 40 + (1 + 12 + 3 \times 16)(2) = 162$$
$$Ca(OH)_2 = 40 + (16 + 1)(2) = 74$$
$$CaCO_3 = 40 + 12 + (16)(3) = 100$$
$$H_2O = 2(1) + 16 = 18$$

From the balanced chemical equation we read that 1 mole of $Ca(HCO_3)_2$ plus 1 mole of $Ca(OH)_2$ combines to produce 2 moles of $CaCO_3$ and 2 moles of H_2O. Since a mole is a molecular weight of a substance in any desired weight units, here we will use 1 mole = 1 lb-molecular weight (in other problems 1 mole might be a gram-molecular weight or a ton-molecular weight or it might be measured in some other weight unit).

$$Ca(HCO_3)_2 + Ca(OH)_2 \rightarrow 2CaCO_3 + 2H_2O$$
$$162 \text{ lb} + 74 \text{ lb} \rightarrow 2(100) \text{ lb} + 2(18) \text{ lb}$$

Now the problem asks the number of pounds of lime required to remove 1 lb of calcium bicarbonate. From the balanced equation we see that 1 mole combines with 1 mole. Thus 74 lb of lime combine with 162 lb of calcium bicarbonate.

For 1 lb of calcium bicarbonate

$$\frac{74}{162} \text{ or } 0.457 \text{ lb of lime is required} \blacktriangleright$$

COMBUSTION

Combustion is a specialized and somewhat more complex example of a chemical balance calculation. In simple complete combustion a hydrocarbon fuel (one containing carbon and hydrogen) is burned in the presence of the theoretically correct amount of oxygen to produce carbon dioxide (CO_2) and water (H_2O) as the combustion products. The source of the oxygen is air. (Air = 21% oxygen and 79% nitrogen by volume.) Often this type of problem is readily solved by a balanced calculation if two additional items are recalled:

1. One g-mole of any gas at standard conditions occupies a volume of 22.4 liters.
2. For 1 liter of oxygen, 100/21 = 4.76 liters of air are required.

Example 3

Propane (C_3H_8) is completely burned in air with carbon dioxide (CO_2) and water (H_2O) being formed. If 15 lb of propane are burned per hour, how many cubic feet per hour of dry CO_2 are formed at 70°F and atmospheric pressure?

Atomic weights: C = 12   H = 1   O = 16

The gas constant for CO_2 is 35 ft-lb/lb-°R.

Solution

The equation for the reaction is:

$$C_3H_8 + 5O_2 \rightarrow 3CO_2 + 4H_2O \qquad \text{(ignoring the } N_2 \text{ in the air)}$$

Molecular weights:

$$C_3H_8 = 36 + 8 = 44 \qquad CO_2 = 12 + 32 = 44$$

1 mole = 1 lb-molecular weight.

Since 44 lb of propane produce 3(44) lb of carbon dioxide, 15 lb of propane produce 45 lb of carbon dioxide. Using the gas equation

$$PV = WRT$$

where $P = 14.7 \times 144 = 2118 \text{ lb/ft}^2$ abs

$W = 45 \text{ lb } CO_2$

$R = 35 \text{ ft-lb/lb-}°R$

$T = 70°F + 460 = 530°F \text{ abs } (°R)$

$$V = \frac{WRT}{P} = \frac{45 \times 35 \times 530}{2118} = 394 \text{ ft}^3/\text{hr } CO_2 \blacktriangleright$$

Example 4

The heating value of natural gas is 1000 Btu/ft³. If a furnace has an output efficiency of 90%, what should be the supply of air in cubic feet per minute for an output of 100,000 Btu/hr? Assume the oxygen content of air is 21% by volume and that natural gas is essentially methane which combines with the oxygen as follows:

$$CH_4 + 2O_2 \rightarrow CO_2 + 2H_2O$$

Solution

The input of natural gas is $100,000/(0.90 \times 1000) = 111 \text{ ft}^3/\text{hr}$. From the balanced equation we see that 1 mole of CH_4 combines with 2 moles of O_2. For 111 ft³/hr of natural gas, twice this amount or 222 ft³/hr of pure oxygen are required. Since air is only 21% by volume oxygen, the amount required is

$$\frac{222}{0.21} = 1058 \text{ ft}^3/\text{hr} \qquad \frac{1058}{60} = 17.6 \text{ ft}^3 \text{ of air per minute} \blacktriangleright$$

MOLAR SOLUTIONS

A molar solution designation is an important method of describing the relative amount of a component of a solution in quantitative form. A molar solution contains 1 g-mole of the solute in 1 liter of solution.

Example 5

How much NaCl is there in a 1 M solution? How much in a 0.5 M solution?

Atomic weights: $Na = 23 \qquad Cl = 35.5$

Solution

A 1 M solution of NaCl would contain $23 + 35.5 = 58.5$ g of NaCl in 1 liter of solution. Similarly, a 0.5 M solution of NaCl would contain 29.25 g/liter.

pH VALUE

The pH value of a solution is the negative logarithm of the hydrogen ion concentration:

$$pH = -\log_{10} [H^+]$$

where $[H^+]$ means hydrogen ion concentration in moles per liter. In neutral solutions the pH is 7. If the pH is less than 7, the solution is acidic; if greater than 7, the solution is basic.

8-1 Wt. 2

Distinguish between compounds and elements.

Solution

▶ Elements: Substances which we are not able, by the ordinary types of chemical change, to decompose into, or make by chemical union from, other substances.

▶ Compounds: Substances which can be made by chemical combination of, or can be decomposed into, two or more substances.

8-2 Wt. 1

The atomic number of an atom is derived from the number of

(*a*) protons
(*b*) electrons
(*c*) neutrons
(*d*) mesons
(*e*) neutrinos

Solution

The atomic number of an atom is the number of protons in the nucleus ▶

Answer is (*a*)

8-3 Wt. 1

A mu meson particle has a mass approximately

(*a*) that of an electron
(*b*) that of a proton
(*c*) that of a neutron
(*d*) between an electron and a proton
(*e*) between a proton and an alpha particle

Solution

A meson is any elementary particle with mass between that of the electron and the proton ▶

Answer is (*d*)

8-4 Wt. 1

The mass number of an atom is the number of

(*a*) electrons in the atom
(*b*) protons in the nucleus
(*c*) neutrons in the nucleus
(*d*) protons and neutrons in the nucleus
(*e*) electrons and neutrons in the nucleus

Solution

The mass number of an atom is the number of protons and neutrons in the nucleus ▶

Answer is (*d*)

8-5 Wt. 1

The halogen group includes the elements

(*a*) Na, Ca, K
(*b*) F, Cl, Br, I
(*c*) Au, Ag, Pt
(*d*) Mg, Mn, Mo
(*e*) C, O, H

Solution

The halogen group of elements ("the salt producers") contains Fluorine (F), Chlorine (Cl), Bromine (Br), and Iodine (I) ▶

Answer is (*b*)

8-6 Wt. 1

The group of metals which is comprised of lithium, sodium, potassium, rubidium, and cesium forms a closely related family known as

(*a*) the rare earth group
(*b*) the metals of the fourth outer group
(*c*) the alkali metals
(*d*) the elements of the inner group
(*e*) the metals of Group VIII

Solution

These metals are all members of Group I_a—the alkali metals ▶

Answer is (*c*)

8-7 Wt. 3

Define the following chemical terms:

(*a*) catalyst
(*b*) oxidation
(*c*) reduction

Solution

▶ Catalyst: A substance which changes the speed of a chemical reaction without itself suffering any permanent change.
▶ Oxidation: When electrons are removed from a substance.
▶ Reduction: When electrons are added to a substance.

8-8 Wt. 1

The substance which does not belong in the following group is

(*a*) lye
(*b*) digestive fluids
(*c*) sour milk
(*d*) sulfuric acid
(*e*) HCl

Solution

Lye is a base. The other four are acids ▶

Answer is (*a*)

8-9 Wt. 1

The number which expresses the combining capacity of an atom of an element or of a group of atoms is called the

(*a*) indicator
(*b*) displacement factor
(*c*) electrolyte
(*d*) valence
(*e*) conductance

Solution

The combining capacity is called valence ▶

Answer is (*d*)

8-10 Wt. 1

Which of the following statements is false?

(a) The atomic weight of oxygen is 16.
(b) Alcohol has a freezing point of $-40°F$.
(c) 778 ft-lb = 1 Btu.
(d) Absolute zero on the temperature scale is approximately $-460°F$.
(e) A coulomb of electricity can be defined as that quantity which will deposit a stated fraction of a gram of silver from a stated normal solution of silver nitrate under certain conditions.

Solution

The freezing point of methyl alcohol is about $-100°C$; the freezing point of ethyl alcohol is even lower ▶

Statement (b) is incorrect

8-11 Wt. 5

Phosphorus pentoxide is a more effective dehydrating agent than is sulfur trioxide. Write the chemical equations which express this property. (Note: the oxides react with water to form acids.)

Solution

$$P_2O_5 + 3H_2O \rightarrow 2H_3PO_4 \blacktriangleright$$

$$SO_3 + H_2O \rightarrow H_2SO_4 \blacktriangleright$$

8-12 Wt. 1

The chemical formula for ethyl alcohol (Ethanol) is

(a) C_2H_5
(b) CH_4
(c) CH_3OH
(d) C_2H_6OH
(e) C_3H_7OH

Solution

The formulas given are:

(a) ethane
(b) methane
(c) methyl alcohol
(d) ethyl alcohol ▶
(e) propyl alcohol

Answer is (d)

8-13 Wt. 1

The ion responsible for the chemistry of bases in water is

(a) H⁺
(b) Na⁺
(c) H$_3$O⁺
(d) OH⁻
(e) HOH⁻

Solution

Hydroxyl ion (OH⁻) is the characteristic base of aqueous solutions ▶

Answer is (d)

8-14 Wt. 1

In the earth's crust (which includes a 10-mile shell and the atmosphere above it), aside from oxygen and silicon, what is the most abundant element found?

(a) aluminum
(b) iron
(c) calcium
(d) sodium
(e) hydrogen

Solution

Oxygen	49.2%	Calcium	3.4%
Silicon	25.7	Sodium	2.4
Aluminum	7.4 ▶	Hydrogen	1.0
Iron	4.7		

(While different sources show slightly different percentages, they agree with this order of ranking.)

Answer is (a)

8-15 Wt. 1

The principal metal applied to iron or steel in the galvanizing process is

(a) nickel
(b) chromium
(c) cadmium
(d) mercury
(e) zinc

Solution

Zinc is the principal metal applied to iron or steel in the galvanizing process ▶

Answer is (*e*)

8-16 Wt. 1

Aside from nitrogen and oxygen, which of the following elements is the most plentiful in dry air?

(*a*) argon
(*b*) carbon dioxide
(*c*) neon
(*d*) helium
(*e*) krypton

Solution

Nitrogen	78.03%	Neon	0.0015%
Oxygen	20.99	Helium	0.0005
Argon	0.94 ▶	Krypton	0.00011

Carbon dioxide 0.023 to 0.050%

(We should point out that carbon dioxide is not an element, but a compound.)

Answer is (*a*)

8-17 Wt. 1

Oxygen is converted into ozone by

(*a*) great heat
(*b*) high pressure
(*c*) electric discharge
(*d*) a catalyst
(*e*) none of these processes

Solution

Ozone is formed by passing an electric discharge through oxygen ▶

Answer is (*c*)

8-18 Wt. 1

The phenomenon known as corrosion in metals is by nature

(*a*) chemical
(*b*) organic
(*c*) electrochemical
(*d*) inorganic
(*e*) electrical

Solution

The phenomenon is electrochemical. It is often called electrolytic corrosion ▶

Answer is (*c*)

8-19 Wt. 1

Radioactivity is a property of all elements with atomic numbers greater than

(*a*) 48
(*b*) 62
(*c*) 83
(*d*) 88
(*e*) 90

Solution

Radioactivity is the property of elements with atomic numbers greater than 83 ▶

Answer is (*c*)

8-20 Wt. 1

A certain process has three factors which always occur: (1) a current flow; (2) substances that are formed at each electrode; and (3) matter that is transported through a solution or molten salt. This process is called

(*a*) evaporation
(*b*) decomposition
(*c*) hydrolysis
(*d*) condensation
(*e*) electrolysis

Solution

The process is called electrolysis ▶

Answer is (*e*)

8-21 Wt. 1

The process in which a solid changes directly to the gaseous state is called

(*a*) sublimation
(*b*) homogenization
(*c*) crystallization
(*d*) vaporization
(*e*) distillation

Solution

The process is called sublimation. Iodine, camphor, and naphthalene are examples of materials that sublime ▶

Answer is (a)

8-22 Wt. 1

Tetraethyl lead is a compound which is used as an antiknock in gasoline. The formula for tetraethyl lead is

(a) CH_4Pb
(b) C_2H_5Pb
(c) $(C_2H_5)_4Pb_4$
(d) $(C_2H_5)_4Pb$
(e) $(C_2H_4)_4Pb$

Solution

Ethyl is C_2H_5, hence tetraethyl lead would be $(C_2H_5)_4Pb$ or, as it is often written, $Pb(C_2H_5)_4$ ▶

Answer is (d)

8-23 Wt. 1

Carbon dioxide "snow" from a fire extinguisher puts out a fire because it

(a) lowers the kindling temperature of the material
(b) displaces the supply of oxygen
(c) raises the kindling temperature of the material
(d) cools the material below its kindling temperature
(e) acts as a catalyst to induce a chemical reaction

Solution

Carbon dioxide "snow" smothers a fire by keeping the oxygen away from the fire ▶

Answer is (b)

8-24 Wt. 1

Hardness in water supplies is primarily due to the solution in it of

(a) carbonates and sulfates of calcium and magnesium
(b) alum
(c) soda ash
(d) sodium sulfate
(e) sodium chloride

Solution

The compounds of calcium and magnesium cause hardness. These may include carbonates, bicarbonates, sulfates, and chlorides ▶

Answer is (a)

8-25 Wt. 2

One g of a sample of coal selling at $10.00/ton was burned in a calorimeter containing 1200 g of water, and the heat evolved raised the temperature of the water from 16°C to 23°C. The same weight of a sample of another coal selling at $8.00/ton was found to evolve 7300 cal of heat. Other things being equal, which of the two kinds of coal would be the more economical?

Solution

1200 g of water raised 7°C means 8400 cal of heat were liberated by 1 g of the $10.00/ton coal; the $8.00/ton coal gave 7300 cal.

Compare the ratio of heat to cost to determine the more desirable alternative:

$$\frac{8400}{10.00} = 840 \frac{\text{units heat}}{\text{unit cost/ton}} \qquad \frac{7300}{8.00} = 912.5$$

Therefore, select the $8.00/ton coal ▶

8-26 Wt. 6

A gasoline engine burns octane (C_8H_{18}) at a rate of 40 lb/hr. The products of combustion are carbon dioxide (CO_2), water vapor (H_2O), and nitrogen (N_2). Assume that the theoretical amount of dry air is supplied for complete combustion and that air is composed of 21% oxygen and 79% nitrogen.

(a) Write the equation for complete combustion.

(b) Compute the air-fuel ratio $\left(\frac{\text{mass air}}{\text{mass fuel}}\right)$.

Atomic weights: C = 12   H = 1   N = 14   O = 16

Solution

(a) The products of complete combustion of a hydrocarbon will be water and carbon dioxide:

$$2C_8H_{18} + 25O_2 \rightarrow 16CO_2 + 18H_2O \blacktriangleright \qquad \text{(ignoring the } N_2\text{)}$$

(b) This equation tells us that 2 molecular weights of C_8H_{18} (8 × 12 + 18 × 1) or 2 × 114 g react with 25 × 32 or 800 g of oxygen. As 228 g of octane require 800 g of oxygen, then each g of octane requires

$$1 \text{ g octane} \times \frac{800 \text{ g oxygen}}{228 \text{ g octane}} = 3.51 \text{ g O}_2$$

Now 1 volume of air contains

$$0.21 \times 32 = 6.72 \text{ g O}_2$$
$$0.79 \times 28 = 22.12 \text{ g N}_2$$
$$\overline{28.84 \text{ g}}$$

or

$$\frac{6.72 \times 100}{28.84} = 23.3\% \text{ O}_2 \text{ by weight}$$

and

$$\frac{22.12 \times 100}{28.84} = 76.7\% \text{ N}_2 \text{ by weight}$$

Therefore, 3.51 g O_2 will be contained in

$$3.51 \text{ g O}_2 \times \frac{100 \text{ g air}}{23.3 \text{ g O}_2} = 15.1 \text{ g air}$$

So

$$\frac{\text{mass air}}{\text{mass fuel}} = \frac{15.1}{1} \blacktriangleright$$

8-27 Wt. 6

(a) Hydrogen-free carbon in the form of coke is burned with complete combustion, using 26.4% excess air. Assume the air contains 79 mole % nitrogen and 21 mole % oxygen. Calculate the mole % of each gas contained in the flue gases.

(b) The above fuel is burned with just enough air to convert theoretically all of the carbon to carbon dioxide, but actually 15.2% of it was converted to carbon monoxide. Assume the air contains 79 mole % nitrogen and 21 mole % oxygen. Compute the mole % of each gas contained in the flue gases.

Solution

(a)

$$C + O_2 \rightarrow CO_2$$

Basis: 100 moles C burned

$$CO_2 \text{ formed} = 100 \text{ moles}$$
$$O_2 \text{ required} = 100 \text{ moles}$$
$$O_2 \text{ in excess} = 0.264(100) = 26.4 \text{ moles}$$
$$\text{Total } O_2 \text{ supplied} = 126.4 \text{ moles}$$
$$N_2 \text{ entering with } O_2 = \tfrac{79}{21}(126.4) = 476 \text{ moles}$$

Output gases

Component	Moles	Mole %	
CO_2	100	(100/602)(100) =	16.6 ▶
O_2	26.4	(26.4/602)(100) =	4.4 ▶
N_2	476	(476/602)(100) =	79.0 ▶
	602		100.0

Note that, here, 1 mole of product gas, CO_2, is formed for each mole of O_2 reacting, so that the combined number of moles of CO_2 and O_2 leaving the burner must equal the number of moles of O_2 entering the burner, 21. This gives a quick check on the calculations.

(b)
$$C + O_2 \rightarrow CO_2 \qquad (1)$$
$$C + \tfrac{1}{2}O_2 \rightarrow CO \qquad (2)$$

Basis: 100 moles C burned

O_2 supplied = 100 moles
N_2 supplied = $\tfrac{79}{21}$(100) = 376 moles
C forming CO = 15.2 moles
requiring $\tfrac{1}{2}$(15.2) = 7.6 moles O_2 by equation (2)
C forming CO_2 = 100 − 15.2 = 84.8 moles
requiring = 84.8 moles O_2 by equation (1)
Total O_2 used = 7.6 + 84.8 = 92.4 moles
O_2 in excess = 100 − 92.4 = 7.6 moles

Output Gases

Component	Moles	Mole %	
CO_2	84.8	(84.8/484)(100) =	17.5 ▶
CO	15.2	(15.2/484)(100) =	3.1 ▶
O_2	7.6	(7.6/484)(100) =	1.6 ▶
N_2	376	(376/484)(100) =	77.8 ▶
	484		100.0

Note that, here, *more* than 1 mole of product gases, CO_2 and CO, are formed for each mole of O_2 reacting, so that the combined number of moles of CO_2, CO, and O_2 leaving the burner should be greater than the number of moles of O_2 entering the burner, 21. Inspection of the results on this basis shows them to be reasonable.

8-28 Wt. 6

(a) Pure butane, C_4H_{10}, is burned with complete combustion, using 23.7% excess air. Assume the air contains 79.1 mole % nitrogen and 20.9 mole %

oxygen. Compute the mole % of each gas contained in the flue gases, on a dry basis (free of water vapor).

(b) Pure butane is burned in such a way that all the oxygen in the air is used up and 7.85 times as many moles of carbon dioxide are formed as carbon monoxide. Assume the air contains 79.1 mole % nitrogen and 20.9 mole % oxygen. Calculate the mole % of each gas contained in the flue gases.

Solution

(a) $$2C_4H_{10} + 13O_2 \rightarrow 8CO_2 + 10H_2O$$

Each mole of C_4H_{10} requires 6.5 moles of O_2 for complete combustion. Air containing 6.5 moles of O_2 will also contain $6.5 \times 79.1/20.9 = 24.6$ moles of N_2 or a total of $6.5 + 24.6 = 31.1$ moles of air that reacts. As mole % and volume % are equal, then

$$\% \text{ excess air} = \frac{\text{moles excess air} \times 100}{31.1} = 23.7$$

or moles excess air = 7.37, of which $7.37 \times 79.1/100 = 5.83$ moles of N_2 and $7.37 - 5.83 = 1.54$ moles O_2.

MOLES/MOLE BUTANE

	O_2	N_2
Reacting air	6.50	24.6
Excess air	1.54	5.83

Assuming complete combustion, each mole of O_2 reacting will produce $\frac{8}{13}$ mole of CO_2. Therefore 6.5 moles of O_2 will produce $6.5 \times 8/13 = 4.0$ moles of CO_2.

MOLES FLUE GASES/MOLE BUTANE

	O_2	N_2	CO_2	
From reacting air	0	24.6	4.0	
From excess air	1.54	5.83	0	
	1.54	30.43	4.0	Total moles = 35.97

$$\% O_2 = \frac{1.54 \times 100}{36.0} = 4.3\% \blacktriangleright$$

$$\% N_2 = \frac{30.43 \times 100}{36.0} = 84.6\% \blacktriangleright$$

$$\% CO_2 = \frac{4.0 \times 100}{36.0} = 11.1\% \blacktriangleright$$

$$100.0\%$$

(b) Each mole of butane that burns will produce a total of 4 moles of CO_2 and CO

$$N_{CO} = X$$

$$N_{CO_2} = 7.85X$$

$7.85X + X = 4.0 \qquad X = 0.452 \text{ mole CO}$

$$7.85X = 3.55 \text{ moles } CO_2$$

One mole CO requires 0.5 mole O_2, so 0.452 mole CO will require $0.452 \times 0.5 = 0.226$ mole O_2. Also, 1 mole CO_2 requires 1 mole O_2, so 3.55 moles O_2 are required for CO_2. Each mole of butane will produce 5 moles of condensable H_2O, but requires 2.50 moles of O_2. The total moles of O_2 required are $2.50 + 3.55 + 0.226 = 6.28$. This means that $6.28 \times 100/20.9 = 30.0$ moles of air are needed per mole of butane, and this air contains 23.7 moles N_2.

MOLES FLUE GASES/MOLE BUTANE

CO	0.452
CO_2	3.55
N_2	23.7
Total moles	27.7

$$\% \text{ CO} = \frac{0.452 \times 100}{27.7} = 1.6\% \blacktriangleright$$

$$\% \text{ CO}_2 = \frac{3.55 \times 100}{27.7} = 12.8\% \blacktriangleright$$

$$\% \text{ N}_2 = \frac{23.7 \times 100}{27.7} = \underline{85.6\%} \blacktriangleright$$

$$100.0\%$$

8-29 Wt. 6

The flue gases from an oil-fired furnace analyze by the Orsat method 10.8% CO_2, 1.0% CO, 4.4% O_2, 83.8% N_2 by volume. The oil fuel contains only carbon and hydrogen. Assuming the air contains 79.1 vol. % nitrogen and 20.9 vol. % oxygen, calculate

(a) the percent excess air used in the combustion
(b) the composition of the fuel in weight % carbon and hydrogen

Atomic weights: $H = 1 \qquad C = 12 \qquad O = 16 \qquad N = 14$

Solution

(a) In combustion of hydrocarbon fuels all hydrogen is burned to water, and then carbon is burned to CO or CO_2 depending upon the O_2 supply in the combustion zone.

Basis: 100 lb-moles dry flue gas

Orsat analysis does not reveal water content. The pound-moles of each gas are directly related to the volume % as reported by Orsat.

Dry flue gas composition: 10.8 lb-moles CO_2
1.0 lb-moles CO
4.4 lb-moles O_2
83.8 lb-moles N_2

100 lb-moles total

The oxygen in the flue gas comes from excess air (above that required to burn all C to CO_2 and all H to H_2O) and from air that (due to improper mixing in the combustion zone) did not burn the CO to CO_2.

4.4 lb-moles O_2 in flue gas

less 0.5 lb-moles O_2 required to burn 1 lb-mole CO to CO_2

equals 3.9 lb-moles "excess" O_2 in flue gas

(over and above that required for stoichiometric or complete combustion of all fuel H to H_2O and C to CO_2).

The percent excess O_2 cannot be calculated directly because we do not yet know how much water is in the flue gases. Therefore, we determine excess air by nitrogen balance.

83.8 lb-moles N_2 in and out of combustion zone

Calculate the pound-moles N_2 associated with 3.9 lb-moles of "excess" O_2.

$$3.9 \times \frac{79.1}{20.9} = 14.8 \text{ lb-moles } N_2$$

(since air is 79.1% N_2 and 20.9% O_2 by volume)

Therefore, $83.8 - 14.8 = 69$ lb-moles N_2 associated with stoichiometric O_2 are required for complete combustion. Stoichiometric O_2 is related to the N_2 associated with it.

$$69 \times \frac{20.9}{79.1} = 18.2 \text{ lb-moles } O_2 \text{ required for complete combustion}$$

Percent excess air (or excess O_2)

$$= \frac{\text{lb-moles "excess" } O_2}{\text{lb-moles } O_2 \text{ required for complete combustion}} \times 100$$

$$= \frac{3.9}{18.2} \times 100 = 21.4\% \blacktriangleright$$

(b) Composition of the fuel to give the 100 lb-moles of dry flue gas is found by determining the weight of carbon and weight of hydrogen in the combustion products: H_2O, CO, and CO_2.

First, calculate the water associated with 100 lb-moles of dry flue gas. This is done by an oxygen balance: 18.2 lb-moles of stoichiometric O_2 less 10.8 lb-moles O_2 needed to produce the CO_2 less 1.0 lb-mole to produce the CO and then burn it to CO_2, leaves 6.4 lb-moles of O_2 that could only produce water by burning the fuel hydrogen.

$$2H_2 + O_2 \rightarrow 2H_2O$$

Then 6.4 lb-moles O_2 burn 12.8 lb-moles (or 25.6 lbs) hydrogen. This is all the hydrogen in the fuel, for hydrogen is always burned completely.

Carbon in the fuel is found from the carbon in the CO and CO_2:

10.8 lb-moles CO_2 comes from 10.8 lb-atoms C

1.0 lb-moles CO comes from 1.0 lb-atoms C

Total: 11.8 lb-atoms C in fuel

Since the atomic weight of C is 12, this is equal to 141.6 lb C.

Summary of fuel composition:

Carbon	141.6 lb	84.7 weight % $\blacktriangleright$
Hydrogen	25.6 lb	15.3 weight % $\blacktriangleright$
Total	167.2 lb	100.0 weight %

8-30 Wt. 5

What is the composition of flue gas and how many pound-moles of combustion products are obtained when 1200 lb of carbon are burned to CO_2 with the theoretical amount of dry air containing by volume 21.0% O_2 and 79.0% nitrogen and inerts?

Atomic weights: C = 12   O = 16   N = 14

406 / Chemistry

Solution

Basis: 1200 lb carbon (100 lb-moles)

$$C + O_2 \rightarrow CO_2$$

This says that 1 lb-mole of C when burned by 1 lb-mole of O_2 forms 1 lb-mole of CO_2. Scaling this equation upward, 100 lb-moles of C require 100 lb-moles of O_2 and form 100 lb-moles CO_2. The air entering must consist of 100 lb-moles O_2 plus $\frac{79}{21}(100)$ lb-moles of N_2.

Since 100 lb-moles CO_2 are formed at the expense of the 100 lb-moles of O_2, the flue gases contain:

100 lb-moles	CO_2	21% by volume ▶
376 lb-moles	N_2	79% by volume ▶
▶ 476		100%

8-31 Wt. 6

A hydrocarbon liquid is subjected to flash vaporization. Equilibrium composition of the vapors and equilibrium ratios ($K = y/x$) at that pressure and various temperatures are given in the table.

	Vapor Mole %	$K = y/x$ 100°F	115°F	125°F
Propane	30.9	1.60	1.88	2.10
Isobutane	37.2	0.74	0.90	1.00
Normal butane	31.9	0.52	0.64	0.72

(a) Using 115°F for the first approximation, estimate the temperature of the equilibrium liquid.

(b) A sample of the above equilibrium vapor is separated from the equilibrium mixture, and the temperature is lowered until the vapor condenses practically completely at the same pressure used in the flash vaporization. Estimate the bubble-point temperature of the condensed vapors.

Solution

For miscible liquids, the mole fraction of each component in the vapor phase, y, is related to the mole fraction of the component in the liquid phase,

x, multiplied by an "equilibrium constant," K. For the general ith component,

$$y_i = K_i x_i$$

Solutions such as this boil over a boiling-point *range*. The temperature at which the total vapor pressure over the liquid becomes equal to the ambient pressure is called the "bubble point," and it is found by assuming temperatures and determining the temperature at which the summation of y_i's = unity. Similarly, the "dew point" is the temperature at which the summation of x_i's = unity. The problem as given here, then, is solved by iteration.

(a) Basis: 1 mole vapor

			First approximation: assume 115°F dew point		Second approximation: assume 116°F dew point interpolated	
Component	y_i Mole fraction	K_i Given	$x_i = \dfrac{y_i}{K_i}$	K_i	$x_i = \dfrac{y_i}{K_i}$	
Propane	0.309	1.88	0.164	1.90	0.162
Isobutane	0.372	0.90	0.413	0.91	0.409
Normal butane	0.319	0.64	0.499	0.65	0.491
	1.000		1.076		1.062

Too high—try 116°F dew point   Again too high—try 125°F dew point

	Third approximation: assume 125°F dew point		Fourth approximation: assume 120°F dew point interpolated	
	K_i Given	$x_i = \dfrac{y_i}{K_i}$	K_i	$x_i = \dfrac{y_i}{K_i}$
	2.10	0.147	1.99	0.155
	1.00	0.372	0.95	0.392
	0.72	0.443	0.68	0.469
		0.962		1.016

Too low—try 120°F dew point   Find dew point by interpolation

From Fig. 8-1 the dew point is found to be 121.5°F ▶

408 / Chemistry

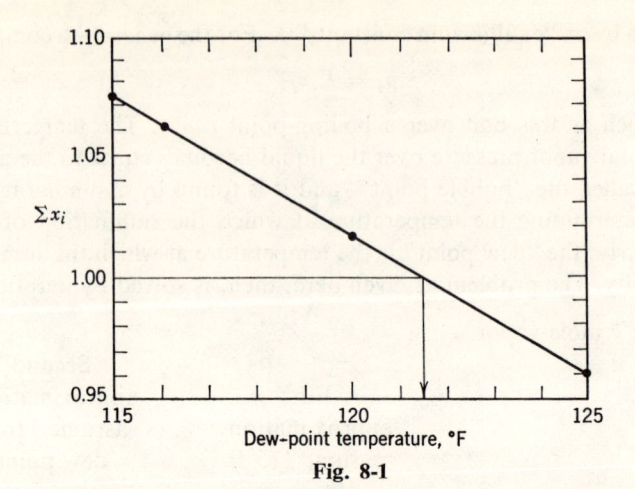

Fig. 8-1

(b) Basis: 1 mole liquid of same composition as vapor in (a).

	x_i Mole fraction	First approximation: assume 115°F bubble point K_i Given	$y_i = K_i x_i$	Second approximation: assume 100°F bubble point K_i Given	$y_i = K_i x_i$
Component					
Propane	0.309	1.88	0.580	1.60	0.495
Isobutane	0.372	0.90	0.335	0.74	0.275
Normal butane	0.319	0.64	0.204	0.52	0.166
	1.000		1.119		0.936

Too high—try 100°F bubble point   Too low—try 105°F bubble point

Third approximation: assume 105°F bubble point interpolated

K_i	$y_i = K_i x_i$
1.69	0.522
0.79	0.294
0.56	0.178
	0.994

Close—get final value by interpolation

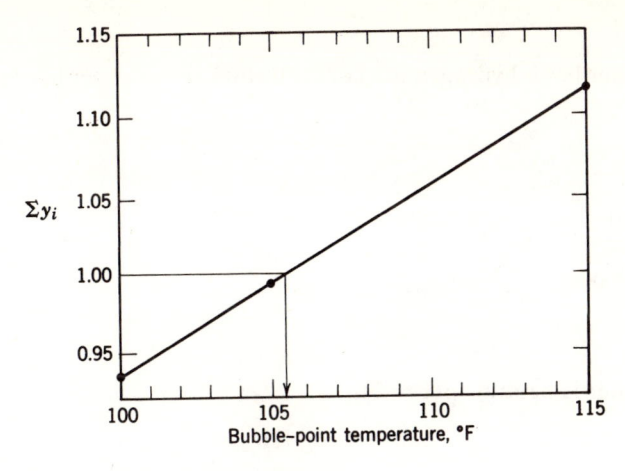

Fig. 8-2

Figure 8-2 establishes that the bubble-point temperature is 105.5°F ▶

8-32 Wt. 1

Helium gas and hydrogen gas are used in lighter-than-air craft. The lifting power of helium as compared to the same volume of hydrogen is approximately

(a) 25%
(b) 50%
(c) 75%
(d) 90%
(e) 100%

Solution

In equal volumes of different gases, at the same temperature and pressure, the number of molecules present are equal. Thus density is proportional to molecular weight.

$$H_2 = 2 \times 1.008 = 2.016$$
$$He = 4.003$$

But the lifting power of gas is proportional to the difference between its density and the density of air. Since air has an average molecular weight of about 29, the lifting power of hydrogen is $29 - 2 = 27$ and of helium $29 - 4 = 25$.

Thus the ratio of lifting power $\dfrac{He}{H_2} = \dfrac{25}{27} = 92\%$ ▶

Answer is (d)

8-33 Wt. 1

Pure water has a hydrogen-ion concentration in moles per liter of approximately

(*a*) 1.0
(*b*) 0.1
(*c*) 0.001
(*d*) 0.00001
(*e*) 0.0000001

Solution

Pure water is neutral (pH = 7).

$$\text{pH} = -\log [\text{H}^+] = 7$$

Since $-(-7) \log 10 = 7$, the hydrogen-ion concentration $[\text{H}^+] = 10^{-7}$ ▶

Answer is (*e*)

8-34 Wt. 2

The common anesthetic chloroform is a colorless liquid boiling at 61.2°C. Its formula is $CHCl_3$ and its molecular weight is approximately 119.5. Give the proportions of each element it contains.

Atomic weights: C = 12   H = 1   Cl = 35.5

Solution

Proportions of each element:

1 atom of carbon
1 atom of hydrogen
3 atoms of chlorine

By weight

$$\% \text{ carbon} = \frac{12}{119.5} \times 100 = 10.03\% \blacktriangleright$$

$$\% \text{ hydrogen} = \frac{1}{119.5} \times 100 = 0.84\% \blacktriangleright$$

$$\% \text{ chlorine} = \frac{3(35.5)}{119.5} \times 100 = 89.13\% \blacktriangleright$$

8-35 Wt. 1

The hydroxyl ion concentration of 1×10^{-11} moles/liter of water can be expressed by what pH number?

Solution

$$pOH = -\log [OH] = -\log 10^{-11} = 11$$

For water, pH + pOH = 14; therefore, pH = 14 − 11 = 3 ▶

8-36 Wt. 3

Potassium cyanide (or sodium cyanide) is used in many practical operations. It is a typical chemical salt, high melting, nonvolatile, soluble in water, and so forth. Careless handling and improper storage can create very hazardous conditions. Explain chemical or physical actions that give rise to these. Express the chemical action(s) by a chemical equation and give the chemical terms applied to the chemical action(s) involved.

Solution

KCN and NaCN are hazardous chemicals as they are very poisonous, and in addition HCN is very poisonous. When solids are used, caution must be exercised in handling and disposing of the cyanides to prevent the contamination of drinking water and food because of the oral toxicity of soluble cyanides.

The greater hazard is associated with the action of acids on these substances.

$$KCN + H^+ \rightarrow HCN\uparrow + K^+$$
$$CN^- + H^+ \rightarrow HCN\uparrow$$

(These reactions are examples of the reaction of an acid, H^+, with an anion, CN^-, to form the weak acid, HCN.) Here, exposure to acid fumes (like HCl) or to acidic liquids will generate the poisonous gas HCN ▶

8-37 Wt. 1

The volume of 1 mole of oxygen at 0°C and 1 atm of pressure is approximately

(a) 1.0 liter
(b) 16.0 liters
(c) 22.4 liters
(d) 1.0 ft³
(e) 62.4 ft³

Solution

The volume of 1 mole of any gas at standard conditions (0°C and 1 atm pressure) is 22.4 liters ▶

Answer is (c)

8-38 Wt. 2

Which one of the following chemical equations is correct? The valences of the various elements or radicals are listed as follows:

$$H^+ \quad Na^+ \quad Ag^+ \quad Ca^{++} \quad Zn^{++} \quad C^{++++}$$
$$(OH)^- \quad Cl^- \quad NO_3^- \quad O^{--} \quad CO_3^{--} \quad SO_4^{--}$$

(a) $H_2 + O_2 \rightarrow H_2O$
(b) $NaCl + 2AgNO_3 \rightarrow NaNO_3 + AgCl$
(c) $CaO + H_2O \rightarrow Ca(OH)_3$
(d) $CaSO_4 + Na_2CO_3 \rightarrow CaCO_3 + Na_2SO_4$
(e) $ZnCO_3 \rightarrow ZnO + 2CO_2$

Solution

The correct equations are:

(a) $2H_2 + O_2 \rightarrow 2H_2O$
(b) $NaCl + AgNO_3 \rightarrow NaNO_3 + AgCl$
(c) $CaO + H_2O \rightarrow Ca(OH)_2$
(d) correct as given ▶
(e) $ZnCO_3 \rightarrow ZnO + CO_2$

Answer is (d)

8-39 Wt. 1

Which one of the following chemical equations is correct? The valences of the various elements or radicals are as follows:

$$Ag^+ \quad H^+ \quad Na^+ \quad Ca^{++} \quad C^{++++} \quad S^{++++} \quad \text{and} \quad S^{--} \quad (OH)^- \quad Cl^-$$
$$NO_3^- \quad O^{--} \quad CO_3^{--} \quad SO_4^{--}$$

(a) $2H_2 + O_2 \rightarrow 2H_2O_2$
(b) $NaCl + 2AgNO_3 \rightarrow NaNO_3 + AgCl$
(c) $2H_2S + 3O_2 \rightarrow 2H_2O + 2SO_2$
(d) $CaO + H_2O \rightarrow Ca(OH)_3$
(e) $CaSO_4 + 2NaCO_3 \rightarrow CaCO_3 + 2NaSO_4$

Solution

The correct equations are:

(a) $2H_2 + O_2 \rightarrow 2H_2O$
(b) $NaCl + AgNO_3 \rightarrow NaNO_3 + AgCl$
(c) correct as given ▶
(d) $CaO + H_2O \rightarrow Ca(OH)_2$
(e) $CaSO_4 + Na_2CO_3 \rightarrow CaCO_3 + Na_2SO_4$

Answer is (c)

8-40 Wt. 2

The action of nitric acid on silver gives silver nitrate, water, and nitric oxide gas. Write and balance the equation for this reaction.

Solution

$$4H^+ + 3e^- + NO_3^- \rightarrow NO + 2H_2O$$
$$3Ag^0 - 3e^- \rightarrow 3Ag^+$$

$$3Ag + 4HNO_3 \rightarrow 3AgNO_3 + NO + 2H_2O \blacktriangleright$$

8-41 Wt. 1

The specific gravity of a substance is the weight of a given volume of the substance divided by the weight of an equal volume of water. If 2 lb of a liquid with a specific gravity of 0.800 are mixed with 1 lb of water, the specific gravity of the mixture is nearest

(a) 0.833
(b) 0.850
(c) 0.857
(d) 0.867
(e) 1.16

Solution

The equivalent volume of liquid $= \dfrac{2}{0.8} + 1 = 3.5$. The weight of liquid $= 2 + 1 = 3$ lb.

$$\text{Specific gravity} = \frac{\text{Wt of liquid}}{\text{Wt of equiv. volume of water}}$$

$$= \frac{3}{3.5(1)} = 0.857 \blacktriangleright$$

Answer is (c)

8-42 Wt. 5

How much water is required to set 100 lb of plaster of Paris? The formulas are

plaster of Paris $CaSO_4 \cdot \tfrac{1}{2}H_2O$

set form of plaster of Paris (gypsum), $CaSO_4 \cdot 2H_2O$

Atomic weights: Ca = 40.1   S = 32.1   O = 16   H = 1

Solution

$$2(CaSO_4 \cdot \tfrac{1}{2}H_2O) + 3H_2O \rightarrow 2(CaSO_4 \cdot 2H_2O)$$

Molecular weights:
$$CaSO_4 \cdot \tfrac{1}{2}H_2O = 40.1 + 32.1 + 4(16) + \tfrac{1}{2}(2 + 16) = 145.2$$
$$H_2O = 2(1) + 16 = 18$$

For 1 molecule of $CaSO_4 \cdot \tfrac{1}{2}H_2O$, $1\tfrac{1}{2}$ molecules of H_2O are needed to form $CaSO_4 \cdot 2H_2O$. Thus for 145.2 lb of plaster of Paris, $\tfrac{3}{2}(18) = 27$ lb of water are needed to form gypsum.

$$\frac{100}{145.2} = \frac{X}{27}$$

The amount of water (X) to set 100 lb of plaster of Paris is 18.6 lb ▶

8-43 Wt. 1

The number of moles of sulfuric acid required to dissolve 10 moles of aluminum according to the reaction
$$2Al + 3H_2SO_4 = Al_2(SO_4)_3 + 3H_2$$
is

(a) 3
(b) 6
(c) 10
(d) 15
(e) 30

Solution

$$\frac{10 \text{ moles Al}}{2Al} = \frac{X \text{ moles } H_2SO_4}{3H_2SO_4} \qquad X = 15 \text{ moles } H_2SO_4 \blacktriangleright$$

Answer is (d)

8-44 Wt. 3

C_7H_{16} burns completely to form water and carbon dioxide. If 10 lb of it are burned, how many liters of carbon dioxide are formed at standard conditions?

Solution

Basis: 1 lb-mole C_7H_{16} burned

Atomic weights: $C = 12 \qquad H = 1$

Molecular weight of $C_7H_{16} = 7(12) + 16(1) = 100$

By inspection: 1 lb-mole of C_7H_{16} will yield, on complete combustion, 7 lb-moles of CO_2 occupying $7(454)(22.4) = 71,300$ liters at standard conditions.

Basis: 10 lb C_7H_{16} burned

Since 100 lb of C_7H_{16} produce 71,300 liters of CO_2, 10 lb will produce one-tenth that amount, or 7130 liters of CO_2 at standard conditions ▶

8-45 Wt. 5

(a) How many pounds of limestone, 80% $CaCO_3$, would be required to make 1 ton of CaO?

(b) How many pounds of water would be required to slake the lime made in part (a)?

Atomic weights: Ca = 40   C = 12   O = 16   H = 1

Solution

(a) $$CaCO_3 \rightarrow CaO + CO_2$$

Thus 1 mole of $CaCO_3$ yields 1 mole of CaO.

Molecular weights:

$$CaCO_3: 40 + 12 + 3(16) = 100$$

$$CaO: 40 + 16 = 56$$

So 100 lb of $CaCO_3$ yields 56 lb of CaO. For 2000 lb of CaO, the quantity of limestone required is

$$\frac{2000}{56} \times \frac{100}{0.8} = 4470 \text{ lb} \blacktriangleright$$

(b) $$CaO + H_2O \rightarrow Ca(OH)_2$$

Molecular weights:

$$CaO = 56 \qquad H_2O = 18$$

Here we see that 56 lb of CaO react with 18 lb of water. For 2000 lb of CaO the quantity of water required would be

$$\frac{2000}{56} \times 18 = 643 \text{ lb of water} \blacktriangleright$$

8-46 Wt. 5

Compute the weights in grams of commercial sodium carbonate and hydrochloric acid required to produce 234 g of pure sodium chloride. The commercial soda ash is 90% pure and the commercial acid contains 30% by weight of dissolved acid.

Molecular weights of pure compounds:

$$Na_2CO_3 = 106 \qquad HCl = 36.5 \qquad NaCl = 58.5$$

Solution

$$Na_2CO_3 + 2HCl \rightarrow 2NaCl + CO_2 + H_2O$$

416 / Chemistry

From the molecular weights and the equation for the reaction,

$$106 \text{ g Na}_2\text{CO}_3 + 2(36.5) \text{ g HCl yields } 2(58.5) \text{ g NaCl}$$

or $\quad$ 106 g Na$_2$CO$_3$ + 73 grams HCl yields 117 g NaCl

Since we want 234 g of NaCl, simply double the quantities above:

$$\text{use 212 g of pure Na}_2\text{CO}_3 \text{ or } \frac{212}{0.9} = 236 \text{ g commercial Na}_2\text{CO}_3 \blacktriangleright$$

$$\text{use 146 g of HCl or } \frac{146}{0.3} = 487 \text{ g commercial HCl} \blacktriangleright$$

8-47 Wt. 2

If silver sells for $1.00/oz (troy), what is the maximum you could pay for a tank containing 1000 lb (avoirdupois) of silver nitrate solution which is 60% water by weight? The cost of recovering the silver from the silver nitrate solution is $0.06/lb of solution as purchased.

$$\text{Atomic weights: Ag} = 108 \quad \text{N} = 14 \quad \text{O} = 16$$

$$175 \text{ lb troy} = 144 \text{ lb av}$$

Solution

$$\text{Pure AgNO}_3 = 1000(0.40) = 400 \text{ lb}$$

$$\text{Molecular weight: AgNO}_3 = 108 + 14 + 3(16) = 170$$

$$\text{AgNO}_3 \rightarrow \text{Ag}\downarrow + \text{NO}_3^-$$

$$170 \text{ lb} \rightarrow 108 \text{ lb}$$

$$\text{Weight of Ag in 400 lb of AgNO}_3 = \frac{400 \times 108}{170} = 254 \text{ lb av}$$

$$\text{Value of pure Ag} = 245 \text{ lb av} \left(\frac{175 \text{ lb t}}{144 \text{ lb av}}\right)\left(\frac{12 \text{ oz t}}{\text{lb t}}\right)\left(\frac{\$1.00}{\text{oz t}}\right) = \$3710$$

The cost of recovering the silver = 1000(0.06) = $60. The maximum that could by paid for AgNO$_3$ solution = 3710 − 60 = $3650 $\blacktriangleright$

8-48 Wt. 3

A company owns a limestone quarry containing 1 million tons of CaCO$_3$. If it costs $12.00 to mine and convert a ton of CaCO$_3$ into quicklime (CaO), and the CaO will sell for $0.015/lb, what is the value of the quarry?

$$\text{Molecular weights: CaCO}_3 = 100 \quad \text{CaO} = 56$$

Solution
$$CaCO_3 \rightarrow CaO + CO_2\uparrow$$

From the equation of the reaction and the molecular weights we see that 100 tons of $CaCO_3$ yield 56 tons of CaO. Therefore, 100×10^4 tons of $CaCO_3$ will yield 56×10^4 tons of CaO.

$$\text{Value of CaO} = 56 \times 10^4 \left(2000 \frac{\text{lb}}{\text{ton}}\right)\left(\frac{0.015}{\text{lb}}\right) = \$16.8 \times 16^6$$

$$\text{Mining cost} = \$12(1 \times 10^6) = \$12 \times 10^6$$

The net value of the quarry $= \$16.8 \times 10^6 - \$12 \times 10^6 = \$4.8 \times 10^6$ ▶

8-49 Wt. 4

The number of liters of oxygen necessary to burn 10 liters of H_2S gas according to the reaction $2H_2S + 3O_2 \rightarrow 2H_2O + 2SO_2$ is

(a) 3
(b) 6
(c) 10
(d) 15
(e) 30

Solution

Using a properly balanced formula we could easily calculate the molecular weights of the gases and of the resulting products. This is not necessary in this problem because we know that equal numbers of molecules of gases occupy equal volumes.

Here we have 2 molecules of hydrogen sulfide and 3 molecules of oxygen reacting. We see, therefore, that 2 volumes of H_2S will react with 3 volumes of O_2.

For 10 liters of H_2S, $\frac{3}{2} \times 10 = 15$ liters of oxygen will be required ▶

Answer is (d)

8-50 Wt. 6

Chlorine may be prepared by oxidizing hydrochloric acid with manganese dioxide, MnO_2. The by-products are water and manganous ion, Mn^{++}. What weight of MnO_2 will be required to produce 10 liters of chlorine, measured at standard temperature and pressure? It is understood that excess hydrochloric acid is present.

Atomic weights: $Mn = 54.9$ $O = 16$ $Cl = 35.5$

Solution

The chemical equation for this process is:

$$4H^+ + 2Cl^- + MnO_2 \rightarrow Cl_2 + Mn^{++} + 2H_2O$$

From this equation it is apparent that 1 mole of MnO_2 is required to produce 1 mole of Cl_2. And we know that 1 g-mole of a gas occupies 22.4 liters at standard conditions.

$$\text{Molecular weight of } MnO_2 = 54.9 + 2(16) = 86.9$$

So 86.9 g of MnO_2 yield 22.4 liters of Cl_2.

For 10 liters of Cl_2, $\dfrac{10}{22.4} \times 86.9 = 38.8$ g of MnO_2 required ▶

8-51 Wt. 6

Ferrous chloride ($FeCl_2$) is oxidized by potassium dichromate ($K_2Cr_2O_7$) in aqueous solution containing an excess of hydrochloric acid (HCl) to form ferric chloride ($FeCl_3$) and chromic chloride ($CrCl_3$).

(*a*) Write the balanced chemical equation for the reaction.
(*b*) How many pounds of ferric chloride would be formed from 3.50 lb of ferrous chloride in the above reaction?

Atomic weights: Cl = 35.5   Cr = 52   H = 1   Fe = 55.8
O = 16   K = 39.1

Solution

(*a*) $$\overset{++}{Fe}Cl_2 + K_2\overset{++++++}{Cr_2}O_7 + HCl \rightarrow \overset{+++}{Fe}Cl_3 + \overset{+++}{Cr}Cl_3 + H_2O$$

Using the valence-change method of balancing equations, we see that the valence of Fe has changed from +2 to +3 and of Cr from +6 to +3. Thus every iron atom adds in positive valence by 1 and every chromium atom loses in positive valence by 3. These changes are associated with the transfer of electrons and we can write the partial equations and their sum:

$$\begin{array}{ll} 3Fe^{++} \quad -3e^- \rightarrow 3Fe^{+++} & \text{oxidation} \\ Cr^{++++++} \quad +3e^- \rightarrow Cr^{+++} & \text{reduction} \\ \hline 3Fe^{++} + Cr^{++++++} \rightarrow 3Fe^{+++} + Cr^{+++} & \end{array}$$

We can now start to balance the equation:

$$3FeCl_2 + K_2Cr_2O_7 + HCl \rightarrow 3FeCl_3 + CrCl_3 + KCl + H_2O$$

Since there are 2 Cr atoms on the left, there must be 2 on the right and we must double the Fe atoms to keep our partial equations balanced. With 2K on the left, there must be 2K on the right. The equation at this point is:

$$6FeCl_2 + K_2Cr_2O_7 + HCl \rightarrow 6FeCl_3 + 2CrCl_3 + 2KCl + H_2O$$

That makes 26Cl on the right so there must be $26 - 12 = 14$HCl on the left. 14H on the left would yield $7H_2O$ on the right.

The balanced equation is:

$$6FeCl_2 + K_2Cr_2O_7 + 14HCl \rightarrow 6FeCl_3 + 2CrCl_3 + 2KCl + 7H_2O \blacktriangleright$$

(b) Molecular weights:

$$FeCl_2: 55.8 + 2(35.5) = 126.8$$

$$FeCl_3: 55.8 + 3(35.5) = 162.3$$

From the balanced equation we see that 6 moles of $FeCl_2$ yield 6 moles of $FeCl_3$. Thus 6(126.8) lb of $FeCl_2$ yield 6(162.3) lb of $FeCl_3$.

For 3.5 lb of $FeCl_2$ the yield is $\dfrac{3.5}{6(126.8)} \times 6(162.3) = 4.48$ lb of $FeCl_3$ $\blacktriangleright$

8-52 Wt. 2

A compound was found to contain 42.12 wt % carbon, 51.45 wt % oxygen, and 6.43 wt % hydrogen. Its molecular weight was determined to be approximately 340. What is the chemical formula of the compound?

Atomic weights: $C = 12$ $H = 1$ $O = 16$

Solution

One molecular weight of the compound contains:

```
C   42.12 × 340 = 143.21 ÷ 12 = 11.93   or   12 at. wt of C
O   51.45 × 340 = 174.93 ÷ 16 = 10.93   or   11 at. wt of O
H    6.43 × 340 =  21.86 ÷  1 = 21.86   or   22 at. wt of H
    ──────                ──────
    100.00                340.00
```

The formula of the compound contains:

12C   11O   22H

We would recognize this as the formula for plain sugar, which would be correctly written $C_{12}H_{22}O_{11}$ $\blacktriangleright$

8-53 Wt. 3

A sample of air is found to have a pressure P_1, volume V, and temperature T_1. After the oxygen has been removed, the remaining nitrogen is found to have a pressure P_2, volume V, and temperature T_2. Express the ratio of the number of moles of nitrogen to the number of moles of oxygen in this sample in terms of the given pressures, volume, and temperatures.

420 / Chemistry

Solution

The ideal gas equation states:

$$PV = NRT$$

where P = pressure
V = volume
N = total moles of gas
R = gas constant
T = absolute temperature

So for the sample of air

$$P_1V = NRT_1 = (N_{O_2} + N_{N_2})RT_1 = N_{O_2}RT_1 + N_{N_2}RT_1$$

Solving,

$$N_{O_2} = \frac{P_1V - N_{N_2}RT_1}{RT_1}$$

Also,

$$P_2V = N_{N_2}RT_2 \qquad N_{N_2} = \frac{P_2V}{RT_2}$$

Therefore

$$N_{O_2} = \frac{P_1V - \left(\frac{P_2V}{RT_2}\right)RT_1}{RT_1} = \frac{V\left(P_1 - \frac{T_1}{T_2}P_2\right)}{RT_1}$$

$$\frac{N_{N_2}}{N_{O_2}} = \frac{P_2V}{RT_2} \cdot \frac{RT_1}{V\left(P_1 - \frac{T_1}{T_2}P_2\right)} = \frac{P_2T_1}{\left(P_1 - \frac{T_1}{T_2}P_2\right)T_2}$$

$$= \frac{P_2T_1}{P_1T_2 - P_2T_1} = \frac{1}{\left(\frac{P_1}{P_2}\right)\left(\frac{T_2}{T_1}\right) - 1} \blacktriangleright$$

8-54 Wt. 6

Acetylene (C_2H_2) is made by the action of water upon calcium carbide according to the following reaction:

$$CaC_2 + 2H_2O \rightarrow Ca(OH)_2 + C_2H_2$$

How many cubic feet of acetylene at atmospheric pressure of 14.7 psi and 60°F will be produced from 1 ton (2000 lb) of calcium carbide? The gas constant for acetylene is 59.4 ft-lb/lb-°R.

Atomic weights: $C = 12$ $\quad$ $Ca = 40.1$ $\quad$ $H = 1$ $\quad$ $O = 16$

Solution

Molecular weights:
$$CaC_2: 40.1 + 2(12) = 64.1$$
$$C_2H_2: 2(12) + 2(1) = 26$$

Hence 64.1 lb of CaC_2 will produce 26 lb of C_2H_2. So 2000 lb of CaC_2 will produce $2000 \times 26/64.1 = 812$ lb of C_2H_2. Using the ideal gas equation $PV = WRT$, where P is in lb/ft² abs and T is in °R (abs),

$$14.7 \times 144 \times V = 812 \times 59.4(60 + 460)$$

$$V = \frac{812 \times 59.4 \times 520}{14.7 \times 144} = 11{,}840 \text{ ft}^3 \blacktriangleright$$

8-55 Wt. 5

How many cubic feet of SO_2, measured at 760 mm mercury and a temperature 500°F, can be made by roasting 2000 lb. of ZnS with oxygen?

Atomic weights: Zn = 65.4   S = 32.1   O = 16

The gas constant R is 1544 ft-lb/lb-mole-°R.

Solution

$$2ZnS + 3O_2 \rightarrow 2SO_2 + 2ZnO$$

Molecular weights:
$$ZnS = 65.4 + 32.1 = 97.5$$
$$SO_2 = 32.1 + 2(16) = 64.1$$

2 moles of ZnS yield 2 moles of SO_2, or 97.5 lb of ZnS yield 1 lb-mole of SO_2. For 2000 lb of ZnS, $2000/97.5 = 20.5$ lb-moles of SO_2 are formed. Applying the equation $PV = WRT$,

$$V = \frac{WRT}{P} = (20.5 \text{ lb-moles})\left(1544 \frac{\text{ft-lb}}{\text{lb-mole °R}}\right)(500 + 460)\left(\frac{1}{14.7 \text{ psia} \times 144}\right)$$

$$= 14{,}350 \text{ ft}^3 \blacktriangleright$$

8-56 Wt. 6

Methane gas, CH_4, and aluminum hydroxide, $Al(OH)_3$, are produced by treating aluminum carbide, Al_4C_3, with water, H_2O.

(a) Write the balanced chemical equation.

(b) What is the volume of methane at atmospheric pressure (14.7 psia) and 40°F that can be produced from 10 lb of aluminum carbide?

Atomic weights: Al = 27   C = 12   H = 1   O = 16

The gas constant for CH_4 is 96.4 ft-lb/lb-°R.

Solution

(a) The balanced equation is

$$Al_4C_3 + 12H_2O \rightarrow 4Al(OH)_3 + 3CH_4 \blacktriangleright$$

(b) Molecular weights:

$$Al_4C_3 = 4(27) + 3(12) = 144 \qquad CH_4 = 12 + 4(1) = 16$$

Thus 144 lb of aluminum carbide will produce $3(16) = 48$ lb of methane. So, 10 lb of Al_4C_3 will produce 3.33 lb of CH_4. Using the ideal gas equation $PV = WRT$ and substituting in absolute units (note also that pressure must be in pounds per square foot):

$$14.7 \times 144 \times V = 3.33 \times 96.4(40 + 460)$$

$$V = \frac{3.33 \times 96.4 \times 500}{14.7 \times 144} = 75.8 \text{ ft}^3 \blacktriangleright$$

8-57 Wt. 6

A solution contains 5.56 lb of potassium carbonate (K_2CO_3) per gallon at 68°F. The specific gravity of the solution is 1.477. An acid is added to the solution, and carbon dioxide and water vapor are evolved at 745 mm Hg and 110°F. How many cubic feet of gas are obtained from 100 lb of the initial solution, assuming that an excess of acid is added and that the carbon dioxide has negligible solubility in the reaction mixture?

Data: 1 lb-mole of gas occupies 359 ft³ at 32°F and 14.7 psia
Vapor pressure of water at 110°F = 1.275 psia
Vapor pressure of the reaction mixture at 110°F = 1.078 psia
Atomic weights: $C = 12$ $H = 1$ $O = 16$ $K = 39.1$

Solution

The balanced partial equation representing the reaction is:

$$K_2CO_3 + 2H^{++} \rightarrow CO_2 \cdot H_2O\uparrow + 2K^+$$

100 lb of solution at a specific gravity of 1.477 equals

$$\frac{100}{1.477 \times 8.33} = 8.13 \text{ gal of solution}$$

At 5.56 lb/gal the weight of K_2CO_3 in solution is $5.56 \times 8.13 = 45.2$ lb.

Molecular weight of $K_2CO_3 = 2(39.1) + 12 + 3(16) = 138.2$

Hence

$$\frac{45.2}{138.2} = 0.326 \text{ lb-mole of } K_2CO_3$$

Thus there is 0.326 lb-mole of CO_2 produced.

Partial volume of $CO_2 = 0.326 \times 359 = 117 \text{ ft}^3$ at 32°F and 14.7 psia
(standard conditions)

But the total pressure of the gas is 745 mm of Hg, or

$$745 \times \frac{14.7}{760} = 14.4 \text{ psia at } 110°F$$

The vapor pressure of the reaction mixture is 1.078 psia. Since total pressure equals the sum of the partial pressures, the partial pressure of CO_2 is

$$14.4 - 1.1 = 13.3 \text{ psia}$$

We want the volume of CO_2 at 13.3 psia and 110°F:

$$\frac{P_1 V_1}{T_1} = \frac{P_2 V_2}{T_2}$$

$P_1 = 14.7 \text{ psia} \quad T_1 = 32° + 460° = 492° \text{ abs}$
$P_2 = 13.3 \text{ psia} \quad T_2 = 110° + 460° = 570° \text{ abs}$
$V_1 = 117 \text{ ft}^3 \quad V_2 = ?$

$$V_2 = \frac{T_2 P_1 V_1}{T_1 P_2} = \frac{570}{492} \times \frac{14.7}{13.3} \times 117 = 150 \text{ ft}^3 \ CO_2 \text{ at 13.3 psia and 110°F} \blacktriangleright$$

8-58 Wt. 5

Calculate the molecular weight of a gas, 1.13 liters of which, collected over water at 24°C and 754 mm pressure, weighed 1.251 g when deprived of the aqueous vapor. The vapor pressure of water at 24°C is 22.2 mm. The gas constant $R = 0.082$ liter-atm/g-mole-°K.

Solution

To apply the equation $PV = WRT$, we first must correct for the water vapor present:

$$P_{\text{total}} = P_{\text{gas}} + P_{\text{water}}$$

$$P_{\text{gas}} = P_t - P_w = (754 - 22.2) = 731.8 \text{ mm Hg} = \frac{731.8}{760} \text{ atm}$$

The number of moles of the gas, W, is $\dfrac{\text{weight of gas }(m)}{\text{molecular weight }(M)}$.

$$PV = WRT = \frac{m}{M} RT$$

$$M = \frac{mRT}{PV} = \frac{1.251 \text{ g} \times 0.082(24 + 273)}{(731.8/760) \times 1.13} = 28 \text{ g/mole}$$

Thus the molecular weight of the gas is 28 ▶

8-59 Wt. 4

In many petroleum refineries hydrogen sulfide gas (H_2S) is burned with the stoichiometric amount of air to produce elementary sulfur (S).

(a) Write the balanced chemical equation for this reaction.

(b) If a certain plant produces 20 long tons of S per day with an overall recovery of 97%, how many cubic feet per hour at 80°F and 15 psig of H_2S gas must have been burned?

Data: Air contains 20.9 vol. % oxygen, 79.1 % nitrogen
 1 long ton contains 2240 lb
 1 lb-mole of gas occupies 359 lb/ft³ at 32°F and 14.7 psia

 Atomic weights: H = 1   O = 16   N = 14   S = 32.1

Solution

(a) $$H_2S + O_2 \rightarrow S + H_2O$$

In balancing the equation we see that the valence of S has changed from -2 to 0 and the valence of O has changed from 0 to -2. The partial equations of the reaction are:

$$S^{--} - 2e^- \rightarrow S$$
$$\tfrac{1}{2}O_2 + 2e^- \rightarrow O^{--}$$

Their sum: $S^{--} + \tfrac{1}{2}O_2 \rightarrow S + O^{--}$

or $2S^{--} + O_2 \rightarrow 2S + 2O^{--}$

Now, writing the equation, we get:

$$2H_2S + O_2 \rightarrow 2S + 2H_2O \blacktriangleright$$

We find that the equation balances and therefore is the answer to part (a)

(b)

$$20 \text{ long tons S per day} = \frac{20 \times 2240}{24} = 1867 \text{ lb S produced per hour}$$

The theoretical amount that should have been produced is:

$$\frac{1867}{0.97} = 1925 \text{ lb S per hour}$$

Molecular weights:

$$H_2S = 2(1) + 32.1 = 34.1 \qquad S = 32.1$$

The pound-moles of S theoretically produced are

$$\frac{1925}{32.1} = 60 \text{ lb-moles/hr}$$

From the balanced equation:

$$2 \text{ lb-moles of } H_2S \text{ yield 2 lb-moles of S}$$

Hence we see that 60 lb-moles of H_2S are required per hour. The volume of gas is:

$$60 \times 359 = 21{,}540 \text{ ft}^3 \text{ } H_2S \text{ at } 32°F \text{ and } 14.7 \text{ psia}$$

But the problem asks for the volume at 80°F and 15 psig. The universal gas law is:

$$\frac{P_1 V_1}{T_1} = \frac{P_2 V_2}{T_2}$$

(absolute temperatures and pressures must be used in this equation)

$P_1 = 14.7$ psia $\qquad P_2 = 15 + 14.7 = 29.7$ psia
$T_1 = 32° + 460° = 492°$ abs $\qquad T_2 = 80° + 460° = 540°$ abs
$V_1 = 21{,}540$ ft^3 $\qquad V_2 = ?$

$$V_2 = \frac{T_2 P_1 V_1}{T_1 P_2} = \frac{540}{492} \times \frac{14.7}{29.7} \times 21{,}540 = 11{,}700 \text{ ft}^3 \text{ } H_2S \blacktriangleright$$

8-60 Wt. 6

One lb-mole of a gas at 745 mm Hg pressure and 450°F contains the following components:

O_2 21 wt % $\qquad CO_2$ 30 wt % $\qquad N_2$ 25 wt % $\qquad H_2O$ 24 wt %

(a) Calculate its density in pounds per cubic foot.
(b) At what temperature will condensation of water begin when the gas is cooled at constant pressure?
(c) How many pounds of water will be condensed when the gas is cooled to 100°F?

426 / Chemistry

Data: The gas law constant R is 1545 ft-lb/lb-mole-°R.

Molecular weights: $O_2 = 32$ $CO_2 = 44$ $N_2 = 28$ $H_2O = 18$

VAPOR PRESSURE OF WATER

Temperature (°F)	Vapor Pressure (Inches of Hg)
100	1.933
120	3.446
140	5.881
160	9.652
180	15.291
200	23.467
220	34.992

Solution

(a) From the ideal gas equation

$$PV = NRT = \frac{\text{wt } RT}{\text{lb/mole}}$$

where P = absolute pressure
V = volume
R = gas constant
T = absolute temperature
N = total number of moles
wt = total weight of gas

or

$$P = \frac{\text{wt}}{V} \frac{RT}{\text{lb/mole}} = D \frac{RT}{\text{lb/mole}}$$

or

$$\text{density } D = \frac{P \times \text{lb/mole}}{RT}$$

For the gas in question, if we take 100 lb, we will have

$$\frac{21 \text{ lb}}{32 \text{ lb/mole}} = 0.656 \text{ mole } O_2 \qquad \frac{25 \text{ lb}}{28 \text{ lb/mole}} = 0.893 \text{ mole } N_2$$

$$\frac{30 \text{ lb}}{44 \text{ lb/mole}} = 0.682 \text{ mole } CO_2 \qquad \frac{24 \text{ lb}}{18 \text{ lb/mole}} = 1.333 \text{ mole } H_2O$$

Therefore, for the gas the value of pounds/mole is

$$\frac{100 \text{ lb}}{(0.656 + 0.893 + 0.682 + 1.333) \text{ moles}} = \frac{100}{3.564} = 28 \text{ lb/mole}$$

$$D = \frac{745 \text{ mm}}{760 \text{ mm/atm}} \times \frac{14.7 \text{ lb}}{\text{in}^2\text{-atm}} \times \frac{144 \text{ in}^2}{1 \text{ ft}^2} \times \frac{28 \text{ lb}}{\text{lb/mole}} \times \frac{\text{lb-mole-}°\text{R}}{1545 \text{ ft-lb}} \times \frac{1}{910°\text{R}}$$

$D = 0.0414 \text{ lb/ft}^3$ ▶

(b) The mole fraction of water in the original gaseous mixture will be:

$$\frac{1.333}{3.564} = 0.374$$

As the pressure is held constant at $\frac{745 \text{ mm Hg}}{25.4 \text{ mm/in}} = 29.33$ in. Hg, condensation will occur at that temperature when saturation pressure is reached, which will be at 0.374×29.33 in. Hg = 11.0 in. Hg. From the table and by interpolation we arrive at a value of about 165°F. A graphical plot, Fig. 8-3, produces better results, giving a value of 166°F ▶

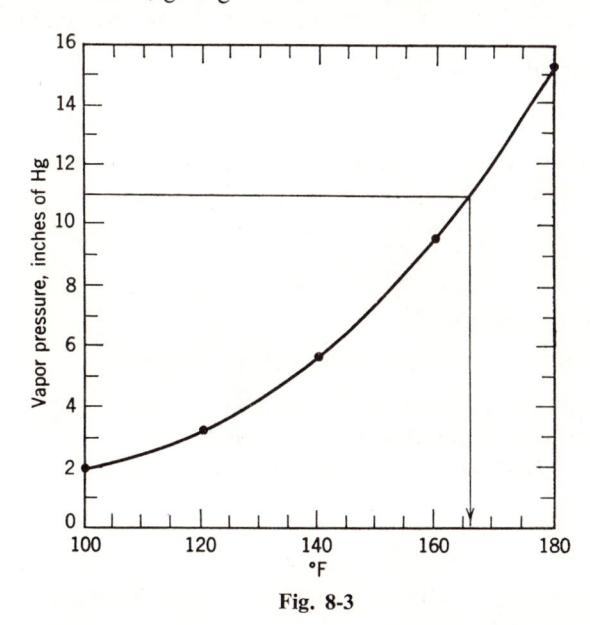

Fig. 8-3

(c) The mole fraction of water left in the gas at constant pressure and at 100°F will be

$$N_{\text{H}_2\text{O}} \times \frac{745 \text{ mm}}{25.4 \text{ mm/in.}} = 1.933 \text{ in.}$$

or

$$\text{mole fraction } N_{H_2O} = 0.0659 \text{ at } 100°F$$

Therefore

$$0.0659 = \frac{\text{moles } H_2O \text{ left}}{2.231 + \text{moles } H_2O \text{ left}}$$

where 2.231 is the total moles other than water originally in the gaseous mixture.

$$\frac{1}{0.0659} - 1 = \frac{2.231 + n_{H_2O}}{n_{H_2O}} - \frac{n_{H_2O}}{n_{H_2O}} = \frac{2.231}{n_{H_2O}}$$

$$14.17 = \frac{2.231}{n_{H_2O}}$$

$$n_{H_2O} = 0.158 \text{ mole left}$$

$$1.333 - 0.158 = 1.175 \text{ moles condensed}$$

1 lb-mole of the gas or 28 lb originally contained 28×0.24 or 6.72 lb of water. $\frac{1.175 \times 100}{1.333} = 88.1\%$ condenses, so we know that $0.881 \times 6.72 = 5.92$ lb of water will condense ▶

8-61 Wt. 3

A dry mixture of clay and barium sulfate has a density of 3.45 g/cc. The respective densities of the constituents are 2.67 g/cc and 4.10 g/cc. Compute the weight percent and volume percent of each constituent in the mixture.

Solution

Basis: 100 cc of mixture

Let x = volume clay and $100 - x$ = volume of $BaSO_4$.

$$\text{Wt clay} + \text{wt } BaSO_4 = \text{total wt}$$
$$2.67x + 4.10(100 - x) = 3.45(100)$$
$$2.67x + 410 - 4.10x = 345$$
$$-1.43x = -65$$

Therefore

$$x = 45.5 = \% \text{ vol. clay ▶}$$
$$100 - x = 54.5 = \% \text{ vol. } BaSO_4 \text{ ▶}$$

Weight = volume × density
Wt clay = $(45.5)(2.67) = 121.5 \text{ g} = 35.2\%$ wt clay ▶
Wt $BaSO_4$ = $(54.5)(4.10) = 223.5 \text{ g} = 64.8\%$ wt $BaSO_4$ ▶

total wt = $345.0 \text{ g} = 100.0\%$

8-62 Wt. 6

Compute the number of pounds of barium sulfate that must be added per barrel of mud to balance a formation pressure of 2860 psig at a depth of 6080 ft. The density of the mud is 65.2 lb/ft³ and the density of the barium sulfate is 4.05 g/cc. (Assume 1 barrel is equivalent to 42 gal.)

Solution

Since gage pressure is given, this represents the pressure above atmospheric pressure which must be balanced. As a result the column of mud in the hole must produce a pressure of 2860 psi at a depth of 6080 ft.

$$\text{Required mud density} = \frac{2860 \times 144}{6080} = 67.7 \text{ lb/ft}^3$$

$$\text{Density of barium sulfate} = 4.05 \times 62.45 = 253 \text{ lb/ft}^3$$

Let x = % mud expressed as a decimal and $1 - x$ = % barium sulfate expressed as a decimal,

$$65.2x + 253(1 - x) = 67.7$$

$$253 - 67.7 = (253 - 65.2)x$$

$$x = \frac{185.3}{187.8} = 0.9867$$

% barium sulfate by volume = $1.00 - 0.9867 = 0.0133 = 1\frac{1}{3}\%$

Amount of barium sulfate per barrel = $0.0133 \times 42 \times 8.33 \times 4.05$

$$= 18.9 \text{ lb} \blacktriangleright$$

8-63 Wt. 4

The crude oil feed to a petroleum refinery is 25,000 bbl/day of petroleum having a specific gravity of 0.88 and containing 1.45 wt % of sulfur. On an overall refinery basis 75% of this sulfur is converted to H_2S, which is partially oxidized to elemental sulfur S with 97% recovery. The elemental S is sent to a sulfuric acid plant where there is 99% recovery in the form of 98 wt % H_2SO_4 solution. How many pounds of 98 wt % H_2SO_4 can be produced per day?

$$1 \text{ bbl} = 42 \text{ U.S. gallons}$$

Atomic weights: H = 1   O = 16   S = 32.1

Solution

$$25,000 \frac{\text{bbl}}{\text{day}} \times 42 \frac{\text{gal}}{\text{bbl}} \times 8.34 \frac{\text{lb}}{\text{gal}} \times 0.88 \times 0.0145 = 112,000 \text{ lb S/day}$$

$$S \xrightarrow{75\%} H_2S \xrightarrow{97\%} S \xrightarrow{99\%} H_2SO_4$$

So 112,000 lb S/day × 0.75 × 0.97 × 0.99 = 80,500 lb S/day into 100% H_2SO_4. From the equation 32.1 lb S will produce 2(1) + 32.1 + 4(16) = 98.1 lb of 100% H_2SO_4 or 98.1/0.98 = 100 lb of 98% H_2SO_4. Therefore

$$\frac{80{,}500 \text{ lb S/day} \times 100 \text{ lb } 98\% \text{ H}_2\text{SO}_4}{32.1 \text{ lb S}} = 251{,}000 \text{ lb } 98\% \text{ H}_2\text{SO}_4/\text{day}$$

Acid production = 251,000 lb 98% H_2SO_4/day ▶

8-64 Wt. 6

An ammonia synthesis reactor received a total feed containing H_2 and N_2 in a 3:1 ratio as required by the reaction $3H_2 + N_2 \rightarrow 2NH_3$. The total feed to the reactor (T moles) is made up of fresh (F moles) and a recycle stream (R moles). The reactor effluent (E moles) contains 18.0 mole % NH_3, the remainder being unreacted H_2 and N_2. The reactor effluent goes to a separator where substantially all of the ammonia is removed as product (P moles), and the remainder is the recycle stream (R moles) returned to the reactor. If F is 100 moles, determine the numerical values of T, R, E, and P expressed as moles, when the plant is operating in a continuous steady-state manner.

Solution

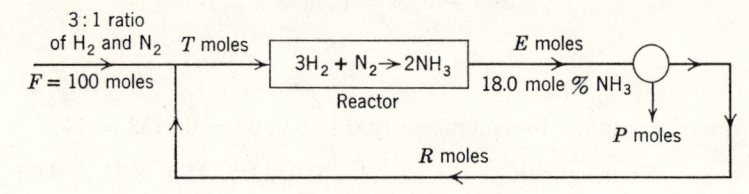

Fig. 8-4

$$3H_2 + N_2 \rightarrow 2NH_3$$

Basis: 1 mole N_2 entering reactor

At the exit of the reactor, let x = the moles of N_2 that have reacted and let N_{N_2} = number of moles of N_2 present, etc. Then

$$N_{N_2} = 1 - x \text{ moles}$$
$$N_{H_2} = 3(1 - x) \text{ moles}$$
$$N_{NH_3} = 2x \quad \text{moles}$$
$$\overline{N_{total} = 1 - x + 3 - 3x + 2x = 2(2 - x) \text{ moles}}$$

But $\dfrac{N_{NH_3}}{N_{total}}(100) = 18.0$. Thus $\dfrac{2x}{2(2-x)} = 0.180$. Solving, $x = 0.305$. Thus

$$N_{N_2} = 1 - x = 0.695 \text{ mole}$$
$$N_{H_2} = 3(1 - x) = 2.085 \text{ moles}$$
$$N_{NH_3} = 2x = 0.610 \text{ mole}$$
$$\overline{N_{total} = 3.390 \text{ moles}}$$

The compositions of the various streams may now be tabulated:

moles	T	E	P	R	F*
N_2	1	0.695	0	0.695	0.305
H_2	3	2.085	0	2.085	0.915
NH_3	0	0.610	0.610	0	0
Total	4	3.390	0.610	2.780	1.220

* By a material balance around input to reactor, $F = T - R$.

Basis: 100 moles of feed

The calculations above correspond to 1.22 moles of feed. The numbers in the table, therefore, need to be scaled by the factor $(100/1.22) = 82$.

$$T = 82 \times 4 = 328 \text{ moles} \blacktriangleright$$
$$R = 82 \times 2.78 = 228 \text{ moles} \blacktriangleright$$
$$E = 82 \times 3.39 = 278 \text{ moles} \blacktriangleright$$
$$P = 82 \times 0.61 = 50 \text{ moles} \blacktriangleright$$

8-65 Wt. 1

Which of the following metals has the slowest rate of combining with oxygen to form metallic oxide?

(a) aluminum
(b) copper
(c) iron
(d) magnesium
(e) zinc

Solution

These metals when placed in order of activity as indicated by the emf scale would be:

> magnesium
> aluminum
> zinc
> iron
> copper

with the most active metal placed first.

Metals of the copper subgroup (Cu, Ag, and Au) are known for their inactivity as a result of their low oxidation potentials. These are the coinage metals. As the oxidation potential of Cu is low and nearest to that for the reduction potential of oxygen in the presence of water, it will thus have the slowest rate of reaction ▶

Answer is (b)

8-66 Wt. 1

Which of the following is the most chemically active (highest in electromotive series)?

(a) cesium
(b) potassium
(c) sodium
(d) strontium
(e) magnesium

Solution

Electromotive Series

1. lithium	9. aluminum
2. potassium ▶	10. manganese
3. cesium	11. zinc
4. barium	·
5. strontium	·
6. calcium	·
7. sodium	silver
8. magnesium	gold

Answer is (b)

8-67 Wt. 1

The concentration of H_2SO_4 is given in terms of moles per liter as 0.2 M—H_2SO_4. The normal concentration of this solution would be designated as

(a) 0.05 N—H_2SO_4
(b) 0.1 N—H_2SO_4
(c) 0.2 N—H_2SO_4
(d) 0.4 N—H_2SO_4
(e) 0.5 N—H_2SO_4

Solution

A molar solution contains a gram-molecular weight of the solute in 1 liter of solution. A normal solution contains a gram-equivalent weight of the solute in 1 liter of solution.

Since valence = $\dfrac{\text{atomic wt}}{\text{equivalent wt}}$ and in H_2SO_4 the radical is bivalent, it follows that a 1 M solution of H_2SO_4 is also a 2 N solution. In this problem the 0.2 M solution could also be called a 0.4 N solution ▶

Answer is (d)

8-68 Wt. 6

The rate of flow of a certain stream of water is to be found by the "injection" method. A solution containing 100 g of NaCl in 400 g of water is added at constant rate over a 2-min period. Titration with standard silver nitrate solution shows the chloride content of the water before injection to be 50 ppm, and after injection the chloride content of the water downstream was found to be 250 ppm. Find the flow rate of the water stream in gallons per minute.

Atomic weights: $Cl = 35.5 \quad H = 1 \quad N = 14 \quad O = 16$
$Ag = 107.9 \quad Na = 23$

Solution

ppm = parts per million by weight. From the problem we can see that the addition of 50 g of NaCl per minute raised the chloride content of the stream by 200 ppm. The chloride added per minute = $\dfrac{\text{wt Cl}^-}{\text{wt NaCl}} \times 50 = \dfrac{35.5}{23 + 35.5} \times 50 = 30.3$ g.

$$30.3 \text{ g} = \dfrac{200}{1,000,000} \times \text{total wt of liquid flowing}$$

$$\text{Total wt of liquid} = \dfrac{30.3}{0.0002} = 151,500 \, \dfrac{g}{\min} = 151.5 \, \dfrac{kg}{\min}$$

But 1 kg = 2.205 lb, so the weight of flow is

$$151.5 \times 2.205 = 334 \text{ lb/min}$$

Since 1 gal of water weighs 8.33 lb,

$$\text{Flow rate} = \dfrac{334}{8.33} = 40.1 \, \dfrac{\text{gal}}{\min}$$

To be strictly correct and find the true stream flow we would have to deduct our 200 g/min addition or approximately 0.5 gal/min from the total flow.

$$\text{Stream flow} = 40.1 - 0.5 = 39.6 \text{ gal/min} \blacktriangleright$$

8-69 Wt. 1

A solution of 0.1 N hydrochloric acid is required. The number of liters of water that should be added to 1 liter of 2.0 N hydrochloric acid in order to obtain the desired solution is nearest

(a) 19
(b) 20
(c) 21
(d) 22
(e) 23

Atomic weights: H = 1   O = 16   Cl = 35.5

Solution

A normal solution contains a gram-equivalent weight of the solute in 1 liter of solution.

$$x = \text{total quantity}$$

$$2.0(1 \text{ liter}) = 0.1(x \text{ liters})$$

$$x = \frac{2.0}{0.1} = 20 \text{ liters}$$

Therefore 1 liter of 2.0 N HCl yields 20 liters of 0.1 N HCl. So add $20 - 1 = 19$ liters of water ▶

Answer is (a)

8-70 Wt. 4

What volumes of 2 N H_2CO_3 and 6 N H_2CO_3 must be mixed to yield 700 ml of a 3 N H_2CO_3 solution? Assume all volumes are additive.

Solution

Let $x =$ ml of 2 N H_2CO_3 and $y =$ ml of 6 N H_2CO_3. Now we can write two expressions:

Total volume $x + y = 700$ (1)

Normality × volume $2x + 6y = 3(700)$ (2)

$$x = 700 - y \quad (1)$$

Substitute into equation (2)

$$2(700 - y) + 6y = 2100$$
$$1400 - 2y + 6y = 2100$$
$$y = \tfrac{700}{4} = 175 \text{ ml of 6 } N \text{ } H_2CO_3 \blacktriangleright$$
$$x = 700 - 175 = 525 \text{ ml of 2 } N \text{ } H_2CO_3 \blacktriangleright$$

8-71 Wt. 3

The common commercial preparation of magnesium metal is said to be by electrolysis of a mixture of 70% magnesium chloride and 30% sodium chloride.

Atomic weights: Mg = 24.32   Cl = 35.45

(a) Would 5 kg of this mixture be enough to yield 1 kg of magnesium? Justify your answer.

(b) During an electrolytic production of 1 kg of magnesium metal, what volume of chlorine would be formed as a by-product?
Assume a temperature of 127°C and a pressure of 740 mm Hg.

Solution

(a) 5 kg of mixture contains $0.70 \times 5 = 3.5$ kg of $MgCl_2$. The weight of Mg in 3.5 kg of $MgCl_2 = \dfrac{24.32}{24.32 + 2(35.45)} \times 3.5 = 0.892$ kg.

Thus 5 kg of the mixture could not yield 1 kg of magnesium ▶

(b) $$MgCl_2 \rightarrow Mg + Cl_2$$

Thus 1 mole of $MgCl_2$ yields 1 mole of Mg metal and 1 mole of Cl_2. Gram-moles of Mg metal = $\dfrac{1000}{24.32} = 41.1$. 41.1 g-moles of Cl_2 are also produced. At standard conditions 1 mole of gas equals 22.4 liters. So at standard conditions the 41.1 moles of Cl_2 occupy $41.1 \times 22.4 = 922$ liters. At 127°C and 740 mm Hg it would occupy

$$922 \left(\frac{273 + 127}{273}\right)\left(\frac{760}{740}\right) = 1386 \text{ liters } \blacktriangleright$$

8-72 Wt. 6

Iodine reacts with sodium thiosulfate in aqueous solution according to the following unbalanced equation:

$$I_2 + Na_2S_2O_3 \rightarrow NaI + Na_2S_4O_6$$

(a) Balance the above equation.

(b) A 0.333-g sample of impure iodine required 16.7 ml of $N/10$ sodium thiosulfate solution to react with the iodine in the sample. Calculate the weight percent of iodine in the sample.

Atomic weights: H = 1   I = 126.9   O = 16   Na = 23   S = 32.1

Solution

(a) Using the valence-change method of balancing equations, we note that the valence of I has changed from 0 to -1 and 2S have changed from $+4$ to $+5$. The partial equations for the transfer of electrons are:

$$2I^0 + 2e^- \rightarrow 2I^-$$
$$4S^{++} - 2e^- \rightarrow 4S^{++\frac{1}{2}}$$

$$\overline{I_2 + 2S_2 \rightarrow 2I + S_4}$$

Hence the equation for the reaction is

$$I_2 + 2Na_2S_2O_3 \rightarrow 2NaI + Na_2S_4O_6$$ ▶

(b) A normal solution contains 1 g-equivalent weight of a substance per liter of solution; therefore, liters × normality equals the number of gram-equivalents, or

$$\frac{16.7}{1000} \times 0.1 = 0.00167 \text{ g-equivalents of sodium thiosulfate}$$

Solutions react such that equivalents react so that we also have 0.00167 g-equivalents of I_2.

A gram-equivalent of a reductant is that amount of material that uses up Avogadro's number of electrons, or for I_2 it will be

$$\frac{2 \times 126.9}{2} = 126.9 \text{ g}$$

The number of equivalents of I_2 will be the weight of I_2 in grams divided by its gram-equivalent weight. Therefore,

$$0.00167 = \frac{wt}{126.9} \text{ or wt. of } I_2 = 0.212 \text{ g}$$

$$\% I_2 = \frac{0.212 \times 100}{0.333} = 63.6\%$$ ▶

8-73 Wt. 6

A total of 30.6 ml of silver nitrate solution was used in titrating 3.62 ml of formation water for chloride content. The silver nitrate solution contained 14.796 g of $AgNO_3$ per liter. Compute for the formation water:

(a) parts per million of Cl
(b) grains per gallon of Cl
(c) grains per gallon of NaCl

Note: 1 grain/gal = 17.12 ppm

Atomic weights: Cl = 35.5   O = 16   Na = 23   N = 14   Ag = 107.9

Solution

(a) $$Ag^+NO_3^- + Cl^- \rightarrow AgCl\downarrow + NO_3^-$$

or

$$Ag^+ + Cl^- \rightarrow AgCl\downarrow$$

The gram-equivalent weight of AgNO$_3$ is its formula weight divided by its valence of 1 or $107.9 + 14 + 3(16) = 169.9$ g/liter. The normality of the silver nitrate solution is the number of gram-equivalents per liter of solution, or

$$N = \frac{14.796}{169.9}$$

Now, liters × N = number of gram-equivalents or ml × N = number of mg-equivalents.

$$30.6 \times \frac{14.796}{169.9} = \text{number of mg-equivalents of AgNO}_3$$

In reactions, equivalents react with equivalents; therefore

$$30.6 \times \frac{14.796}{169.9} = \text{number of mg-equivalents of Cl}^-$$

But the number of equivalents is also equal to the weight of Cl$^-$ divided by its equivalent weight of 35.5, or $\frac{wt}{35.5 \text{ g/g-equiv.}}$, and $\frac{wt \times 1000}{35.5}$ is equal to the number of mg-equivalents. Therefore,

$$30.6 \times \frac{14.796}{169.9} = \frac{\text{wt of Cl}^- \times 1000}{35.5} \quad \text{or wt of Cl}^- = 0.0945 \text{ g}$$

As this weight of Cl$^-$ was contained in 3.62 ml or 3.62 g, then the amount in 10^6 g or ppm would be:

$$\text{ppm Cl}^- = \frac{0.0945}{3.62} \times 10^6 = 26{,}100 \text{ ppm} \blacktriangleright$$

(b) $\qquad 26{,}100 \times \dfrac{1 \text{ grain/gal}}{17.12 \text{ ppm}} = 1530 \text{ grains Cl}^-/\text{gal} \blacktriangleright$

(c) $\quad 1530 \text{ grains Cl}^-/\text{gal} \times \dfrac{58.5 \text{ NaCl}}{35.5 \text{ Cl}^-} = 2520 \text{ grains NaCl/gal} \blacktriangleright$

8-74 Wt. 4

It is desired to produce 100 g-moles of compound AD by reacting compounds A$_2$B and CD$_2$ at a constant temperature of 25°C.

$$A_2B + CD_2 \rightarrow BC + 2AD \pm Q \text{ kcal}$$

The heats of formation of these compounds from the elements A, B, C, and D at 25°C are as follows:

A_2B 100 kcal/g-mole
CD_2 200 kcal/g-mole
BC 300 kcal/g-mole
AD 150 kcal/g-mole

Assume that the reactants react completely to form the products. How much heat must be added or removed during the reaction to maintain the desired temperature of 25°C? Use − sign if heat is removed and + sign if heat is added.

Solution

50 g-moles A_2B + 50 g-moles CD_2 → 50 g-moles BC + 100 g-moles AD

50(100 kcal) + 50(200 kcal) → 50(300 kcal) + 100(150 kcal) ± Q kcal

5000 kcal + 10,000 kcal → 15,000 kcal + 15,000 kcal ± Q kcal

For a heat balance

15,000 kcal → 30,000 kcal − 15,000 kcal

Therefore

$Q = -15{,}000$ kcal ▶

8-75 Wt. 6

The solubility of anhydrous sodium sulfate (Na_2SO_4) in water is determined experimentally to be

42.5 lb/100 lb water at 100°C

5.38 lb/100 lb water at 0°C

100 lb of water is saturated at 100°C with anhydrous Na_2SO_4. The solution is cooled, with an evaporation loss of 6.2 lb, to 0°C. During this cooling, crystallization of the solid decahydrate ($Na_2SO_4 \cdot 10H_2O$) takes place. How many pounds of the solid crystals of the decahydrate can be obtained?

Atomic weights: Na = 23   S = 32.07   O = 16   H = 1.01

Solution

$M_{Na_2SO_4} = 2(23) + 32.07 + 4(16) = 142.07$

$M_{Na_2SO_4 \cdot 10H_2O} = 2(23) + 32.07 + 4(16) + 20(1.01) + 10(16) = 322.27$

Basis: 100 lb of water
Let

X = weight ↓ $Na_2SO_4 \cdot 10H_2O$ crystallized out

Overall material balance:

$$142.5 \text{ lb initially}$$
$$6.2 \text{ lb evaporated}$$
$$X \text{ lb Na}_2\text{SO}_4 \cdot 10\text{H}_2\text{O crystallized}$$

The solution left is then $(142.5 - 6.2 - X) = (136.3 - X)$ lb

Na_2SO_4 balance:

$$42.5 = \frac{5.38}{105.38}(136.3 - X) + \frac{142.07}{322.27}X$$

$$42.5 = 0.0511(136.3 - X) + 0.441X = 6.96 + 0.39X$$

Thus

$$X = \frac{42.5 - 6.96}{0.39} = \frac{35.54}{0.39} = 91.0 \text{ lb Na}_2\text{SO}_4 \; 10\text{H}_2\text{O crystallized} \blacktriangleright$$

(As a check the H_2O balance can be calculated. From it a value of $X = 91.3$ lb was obtained.)

8-76 Wt. 6

A binary mixture consists of 65 mole % nonane and 35 % decane. Assuming equilibrium conditions and ideal solutions, calculate

(a) the bubble-point pressure at 360°F
(b) the dew-point pressure at 360°F

Vapor pressures of pure components, psia.

Temperature (°F)	Nonane	Decane
310	16.1	8.9
320	18.3	10.3
330	20.8	11.9
340	23.6	13.6
350	26.5	15.6
360	30.0	17.8
370	33.7	20.3
380	37.5	22.8

Solution

Nomenclature: X_i = mole fraction of component i in the liquid phase
Y_i = mole fraction of component i in the vapor phase
p_i = partial pressure of component i at 360°F
p_i° = vapor pressure of component i at 360°F
P = total pressure

(a) Bubble point. All mixture is in liquid phase, so that

$$X_A = 0.65$$
$$X_B = 0.35$$
$$X_A + X_B = 1.0$$

Basis: 1 mole of mixture
(Raoult's law)

$$p_A = X_A p_A^\circ = 0.65(30.0) = 19.5 \text{ psia}$$
$$p_B = X_B p_B^\circ = 0.35(17.8) = \underline{6.24} \text{ psia}$$

so that the total pressure = 25.74 psia ▶

(b) Dew point. All mixture is in vapor phase, so that

$$Y_A = 0.65$$
$$Y_B = 0.35$$

For the first bit of liquid condensed,

$$X_A + X_B = 1$$

Thus

$$p_A = Y_A P = 0.65P$$
$$p_B = Y_B P = 0.35P$$

so that

$$\left. \begin{array}{l} X_A = \dfrac{p_A}{p_A^\circ} = \dfrac{0.65}{30.0} P \\[1em] X_B = \dfrac{p_B}{p_B^\circ} = \dfrac{0.35}{17.8} P \end{array} \right\} \text{ sum is equal to 1}$$

Then

$$\left(\frac{0.65}{30.0} + \frac{0.35}{17.8} \right) P = 1$$

$$(0.0216 + 0.0197)P = 1$$

$$0.0413P = 1 \quad \text{or} \quad P = 24.2 \text{ psia} \blacktriangleright$$

8-77 Wt. 6

Dissolving a petroleum product in a particular hydrocarbon solvent is known to lower the freezing point by an amount which is proportional to the mole percent solute. Properties are given as follows:

Molecular weight of hydrocarbon solvent, 78.1 g
Density of hydrocarbon solvent, 0.879 g/ml
Freezing point lowering for 1 g-mole in 1000 g hydrocarbon solvent, 5.00°C
Density of petroleum product, 0.783 g/ml

A mixture containing 10.03 ml of the hydrocarbon solvent and 1.46 ml of the petroleum product freezes at 4.34°C below the freezing point of the pure hydrocarbon solvent. Calculate for the mixture:

(a) mole percent petroleum product
(b) number of moles of petroleum product
(c) average molecular weight of petroleum product

Solution

$$\text{Gram-mole of solvent} = \frac{1000}{78.1} = 12.81$$

$$\frac{1 \text{ g-mole solute}}{12.81 \text{ g-mole solvent}} = 0.078 = \frac{5.00°C}{100\% \text{ reduction}}$$

$$100\% \text{ reduction of freezing point} = \frac{5.00}{0.078} = 64.1°C$$

Mixture:

$$10.03 \text{ ml solvent} \times 0.879 = 8.82 \text{ g solvent}$$
$$1.46 \text{ ml solute} \times 0.783 = 1.15 \text{ g solute}$$

$$\text{Gram-mole of solvent} = \frac{8.82}{78.1} = 0.113$$

(a) $$\frac{4.34°C}{64.1°C} = 0.0678 \frac{\text{mole solute}}{\text{mole solvent}}$$

$$\text{mole \% petroleum product (solute)} = \frac{0.0678 \times 10^2}{1 + 0.0678} = 6.36\% \blacktriangleright$$

(b) $$0.0678 \frac{\text{mole solute}}{\text{mole solvent}} \times 0.113 \text{ g-mole solvent}$$
$$= 0.00766 \text{ moles of petroleum product} \blacktriangleright$$

(c) average molecular weight = grams/mole

$$= \frac{1.15}{0.00766} = 150 \blacktriangleright$$

Chapter 9

Electricity

The study and use of electrical energy in an ever-increasing variety of ways is the goal of an entire engineering discipline, that of electrical engineering. And as this field continues its rapid growth, the importance of a fundamental knowledge of electric circuits and electromechanical energy conversion grows with it. The problems collected here primarily emphasize a few basic principles of direct-current (d-c) and alternating-current (a-c) circuits and motors.

D-C CIRCUITS

In d-c circuitry a potential difference or voltage e across the terminals of some resistive device causes an electric current i to flow through it. According to Ohm's law,

$$e = iR \tag{9-1}$$

where R is the resistance, in ohms, of the element to current flow, and the current is measured in amperes. Furthermore, power must be expended in sustaining this current flow in the amount

$$p = ei = i^2 R \tag{9-2}$$

The units of this equation are volt-amperes or watts. The driving force for such current flows is called an electromotive force (emf). Typical sources of emf are batteries, generators, and motors.

For resistors of uniform geometric shape, the resistance R is a function of material properties and the size of the element. For a uniform wire, for example,

$$R = \rho \frac{L}{A} \tag{9-3}$$

where ρ is the specific resistance or resistivity (a material property) and L and A are, respectively, the length and cross-sectional area of the wire.

Kirchhoff's two laws define the relations which exist when individual elements are combined to form an electric circuit:

1. The sum of all currents flowing into a junction must equal the sum of all currents flowing away from the junction.
2. The algebraic sum of the emf's and the voltage drops around a closed circuit is equal to zero.

Rule one assures that charge is conserved at a point. In a large circuit or network containing n junctions, this rule must be applied $n - 1$ times; the rule is then automatically satisfied at the last junction. Rule two is in a way a generalized application of Ohm's law to a closed circuit. For larger networks the principle should be applied once to each elementary loop or mesh. In the use of these rules care must be taken so that a chosen sign convention is used consistently.

In d-c circuits resistors and emf's may be connected in a seemingly bewildering variety of ways for different purposes. Most, but not all, resistor combinations are sequences of either the basic series or parallel arrangements of individual resistors. In these cases the single equivalent series resistance R_e is

$$R_e = \sum_{n=1}^{N} R_n \qquad (9\text{-}4)$$

for a series of N individual resistors and

$$\frac{1}{R_e} = \sum_{n=1}^{N} \frac{1}{R_n} \qquad (9\text{-}5)$$

for a parallel combination of N resistors. The Y-Δ transformation, an important exception, is treated in the problems.

Example 1

Solve for the five currents shown, Fig. 9-1, of the diagram of a d-c network.

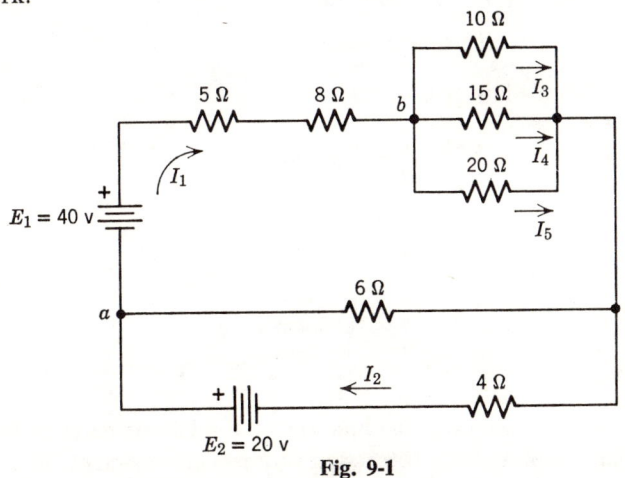

Fig. 9-1

Solution

We begin by applying Kirchhoff's first rule. This network may be considered to have only two junctions if the parallel combination of resistors is replaced by one equivalent resistor R_e in series. Examining the junction marked a, the current through the 6-ohm resistor I_6 to the left is

$$I_6 = I_1 - I_2$$

Next we replace the three resistors by the one equivalent series resistor R_e:

$$\frac{1}{R_e} = \frac{1}{10} + \frac{1}{15} + \frac{1}{20} = \frac{13}{60}$$

$$R_e = \frac{60}{13} \text{ ohms}$$

Now we can apply Kirchhoff's second rule to the two loops to obtain

$$40 - (5 + 8 + \tfrac{60}{13})I_1 - 6(I_1 - I_2) = 0 \qquad (1)$$

$$20 - 6(I_2 - I_1) - 4I_2 = 0 \qquad (2)$$

From equation (1), $I_2 = 0.1(20 + 6I_1)$. Substituting the relation into equation (2) gives $I_1 = 2.6$ amp ▶

Then we find $I_2 = 3.56$ amp ▶

Returning to the parallel combination of resistors, the voltage drop across this combination is, by use of Ohm's law,

$$V_e = I_1 R_e = (2.6)(\tfrac{60}{13}) = 12 \text{ volts}$$

This voltage drop occurs across each of these three resistors, while the use of Kirchhoff's first rule at point b shows that

$$I_1 = I_3 + I_4 + I_5$$

Repeated use of Ohm's law gives

$$I_3 = \frac{V_e}{10} = 1.2 \text{ amp} \blacktriangleright$$

$$I_4 = \frac{V_e}{15} = 0.8 \text{ amp} \blacktriangleright$$

$$I_5 = \frac{V_e}{20} = 0.6 \text{ amp} \blacktriangleright$$

A-C CIRCUITS

Ohm's law expresses only one of three possible relations between voltage and current, that resulting from the action of a resistive element. Two other relations exist and are characterized by two electrical elements: the inductance and the capacitance.

For an inductance

$$e = L \frac{di}{dt} \tag{9-6}$$

where L is the inductance in henries when i is in amperes and e is in volts. For a sinusoidal alternating current the current lags 90° behind the voltage in a purely inductive circuit, as can be seen by assuming a sinusoidal form for i and computing e.

Capacitance, the third electrical element, directly relates the charge q to the voltage e. Thus the current-voltage relation is

$$i = \frac{dq}{dt} = C \frac{de}{dt} \tag{9-7}$$

and C is measured in farads. In contrast to the inductive circuit, the current leads the voltage by 90° in a purely capacitive a-c circuit. Each phase relation contrasts with the purely resistive element which in itself causes no phase shift.

In a-c circuits we are usually interested in effective or rms values of the voltage or current rather than the maximum value of the quantity over a cycle. The relations are

$$I = \frac{I_m}{\sqrt{2}} \quad \text{and} \quad E = \frac{E_m}{\sqrt{2}}$$

where the subscript m stands for the maximum value over a cycle. The effective values are unsubscripted since they are the commonly used quantities.

Resistance, inductance, and capacitance are all generally present in a-c circuits. When elements are connected in series, a voltage drop between two points is the sum of the individual *phasor* voltage drops; for a parallel combination of elements, the total current at a junction is the *phasor* sum of individual currents through the parallel elements. Phasor algebra is the same as vector algebra since phasors have a magnitude and orientation, but we must remember that the angle is a phase angle between two sinusoidally varying quantities and not a definite direction in space.

The relation between current and voltage is conveniently given by

$$E = IZ \tag{9-8}$$

where these quantities are all phasor quantities, and Z is the impedance. If the inductive and capacitive reactances X_L and X_C are defined in terms of frequency f, inductance L, and capacitance C as

$$X_L = 2\pi f L \qquad X_C = -\frac{1}{2\pi f C} \qquad (9\text{-}9)$$

then the relations between impedance Z, reactance X, and resistance R, as shown in Fig. 9-2 for $X > 0$, are $Z^2 = R^2 + X^2$, $\tan \theta = X/R$ with the

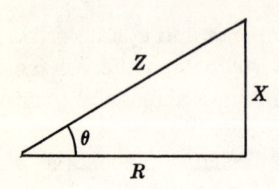

Fig. 9-2

reactance $X = X_L + X_R$. Note however that X may be either positive or negative.

An alternative to the use of the impedance triangle is the volt-ampere method which is based on a-c power considerations. Most a-c equipment is rated in volt-amperes VA just as d-c power is the product EI. Because of the phase shift θ, however, the useful average power in an a-c circuit is $I^2 R = P = EI \cos \theta$. $\cos \theta$ is called the power factor. The reactive power is $Q = I^2 X = EI \sin \theta$. These relations are summarized in Fig. 9-3, the volt-ampere triangle.

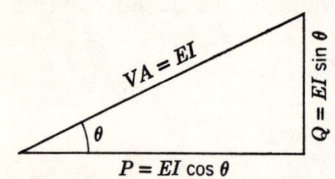

Fig. 9-3

Equations (9-9) show that reactance is a function of frequency. By selecting this frequency so that $X = 0$, one can maximize the current in the system and at the same time maximize the power by operating with a power factor of unity. This is called the resonant frequency; at this frequency the response to a given voltage input will also be greatest.

Example 2

In a 440-volt a-c circuit there is a resistance and an inductance load which requires 30 amp at 0.80 power factor. Calculate

 (a) volt-amperes
 (b) average power
 (c) reactive volt-amperes
 (d) resistance R
 (e) impedance Z
 (f) reactance X_L

Solution

We apply the volt-ampere method here.

(a) Volt-amperes $VA = EI = (440)(30) = 13{,}200$ va ▶
(b) Average power $P = EI \cos \theta = (440)(30)(0.8) = 10{,}560$ watts ▶
(c) Reactive volt-amperes $Q = EI \sin \theta = (440)(30)(0.6) = 7920$ vars ▶
(d) The resistance R is given by the relation $P = I^2R$, or

$$R = \frac{P}{I^2} = \frac{10{,}560}{(30)^2} = 11.73 \text{ ohms} \blacktriangleright$$

(e) The impedance relation is $I^2Z = VA = EI$. The impedance magnitude is

$$Z = \frac{EI}{I^2} = \frac{E}{I} = \frac{440}{30} = 14.67 \text{ ohms} \blacktriangleright$$

with a phase angle $\theta = \cos^{-1} 0.8 = 37°$.

(f) The expression I^2X_L is an alternative expression for reactive volt-amperes. Hence

$$X_L = \frac{Q}{I^2} = \frac{7920}{(30)^2} = 8.8 \text{ ohms} \blacktriangleright$$

Example 3

A coil with an inductance of 1 henry and a resistance of 50 ohms is connected in series with a 0.01-μf capacitor and a variable frequency supply, as in Fig. 9-4.

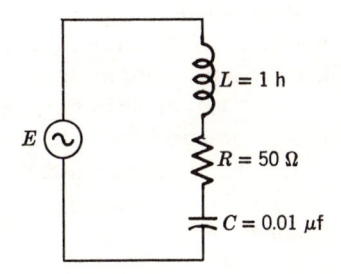

$L = 1$ h
$R = 50 \, \Omega$
$C = 0.01 \, \mu f$

Fig. 9-4

(a) At what frequency f_0 will the maximum voltage appear across the capacitor if the supply voltage is held constant as the frequency is varied?
(b) Using the frequency f_0, what is the maximum supply voltage which may be used in this circuit if the capacitor voltage rating is 200 volts?

Solution

(a) The voltage is maximized at resonance when $X = 0$:

$$X = 2\pi f_0 L - \frac{1}{2\pi f_0 C} = 0$$

Solving,

$$2\pi f_0 = \frac{1}{(LC)^{1/2}} = \frac{1}{[1(10^{-8})]^{1/2}} = 10^4 \text{ rad/sec}$$

$$f_0 = \frac{10^4}{2\pi} = 1592 \text{ cps} \blacktriangleright$$

(b) At the frequency f_0 the current $I_0 = E/R$. The maximum capacitor voltage $E_c = 200 = I_0 X_C$, where $X_C = -(2\pi f_0 C)^{-1}$. Combining these relations and solving for E yields

$$E = I_0 R = \frac{E_c}{X_c} R = 2\pi f_0 C E_c R$$

$$E = (10^4)(10^{-8})(200)(50) = 1.00 \text{ volt} \blacktriangleright$$

ELECTRICAL MACHINERY

Both electric motors and generators utilize the same fundamental principles in their operation. For this reason the analysis of either type of machine is essentially the same. All these machines are basically usable as *either* a motor or a generator; it is only our optimization of the primary and secondary operating characteristics of a given machine so that it performs better in one role than in the other that has led us to separate the two.

Both a-c and d-c machines are widely used. By various means electric and magnetic fields, fields set up by the current flow through the machine, create a mechanical torque, or vice versa. However, there are sufficiently many different variations in the actual design of the major types of machinery that it is impractical to outline them here. One sample case is presented next; others may be found in the problems.

Example 4

A d-c shunt motor running at no load draws 5.0-amp armature current and 1.0-amp field current from a 120-volt line. Its speed is 1190 rpm. When loaded, the armature current increases to 50 amp without adjustment of the field. Neglecting the effect of armature reaction, what will be the speed under this load condition? The armature resistance is 0.2 ohm.

Solution

One primary characteristic of this motor is the relative constancy of the armature voltage over a wide speed range.

Letting E_a = armature voltage
I_a = armature current
R_a = armature resistance
V = terminal voltage

$$E_a = V - I_a R_a$$

At no load,
$$E_a = 120 - (5.0)(0.2) = 119 \text{ volts}$$

Under load,
$$E_a = 120 - (50)(0.2) = 110 \text{ volts}$$

For a given machine, the armature voltage E_a is proportional to the product of the field strength ϕ and the armature speed n in rpm, or

$$E_a = K\phi n$$

where K is a proportionality constant dependent on the machine. Here ϕ is also constant so that

$$\frac{E_{a_1}}{E_{a_2}} = \frac{n_1}{n_2}$$

$$n_2 = \tfrac{110}{119}(1190) = 1100 \text{ rpm} \blacktriangleright$$

9-1 Wt. 1

A 0.02 µf capacitor consists of two parallel plates of 100 in.² each, with an air gap of 0.001 in. between them. If the air gap is increased to 0.0015 in., what will be the resulting capacitance?

(a) 0.0133 µf
(b) 0.030 µf
(c) 0.009 µf
(d) 0.045 µf
(e) 0.010 µf

Solution

Capacitance is proportional to A/d. Hence

$$\frac{C_2}{C_1} = \left(\frac{A_2}{d_2}\right)\left(\frac{d_1}{A_1}\right) \quad \text{and} \quad C_2 = C_1\left(\frac{d_1}{d_2}\right) \quad \text{since } A \text{ is constant}$$

$$C_2 = (0.02)\left(\frac{0.001}{0.0015}\right) = 0.0133 \text{ µf} \blacktriangleright$$

Answer is (a)

9-2 Wt. 1

A condenser has a capacitance of 4000 farads and is given a charge of 80 coulombs. The potential difference between the plates is

(a) 5 volts
(b) 50 volts
(c) 0.04 volt
(d) 0.02 volt
(e) 0.2 volt

Solution

$$C = \frac{Q}{V}$$

where C = capacitance in farads
Q = quantity of electricity in coulombs
V = difference in potential in volts

$$V = \frac{Q}{C} = \frac{80}{4000} = 0.02 \text{ volt} \blacktriangleright$$

Answer is (d)

9-3 Wt. 1

The following equation represents a basic law of electricity. Select the man to whom the law is attributed.

$$W = RI^2t$$

where W = energy
R = resistance
I = current
t = time

(a) Ohm
(b) Faraday
(c) Lenz
(d) Kirchhoff
(e) Joule

Solution

In 1845 J. P. Joule demonstrated that the energy expended in "driving the current" through a resistor was converted into heat. He showed experimentally that the heat generated in a conductor per unit time was proportional to the square of the current ▶

Answer is (e)

9-4 Wt. 1

The following equation represents a basic law of electricity. Select the man to whom the law is attributed.

$$e_L = -L\frac{di}{dt}$$

where e_L = emf of self-induction

$\frac{di}{dt}$ = time rate of change of current

L = coefficient of self-induction

(a) Ohm
(b) Faraday
(c) Lenz
(d) Kirchhoff
(e) Joule

Solution

This is Faraday's law of electromagnetic induction ▶
The equation is sometimes written

$$e_L = -N\frac{d\phi}{dt}$$

Answer is (b)

9-5 Wt. 6

A copper rod 1 in. in diameter and 10 ft long is found to have a resistance of 1.02×10^{-4} ohm. If the rod is drawn into a wire with a uniform diameter of 0.05 in., what would be the resistance of the wire?

Solution

The general equation is

$$R = \rho\frac{L}{A}$$

where R = resistance in ohms
ρ = specific resistance in ohm-circular mils per foot
L = length of wire in feet
A = area in circular mils

452 / Electricity

For the rod,
$$1 \text{ in. in diameter} = 1000 \text{ mils in diameter}$$
$$A = 10^3 \times 10^3 = 10^6 \text{ cir mils}$$
$$\rho = \frac{RA}{L} = \frac{1.02 \times 10^{-4} \times 10^6}{10} = 10.2 \text{ ohm-cir mils/ft}$$

For the drawn wire,
$$0.05 \text{ in. in diameter} = 50 \text{ mils in diameter}$$
$$A = 50 \times 50 = 2500 \text{ cir mils}$$
$$\text{Volume of wire: } \frac{\pi}{4} 1^2 \times 10 \times 12 = 30\pi \text{ in}^3$$

$$\text{Length} = \frac{\text{volume}}{\text{area}} = \frac{30\pi}{\frac{\pi}{4}(0.05)^2} = \frac{120}{0.0025} = 48{,}000 \text{ in.} = 4000 \text{ ft}$$

$$\text{New } R = \rho \frac{L}{A} = 10.2 \frac{4000}{2500} = 16.3 \text{ ohms} \blacktriangleright$$

9-6 Wt. 1

A device that produces coherent light with predictable properties that can be controlled in a manner comparable to signals at radio and microwave frequencies is called a

(a) radar
(b) capacitor
(c) photoelectric cell
(d) reflector
(e) laser

Solution

The laser (*L*ight *A*mplification by *S*timulated *E*mission of *R*adiation) emits coherent light (a narrow, intense beam whose waves are all nearly parallel, in phase, and of the same wavelength) ▶

Answer is (*e*)

9-7 Wt. 1

Cathode rays are

(a) high-energy X rays
(b) alpha particles
(c) protons
(d) electrons
(e) neutrons

Solution

Here we are dealing with a flow of electrons ▶

Answer is (d)

9-8 Wt. 1

A kilowatt-hour is a unit of

(a) momentum
(b) power
(c) acceleration
(d) energy
(e) impulse

Solution

A kilowatt-hour is a measure of work or energy. Energy = Power × Time. Power, on the other hand, is work per unit time. It might be measured in watts or kilowatts, not kilowatt-hours ▶

Answer is (d)

9-9 Wt. 3

List three ways in which a current or voltage may be developed.

Solution

Sources of voltage:

▶ (1) Generators (a-c or d-c). Electromechanic conversion. Rotate coil through the field of a magnet.

▶ (2) Battery (d-c). Electrochemical conversion. Use oxidation-reduction in the cells.

▶ (3) Junction (d-c). Heat a junction of dissimilar metals and complete the circuit at a different temperature. A thermocouple uses this principle.

9-10 Wt. 1

Permittivity is expressed by the Greek letter ε, and its defining equation is Coulomb's law:

$$\varepsilon = \frac{QQ'}{4\pi F S^2}$$

where Q and Q' = two point charges
F = force between these charges
S = distance between these charges

454 / Electricity

Permittivity in the mks system is

(a) coulombs²/newton-meter²
(b) coulombs²/dyne-meter²
(c) statcoulombs²/dyne-meters²
(d) coulombs²/newton-centimeter²
(e) statcoulombs²/gram-centimeter²

Solution

Permittivity in the mks system is coulombs²/newton-meter² ▶

Answer is (a)

9-11 Wt. 5

Given the Δ and Y circuits shown, determine the resistance of x, y, and z, so that the two circuits are equivalent.

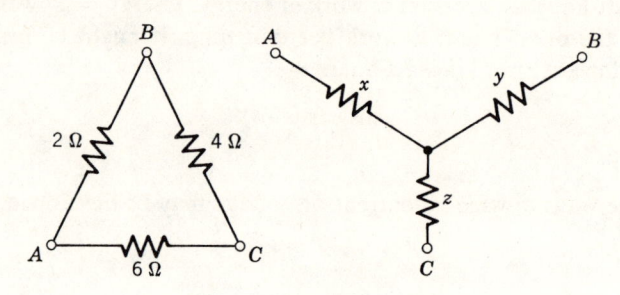

Fig. 9-5

Solution

First, determine the line-to-line resistance for the Δ circuit:

$$AB: \quad \frac{1}{R_{AB}} = \frac{1}{2} + \frac{1}{6+4} = 0.6 \qquad R_{AB} = \frac{1}{0.6} = 1.67 \text{ ohms}$$

$$BC: \quad \frac{1}{R_{BC}} = \frac{1}{4} + \frac{1}{2+6} = 0.375 \qquad R_{BC} = \frac{1}{0.375} = 2.67 \text{ ohms}$$

$$CA: \quad \frac{1}{R_{CA}} = \frac{1}{6} + \frac{1}{4+2} = 0.333 \qquad R_{CA} = \frac{1}{0.333} = 3.0 \text{ ohms}$$

Similarly, determine the line-to-line resistance for the Y circuit:

$$AB: \quad R_{AB} = R_x + R_y$$
$$BC: \quad R_{BC} = R_y + R_z$$
$$CA: \quad R_{CA} = R_z + R_x$$

For the circuits to be equivalent, the line-to-line resistances must be equal. Therefore,

$$R_{AB} = R_x + R_y \qquad = 1.67 \qquad (1)$$
$$R_{BC} = \qquad R_y + R_z = 2.67 \qquad (2)$$
$$R_{CA} = R_x \qquad + R_z = 3.0 \qquad (3)$$

Solving these three equations simultaneously,

(1) $\quad R_x + R_y \qquad = 1.67$

$-$(3) $\quad \underline{-R_x \qquad\quad - R_z = -3.0}$

$\qquad\qquad R_y - R_z = -1.33$

(2) $\qquad R_y + R_z = 2.67$

$\qquad\quad \overline{2R_y \qquad\quad = 1.34} \qquad R_y = \dfrac{1.34}{2} = 0.67 \text{ ohm} \blacktriangleright$

(2) $\qquad 0.67 + R_z = 2.67 \qquad R_z \qquad = 2.00 \text{ ohms}\blacktriangleright$

(1) $\qquad R_x + 0.67 \quad = 1.67 \qquad R_x \qquad = 1.00 \text{ ohm} \blacktriangleright$

9-12 Wt. 1

In a Y-connected circuit the line current equals

(a) the phase current

(b) $\dfrac{1}{\sqrt{3}}$ times the phase current

(c) $\sqrt{3}$ times the phase current

(d) $\dfrac{\sqrt{3}}{2}$ times the phase current

(e) $\dfrac{2}{\sqrt{3}}$ times the phase current

Solution

For the balanced three-phase Y connection, line current equals phase current $\blacktriangleright$

Answer is (a)

9-13 Wt. 3

A Δ and a Y capacitance configuration are shown in Fig. 9-6 Find C_Y, in terms of C_Δ, so that the capacitances between any corners of the Δ

456 / Electricity

circuit are equal to the capacitances between corresponding points in the Y circuit.

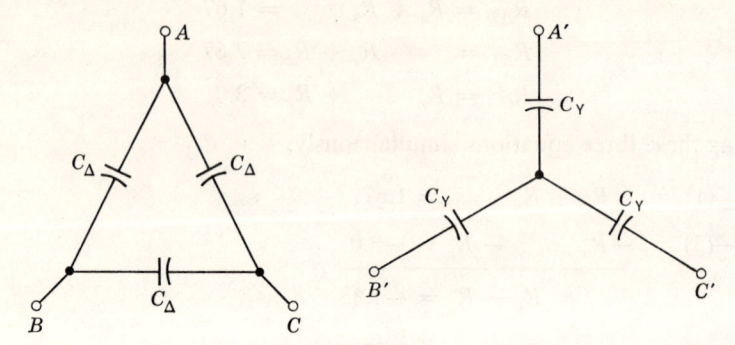

Fig. 9-6

Solution

To find C_Y in terms of C_Δ, we must find the equivalent capacitances between corresponding terminals of the Y and Δ connections:

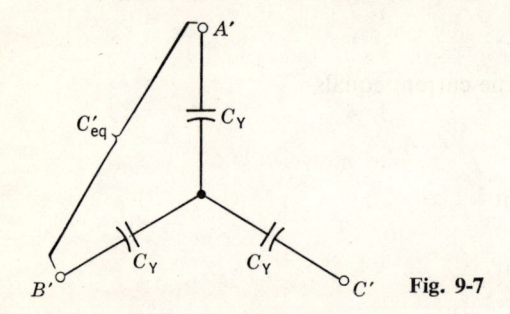

Fig. 9-7

For the Y connection with C' open: Between A' and B',

$$C'_{eq} = \frac{(C_Y)(C_Y)}{2C_Y}$$

since the capacitors are in series, and

$$C'_{eq} = \frac{C_a C_b}{C_a + C_b}$$

for C_a and C_b in series. Hence

$$C'_{eq} = \frac{C_Y}{2}$$

For the Δ connection, with C open as before,

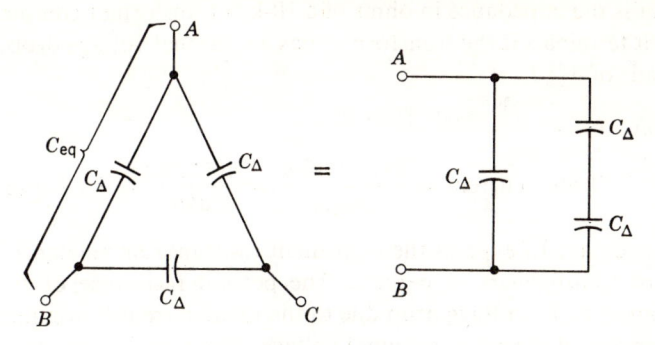

Fig. 9-8

Consequently,

$$C_{eq} = C_\Delta + \frac{C_\Delta^2}{2C_\Delta} = \frac{3C_\Delta}{2}$$

But $C'_{eq} = C_{eq}$ if the two configurations are equivalent between terminals AB and $A'B'$ or between any other terminals, and

$$C'_{eq} = C_{eq}, \quad \frac{C_Y}{2} = \frac{3C_\Delta}{2}, \quad \text{or} \quad C_Y = 3C_\Delta \blacktriangleright$$

9-14 Wt. 4

A bank of three transformers is to be connected to reduce the voltage from a three-phase, 12,000-volt (line-to-line) distribution line to supply power for a small irrigation pump driven by a 440-volt induction motor. A Y connection will be used for the primary and a Δ connection for the secondary.

(a) What should be the primary voltage rating of each transformer?
(b) What should be the secondary voltage rating of each transformer?

Solution

(a) In a Y connection

$$V_{line} = \sqrt{3}\, V_{phase}$$

Hence

$$V_{phase} = \frac{12,000}{\sqrt{3}} = 6920 \text{ volts} = \text{primary rating} \blacktriangleright$$

(b) In a Δ connection

$$V_{line} = V_{phase} = 440 \text{ volts} = \text{secondary rating} \blacktriangleright$$

458 / Electricity

9-15 Wt. 5

What is the impedance in ohms of a 10-kva transformer computed from its 480-volt terminals if the transformer has an internal voltage drop, no load to full load, of 5%?

Solution

$$\text{Rated current} = \frac{\text{kva} \times 1000}{\text{volts}} = \frac{10(1000)}{480} = 20.8 \text{ amp}$$

It is customary to express the equivalent resistance or reactance (or impedance) of transformers in percent. The percent resistance or reactance (or impedance) is the voltage drop due to the rated current flowing, expressed as a percentage of the rated terminal voltage. What might have been requested was the value of the internal impedance if the percent impedance was 5%. Then the answer would be

$$Z_{eq} = \frac{5\% \, E_{rated}}{I_{rated}} = \frac{(0.05)(480)}{20.8} = 1.15 \text{ ohms}$$

The problem as stated (and worth 5 points) is more complex. One must recall the equivalent circuit of the transformer:

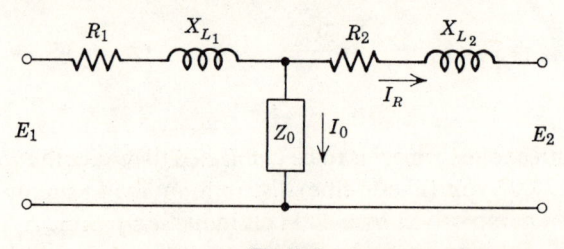

Fig. 9-9

Since the magnetizing current I_0 is much smaller than the rated current I_R, one may neglect the effect of the shunt impedance Z_0. Also, since the resistance-reactance ratio is not given, one could assume that the X_L's are much larger than the R's. Then the very approximate equivalent circuit is:

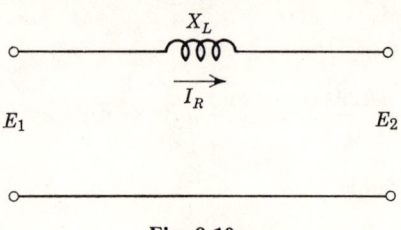

Fig. 9-10

where
$$L = L_1 + L_2 \quad \text{or} \quad X_L = X_{L_1} + X_{L_2}$$

From the voltage triangle one may now compute the series reactance voltage drop.

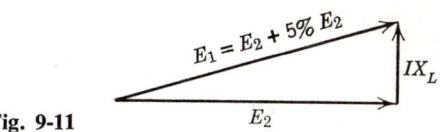

Fig. 9-11

$$IX_L = \sqrt{E_1^2 - E_2^2}$$

where
$$E_1 = 480 + 24 = 504 \text{ volts}$$
$$IX_L = \sqrt{504^2 - 480^2} = 153 \text{ volts}$$

Therefore
$$Z \approx X_L = \frac{153 \text{ volts}}{20.8 \text{ amp}} \approx 7.3 \text{ ohms} \blacktriangleright$$

9-16 Wt. 3

What is the combined capacitance of five condensers, each of 5-μf capacity, when connected in

(a) series?
(b) parallel?

Solution

(a) Condensers in series:
$$\frac{1}{C_{eq}} = \frac{1}{C_1} + \frac{1}{C_2} + \frac{1}{C_3} + \frac{1}{C_4} + \frac{1}{C_5}$$
$$\frac{1}{C_{eq}} = 5\left(\frac{1}{5\,\mu\text{f}}\right) \qquad C_{eq} = 1\,\mu\text{f} \blacktriangleright$$

(b) Condensers in parallel:
$$C_{eq} = C_1 + C_2 + C_3 + C_4 + C_5 = 5(5)$$
$$= 25\,\mu\text{f} \blacktriangleright$$

9-17 Wt. 3

Show that if n identical cells of emf E and internal resistance r are connected in series, they will produce a current through an external resistance R equal to $\dfrac{nE}{R + nr}$, and if they are connected in parallel the current will be $\dfrac{nE}{nR + r}$.

Solution

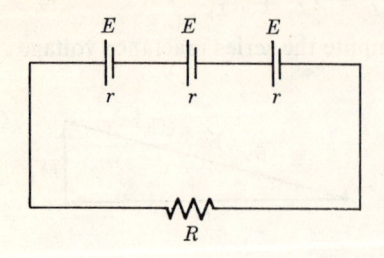

Fig. 9-12

Ohm's law may be applied to this undivided circuit to give

$$I = \frac{E}{R} = \frac{\text{total emf}}{\text{total resistance}}$$

In the circuit total emf $= \sum E = nE$. The total resistance in the circuit $= R + \sum r = R + nr$. Hence

$$I = \frac{nE}{R + nr} \blacktriangleright$$

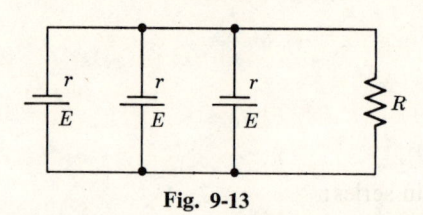

Fig. 9-13

In this situation Ohm's law again applies. Here the emf $= E$ as any electron goes through only one cell—not all of them—hence the total emf of the circuit is the same as the emf of one cell. The resistance for the cells is

$$\frac{1}{R_{\text{cells}}} = \frac{1}{r_1} + \frac{1}{r_2} + \cdots = \frac{n}{r} \qquad R_{\text{cells}} = \frac{r}{n}$$

Then the total circuit resistance is $R + r/n$. Ohm's law now gives

$$I = \frac{\text{total emf}}{\text{total resistance}} = \frac{E}{R + r/n}$$

This may be rewritten as

$$I = \frac{nE}{nR + r} \blacktriangleright$$

9-18 Wt. 3

What is the total equivalent resistance of the circuit between A and B?

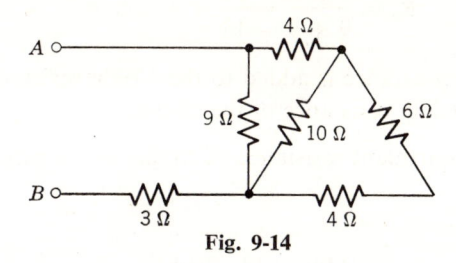

Fig. 9-14

Solution

First redraw the circuit in a more conventional form:

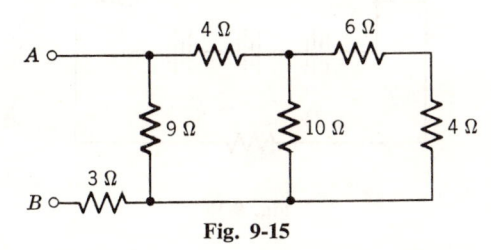

Fig. 9-15

The 6-ohm and 4-ohm resistances in series may be added:

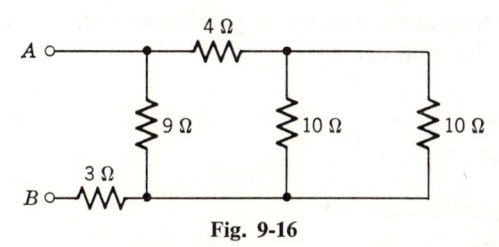

Fig. 9-16

The equivalent resistance of the two 10-ohm resistances in parallel is

$$R_e = \frac{10 \times 10}{10 + 10} = \frac{100}{20} = 5 \text{ ohms}$$

This 5-ohm resistance added to the 4-ohm resistance in series reduces the circuit to:

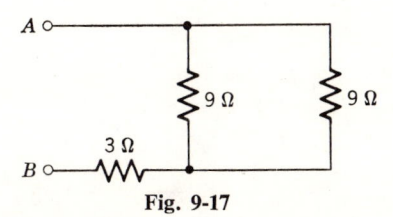

Fig. 9-17

462 / Electricity

The equivalent resistance of the parallel 9-ohm resistances is

$$R_e = \frac{9 \times 9}{9 + 9} = \frac{81}{18} = 4.5 \text{ ohms}$$

When this 4.5-ohm resistance is added to the 3-ohm resistance in series, we have the total equivalent resistance of the circuit.

Total equivalent resistance of circuit = 7.5 ohms ▶

9-19 Wt. 3

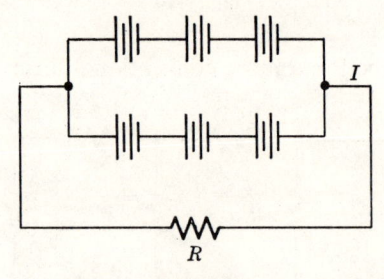

Fig. 9-18

When Ohm's law is applied to three similar cells which are connected in series, and then they are connected in parallel with three other similar cells, as in the figure, the total current is

(a) $I = \dfrac{E}{\dfrac{r}{3} + R}$

(b) $I = \dfrac{3E}{\dfrac{3r}{2} + R}$

(c) $I = \dfrac{E}{\dfrac{3r}{2} + R}$

(d) $I = \dfrac{3E}{3r + R}$

(e) $I = \dfrac{1}{\dfrac{6}{r} + R}$

Note: E is the emf of each cell, r is the internal resistance of each cell, and R is the resistance of the external circuit.

Solution

In a series circuit,
$$E_t = E_1 + E_2 + \cdots$$
$$R_t = R_1 + R_2 + \cdots$$

In a parallel circuit,
$$E_1 = E_2 = E_3 \cdots$$
$$R_t = \frac{R_1 R_2}{R_1 + R_2}$$

Total E in circuit $= E_1 + E_2 + E_3 = 3E$

R in cells $= \dfrac{R_1 R_2}{R_1 + R_2} = \dfrac{(3r)(3r)}{(3r) + (3r)} = \dfrac{9r^2}{6r} = 1.5r$

$E = IR$ or $I = \dfrac{E}{R} = \dfrac{3E}{1.5r + R}$ or $\dfrac{3E}{\dfrac{3r}{2} + R}$ ▶

Answer is (b)

9-20 Wt. 3

Determine the current I in the circuit shown.

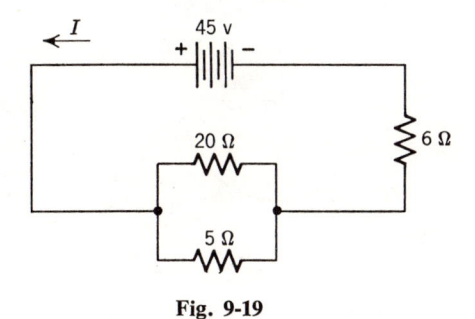

Fig. 9-19

Solution

Finding the equivalent resistance of the circuit,

$$R_e = \frac{1}{\frac{1}{5} + \frac{1}{20}} + 6 = \frac{1}{0.2 + 0.05} + 6 = \frac{1}{0.25} + 6 = 4 + 6 = 10 \text{ ohms}$$

$$I = \frac{E}{R} = \frac{45}{10} = 4.5 \text{ amp} \blacktriangleright$$

9-21 Wt. 4

A wire having a uniform resistance of 5 ohms/ft is formed into the configuration shown in Fig. 9-20. What is the resistance measured between the points O and A?

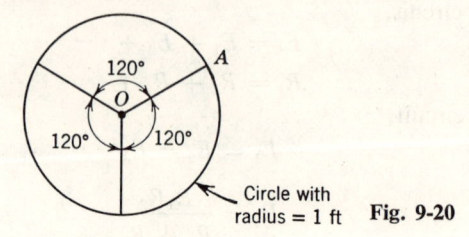

Fig. 9-20

Solution

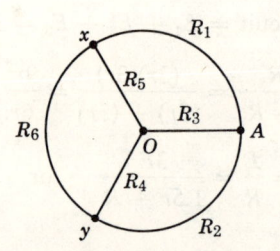

Fig. 9-21

Since R_5 is equal to R_4, there is no unbalanced current that will flow between x and y. As a result, R_6 will have no effect on the resistance between O and A and can be deleted.

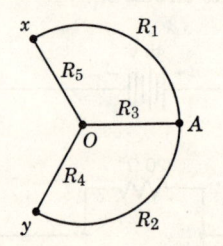

Fig. 9-22

The resistance values are

$$R_1 = R_2 = \tfrac{1}{3}\pi \times 2 \times 5 = 10.47 \text{ ohms}$$
$$R_3 = R_4 = R_5 = 1 \times 5 = 5.0 \text{ ohms}$$

The circuit can now be drawn in a more conventional form.

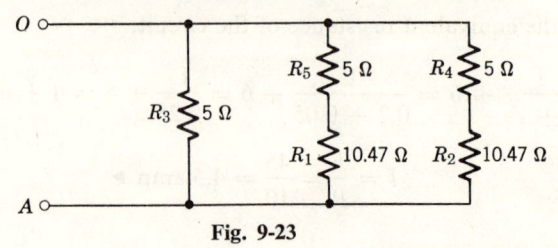

Fig. 9-23

The resistors in series can be replaced by an equivalent resistance equal to the sum of individual resistances.

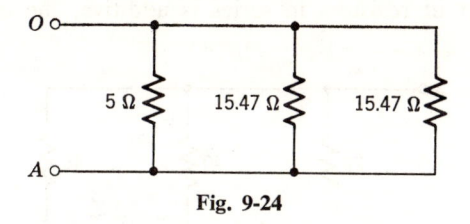

Fig. 9-24

For resistors in parallel, the equivalent resistance is

$$\frac{1}{R_e} = \frac{1}{R_3} + \frac{1}{R_5 + R_1} + \frac{1}{R_4 + R_2}$$

$$= \frac{1}{5} + \frac{1}{15.47} + \frac{1}{15.47}$$

$$= 0.2 + 2(0.0647) = 0.33$$

$$R_e = \frac{1}{0.33} = 3.0 \text{ ohms} = \text{resistance between } O \text{ and } A \blacktriangleright$$

9-22 Wt. 4

A tetrahedron $ABCD$ is constructed from wire of resistance 5 ohms/ft. Each side is 4 ft in length. Find the resistance from A to B.

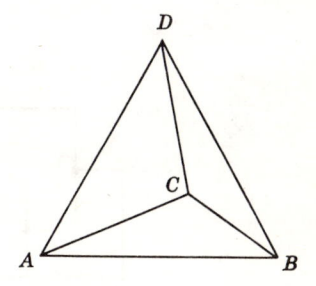

Fig. 9-25

Solution

The equivalent circuit would be:

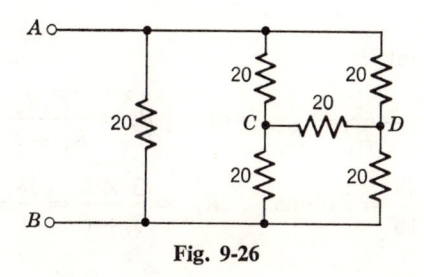

Fig. 9-26

466 / Electricity

For equal resistances, the current at C is equal to that at D; consequently no current will flow between C and D and the connection can be ignored.

Since the effect of resistors in series is additive, the equivalent circuit becomes:

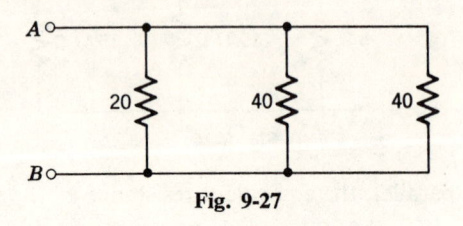

Fig. 9-27

For these parallel resistors, the equivalent resistance is

$$\frac{1}{R_{AB}} = \frac{1}{R_{20}} + \frac{1}{R_{40}} + \frac{1}{R_{40}}$$

$$= \tfrac{1}{20} + \tfrac{1}{40} + \tfrac{1}{40} = 0.05 + 2(0.025)$$

$$= 0.10$$

$$R_{AB} = \frac{1}{0.10} = 10 \text{ ohms} \blacktriangleright$$

9-23 Wt. 3

What is the resistance of the circuit shown?

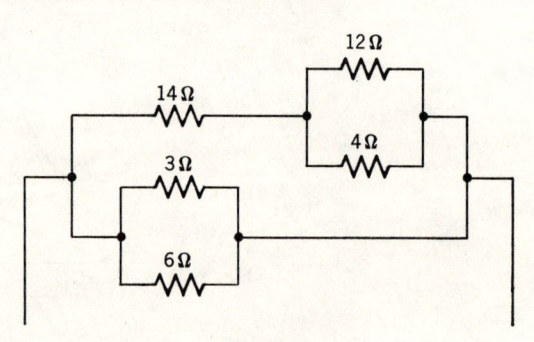

Fig. 9-28

Solution

For a parallel circuit,

$$\frac{1}{R_e} = \frac{1}{R_1} + \frac{1}{R_2} \quad \text{or} \quad R_e = \frac{R_1 R_2}{R_1 + R_2}$$

$$R_{e_A} = \frac{12 \times 4}{12 + 4} = \frac{48}{16} = 3 \text{ ohms} \qquad R_{e_B} = \frac{3 \times 6}{3 + 6} = \frac{18}{9} = 2 \text{ ohms}$$

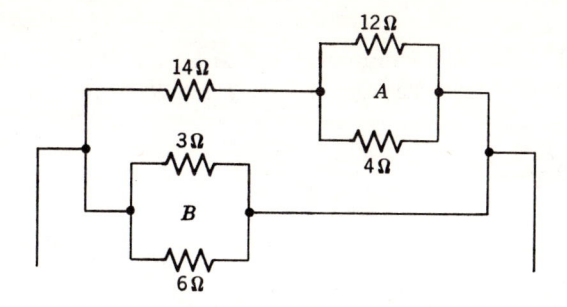

Fig. 9-29

Thus the circuit reduces to:

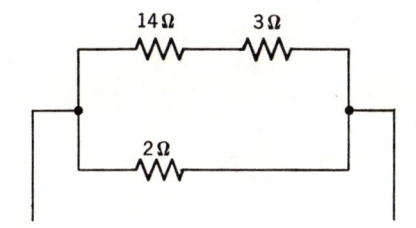

Fig. 9-30

For a series circuit

$$R_e = R_1 + R_2$$
$$R_e = 14 + 3 = 17 \text{ ohms}$$

So the circuit further reduces to:

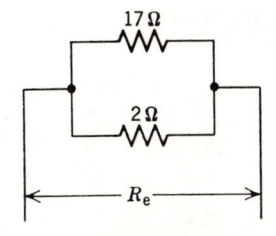

Fig. 9-31

$$R_e = \frac{17 \times 2}{17 + 2} = \frac{34}{19} = 1.79 \text{ ohms} \blacktriangleright$$

9-24 Wt. 3

The circuit shown is composed of resistors which have equal resistances R. What is the resistance between points A and B?

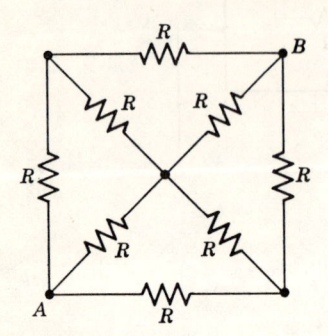

Fig. 9-32

Solution

We can redraw the same circuit in a more conventional fashion:

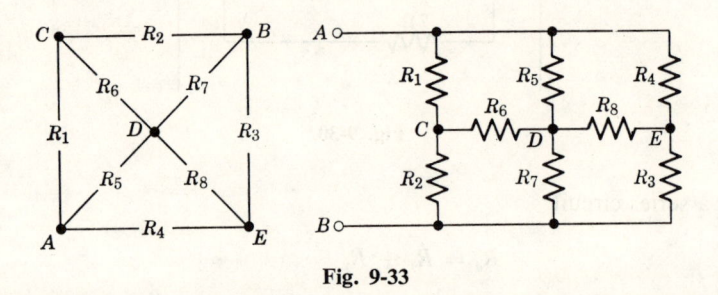

Fig. 9-33

Since all resistors have equal resistances R, we see that the current at C is equal to the current at D. Consequently no current will flow in R_6, and the connection C-D (and R_6) can be ignored. The same is true for connection D-E (and R_8). Thus for equal resistances the equivalent circuit reduces to:

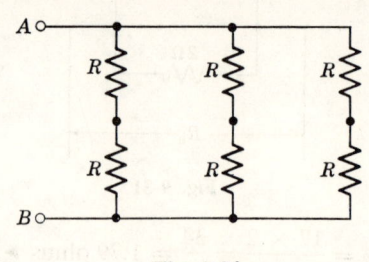

Fig. 9-34

We can replace series resistances by an equivalent resistor equal to the sum of the resistances and further reduce the circuit to:

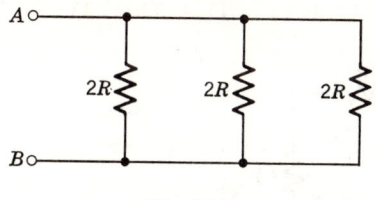

Fig. 9-35

For parallel resistors the equation for equivalent resistance is

$$\frac{1}{R_{AB}} = \frac{1}{R_1} + \frac{1}{R_2} + \frac{1}{R_3} \qquad \frac{1}{R_{AB}} = \frac{1}{2R} + \frac{1}{2R} + \frac{1}{2R} = \frac{3}{2R}$$

Therefore, the resistance R_{AB} between A and B is

$$R_{AB} = \frac{2R}{3} = \frac{2}{3}R \text{ ohm} \blacktriangleright$$

9-25 Wt. 4

Find the current input to the following circuit:

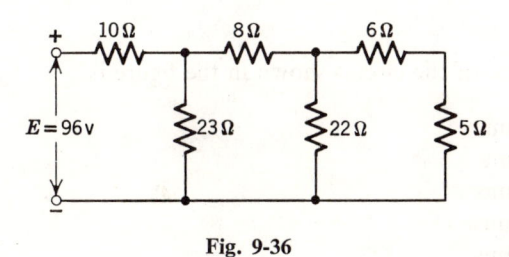

Fig. 9-36

Solution

Add the resistors in series: $6 + 5 = 11$ ohms:

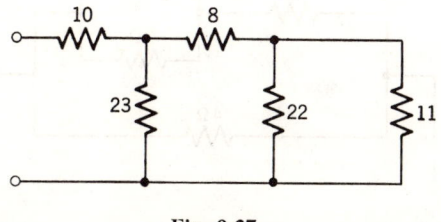

Fig. 9-37

For the 11- and 22-ohm parallel resistors,

$$R_e = \frac{1}{\frac{1}{11} + \frac{1}{22}} = \frac{11 \times 22}{11 + 22} = \frac{242}{33} = 7\frac{1}{3} \text{ ohms}$$

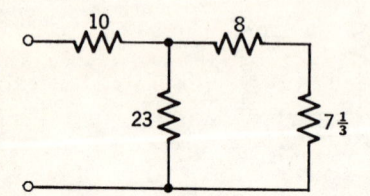

Fig. 9-38

Again adding the resistors in series, $8 + 7\frac{1}{3} = 15\frac{1}{3}$ ohms

$$R_e = \frac{15\frac{1}{3} \times 23}{15\frac{1}{3} + 23} = \frac{352.7}{38.3} = 9.2 \text{ ohms}$$

for the last pair of parallel resistors. Finally we add the last resistors in series: $10 + 9.2 = 19.2$ ohms = equivalent circuit R.

$$I = \frac{E}{R} = \frac{96}{19.2} = 5 \text{ amp} \blacktriangleright$$

9-26 Wt. 3

The resistance of the circuit shown in the figure is

(a) 0.43 ohm
(b) 0.80 ohm
(c) 2.28 ohms
(d) 5.50 ohms
(e) 14.00 ohms

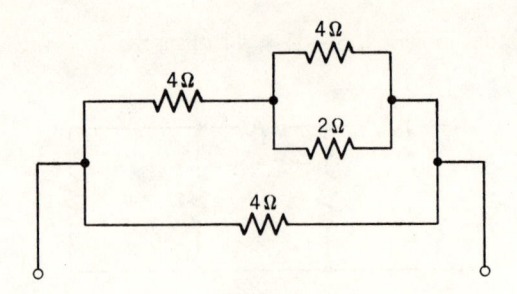

Fig. 9-39

Solution

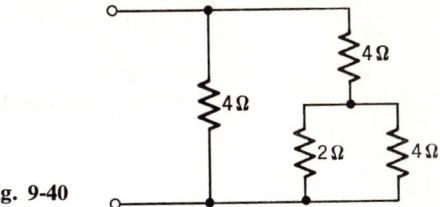

Fig. 9-40

For equivalent resistances, in series,

$$R_e = R_1 + R_2 + R_3 + \cdots$$

and in parallel,

$$\frac{1}{R_e} = \frac{1}{R_1} + \frac{1}{R_2} + \frac{1}{R_3} + \cdots$$

For the 2- and 4-ohm resistors in parallel,

$$\frac{1}{R_{2+4}} = \frac{1}{4} + \frac{1}{2} = \frac{3}{4}$$

$$R_{2+4} = \tfrac{4}{3} \text{ ohm}$$

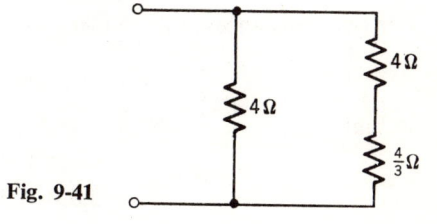

Fig. 9-41

In series,

$$R_e = 4 + \tfrac{4}{3} = 5\tfrac{1}{3} \text{ ohms}$$

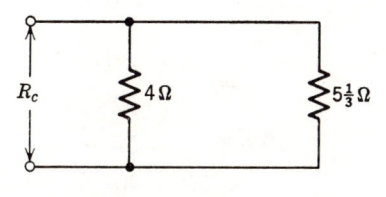

Fig. 9-42

472 / Electricity

The $5\frac{1}{3}$- and 4-ohm resistors in parallel add to give

$$\frac{1}{R_e} = \frac{1}{4} + \frac{1}{5\frac{1}{3}} = 0.25 + 0.188 = 0.438$$

$$R_e = \frac{1}{0.438} = 2.28 \text{ ohms} \blacktriangleright$$

Answer is (c)

9-27 Wt. 2

Solve for R in the circuit shown.

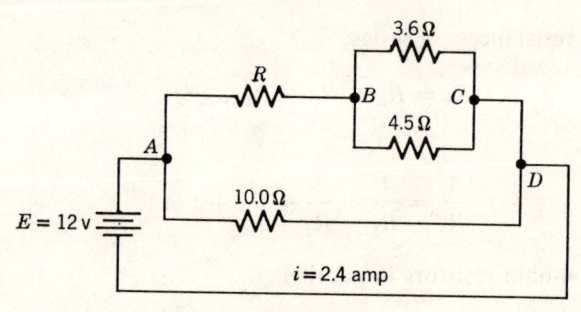

Fig. 9-43

Solution

For resistances in parallel,

$$\frac{1}{R_e} = \frac{1}{R_1} + \frac{1}{R_2} + \frac{1}{R_3} + \cdots$$

For two resistances, this reduces to

$$R_{1-2} = \frac{R_1 R_2}{R_1 + R_2}$$

so that

$$R_{BC} = \frac{(3.6)(4.5)}{3.6 + 4.5} = \frac{16.2}{8.1} = 2 \text{ ohms}$$

$$R_e = R_{AD} = \frac{(R + R_{BC})(10)}{R + R_{BC} + 10} = \frac{10(R + 2)}{R + 12}$$

$$R_e = \frac{E}{I} = \frac{12}{2.4} = 5$$

$$5 = \frac{10(R + 2)}{R + 12}$$

$$10R + 20 = 5R + 60$$

$$5R = 40$$

$$R = 8 \text{ ohms} \blacktriangleright$$

9-28 Wt. 6

Solve for the current I through the resistor R shown in the circuit. What should the power rating of resistor R be?

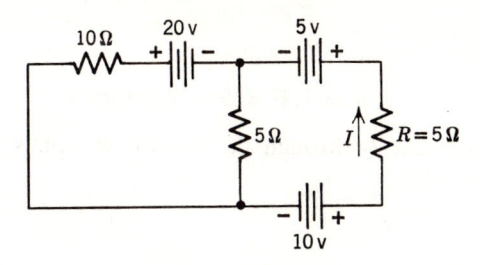

Fig. 9-44

Solution

The problem can be solved by using mesh equations. Assume a counter-clockwise direction for the current.

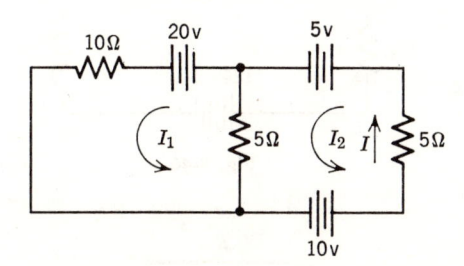

Fig. 9-45

Writing the equation for I_1,

$$20 \text{ volts} - (10 + 5)I_1 + 5I_2 = 0, \tag{1}$$

and for I_2,

$$-5 \text{ volts} - 5(I_2 - I_1) + 10 \text{ volts} - 5I_2 = 0$$

or

$$5 + 5I_1 - 10I_2 = 0 \tag{2}$$

$2 \times$ equation (1)

$$\frac{40 - 30I_1 + 10I_2 = 0}{45 - 25I_1 \qquad = 0}$$

$$I_1 = \tfrac{45}{25} = 1.8 \text{ amp}$$

474 / Electricity

Then, equation (2) becomes

$$5 + 5(1.8) - 10I_2 = 0$$

$$I_2 = \frac{5+9}{10} = 1.4 \text{ amp} \blacktriangleright$$

The power rating of resistor R is equal to

$$I^2R = 1.4^2 \times 5 = 9.8 \text{ watts} \blacktriangleright$$

Therefore, the current I_2 through the resistor R equals 1.4 amp, and the power loss is 9.8 watts.

9-29 Wt. 5

In the circuit shown, compute the voltage drop across the 7-ohm resistor. Assume the inductance has negligible resistance.

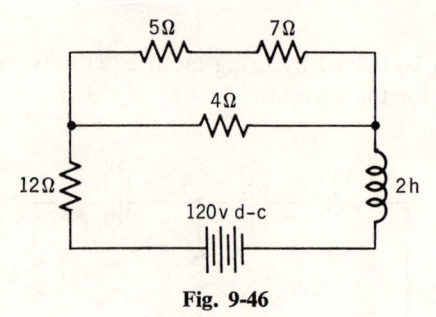

Fig. 9-46

Solution

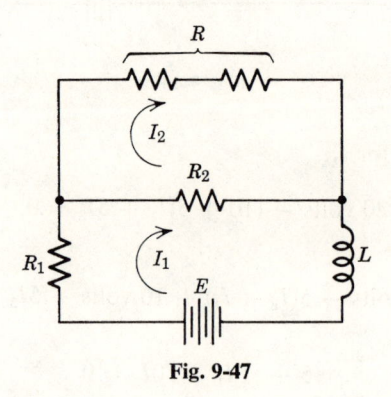

Fig. 9-47

Here we use the notation $p \equiv d/dt$. We now write loop equations:

Loop 1: $E = R_1 i_1 + R_2(i_1 - i_2) + pLi_1$
Loop 2: $0 = R_2(i_2 - i_1) + Ri_2$

Hence
$$i_1 R_2 = (R_2 + R)i_2 \quad \text{and} \quad i_1 = \frac{(R_2 + R)i_2}{R_2}$$

Loop 1: $\quad E = \dfrac{R_1}{R_2}(R_2 + R)i_2 + \dfrac{R_2}{R_2}(R_2 + R)i_2 - R_2 i_2 + p\dfrac{L}{R_2}(R_2 + R)i_2$

Collecting terms and inserting actual values:
$$E = \tfrac{12}{4}(4 + 12)i_2 + 1(4 + 12)i_2 - 4i_2 + p\tfrac{2}{4}(4 + 12)i_2$$
$$= 48i_2 + 16i_2 - 4i_2 + p8i_2 = 60i_2 + p8i_2$$

There are two possible modes of operation of this circuit:

Condition No. 1 (if the circuit has been "on" for a long period of time):
$$E = 60i_2 + 8(pi_2)^0$$
since the driving function is a constant and all transients will have died out. Hence
$$i_2 = \frac{E}{60} = \frac{120}{60} = 2 \text{ amp} \quad \text{and} \quad V \text{ across } R_7 = i_2 R_7 = 2(7) = 14 \text{ volts} \blacktriangleright$$

Condition No. 2 (if the battery has been placed in the circuit at $t = 0$):
$$pi_2 + \frac{60}{8}i_2 = \frac{E}{8}$$

This differential equation has the solution
$$i_2 = \frac{E}{60}(1 - e^{-60t/8}) = \frac{E}{60}(1 - e^{-7.5t})$$

Then the voltage drop is
$$V \text{ across } R_7 = i_2 R_7 = \frac{7E}{60}(1 - e^{-7.5t})$$
$$= 14.0(1 - e^{-7.5t}) \text{ volts} \blacktriangleright$$

9-30 Wt. 3

A radio circuit oscillates at a frequency $f = \dfrac{1}{2\pi(LC)^{1/2}}$. A given circuit has an inductance L of 0.1 henry, but its capacitance is not exact, being given as 0.1 μf with a 10% tolerance. Calculate the resonant frequency and, by methods of calculus, the variation in this frequency caused by the condenser tolerance.

476 / Electricity

Solution

(a) $f = \dfrac{1}{2\pi(LC)^{1/2}} = \dfrac{1}{2\pi(0.1 \times 0.1 \times 10^{-6})^{1/2}} = \dfrac{1}{2\pi(1 \times 10^{-8})^{1/2}} = \dfrac{10^4}{2\pi}$

$f = 1591$ cps ▶

(b) $f = \dfrac{1}{2\pi(L)^{1/2}} C^{-1/2}$ so $\dfrac{df}{dC} = \dfrac{1}{2\pi L^{1/2}}(-\tfrac{1}{2}C^{-3/2})$

$= -\dfrac{1}{2}\dfrac{1}{2\pi L^{1/2}}\dfrac{1}{C(C)^{1/2}} = \dfrac{-1}{2C}\dfrac{1}{2\pi(LC)^{1/2}}$

$= -\dfrac{1}{2C}f$

For finite increments,

$\left|\dfrac{\Delta f}{\Delta C}\right| = \dfrac{1}{2C}f$ or $|\Delta f| = \dfrac{f}{2}\dfrac{\Delta C}{C}$

Then the frequency variation Δf is

$\Delta f = (\tfrac{1591}{2})(0.1) = 79.5$ cps

The frequency variation is about 80 cps for a 10% tolerance ($\pm 5\%$) ▶

9-31 Wt. 1

In a series circuit containing resistance, capacitance, and inductance, to increase the resonant frequency it is necessary to

(a) increase the resistance
(b) decrease the inductance
(c) increase the capacitance
(d) decrease the resistance
(e) increase the voltage

Solution

At resonance, $2\pi fL = \dfrac{1}{2\pi fC}$,

$2\pi f = \dfrac{1}{(LC)^{1/2}}$ and $f = \dfrac{1}{2\pi(LC)^{1/2}}$

Therefore, to increase f, L or C must be decreased ▶

Answer is (b)

9-32 Wt. 3

A 12-volt car battery, with an internal resistance of 4 ohms, supplies a bank of 25 lights that are connected in parallel. Each light has an effective resistance of 500 ohms. What is the wattage supplied to the bank by the battery?

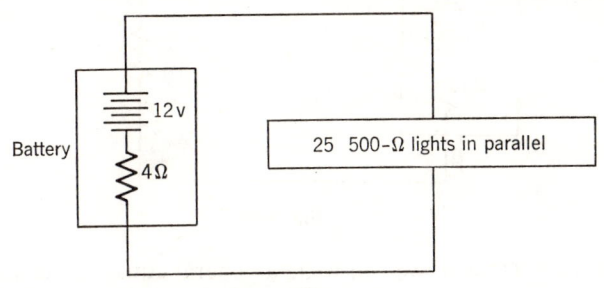

Fig. 9-48

Solution

The equivalent resistance of 25 lights in parallel is

$$R_e = \frac{1}{\dfrac{1}{R_1} + \dfrac{1}{R_2} + \cdots + \dfrac{1}{R_n}} = \frac{1}{25(\frac{1}{500})} = \frac{500}{25} = 20 \text{ ohms}$$

$$\text{Current } I = \frac{E}{R} = \frac{12}{20 + 4} = 0.5 \text{ amp}$$

The wattage *supplied to the bank* is

$$\text{Power} = I^2 R = 0.5^2 \times 20 = 5 \text{ watts} \blacktriangleright$$

Note: we are asked for the power supplied to the bank of lights; this is 5 watts, *not* 6 watts.

9-33 Wt. 6

A "black box" has within it several batteries and resistors. An external resistor R and an ammeter are placed in series between the terminals A-B of the box as shown. When $R = 3.0$ ohms, the current flow indicated by the meter is 1.0 amp. When $R = 9.0$ ohms, the indicated flow of current is 0.50 amp. What is the effective voltage and resistance between the terminals A-B?

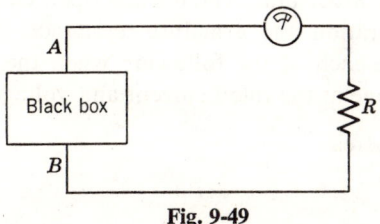

Fig. 9-49

Solution

We can draw the circuit diagram:

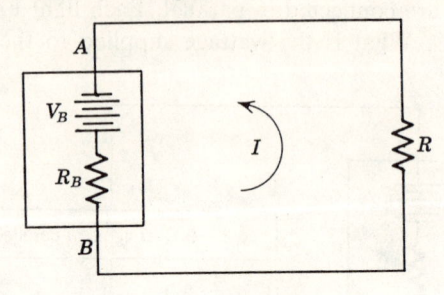

Fig. 9-50

Since $\sum V - \sum IR = 0$, we can write a general equation for the circuit:

$$V_B - R_B I - RI = 0$$

When $R = 3.0$ ohms and $I = 1.0$ amp,

$$V_B - 1.0 \times R_B - 3.0 \times 1.0 = 0 \tag{1}$$

When $R = 9.0$ ohms and $I = 0.50$ amp,

$$V_B - 0.50 \times R_B - 9.0 \times 0.50 = 0 \tag{2}$$

$$\begin{array}{ll}(1) & V_B - R_B - 3.0 = 0 \\ -(2) & -V_B + 0.5 R_B + 4.5 = 0 \end{array}$$

$$-0.5 R_B + 1.5 = 0 \qquad R_B = \frac{1.5}{0.5} = 3.0 \text{ ohms} \blacktriangleright$$

Then equation (1) becomes

$$V_B - 3.0 - 3.0 = 0 \qquad V_B = 6.0 \text{ volts} \blacktriangleright$$

9-34 Wt. 6

A three-phase, four-pole, 60-cycle, cylindrical-rotor synchronous machine is rated at 10,000 kw, 5000 volts, and 0.8 leading power factor. The synchronous reactance is 0.8 per unit. The normal open circuit field current is 200 amp. Neglect saturation and armature resistance. Assume a constant field current. Calculate each of the following when the machine is taking rated power from the line at the rated current and voltage:

(a) reactive volt-amperes
(b) field current
(c) torque angle

Solution

(a) $$P = EI \cos \theta \qquad EI = \frac{P}{\cos \theta} \qquad \text{Vars} = EI \sin \theta$$

$$\text{Vars} = \frac{P \sin \theta}{\cos \theta} = \frac{10{,}000 \text{ kw} \times 0.6}{0.8} = 7500 \text{ kvars} \blacktriangleright$$

(b) $$P = \sqrt{3} \, V_{L-L} I_L \cos \theta$$

with $V_{L-L} = 5000$ volts, $\cos \theta = 0.8$, and $P = 10{,}000$ kw.

$$|I_L| = \frac{10{,}000{,}000 \text{ watts}}{\sqrt{3} \times 5000 \times 0.8} = 1440 \text{ amp}$$

The per unit reactance $= 0.8 = \dfrac{IX_e}{V/\sqrt{3}}$, so

$$X_e = \frac{5000 \times 0.8}{\sqrt{3} \times 1440} = 1.6 \text{ ohms}$$

$$V_\phi = E_g + jIX_e$$

$$\frac{5000}{\sqrt{3}} = E_g + (j1.6)(1440 \angle{+37°})$$

$$2890 = E_g + 2300(-0.6 + j0.8)$$

$$E_g = 4270 - j1840$$

$$|E_g| = \sqrt{(4270)^2 + (1840)^2} = 4650 \text{ volts}$$

$$\frac{4650 \text{ volts}}{I_f} = \frac{2890 \text{ volts}}{200 \text{ amp}}$$

so the field current $I_f = 322$ amp $\blacktriangleright$

(c) The torque angle is

$$\delta = \tan^{-1} \tfrac{1840}{4270} = 23.3° \blacktriangleright$$

9-35 Wt. 1

For a d-c shunt motor, operating from a constant potential supply, one of the following will happen:

(a) Increasing the shunt field resistance will cause the motor speed to increase.

(b) Increasing the shunt field resistance will cause the motor speed to decrease.

(c) Increasing the shunt field resistance will cause the shunt field current to increase.

(d) Decreasing the shunt field resistance will cause the shunt field current to decrease.

(e) Increasing the shunt field resistance will cause the armature current to decrease.

Solution

The practical way to control the speed of a d-c shunt motor is to place a rheostat in series with the shunt field. Increasing the resistance will decrease the field current i_f, which is proportional to the flux ϕ_f. This can be seen in the equation for motor speed,

$$\text{Speed} = \frac{V - I_a R_a}{K \phi_f} \blacktriangleright$$

Answer is (a)

9-36 Wt. 1

A four-pole synchronous motor operating from a 50-cps supply will have a synchronous speed of

(a) 3600 rpm
(b) 3000 rpm
(c) 1800 rpm
(d) 1500 rpm
(e) 1200 rpm

Solution

$$\text{Frequency } f = \frac{\text{no. of poles}}{2} \times \frac{\text{rpm}}{60} \qquad 50 = \frac{4}{2} \times \frac{\text{rpm}}{60} \qquad \text{rpm} = 1500 \blacktriangleright$$

Answer is (d)

9-37 Wt. 4

Two wattmeters are properly connected in a three-phase circuit to read the power supplied to a 220-volt, 15-hp, Δ-connected, three-phase induction motor running at no load. Voltmeters and ammeters are also properly connected in the circuit to supply lines A, B, and C. The readings are:

Voltmeter (Volts)	Ammeter (Amperes)	Wattmeter (Watts)
$V_{AB} = 221$	$I_A = 4.04$	810
$V_{BC} = 221$	$I_C = 4.04$	(−)200

(a) What is the power input to the motor?
(b) What is the power factor of the motor?

Solution

(a) The power in a three-phase circuit can be determined by the use of two wattmeters. The total power is the algebraic sum of the two meter readings.

$$P_{\text{total}} = W_1 + W_2 = 810 + (-200) = 610 \text{ watts} \blacktriangleright$$

(b) $$P_t = (3)^{1/2} VI \cos \theta$$

(Power factor will be lagging with an inductive load.)

$$\cos \theta = \frac{P_t}{(3)^{1/2} VI} = \frac{610}{(3)^{1/2} \times 221 \times 4.04} = 0.394$$

Power factor = 0.394 current lagging ▶

9-38 Wt. 6

The armature resistance of a 50-hp, 550-volt, d-c shunt wound motor is 0.35 ohm. The full-load armature current of this motor is 76 amp.

(a) What should the resistance of the starter be so that the initial starting current is 150% of the full-load value?

(b) If the field current under full load is 3 amp, what is the overall efficiency of the motor?

(c) Compute the stray power losses.

Solution

(a) The starting current is to be limited to 1.50(76) = 114 amp. The total resistance is then

$$R = \frac{V}{I_s} = \frac{550}{114} = 4.82 \text{ ohms}$$

and the resistance of the starter is 4.82 − 0.35 = 4.47 ohms ▶

(b) Line current = armature current + field current = 76 + 3 = 79 amp

$$\text{Efficiency} = \frac{\text{output}}{\text{input}} = \frac{50 \times 746}{79 \times 550} = 0.859 = 85.9\% \blacktriangleright$$

(c) Input = output + losses
Input = output + $I_a^2 R_a + I_f^2 R_f$ + stray power losses

$$R_f = \frac{550 \text{ volt}}{3 \text{ amp}} = 183.3 \text{ ohms}$$

79 × 550 = 50 × 746 + 76²(0.35) + 3²(183.3) + stray power losses
43,450 = 37,300 + 2020 + 1650 + stray power losses
Stray power losses = 43,450 − 40,970 = 2480 watts ▶

482 / Electricity

9-39 Wt. 1

The rpm of an a-c electric motor

(a) varies directly as the number of poles
(b) varies inversely as the number of poles
(c) is independent of the number of poles
(d) is independent of the frequency
(e) is directly proportional to the square of the frequency

Solution

The speed varies directly as the frequency and inversely as the number of poles ▶

Answer is (b)

9-40 Wt. 6

Prove *mathematically* the following statement: Maximum power transfer is obtained by making the load resistance R_L equal to the generator resistance R_g and the load reactance X_L equal and opposite to the generator reactance X_g.

Nomenclature: E = generator voltage
Z_g = generator impedance (fixed)
Z_L = load impedance (variable)
P_L = power to load
I = line current

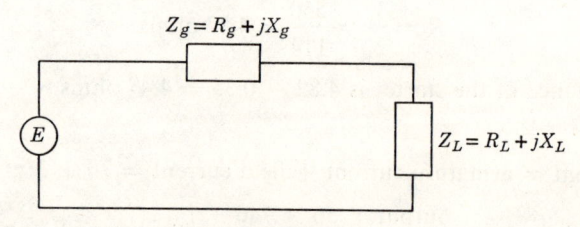

Fig. 9-51

Solution

Proof: The line current is, from Ohm's law,

$$I = \frac{E}{Z} = \frac{E}{[(R_g + R_L)^2 + (X_g + X_L)^2]^{1/2}}$$

The power to the load is

$$P_L = I^2 R_L = \frac{E^2 R_L}{(R_g + R_L)^2 + (X_g + X_L)^2}$$

P_L is the largest for constant R_L and R_g if the denominator is smallest, that is, if $X_L = -X_g$. Then

$$P_L = \frac{E^2 R_L}{(R_g + R_L)^2}$$

The maximum of a function is found by taking its derivative and equating it to zero:

$$\frac{dP_L}{dR_L} = \frac{(R_g + R_L)^2 E^2 - E^2 R_L (2)(R_g + R_L)}{(R_g + R_L)^4} = 0$$

$$(R_g + R_L)^2 - 2R_L(R_g + R_L) = 0 \quad \text{or} \quad R_g + R_L - 2R_L = 0$$

which gives

$$R_L = R_g \quad \text{Q.E.D.} \blacktriangleright$$

These conditions ($R_L = R_g$, $X_L = -X_g$) are thus required if maximum power is to be obtained.

9-41 Wt. 3

A metal transport plane has a wing spread of 88 ft. What difference of potential exists between the extremities of the wings when the plane moves in a horizontal plane with a speed of 150 mph? The value of the vertical component of the earth's magnetic field is 0.65 gauss at the plane.

Solution

An emf is induced in a circuit whenever any of its conductors cuts magnetic flux. The equation is

$$e_i = BLv$$

where e_i = induced voltage (in units of 10^{-8} volt)
B = flux density in gausses
L = conductor length in centimeters
v = velocity of conductor in centimeters per second

$$e_i = 0.65(88 \times 12 \times 2.54)\left(\frac{150 \times 5280 \times 12 \times 2.54}{60 \times 60}\right)$$

$$e_i = 0.117 \text{ volt} \blacktriangleright$$

9-42 Wt. 6

A circuit consisting of $R = 3$ ohms and $X_L = 4$ ohms in series is connected to a 100-volt, 60-cycle source as shown. Determine the following:

(a) real volt-amperes
(b) reactive volt-amperes
(c) total volt-amperes

484 / Electricity

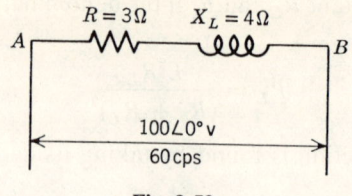

Fig. 9-52

Solution

$$\text{Impedance } \bar{Z} = R + jX_L = 3 + j4$$

In polar form,

$$\bar{Z} = (3^2 + 4^2)^{1/2} \tan^{-1}(\tfrac{4}{3}) = 5 \angle 53.1° \text{ ohms}$$

$$\text{Current } \bar{I} = \frac{\bar{E}}{\bar{Z}} = \frac{100 \angle 0°}{5 \angle 53.1°} = 20 \angle -53.1°$$

(*a*) Real volt-amperes

$$P = \bar{E} \times \bar{I} = 100 \angle 0° \times 20 \angle -53.1° = 100 \times 20 \cos(-53.1°)$$
$$P = 2000 \times 0.6 = 1200 \text{ va} \blacktriangleright$$

(*b*) Reactive volt-amperes

$$\text{vars} = 1200 \tan(-53.1) = 1200 \times (-\tfrac{4}{3}) = -1600 \text{ vars} \blacktriangleright$$

(*c*) Apparent or total volt-amperes $= EI = 100 \times 20 = 2000 \text{ va} \blacktriangleright$

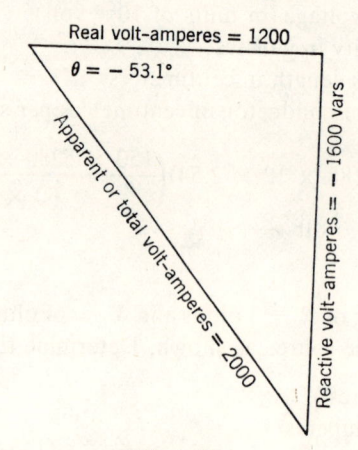

Fig. 9-53

9-43 Wt. 6

Determine the value of e_0 in the circuit shown, where

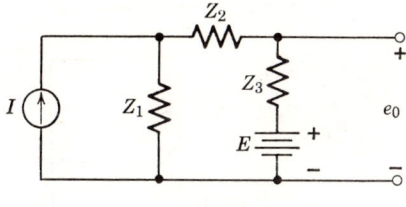

$$Z_1 = 3 + j4 \qquad I = 0 - j6 \text{ amp}$$
$$Z_3 = 2 \qquad E = 5 \text{ volts}$$
$$Z_2 = 5 - j4$$

Fig. 9-54

Solution

Convert I and Z_1 to a voltage source.

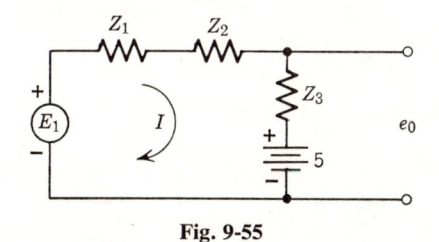

Fig. 9-55

$$E_1 = IZ_1 = (-j6)(3 + j4) = 24 - j18 = 30 \angle -37°$$

If we assume that

$$Z_2 = 5 - j4 =$$

Fig. 9-56

then no d-c path exists, and the e_0 will then be given by superposition as the sum of 5 volts d-c due to the battery and the IZ_3 due to the voltage source E_1.

$$IZ_3 = \frac{E_1 Z_3}{Z_1 + Z_2 + Z_3} = \frac{(24 - j18)(2)}{3 + j4 + 5 - j4 + 2} = \frac{24 - j18}{5} = 6 \angle -37°$$

The instantaneous voltage drop across Z_3 therefore is $6\sqrt{2} \sin(\omega t - 37°)$, and the total instantaneous voltage is the sum of individual values $e_0 = 5 + 8.5 \sin(\omega t - 37°)$ volts ▶

9-44 Wt. 6

For the parallel circuit shown:

(a) Find the resonant frequency f_0.
(b) What is the figure of merit Q of the circuit at f_0?
(c) What is the wavelength at resonant frequency?

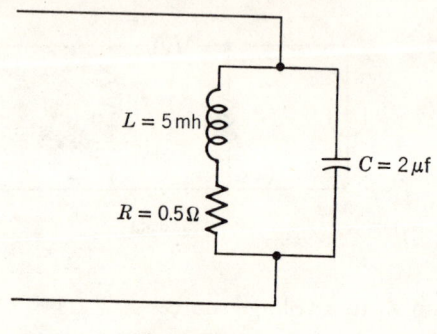

Fig. 9-57

Solution

(a) The resonant frequency occurs when $X_L = X_C$, assuming $X_L \gg R$

$$2\pi f_0 L = \frac{1}{2\pi f_0 C} \qquad 2\pi f_0 \times 0.005 = \frac{1}{2\pi f_0 \times 2 \times 10^{-6}}$$

$$f_0 = \left(\frac{10^6}{8\pi^2 \times 0.005}\right)^{1/2} = (2.525 \times 10^6)^{1/2} = 1590 \text{ cps} \blacktriangleright$$

(b) $$X_C = X_L = 2\pi \times 1590 \times 0.005 = 50 \text{ ohms}$$

$$Q = \frac{X}{R} = \frac{50}{0.5} = 100 \blacktriangleright$$

(c) Using the value of c = velocity of light = 300×10^6 m/sec, the wavelength is

$$\lambda = \frac{c}{f_0} = \frac{300 \times 10^6}{1590} = 1.89 \times 10^5 \text{ m} \blacktriangleright$$

9-45 Wt. 6

A 10-kw, 60-cps, 100-volt load has a power factor of 0.8 lead. It is desired to correct the power factor to 0.9 lag while keeping the load voltage constant. What electrical element should be added? Where should the element be added, and what is its value?

Solution

An inductance coil should be added in parallel with the load ▶

0.8 leading power factor

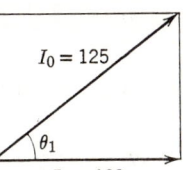

I_C, $I_0 = 125$, θ_1, $I_R = 100$

0.9 lagging power factor

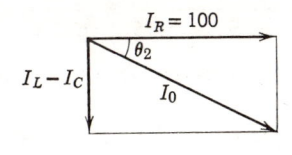

$I_R = 100$, θ_2, $I_L - I_C$, I_0

Fig. 9-58

$$P = EI \cos \theta_1$$

$$10{,}000 = 100 I_0 \times 0.8 \qquad I_0 = 125 \text{ amp}$$

$$I_R = \frac{P}{E} = \frac{10{,}000}{100} = 100 \text{ amp}$$

$$I_C = (125^2 - 100^2)^{1/2} = 75 \text{ amp}$$

$$\cos \theta_2 = \frac{100}{I_0} = 0.9 \qquad I_0 = 111 \text{ amp}$$

$$I_L - I_C = (111^2 - 100^2)^{1/2} = 48.2 \text{ amp}$$

$$I_L = 48.2 + 75 = 123.2 \text{ amp}$$

$$I_L = \frac{E}{2\pi f L} \qquad L = \frac{E}{2\pi f I_L}$$

$$L = \frac{100}{2\pi \times 60 \times 123.2} = 0.00215 \text{ henry} \blacktriangleright$$

9-46 Wt. 6

(a) What value of capacitance C, connected across a 50-cycle source, will give the same reactance as a choke coil of 150 mh (millihenrys) connected across a 60-cycle source?

(b) What is the series resonant frequency of L and C in part (a)? Assume the capacitor has negligible resistance.

Solution

(a) Inductive reactance $X_L = 2\pi f L = 2\pi \times 60 \times 0.150 = 56.5$ ohms

Capacitive reactance $X_C = \dfrac{1}{2\pi f C} = \dfrac{1}{2\pi \times 50 C}$

488 / Electricity

For equal reactance,

$$\frac{1}{2\pi \times 50C} = 56.5 \text{ ohms}$$

$$C = \frac{1}{2\pi \times 50 \times 56.5} = \frac{1}{17{,}750} = 56.3 \times 10^{-6} \text{ farad} = 56.3 \ \mu f \blacktriangleright$$

(b) The series resonant frequency can be found by equating X_L and X_C.

$$2\pi f L = \frac{1}{2\pi f C} \qquad 4\pi^2 f^2 = \frac{1}{LC} \qquad f = \frac{1}{2\pi(LC)^{1/2}}$$

$$f = \frac{1}{2\pi(0.15 \times 56.3 \times 10^{-6})^{1/2}} = \frac{10^3}{2\pi(8.45)^{1/2}}$$

$$f = 54.6 \text{ cycles} \blacktriangleright$$

9-47 Wt. 4

What is the power in watts, reactive power in vars, the total volt-amperes input, and the power factor of the circuit below?

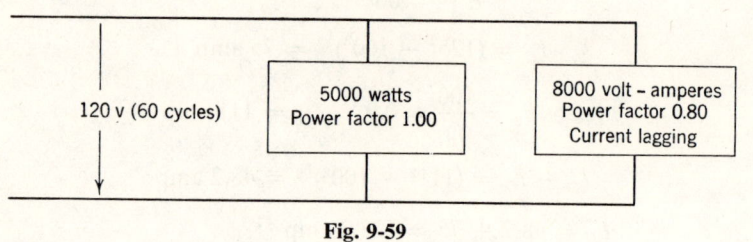

Fig. 9-59

Solution

Power factor = $\cos \theta$ Power = $EI \cos \theta$
Reactive volt-amperes = $EI \sin \theta$

Load	Power	(= $\cos \theta$ ×)	Apparent Power	(× $\sin \theta$ =)	Reactive Volt-Amperes
(1)	5000 watts	1	5000 watts	0	0
(2)	6400 watts	0.8 lag	8000 va	0.6	4800 vars

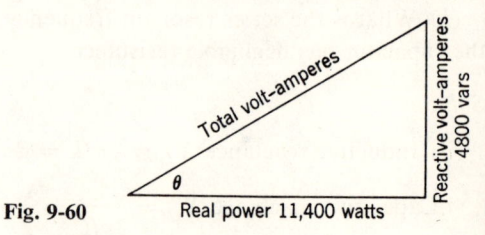

Power = 11,400 watts ▶   Reactive power = 4800 vars ▶

Fig. 9-60

Total volt-amperes $= [(11{,}400)^2 + (4{,}800)^2]^{1/2} = 12{,}380$ va ▶

$$\text{Power factor} = \cos\theta = \frac{11{,}400}{12{,}380} = 0.92 \text{ current lagging} \blacktriangleright$$

9-48 Wt. 4

A rectifier circuit connected to a 60-cps a-c supply is shown below. Each secondary winding of the transformer has a voltage of 120 volts rms. Selenium disks are used as the rectifying element; their voltage drop is negligible.

(a) Sketch the voltage across the load, indicating magnitudes of voltage and time.

(b) What is the average or d-c voltage across the load?

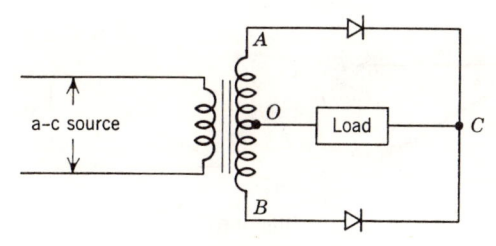

Fig. 9-61

Solution

$$E_{AO} = E_{OB} = 120 \text{ volts rms}$$

$$V = 170 \sin 377t \qquad 0 < t < \tfrac{1}{120}$$

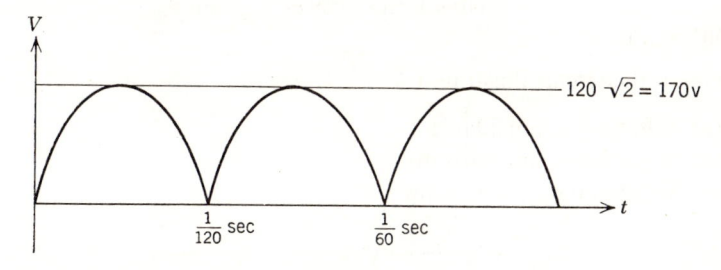

Fig. 9-62

$$E_{av} = \frac{170}{\pi} \int_0^{\pi} \sin \omega t \, d(\omega t) = \frac{170}{\pi}\Big[-\cos \omega t\Big]_0^{\pi} = \frac{170}{\pi}(1+1) = \frac{2}{\pi}(170)$$

$$E_{av} = E_{d\text{-}c} = 108 \text{ volts}$$

490 / Electricity

9-49 Wt. 1

In a series a-c circuit with a lagging power factor, to increase the power factor you should increase the

(a) current
(b) voltage
(c) inductance
(d) capacitance
(e) frequency

Solution

Lagging power factor means the current is lagging the voltage by $\cos \theta$.

$$I \angle -\theta = \frac{E \angle 0°}{Z \angle \theta}$$

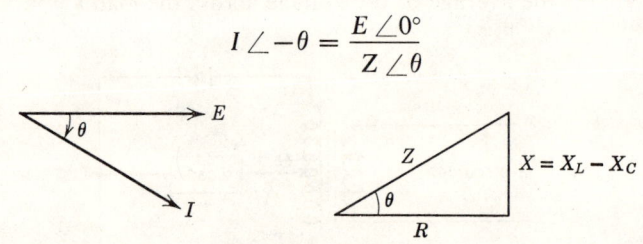

Fig. 9-63

To *increase* the power factor (to a maximum of unity—or 0°) one must make $(X_L - X_C)$ smaller; therefore either the frequency f, or L or C must be *decreased*, since

$$X_L = 2\pi f L \quad \text{and} \quad X_C = \frac{1}{2\pi f C}$$

No correct answer is given ▶

9-50 Wt. 5

Given the circuit illustrated.

(a) What is the impedance?
(b) What is the line current?
(c) What is the power factor?

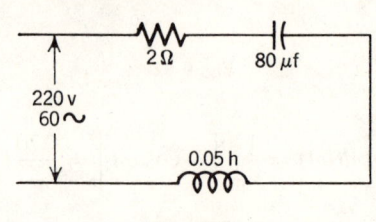

Fig. 9-64

Solution

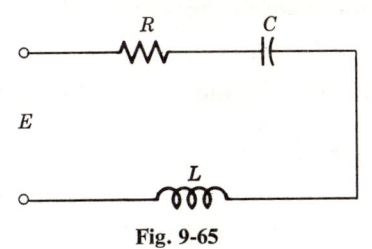

Fig. 9-65

$$I\angle-\theta = \frac{E\angle 0°}{Z\angle\theta}$$

(a) The impedance Z is
$$Z = R + jX_L - jX_C$$

where R = resistance (ohms)

$X_L = 2\pi f L$ (ohms) (inductive reactance)

$X_C = \dfrac{1}{2\pi f C}$ (ohms) (capacitive reactance)

$$Z = 2 + j(2\pi 60)(0.05) - j\left[\frac{1}{(2\pi 60)(80 \times 10^{-6})}\right]$$

$$= 2 + j(18.85) - j(33.25)$$

$$\simeq 2 - j14.4 \text{ ohms (in rectangular form)}$$

$$\simeq 14.7\angle-82.2° \blacktriangleright$$

(b) $$\bar{I} = \frac{E\angle 0°}{\bar{Z}} = \frac{220\angle 0°}{14.7\angle-82.2°} = 14.95\angle+82.2°$$

Hence
$$|I| \simeq 15 \text{ amp} \blacktriangleright$$

(c) Power factor = $\cos 82.2° = 0.135$ (leading) ▶

9-51 Wt. 3

Given the circuit and voltage measurements shown.

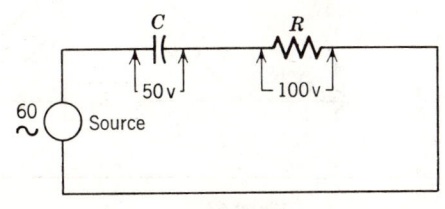

Fig. 9-66

(a) What is the voltage of the 60-cycle source?
(b) What is the power factor of the circuit?

Solution

$$IR = 100 \text{ volts} \quad \text{and} \quad I(-jX_C) = 50 \text{ volts}$$

Therefore

$$\bar{E} = IR - jIX_C = 100 - j50 \text{ volts (rectangular form)}$$
$$\bar{E} = 112 \angle -26.5°$$

(a) $\quad |\bar{E}| = 112 \text{ volts}$ ▶

(b) $\quad$ Power factor $= \cos 26.5° = 0.89$ (leading) ▶

9-52 Wt. 4

It is desired, by discharging a capacitance, to send an electric impulse through a 1000-ohm resistor so that the initial current will be 1.00 amp. At the end of 0.010 sec (10 msec), the voltage across the resistance is to be 368 volts. What size capacitance should be used?

Solution

The time constant T is the time in seconds needed for a condenser to lose 63% of its voltage, that is, for the voltage to drop to 37% of its initial value. In this case $T = 0.01$ sec. Thus

$$T = RC \quad \text{or} \quad C = \frac{T}{R} = \frac{10 \times 10^{-3}}{10^3} = 10 \times 10^{-6} \text{ farad} = 10 \ \mu\text{f} \ \blacktriangleright$$

9-53 Wt. 6

The 208-volt, three-phase line and Δ-connected load shown has impedances in ohms:

$$Z_a = 4 - j3$$
$$Z_b = 7 + j3$$
$$Z_c = 2 + j5$$

If $E_{cb} = 208 \angle 0°$ volts, find the current I_c in line C.

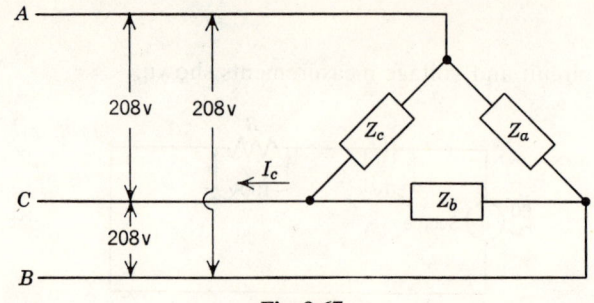

Fig. 9-67

Solution

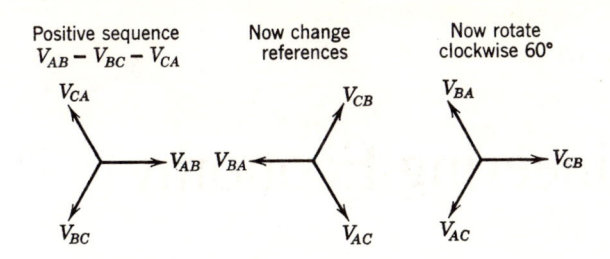

Positive sequence
$V_{AB} - V_{BC} - V_{CA}$

Now change references

Now rotate clockwise 60°

Fig. 9-68

We have

$$V_{CB} = 208 \angle 0° \qquad Z_{CB} = Z_b = 7 + j3$$
$$V_{AC} = 208 \angle -120° \qquad Z_{AC} = Z_c = 2 + j5$$
$$V_{BA} = 208 \angle 120° \qquad Z_{BA} = Z_a = 4 - j3$$

Fig. 9-69

Now $I_C = I_{AC} - I_{CB}$.

$$I_{AC} = \frac{V_{AC}}{Z_{AC}} = \frac{208 \angle -120°}{2 + j5} = \frac{208 \angle -120°}{5.38 \angle 68.2°}$$
$$= 38.7 \angle -188.2°$$

$$I_{CB} = \frac{V_{CB}}{Z_{CB}} = \frac{208 \angle 0°}{7 + j3} = \frac{208 \angle 0°}{7.6 \angle 23.2°}$$
$$= 27.3 \angle -23.2°$$

$$I_C = 38.7 \angle -188.2° - 27.3 \angle -23.2°$$
$$= -38.2 + j5.5 - (25.5 - j10.8)$$
$$= -63.7 + j16.3 = 65.8 \angle 14.3°$$

Current I_C in line $C = 65.8$ amp ▶

Chapter 10

Engineering Economy

Engineering economy is the title given a group of techniques for the systematic economic analysis of alternative courses of action. Being money-based, the analysis gives guidance for economically efficient decision making.

TIME VALUE OF MONEY

Most economy problems involve determining what is economical in the long run, that is, over a considerable period of time. It is necessary to recognize that there is a time value to money. The general rule is: *A dollar now is worth more than the prospect of a dollar at any future time.* Thus $100 today is more valuable to us than someone's guarantee to give us $100 a year from now. The reason is simple; we could deposit our present $100 in a bank at 5% interest, and at the end of the year the bank would return our $100 together with $5 interest. So $100 today *is* worth more than $100 one year hence, and at 5% we know that $100 today is really the same as $105 one year hence.

EQUIVALENCE

If we were unconcerned whether we had the money now or a year from now and if we believed that 5% was a suitable interest rate, we would say that $100 today is equivalent to $105 one year hence. This illustrates an important concept: *Different amounts of money paid at different dates may be equivalent to one another.*

COMPOUND INTEREST FORMULAS

The generally accepted notation is:

$i =$ interest rate per interest period. In the formulas the interest rate is stated as a decimal; for example, 5% interest is entered in formulas as 0.05.

n = number of interest periods. Frequently the interest period is 1 year, but it might be 3 months, 6 months, or some other time period.
P = a present sum of money
F = a sum of money n periods from the present date that is equivalent to P with interest rate i
A = the end-of-period payment or disbursement in a uniform series continuing for n periods, the entire series being equivalent to P at interest rate i

If a present sum P is invested at interest i, the interest for the first period is iP, and the total amount at the end of the first period is $P + iP$ or $P(1 + i)$. For the second period the interest on the amount $P(1 + i)$ is $iP(1 + i)$, and the total amount at the end of the second period is

$$P(1 + i) + iP(1 + i) = P(1 + i)(1 + i) = P(1 + i)^2$$

Similarly, at the end of the third period the total amount is $P(1 + i)^3$, and at the end of n periods it is $P(1 + i)^n$. This is the formula for the future sum F obtainable n periods hence from a present sum P. From this we have the first of six compound interest formulas:

Single payment compound amount factor $CA = (1 + i)^n$

$$F = P(1 + i)^n$$

Solving for P, we obtain the

Single payment present worth factor $PW = (1 + i)^{-n}$

$$P = F \frac{1}{(1 + i)^n}$$

Example 1

An individual wishes to deposit a certain quantity of money now so that at the end of 5 years he will have $500. With interest at 4% compounded annually, how much must he deposit now?

Solution

$$n = 5 \qquad F = \$500$$
$$i = 0.04 \qquad P = ?$$

$$P = F \frac{1}{(1 + i)^n} = 500 \left[\frac{1}{(1 + 0.04)^5} \right] = 500(0.8219) = \$410.95$$

The usual method of solving this problem is not as just demonstrated but by the use of tables of compound interest factors. For solution by tables the problem would be written:

$$P = F(PW\text{-}4\%\text{-}5 \text{ yr}) = 500(PW\text{-}4\%\text{-}5 \text{ yr})$$

Then from compound interest tables (see Appendix B) the single payment present worth factor for $i = 0.04$ and $n = 5$ periods is $(PW\text{-}4\%\text{-}5 \text{ yr}) = 0.8219$. Hence, as before, we find the solution to be

$$P = 500(0.8219) = \$410.95$$

From the calculation we know that if $410.95 is deposited now at 4% interest compounded annually it will increase to $500 by the end of 5 years ▶

Example 2

If you were to deposit $2000 in a bank whose present interest policy is "4% interest, compounded quarterly," how much would be in your account at the end of 2 years?

Solution

We must compute the number of interest periods and the interest rate per interest period. We have

$$n = 8 \qquad P = \$2000$$
$$i = 0.01 \qquad F = ?$$
$$F = P(1 + i)^n \quad \text{or} \quad F = P(CA\text{-}i\%\text{-}n)$$
$$F = 2000(CA\text{-}1\%\text{-}8)$$

From the compound interest tables (Appendix B) we find $(CA\text{-}1\%\text{-}8) = 1.083$.

$$F = 2000(1.083) = \$2166 \blacktriangleright$$

Where there are a series of end-of-period payments or disbursements, the following four compound interest factors are used:

Series compound amount factor $SCA = \dfrac{(1 + i)^n - 1}{i}$

$$F = A\dfrac{(1 + i)^n - 1}{i} = A(SCA\text{-}i\%\text{-}n)$$

Sinking fund factor $SF = \dfrac{i}{(1 + i)^n - 1}$

$$A = F\dfrac{i}{(1 + i)^n - 1} = F(SF\text{-}i\%\text{-}n)$$

Capital recovery factor $CR = \dfrac{i(1 + i)^n}{(1 + i)^n - 1}$

$$A = P\dfrac{i(1 + i)^n}{(1 + i)^n - 1} = P(CR\text{-}i\%\text{-}n)$$

Series present worth factor $SPW = \dfrac{(1 + i)^n - 1}{i(1 + i)^n}$

$$P = A\dfrac{(1 + i)^n - 1}{i(1 + i)^n} = A(SPW\text{-}i\%\text{-}n)$$

Example 3

A man on January 1 deposits $1000 in a credit union that pays 5% interest, compounded annually. He wishes to make five equal end-of-year withdrawals beginning December 31 of the first year. How much should he withdraw each year?

Solution

$$n = 5 \qquad P = 1000$$
$$i = 0.05 \qquad A = ?$$
$$A = P(CR\text{-}i\%\text{-}n) = 1000(CR\text{-}5\%\text{-}5) = 1000(0.2310) = 231.00$$

Therefore, the amount to be withdrawn each year is $231.00 ▶

ECONOMY STUDIES

Several fundamental elements are involved in economy studies:

1. The economy study is usually made from the viewpoint of the owner or owners of the enterprise.
2. The study is a comparison of alternatives and deals with prospective differences between the alternatives.
3. As far as possible, the differences are reduced to differences in money receipts and disbursements.
4. A minimum attractive rate of return or suitable interest rate is selected or will be calculated.
5. The analysis is made by comparing the alternatives by means of (*a*) annual cost, (*b*) present worth, or (*c*) rate of return.
6. The decision concerning the best alternative considers not only the monetary comparison but also the prospective differences between alternatives that are not reduced to money terms.

ANNUAL COST

To compare nonuniform series of money disbursements, it is necessary somehow to make them comparable. One way to do this is by reducing each to an equivalent uniform annual series of payments (or receipts).

Factors in the annual cost method are:

1. A uniform rate of interest is charged on all money, no matter whether borrowed or not.
2. If a net money receipt from salvage value is in prospect at the end of the life of the asset, this reduces the annual cost.

3. Differences in estimated lives of alternatives create no special problem when making an annual cost comparison.

4. The conversion of variable annual disbursements to equivalent uniform annual disbursements requires calculating the present worth of each of the variable disbursements. Then the present worth sum can be multiplied by a suitable capital recovery factor to obtain the uniform equivalent annual cost.

PRESENT WORTH

The present worth of a prospective series of future money receipts has been defined as the present investment that the future receipts would just repay with interest. Comparisons on the basis of present worth may be thought of as comparisons of the sums necessary to endow a given number of years of service of the facility being examined. The valuation of any property that represents a right to future net money receipts is a present worth calculation.

RATE OF RETURN

Present worth calculations start with the assumption of an interest rate or minimum attractive rate of return. Sometimes we want to know the prospective rate of return on an investment rather than simply to determine whether or not it meets some fixed standard of attractiveness. The rate of return computation usually involves a trial-and-error solution involving several present worth calculations, which is solved by assuming different interest rates. The rate of return is the interest rate that equates the present worth of cost and the present worth of benefit. This is illustrated in Problem 10-31.

Example 4

A manufacturing firm is considering replacing an old machine with a new machine which would perform the same task. The old machine had cost $5000 6 years ago and has an estimated useful life of an additional 10 years. The salvage value at the present time is $2400, but no salvage value will remain after 10 years. The annual operating costs, exclusive of depreciation and interest, are $6500. The proposed new machine would cost $7000, and the salvage value at the end of its useful life, estimated to be 10 years, would be zero. The annual operating costs, exclusive of depreciation and interest, are $5400. If the firm uses an interest rate of 10%, what should it do?

Solution

	Old Machine	New Machine
Current value	2400	7000
Useful life	10 more years	10 years
Salvage value	0	0
Annual operating cost	6500	5400

The problem will be solved by the annual cost method.

Old machine

$2400(CR\text{-}10\%\text{-}10) = 2400(0.1627) =$ 390
Annual operating cost $\qquad = 6500$

Equivalent uniform annual cost $= \$6890$

New machine

$7000(CR\text{-}10\%\text{-}10) = 7000(0.1627) =$ 1140
Annual operating cost $\qquad = 5400$

Equivalent uniform annual cost $= \$6540$

Based on the available data, we would choose the alternative with the smaller equivalent uniform annual cost. We choose to purchase the new machine ▶

The problem also may be solved by the present worth method.

Old machine

Current value $\qquad = 2400$
Present worth of annual operating cost
$6500(SPW\text{-}10\%\text{-}10) = 6500(6.145) \quad = 39{,}900$

Present worth of cost $= \$42{,}300$

New Machine

Installed cost $\qquad 7000$
Present worth of annual operating cost
$5400(SPW\text{-}10\%\text{-}10) = 5400(6.145) \quad = 33{,}200$

Present worth of cost $= \$40{,}200$

This time we select the alternative with the lower present worth. As before, we choose the new machine ▶

This problem may be solved by still another method. It will be examined by the approximate method of straight-line depreciation plus average interest. In this approximate method the decline in value of the asset, over its useful life, is distributed equally to each year of asset life. This amounts to straight-line depreciation:

$$\frac{P - L}{n}$$

where P = present value
L = end of useful life salvage value
n = useful life in years

In addition, there must be a charge to reflect the investment in the unrecovered cost of the asset. This is an interest charge on the money still invested in the asset. In the first year the interest is $(P - L)i + Li$, and in the last year of the useful life it is $(P - L)i/n + Li$. The average interest charge is half the sum of the first and last years interest charges, or:

$$\frac{(P - L)i + Li + \left(\frac{P - L}{n}\right)i + Li}{2} = (P - L)\left(\frac{i}{2}\right)\left(\frac{n + 1}{n}\right) + Li$$

Thus the straight-line depreciation plus average interest method is based on two equations:

$$\text{Straight-line depreciation} = \frac{P - L}{n}$$

$$\text{Average interest} = (P - L)\left(\frac{i}{2}\right)\left(\frac{n + 1}{n}\right) + Li$$

For Example 4 the computation is as follows:

Old machine

$$\frac{P - L}{n} = \frac{2400 - 0}{10} \qquad\qquad\qquad = 240$$

$$(P - L)\left(\frac{i}{2}\right)\left(\frac{n + 1}{n}\right) + Li = (2400 - 0)\left(\frac{0.10}{2}\right)\left(\frac{11}{10}\right) = 132$$

Annual operating cost $\qquad\qquad\qquad\qquad\qquad\qquad = \underline{6500}$

Average annual cost = $6872

New machine

$$\frac{P - L}{n} = \frac{7000 - 0}{10} \qquad\qquad\qquad = 700$$

$$(P - L)\left(\frac{i}{2}\right)\left(\frac{n + 1}{n}\right) + Li = (7000 - 0)\left(\frac{0.10}{2}\right)\left(\frac{11}{10}\right) = 385$$

Annual operating cost $\qquad\qquad\qquad\qquad\qquad\qquad = \underline{5400}$

Average annual cost = $6485

The new machine, with the lower average annual cost, is selected ▶

10-1 Wt. 2

Your company has an outstanding note which calls for a lump sum payment of $12,000 at the end of 7 years. If you can borrow money for 6% interest compounded annually, what can you afford to pay now to have the note canceled?

Solution

You can afford to pay the present worth of $12,000 7 years hence at 6% interest, or

$$P = F(PW\text{-}6\%\text{-}7) = 12,000(0.665) = \$7980 \blacktriangleright$$

10-2 Wt. 4

A subdivider offers lots for sale at $5000; $500 is to be paid down, and $900 is to be paid at the end of each year for the next 5 years with no interest to be charged on the payments. This time purchase will require an extra $100 to be paid with the down payment to cover handling charges. Further negotiation reveals that the identical lot may be obtained for $4500 cash. Compute to the nearest $\frac{1}{10}$th % the actual interest rate the buyer will pay if the time purchase arrangement is employed.

Solution

$$4500 = 100 + 500 + 900(SPW\text{-}i\%\text{-}5)$$

$$(SPW\text{-}i\%\text{-}5) = \frac{4500 - 600}{900} = 4.333$$

From the series present worth column of Appendix B:

i	SPW
4%	4.452
5%	4.329

Interpolating,

$$i = 5\% - \tfrac{4}{123}(1\%) \approx 5.0\% \blacktriangleright$$

10-3 Wt. 1

In evaluating the economics of projects, the capital recovery factor equals the

(a) present worth factor plus the sinking fund factor
(b) sinking fund factor plus the interest rate
(c) compound amount factor plus the present worth factor
(d) series compound amount factor plus the present worth factor

Solution

The capital recovery factor equals the sinking fund factor plus the interest rate ▶

Answer is (b)

10-4 Wt. 1

The uniform annual end-of-year payment to repay a debt (the lender's investment) in n years, with an interest rate of i, is determined by multiplying the capital recovery factor by the

(a) average investment
(b) initial investment plus total interest
(c) average investment plus interest
(d) initial investment plus first year's interest
(e) initial investment

Solution

The correct relation is: Annual payment equals the initial investment multiplied by the capital recovery factor, or $A = P(CR\text{-}i\%\text{-}n)$ ▶

Answer is (e)

10-5 Wt. 2

Find the present worth (value) of $10,000 assuming a 3-year return period, if money can be invested at 5%, compounded yearly. Present worth is defined as the amount invested today that will accumulate to the required value at the end of a given time period.

Solution

$$P = F(PW\text{-}5\%\text{-}3) = 10{,}000(0.8638) = \$8638 \blacktriangleright$$

10-6 Wt. 2

A sum of money invested at 4% interest, compounded semiannually, will double in amount in approximately

(a) $15\frac{1}{2}$ years
(b) $17\frac{1}{2}$ years
(c) $19\frac{1}{2}$ years
(d) $21\frac{1}{2}$ years
(e) $23\frac{1}{2}$ years

Solution

$$P = 1 \quad F = 2 \quad i = 0.02 \quad n = ?$$
$$F = P(1 + i)^n$$
$$2 = 1(1 + 0.02)^n \quad 1.02^n = 2 \quad n = 35$$

Therefore the number of semiannual interest periods for money to double at 4% interest is 35. The time period is then $35/2 = 17\frac{1}{2}$ years ▶

Answer is (b)

10-7 Wt. 2

Sufficient funds are to be deposited in a savings account with a bank to provide a total of $10,000 in the account after a 10-year period. If the nominal interest rate is 4% compounded semiannually, what amount P should be deposited?

Solution

$$F = 10,000 \quad P = ? \quad n = 20 \quad i = 0.02$$

$$P = F(PW\text{-}2\%\text{-}20) = 10,000(0.6730) = \$6730 \blacktriangleright$$

10-8 Wt. 4

A contractor wishes to set up a special fund by making uniform semi-annual deposits for 20 years. The fund is to provide $10,000 per year for each of the last 5 years of the 20-year period. If interest is at 4%, compounded semiannually, how much should the semiannual deposit be?

Solution

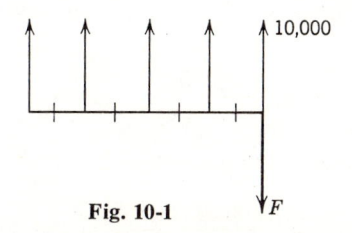

Fig. 10-1

$$F = 10,000 + 10,000(CA\text{-}2\%\text{-}2) + 10,000(CA\text{-}2\%\text{-}4)$$
$$+ 10,000(CA\text{-}2\%\text{-}6) + 10,000(CA\text{-}2\%\text{-}8)$$
$$= 10,000(1 + 1.040 + 1.082 + 1.126 + 1.172)$$
$$= 10,000(5.420) = 54,200$$

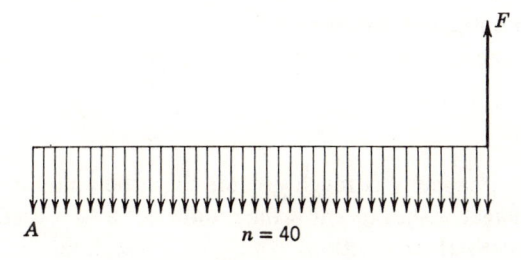

Fig. 10-2

$$A = F(SF\text{-}2\%\text{-}40) = 54,200(0.0166) = \$900 \blacktriangleright$$

504 / Engineering Economy

10-9 Wt. 5

A company which manufactures electric motors has a production capacity of 200 motors per month. The variable costs are $150 per motor. The average selling price of the motors is $275. Fixed costs of the company amount to $20,000 per month, which includes all taxes. Present policy is to pay an annual dividend of $2 per share on each of the 15,000 shares of common stock.

(*a*) By analytical or graphical methods, determine the number of motors that must be sold each month to break even and the sales volume corresponding to the break-even point.

(*b*) If the company can bring about a 20% reduction in both fixed and variable costs, what will be the new break-even point and how much profit will be made if operations are at 90% of capacity?

Solution

(*a*) Fixed cost = $20,000 per month

Variable cost = $150 per motor

Let the number of motor sales per month to break even be x. Then $275x = 150x + 20,000$.

$$x = \frac{20,000}{275 - 150} = 160 \blacktriangleright$$

The required sales volume to break even = $275(160) = \$44,000$ ◄

(*b*) New breakeven point = $\dfrac{0.8(20,000)}{275 - 0.8(150)} = 104$ (actually 103.23) ▶

The profit if operations are at 90% of capacity is:

$$\text{Profit} = 275(180) - 0.8(20,000) - 0.8(150)(180)$$

$$= 49,500 - 16,000 - 21,600 = \$11,900 \blacktriangleright$$

It should be pointed out that the firm's dividend policy cannot be considered a fixed cost to the company. Rather it reflects periodic decisions by the Board of Directors.

10-10 Wt. 6

In a foundry, capacity is much greater than the amount to be produced. The plant manager wishes to know the number of lots of castings to produce, within a given total production, which will give him the least total cost. The cost of one manufacturing setup is $150. The current rate of interest is 6%. Total production for the year is to be 30,000 castings; cost per piece is

$3. What number of manufacturing lots would you recommend, and what would be the total annual cost of production?

Solution

Setup cost $S = \$150$
Interest rate $I = 6\%$
Annual production $P = 30{,}000$ castings
Total cost per piece $T = \$3$
Lot size $= Q$

Assuming that there is a constant demand for the castings and that production capacity is large compared to demand, then

Maximum inventory of castings $= Q$
Variable investment in inventory $= TQ$

Interest on variable investment in maximum inventory $= \dfrac{I}{2} TQ$

Annual setup cost $= \dfrac{SP}{Q}$

Total annual cost which is variable with lot size is $y = \dfrac{I}{2} TQ + \dfrac{SP}{Q}$

For minimum cost, set the derivative (dy/dQ) equal to zero:

$$\frac{dy}{dQ} = \frac{I}{2} T - \frac{SP}{Q^2} = 0 \qquad Q^2 = \frac{SP}{\frac{I}{2} T} \qquad Q = \left(\frac{SP}{\frac{I}{2} T}\right)^{1/2}$$

$$Q = \left(\frac{150(30{,}000)}{\frac{0.06}{2}(3.00)}\right)^{1/2} = (50 \times 10^6)^{1/2} = 7070$$

Number of manufacturing lots per year $= \dfrac{P}{Q} = \dfrac{30{,}000}{7070} = 4.28$

The cost of producing 30,000 castings in four lots, including interest on the investment in inventory, is

$$\text{Cost} = PT + \frac{I}{2} TQ + \frac{SP}{Q} = 30{,}000(3) + \frac{0.06}{2}(3)\left(\frac{30{,}000}{4}\right) + \frac{150(30{,}000)}{\frac{30{,}000}{4}}$$

$$= 90{,}000 + 675 + 600 = \$91{,}275$$

506 / Engineering Economy

Thus the economic lot size is 7070 castings which would require 4.28 manufacturing lots per year. If the annual requirement was manufactured in four lots, the total cost would be $91,275 ▶

10-11 Wt. 3

A 10-hp electric motor is required for a certain plant process; two different models are being considered. Motor A has a first cost of $350 and an overall efficiency of 90%. Motor B has a first cost of $250 and an efficiency of 80%. Both motors have identical lives and will be loaded to rated capacity when in use. The cost of electric energy is 2.5¢/kwhr, and the investment charges are 20% of the first cost. Compute the annual break-even operating time in hours at which the annual cost of operation and ownership of motor A equals that of B.

Solution

$$\text{Efficiency} = \frac{\text{output}}{\text{input}} = \frac{10 \text{ hp}}{\text{input}}$$

$$\text{Difference in power input} = \frac{10 \text{ hp}}{0.80} - \frac{10 \text{ hp}}{0.90} = 12.5 - 11.1 = 1.4 \text{ hp}$$

So the required additional power in kilowatts = 1.4(0.746) = 1.045 kw.

Let the annual incremental investment charge equal the annual incremental power charge:

$$0.20(350 - 250) = 1.045(0.025)(\text{hours of operation})$$

Thus the number of annual hours of operation for break even = 0.20(350 − 250)/1.045(0.025) = 763 hr ▶

10-12 Wt. 3

A manufacturer who produces a single item has a maximum production capacity of 80,000 units. The overall and unit costs for different levels of operation are as follows:

Output (Units)	Total Cost (Dollars)	Total Cost per Unit (Dollars)
0	60,000	
5,000	82,000	16.40
10,000	107,000	10.70
20,000	151,000	7.55
40,000	237,000	5.92
60,000	321,000	5.35
80,000	405,000	5.06

Early in the year orders are received for 20,000 units at $6.00 each. Owing to depressed business conditions it is realized that no further domestic orders can be expected in the current year. Two foreign purchase orders are available, one for 20,000 units at $4.80 each and the other for 40,000 units at $4.70 each. Indicate the reasons why either or both foreign orders should or should not be accepted.

Solution

Output (Units)	Total Cost (Dollars)	Incremental Cost (Dollars)	Incremental Unit Cost (Dollars)
20,000	151,000		
		86,000	4.30
40,000	237,000		
		84,000	4.20
60,000	321,000		
		84,000	4.20
80,000	405,000		

The manufacturer should accept any order where the additional revenue will exceed the incremental cost of supplying the order. In this case he should accept both foreign orders ▶

10-13 Wt. 6

The ABC Corporation seeks to enlarge the market for its product, which is sold under the brand name ExxDRA. Examine the following information and then state which action the corporation should take. Show computations.

At present, 400,000 units of ExxDRA are produced and sold each year. The current selling price is 30¢ each. On the basis of a thorough investigation and analysis, the management believes that it can increase sales by 50,000 units per year if it lowers the price from 30¢ to 29¢; by 150,000 units if the price is reduced to 28¢; and by 300,000 units if the price is lowered to 26¢.

The present capacity of the plant on a one-shift 40-hour week is 600,000 units. Operating at capacity, its cost structure would be:

> Fixed costs = $40,000.00 per year
> Present advertising budget = $20,000.00 per year
> Variable costs = 16¢ per unit of output

To produce in excess of 600,000 units on a one-shift basis would require overtime premium pay. The variable costs for the units over 600,000 would be 20¢ per unit.

If the company should raise its annual advertising budget to $30,000, it could expect to sell 500,000 units without lowering the price from 30¢; if the budget were raised to $45,000, it could expect to sell 600,000 units at 30¢.

Solution

Sales	Sales Price (Dollars)	Total Income (Dollars)	Total Incremental Income (Dollars)	Unit Incremental Income (Dollars)	Incremental Manuf. Cost (Dollars)
400,000	0.30	120,000			
			10,500	0.21	8,000
450,000	0.29	130,500			
			23,500	0.235	16,000
550,000	0.28	154,000			
			28,000	0.187	28,000
700,000	0.26	182,000			

Advertising: Increasing the advertising by $10,000 to $30,000 yields 500,000 sales at 30¢:

Incremental cost: $10,000

Incremental income: $100,000(0.30 - 0.16) = \$14,000$

Further increasing the advertising to $45,000 yields 600,000 sales at 30¢:

Incremental cost: $15,000

Incremental income: $100,000(0.30 - 0.16) = \$14,000$

First we must check that ExxDRA is worthy of continued production: At 400,000 units the income is $120,000. The total cost is

$$40,000 + 20,000 + 0.16(400,000) = \$124,000$$

Thus at a sales volume of 400,000 units, ExxDRA is not recovering its total cost.

Consider increased advertising: Based on the incremental analysis above, the addition of $10,000 of advertising is profitable, for it can be expected to increase income by $14,000. A further addition of $15,000 above the $30,000 advertising level is not profitable, for the incremental cost exceeds the incremental income.

Consider a reduction of the sales price: The reduction of the sales price from 30¢ to 29¢ yields an incremental income of $10,500 and an incremental manufacturing cost of $8,000. This is a favorable result. A further reduction of the sales price to 28¢ results in increased income of $23,500 and increased manufacturing costs of $16,000, which is also a favorable increment. A still further reduction of the sales price to 26¢ produces an incremental cost and income of $28,000 with no net advantage to the ABC Corporation.

Conclusion: The increase of the advertising budget to $30,000 yields a $4000 increment of profit (or reduced loss). The reduction of the sales price to 28¢ can be expected to increase income by $10,000. Based on the available data, the company should reduce its sales price to 28¢ ▶

This is not necessarily the optimal solution. Some other combination of selling price and advertising level might further improve the situation for the ABC Corporation.

10-14 Wt. 6

A state university proposes to purchase a large concrete testing machine for use in their materials testing laboratory. Two different testing machines are available for purchase. Machine A has an initial cost of $30,000 and an estimated life of 10 years. The power requirements to operate the machine amount to $50 per month. Its maintenance cost is $400 per year. The initial cost of Machine B is $20,000 and has an estimated life of 20 years. Power consumption and maintenance costs amount to $250 per month. The salvage value of each machine is zero. Determine the annual cost of each machine. Use 6% sinking fund depreciation and 6% interest on the first cost.

Solution

A sinking fund is a fund established to produce a desired amount at the end of a given period of time by means of a uniform series of payments throughout the period. The sinking fund factor SF is used to compute the value of the individual payments A when the future sum F is known. It should be noted that A is defined as an end-of-year payment while the problem actually calls for various monthly payments. In economy studies this discrepancy is frequently ignored, and payments are assumed to be made at the end of the year or interest period.

	Machine A	Machine B
Initial cost	$30,000	$20,000
Power cost	$50 per month	$250 per month
Maintenance	$400 per year	
Life	10 years	20 years

510 / Engineering Economy

Using an annual cost comparison we have:

	Machine A	Machine B
6% sinking fund depreciation		
$n = 20$ 20,000(0.0272)		= 544
$n = 10$ 30,000(0.0759)	= 2277	
Interest on investment		
0.06(20,000)		= 1200
0.06(30,000)	= 1800	
Power and maintenance		
50(12) + 400	= 1000	
250(12)		= 3000
Equivalent annual cost	= $5077 ▶	$4744 ▶

10-15 Wt. 3

Given the following data on condensers:

	Metal	
Item	Ferrous	Copper
First cost, installed, ready to operate	$800.00	$855.00
Life	5 years	? years
Salvage value at end of useful life	0	0
Operating cost per year	$50.00	$40.00
Interest rate per annum = 5%		

Certain items that are the same for each condenser have been omitted. The length of service of the copper alloy condenser is not known. The ferrous metal condenser may be replaced every 5 years at the same first cost. Compute the life (to the nearest whole year) for the more expensive condenser to be as economical as the cheaper one.

Solution

Ferrous condenser

$$800(CR\text{-}5\%\text{-}5) = 800(0.231) = 184.80$$
Annual operating cost $\quad = \quad 50.00$

Equivalent annual cost $\quad = \$234.80$

Copper condenser

$855(CR\text{-}5\%\text{-}n) \quad = \quad 855(CR\text{-}5\%\text{-}n)$
Annual operating cost $\quad = \quad 40.00$

Equivalent annual cost $\quad = \quad 855(CR\text{-}5\%\text{-}n) + 40.00$

Setting the annual cost of the ferrous condenser equal to the annual cost of the copper condenser, we have

$$234.80 = 855(CR\text{-}5\%\text{-}n) + 40.00$$

or

$$(CR\text{-}5\%\text{-}n) = \frac{234.80 - 40.00}{855.00} = 0.227$$

From the compound interest tables (Appendix B) we see that

$$(CR\text{-}5\%\text{-}6) = 0.197$$
$$(CR\text{-}5\%\text{-}5) = 0.231$$

Thus the life of the copper condenser must be somewhat greater than 5 years for it to be as economical as the ferrous condenser ▶

10-16 Wt. 3

A building costs $50,000. Its estimated life is 20 years. How much should be spent for annual upkeep if this would extend the life to 30 years? Take interest at 10% per annum.

Solution

Assume that we would be willing to spend money on annual upkeep until the point is reached where the equivalent uniform annual cost for the building with its life extended equals the equivalent uniform annual cost for the 20-year life. Then

$$50,000(CR\text{-}10\%\text{-}20) = 50,000(CR\text{-}10\%\text{-}30) + \text{annual upkeep}$$
$$50,000(0.1175) = 50,000(0.1061) + \text{annual upkeep}$$
$$5875 = 5305 + \text{annual upkeep}$$
$$\text{Annual upkeep} = 5875 - 5305 = \$570 \blacktriangleright$$

10-17 Wt. 4

A new snow removal machine costs $50,000. The new machine will operate at a reported saving of $400 per day over the present equipment in terms of time and efficiency. If interest is 5% and the machine's life is assumed to be 10 years with zero salvage, how many days per year must the machine be used to make the investment economical?

Solution

The problem will be solved by the annual cost method. Set the annual saving with the snow removal machine equal to its annual cost and solve for the required utilization.

Let X = annual operation in days per year.

$$\text{Annual saving} = \text{annual cost}$$
$$400X = 50{,}000(CR\text{-}5\%\text{-}10)$$
$$= 50{,}000(0.1295) = 6475$$
$$X = \tfrac{6475}{400} = 16.2 \text{ days per year} \blacktriangleright$$

10-18 Wt. 4

It is desired to determine whether to use a material which is 1 in. thick or one which is 2 in. thick in insulating a steam pipe. The heat loss from this pipe without insulation would cost $2.00 per year per foot. One-inch insulation will eliminate 88% of the loss and cost $0.50 per foot. Two-inch insulation will eliminate 92% of the loss and will cost $1.10 per foot. Compare the annual cost per 1000 ft for the two thicknesses using a life of 10 years with no salvage value. Interest is assumed to be 6%.

Solution

One-inch insulation

Capital recovery of installed cost
$= \$0.50(1000)(CR\text{-}6\%\text{-}10) = 500(0.1359) = 68$
Annual cost of heat loss
$= 0.12(\$2.00)(1000) = 240$

Equivalent uniform annual cost $ = \$308$

Two-inch insulation

Capital recovery of installed cost
$= \$1.10(1000)(CR\text{-}6\%\text{-}10) = 1100(0.1359) = 149$
Annual cost of heat loss
$= 0.08(\$2.00)(1000) = 160$

Equivalent uniform annual cost $ = \$309$

From the calculations we see that there is no significant difference in the cost of the two alternatives ▶

This analysis is incomplete, for there is another important alternative to be considered. It is, of course, the do-nothing alternative—install no insulation. We would find, however, that insulation is the economical decision in this case. The thickness to install should be selected on the basis of other tangible or intangible factors not presented in the problem.

10-19 Wt. 4

A contractor can purchase a heavy-duty truck with a 12-yd^3 dump body for $13,000. Its estimated life is 7 years, and its estimated salvage is $2000.

Maintenance is estimated at $1100 per year. Daily operating expenses are estimated at $53, including the cost of the driver. The contractor can hire a similar unit and its driver for $83 per day. If interest is taken at 8%, how many days per year must the services of a dump truck be required to justify the purchase of a truck?

Solution

The problem is solved by setting the annual cost of truck hire equal to the annual cost of truck ownership. In an annual cost computation where there is a salvage value L, there are two ways to compute the capital recovery of the investment. It is equal to either $(P - L)(CR\text{-}i\%\text{-}n) + Li$ or $P(CR\text{-}i\%\text{-}n) - L(SF\text{-}i\%\text{-}n)$. The results are identical. Here we will use the first computation. Let X = amount of truck utilization in days per year.

$$83X = (13{,}000 - 2000)(CR\text{-}8\%\text{-}7) + 2000(0.08) + 1100 + 53X$$

$$30X = 11{,}000(0.1921) + 160 + 1100$$

$$X = \frac{2{,}110 + 160 + 1{,}100}{30} = 112.3 \text{ days}$$

Thus the truck must be required for more than 112 days per year to justify its purchase ▶

10-20 Wt. 1

The formula for determining average interest is

(a) $(P - L)\left(\dfrac{i}{2}\right)\left(\dfrac{n+1}{n}\right) + Li$

(b) $(P - L)\left[\dfrac{i}{2}\left(\dfrac{n+1}{1}\right)\right]$

(c) $\left(\dfrac{P+L}{2}\right)i$

(d) $(P - L)\left[\dfrac{1}{(1+i)^n - 1}\right] + Li$

(e) $(P - L)\left[\dfrac{1}{(1+i)^n - 1}\right]$

Solution

The formula for average interest is given in (a) above, where P = initial cost, L = salvage value, n = life in years, and i = interest rate stated as a decimal ▶

514 / Engineering Economy

For the derivation of the average interest equation, see Example 4.

Answer is (a)

10-21 Wt. 6

A public utility owns a steam power plant which was built 20 years ago. Its estimated life span when originally built was 40 years. Depreciation on this plant has been written off on the books so that the book value of this plant is now $50,000. The present costs for operating the plant are as follows:

Labor	$2100 per month
Fuel	900 per month
Taxes	1400 per year
Insurance	600 per year

No depreciation sinking fund has been maintained for this steam plant, and its salvage value at the end of its life period is estimated to be $2000.

A new replacement diesel power plant costing $120,000 would reduce the foregoing operating costs by 50%. The life of the new diesel plant is estimated to be 30 years, and its salvage value would be $3000. Using the straight-line depreciation plus average interest method and employing an interest rate of 4%, determine the average annual saving that would result by replacing the steam plant with the diesel power plant.

Solution

The problem does not state a present market value for the steam plant, so we will assume it is equal to the present book value. In reality, however, this is not likely to be the case.

Steam power plant

Straight-line depreciation

$$\frac{P - L}{n} = \frac{50,000 - 2000}{20} = 2,400$$

Average interest (see Problem 10-20)

$$(50,000 - 2000)\left(\frac{0.04}{2}\right)\left(\frac{21}{20}\right) + 2000(0.04) = 1,088$$

Plant operation

Labor	2100 × 12	=	25,200
Fuel	900 × 12	=	10,800
Taxes and insurance		=	2,000
	Average annual cost	=	$41,488

Diesel plant
Straight-line depreciation
$$\frac{120{,}000 - 3000}{30} = 3{,}900$$
Average interest
$$(120{,}000 - 3000)\left(\frac{0.04}{2}\right)\left(\frac{31}{30}\right) + 3000(0.04) = 2{,}538$$

Plant operation
Labor	1050 × 12	= 12,600
Fuel	450 × 12	= 5,400
Taxes and insurance		= 1,000

Average annual cost = $25,438

Average annual saving = 41,488 − 25,438 = $16,050 ▶

10-22 Wt. 6

The cost to rehabilitate a condemned timber highway bridge to place it in usable condition with a life expectancy of 20 years is $11,500. The salvage value of the timber bridge at the end of the 20-year period would be $1500. The annual cost to maintain the bridge is $4200.

The initial or first cost of a replacement steel bridge is $48,000 for a structure that has a life expectancy of 20 years. The salvage value of the bridge at the end of 20 years is $8000. The maintenance cost of the steel bridge is $1810 per year. Using the straight-line depreciation plus average interest method, it is found that the average annual cost for both the rehabilitation of the timber bridge and the construction of a new steel replacement bridge are equal. What annual rate of interest are these average annual costs based upon?

Solution

	Timber Bridge	Steel Bridge
Straight-line depreciation = $\frac{P - L}{n}$		
$\frac{11{,}500 - 1500}{20}$	= 500	
$\frac{48{,}000 - 8000}{20}$		= 2,000
Maintenance	= 4200	= 1,810
Average interest		
$(11{,}500 - 1500)\left(\frac{i}{2}\right)\left(\frac{21}{20}\right) + 1500i$	$= 6750i$	
$(48{,}000 - 8000)\left(\frac{i}{2}\right)\left(\frac{21}{20}\right) + 8000i$		$= 29{,}000i$
Average annual cost =	$4700 + 6750i$	$3810 + 29{,}000i$

516 / Engineering Economy

If the average annual costs are equal, then

$$4700 + 6750i = 3810 + 29{,}000i$$

$$i = \frac{4700 - 3810}{29{,}000 - 6750} = \frac{890}{22{,}250} = 0.04 \qquad i = 4\% \blacktriangleright$$

10-23 Wt. 6

A rancher has been pumping irrigation water for several years with a pump driven by a gasoline engine. The annual fuel cost is $700. It is estimated that the annual repair costs for the next few years will be $600. The salvage value of the pump and engine at the present time is $500.

Consideration is being given to replacing the old system with a new pump and electric motor. The installed cost of the new unit is $3500. The annual power cost is estimated at $500. It is estimated that the average repair costs will be $100. The new unit will be installed if it will pay for itself in 5 years. The interest rate is 6%.

(*a*) Compare the two systems on the basis of annual cost. (Assume the old system to have no salvage value at the end of 5 years. Use straight-line depreciation and average interest.)

(*b*) What are some factors, in addition to the annual cost comparison, that might influence the decision to replace or not to replace the present system?

Solution

(*a*)   Average annual cost—old pump

$$\begin{aligned}
\text{Fuel cost} &= 700 \\
\text{Annual repair costs} &= 600 \\
\text{Depreciation } \tfrac{500}{5} &= 100 \\
\text{Average interest } 500\left(\tfrac{0.06}{2}\right)\left(\tfrac{6}{5}\right) &= 18 \\
\text{Average annual cost} &= \$1418 \blacktriangleright
\end{aligned}$$

Average annual cost—new pump

$$\begin{aligned}
\text{Power cost} &= 500 \\
\text{Average repair costs} &= 100 \\
\text{Depreciation } \tfrac{3500}{5} &= 700 \\
\text{Average interest } 3500\left(\tfrac{0.06}{2}\right)\left(\tfrac{6}{5}\right) &= 126 \\
\text{Average annual cost} &= \$1426 \blacktriangleright
\end{aligned}$$

(b) Other factors which might influence the replacement decision are:
1. Income tax effects. ▶
2. Reliability of electric power supply. ▶
3. Reliability of present pumping unit. ▶
4. Difference in capacity of the two pumps. ▶
5. Other uses of the money. ▶

10-24 Wt. 6

A company is considering the purchase of automatic billing equipment to replace 10 clerks who cost $300.00 per month each to send out 20,000 bills each month. The proposed automatic billing equipment will require two operators costing $400.00 per month each. It is estimated that the operation and maintenance cost of the automatic equipment will average $100 per month, that the life of the equipment is 10 years, and that it will have a scrap value of 10% of its cost. Taxes and insurance (combined) on the equipment will be 3% of the original cost per year. If straight-line depreciation is used and interest is taken as 6%, what is the greatest sum the company can pay for the equipment and break even?

Solution

Annual cost using straight-line depreciation plus average interest:

Annual cost of hand operation: $10 \times 300 \times 12 = \$36,000$

Annual cost of machine operation:

Two operators: $2 \times 400 \times 12$	$= 9600$
Operation and maintenance: 100×12	$= 1200$
Taxes and insurance	$= 0.03P$
Depreciation: $\dfrac{P - 0.1P}{10}$	$= 0.09P$
Average interest: $(P - 0.1P)\left(\dfrac{0.06}{2}\right)\left(\dfrac{11}{10}\right) + 0.1P(0.06)$	$= 0.036P$

$$\text{Average annual cost} = \$10,800 + 0.156P$$

Equating the annual costs of the two alternatives,

$\$36,000 = \$10,800 + 0.156P \quad 0.156P = \$25,200 \quad P = \$161,500$ ▶

10-25 Wt. 6

An existing mine ore treatment plant has become obsolete. It cost $1,100,000 20 years ago. The present-day salvage value is $100,000. During the past 20

518 / Engineering Economy

years the operation and maintenance cost of this plant was $12,500 per year. A new replacement treatment plant is to be constructed at a cost of $800,000. It will have an annual operation and maintenance cost of $13,600 per year. The salvage value of the new plant at the end of life expectancy will be $50,000.

In order to save $30,000 in average annual costs for the new replacement treatment plant, compared to that experienced for the old plant, what would the life expectancy be in years for the new plant? Assume an annual interest rate of 6% was used in computing the average annual costs for both the old and new plant. In calculating the average annual costs, use the straight-line depreciation plus average interest method.

Solution

By equating the annual costs for the two plants, we can solve for the required life of the new plant for the $30,000 annual saving.

Old plant

$$\text{Depreciation} = \frac{P - L}{n} = \frac{1,100,000 - 100,000}{20} = \$50,000$$

Operation and maintenance $= 12,500$

$$\text{Average interest} = (P - L)\left(\frac{i}{2}\right)\left(\frac{n+1}{n}\right) + Li$$

$$= (1,100,000 - 100,000)$$

$$\times \left(\frac{0.06}{2}\right)\left(\frac{21}{20}\right)$$

$$+ 100,000(0.06) = 37,500$$

Annual cost $= \$100,000$

New plant

$$\text{Depreciation} = \frac{800,000 - 50,000}{n} = \frac{750,000}{n}$$

Operation and maintenance $= 13,600$

$$\text{Average interest} = (800,000 - 50,000)\left(\frac{0.06}{2}\right)$$

$$\times \left(\frac{n+1}{n}\right) = 22,500\left(\frac{n+1}{n}\right)$$

$$+ 50,000(0.06) = 3,000$$

For a $30,000 per year saving,

$$\frac{750{,}000}{n} + 13{,}600 + 22{,}500\left(\frac{n+1}{n}\right) + 3000 = 100{,}000 - 30{,}000$$

$$\frac{750{,}000}{n} + 22{,}500\left(\frac{n+1}{n}\right) = 53{,}400$$

Multiplying by n,

$$750{,}000 + 22{,}500n + 22{,}500 = 53{,}400n$$

$$n = \frac{772{,}500}{30{,}900} = 25 \text{ years} \blacktriangleright$$

It must be noted that this calculation does *not* insure that replacement of the old plant is the more economical solution. To find that solution, we must compare the *future* cost of operating the old plant with the cost of operating the new plant.

10-26 Wt. 5

A manufacturing company is faced with the choice of repairing an old machine at a cost of $12,000 or replacing it with a new one at a cost of $50,000. It is estimated that the repaired machine will last 5 years, after which time replacement will be necessary. Present salvage value of the old machine is $10,000, and its salvage value after 5 years will be $5000. It is estimated that the new machine will last 20 years and will have a salvage value of $15,000. The yearly maintenance will be $500 less on the new machine than on the old one. With straight-line depreciation and interest at 6%, determine which is more economical: to replace or to repair the old machine. Indicate clearly the basis for your decision.

Solution

We solve for the average annual cost by the straight-line depreciation plus average interest method:

Old Machine

Straight-line depreciation $\dfrac{22{,}000 - 5000}{5}$ = 3400

Average interest $(22{,}000 - 5000)\left(\dfrac{0.06}{2}\right)\left(\dfrac{6}{5}\right) + 5000(0.06)$ = 912

Additional annual maintenance = 500

Average annual cost = $4812

New Machine

$$\text{Straight-line depreciation } \frac{50{,}000 - 15{,}000}{20} = 1750$$

$$\text{Average interest } (50{,}000 - 15{,}000)\left(\frac{0.06}{2}\right)\left(\frac{21}{20}\right) = 2002$$

$$\text{Average annual cost} = \$3752$$

Therefore replace the old machine, since this results in a reduced average annual cost ▶

10-27 Wt. 6

The state will furnish an automobile to certain employees or will permit them to use their personal car and will pay 11¢ per mile. A car costs \$3350, has a life of 5 years, a trade-in value of \$600, and costs \$5 per month for incidentals. The cost of tires and repairs is \$6 per month, and fuel and oil cost 3¢ per mile. How many miles per year must an employee travel so that the two methods will be equal? Interest is computed at 6%. Use straight-line depreciation plus average interest.

Solution

Average annual fixed cost:

$$\text{Depreciation } \frac{3350 - 600}{5} = 550$$

$$\text{Average interest } (3350 - 600)\left(\frac{0.06}{2}\right)\left(\frac{6}{5}\right) + 600(0.06) = 135$$

$$\text{Incidentals } \$5 \times 12 \text{ months} = 60$$

$$\text{Tires and repairs } \$6 \times 12 \text{ months} = 72$$

$$\text{Average annual fixed cost} = \$817$$

If fuel and oil equal 3¢, then for what distance does 8¢ per mile equal \$817?

$$\text{Distance} = \frac{817}{0.08} = 10{,}210 \text{ miles} \blacktriangleright$$

Thus the employee must travel at least 10,210 miles per year for the 11¢ per mile allowance to pay its pro rata share of the automobile cost.

10-28 Wt. 5

A standard pumping installation costs \$30,000 installed and has an estimated life of 12 years. By the addition of certain auxiliary equipment, an annual saving of \$400 in its operating cost can be obtained, and the estimated life of the installation can be doubled. Neglecting any salvage value for either plan and with interest at 6%, what present expenditure is justified for the auxiliary equipment?

Solution

Assuming the pumping installation is needed for 24 years, the present worth of each of the alternatives may be equated. The value of the auxiliary equipment can then be obtained.

Standard installation: For 24 years of service, a new standard pumping installation (at $30,000) must be installed at the end of 12 years.

$$\text{Present worth} = 30{,}000 + 30{,}000(PW\text{-}6\%\text{-}12)$$
$$= 30{,}000 + 30{,}000(0.497) = \$44{,}910$$

Installation with auxiliary equipment:

$$\text{Present worth} = 30{,}000 + \text{auxiliary equipment} - 400(SPW\text{-}6\%\text{-}24)$$
$$= 30{,}000 + \text{auxiliary equipment} - 400(12.550)$$
$$= \text{auxiliary equipment} + 24{,}980$$

Setting the present worth of the standard installation equal to the present worth of the installation with the auxiliary equipment, we obtain

$$44{,}910 = \text{auxiliary equipment} + 24{,}980$$
$$\text{Auxiliary equipment} = 44{,}910 - 24{,}980 = \$19{,}930$$

An expenditure of up to $19,930 is therefore justified for the auxiliary equipment ▶

10-29 Wt. 6

The Pracdifoil Corporation must decide whether to lease or purchase a piece of manufacturing equipment. The purchase price is $250,000, and installation costs will be $25,000. The equipment can be expected to have a useful life of 5 years, after which it can be sold as scrap with a net realization of $25,000. To maintain the equipment will require an expenditure of $3000 per year; taxes and insurance will be $1200 per year. Assume that the annual outlays occur at the beginning of each year.

The machine can be leased for 5 years at an annual rental fee of $58,500, payable in five equal annual installments at the beginning of each year. The lessor agrees to install the equipment ready to use, to remove it at the end of 5 years, and to see that it is kept in working order. The Pracdifoil Corporation will have to spend about $1000 per year for ordinary maintenance and $500 per year for insurance.

Alternative investment opportunities would produce a 5% return to the Pracdifoil Corporation at this time. What should the corporation management decide to do? Use the present value comparison method.

Solution

The computation should determine the present worth (present value) of the cost of the manufacturing equipment for 5 years. It is important to note the assumption that annual outlays are made at the beginning of each year. The compound interest tables, on the other hand, are based on the assumption of end-of-period payments. A cash flow tabulation will clarify the situation:

	Purchase Equipment	Lease Equipment
Now	−275,000 − 4200	−58,500 − 1500
1 year hence	− 4200	−58,500 − 1500
2 years hence	− 4200	−58,500 − 1500
3 years hence	− 4200	−58,500 − 1500
4 years hence	− 4200	−58,500 − 1500
5 years hence	+25,000	

The present worth (present value) comparison can now be readily computed.

For purchased equipment:

$$\text{Present worth} = -275{,}000 - 4200 - 4200(SPW\text{-}5\%\text{-}4)$$
$$+ 25{,}000(PW\text{-}5\%\text{-}5)$$
$$= -279{,}200 - 4200(3.546) + 25{,}000(0.7835)$$
$$= -279{,}200 - 14{,}890 + 19{,}590 = -274{,}500$$

For leased equipment:

$$\text{Present worth} = -60{,}000 - 60{,}000(SPW\text{-}5\%\text{-}4)$$
$$= -60{,}000 - 60{,}000(3.546)$$
$$= -60{,}000 - 212{,}760 = -272{,}760$$

The present worth of purchasing the equipment is slightly higher than the present worth of leasing the equipment. A very careful analysis should be made of these data plus other tangible and intangible consequences to insure that a correct decision is made. If all other things are equal, however, there is a small advantage to leasing rather than purchasing the equipment ▶

10-30 Wt. 5

A machine is to be purchased for $15,500; it has an estimated life of 8 years and a salvage value of $600. A sinking fund is to be established in order that monies will be available to purchase a replacement machine when the first machine wears out at the end of 8 years. An amount of $1303 is to be deposited annually (at the end of each year) during the lifetime of the first machine into this sinking fund. With compound interest tables determine the annual rate of interest this fund must earn to produce sufficient funds to purchase the replacement machine at the end of 8 years.

Solution

The amount of money that must be accumulated in the sinking fund by the end of 8 years equals the cost of the new machine minus the salvage value of the old machine:

$$P - L = 15{,}500 - 600 = 14{,}900$$

So the annual payment A is $1303, and the future sum F is $14,900. We may solve for the interest rate i using either of two compound interest factors: First,

$$A = F(SF\text{-}i\%\text{-}n)$$

so that

$$(SF\text{-}i\%\text{-}8) = \frac{A}{F} = \frac{1303}{14{,}900} = 0.0874$$

From the interest tables we see that this is exactly the value of the sinking fund factor when the interest rate i is 10%. Alternatively,

$$F = A(SCA\text{-}i\%\text{-}n)$$

or

$$(SCA\text{-}i\%\text{-}8) = \frac{F}{A} = \frac{14{,}900}{1303} = 11.436$$

From the tables the interest rate again is 10% ▶

10-31 Wt. 5

An investor purchased 10 shares of C, B and A Company stock for $10,000. He held this stock for 15 years. For the first 4 years he received annual dividends of $500. For the next 3 years he received annual dividends of $400. For the final 8 years he received annual dividends of $300. At the end of the 15th year he sold his stock for $7000. What approximate rate of return did he make on his investment?

Solution

First, we list the cash flow:

Year	Cash Flow	Year	Cash Flow
0	−10,000	8	+300
1	+500	9	+300
2	+500	10	+300
3	+500	11	+300
4	+500	12	+300
5	+400	13	+300
6	+400	14	+300
7	+400	15	+7300

Then, we write an equation to set the present worth of the cash flow equal to zero. There are, of course, several correct ways of setting up the equation. One way is

$$0 = -10{,}000 + 300(SPW\text{-}i\%\text{-}15) + 100(SPW\text{-}i\%\text{-}7) + 100(SPW\text{-}i\%\text{-}4)$$
$$+ 7000(PW\text{-}i\%\text{-}15)$$

Another correct equation would be

$$0 = -10{,}000 + 500(SPW\text{-}i\%\text{-}4) + 400(SPW\text{-}i\%\text{-}3)(PW\text{-}i\%\text{-}4)$$
$$+ 300(SPW\text{-}i\%\text{-}8)(PW\text{-}i\%\text{-}7) + 7000(PW\text{-}i\%\text{-}15)$$

Now we solve by trial and error for the unknown interest rate. (Use the first equation.)

Assume $i = 4\%$:
$0 = -10{,}000 + 300(11.118) + 100(6.002) + 100(3.630) + 7000(0.5553)$
$0 = -10{,}000 + 3335 + 600 + 363 + 3887$
$0 = -10{,}000 + 8185$

We see that the present worth of the subsequent receipts is too low, indicating that our trial interest rate is too high.

Try $i = 2\%$:
$0 = -10{,}000 + 300(12.849) + 100(6.472) + 100(3.808) + 7000(0.7430)$
$0 = -10{,}000 + 3855 + 647 + 381 + 5201$
$0 = -10{,}000 + 10{,}084$

Thus the rate of return on the investment is slightly greater than 2%.

By interpolation, the rate of return is $0.02 + [84/(84 + 1815)] = 0.0204 = 2.04\%$ ▶

10-32 Wt. 4

A bond issue of $100,000 of 10-year bonds in $1000 units, paying 6% interest in semiannual payments, must be retired by the use of a sinking fund which earns 4%, compounded semiannually. What is the total cost of retirement of the entire bond issue after 10 years?

Solution

$$\text{Semiannual interest payment} = 100{,}000(0.03) = \$3000$$

Assuming there are semiannual payments into the sinking fund, the semiannual sinking fund payment A is

$$A = 100{,}000(SF\text{-}2\%\text{-}20)$$
$$= 100{,}000(0.0412) = \$4120$$

$$\text{Total cost} = \text{total sinking fund payments} + \text{total interest payments}$$
$$= 20(4120) + 20(3000) = \$142{,}400 \blacktriangleright$$

10-33 Wt. 3

A community wishes to purchase an existing utility valued at $500,000 by selling 5% bonds that will mature in 30 years. The money to retire the bonds will be raised by paying equal annual amounts into a sinking fund that will earn 4%. What will be the total annual costs of the bonds until they mature? Compute the answer to the nearest $100.

Solution

The total annual cost of the bonds will equal the annual interest payment on the bonds plus the annual payment into the 4% sinking fund.

Annual interest payment on bonds = $0.05(500{,}000)$ = 25,000
Annual payment into 4% sinking fund A
$$= 500{,}000(SF\text{-}4\%\text{-}30) = 500{,}000(0.0178) = \underline{8{,}900}$$

$$\text{Annual cost of the bonds} = \$33{,}900 \blacktriangleright$$

Chapter 11

Other Problems

Collected here are problems from diverse engineering fields: surveying, engineering graphics, assorted principles of physics (those not covered in earlier chapters), engineering uses of materials, and dimensions of some engineering quantities. Owing to the variety of fundamentals presented in this chapter, we depart from the practice of summarizing the important points of each field. Within a given area of interest, however, the number of problems on a given subject serves as a rough guide to the subject's relative importance.

Some principles of plane surveying practice and traverse computations are the subject of the first group of problems. They review several surveying terms, practices, and the function of some surveying instruments. Traverse computations and correction procedures are then briefly considered.

The first problems on engineering graphics are concerned with orthographic projections, especially the correct completion of a partly finished set of views of an object. Descriptive geometry is then touched upon lightly.

Several topics are treated in the other problems. Miscellaneous topics from physics, such as the properties of light and some fundamentals of optics and acoustics are surveyed. These are followed by problems designed to refresh one's knowledge of the engineering uses of materials such as concrete, wood, and steel. Finally, we review the dimensions of many of the fundamental engineering quantities we have encountered in earlier chapters.

11-1 Wt. 1

The system of linear measurement used in the United States is the

(*a*) English
(*b*) metric
(*c*) Arabic
(*d*) quadratic
(*e*) Roman

Solution

In the United States the English system of linear measurement is used ▶

Answer is (*a*)

11-2 Wt. 1

The angle that a compass needle deviates from true North is called the angle of _____ .

Solution

The angle that magnetic North makes with true North is called the angle of declination ▶

Answer is declination

11-3 Wt. 1

Which measurement listed below is the most precise?

(*a*) a mile measured to the nearest foot
(*b*) a degree measured to the nearest second
(*c*) a kilogram measured to the nearest gram
(*d*) 1.004 in. measured to the nearest thousandth of an inch
(*e*) a second measured to the nearest millisecond

Solution

The most precise measurement is the one with the smallest ratio of measurement error to total measurement, in this case, a mile to the nearest foot ▶

Answer is (*a*)

11-4 Wt. 1

If the coordinates of point *A* are N 1025, W 1575, and those of point *B* are N 975, W 1625, the bearing of course *AB* is

(*a*) N 45° E
(*b*) S 45° E
(*c*) S 45° W
(*d*) N 45° W
(*e*) S 30° E

Solution

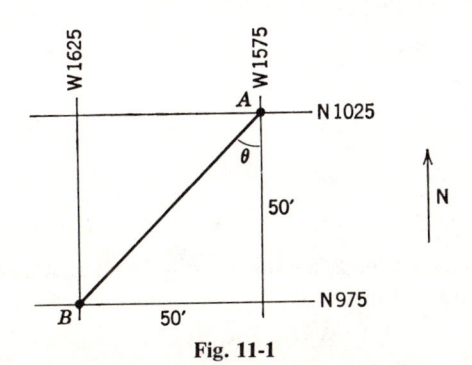

Fig. 11-1

By inspection, $\theta = 45°$ so the bearing of course AB is S 45° W ▶

Answer is (c)

11-5 Wt. 1

The D.M.D. (double meridian distance) method is used primarily to

(a) determine stresses in an airport pavement section
(b) determine the coefficient of compaction of an earth fill
(c) calculate the spacing of spiral ties in a round column of reinforced concrete
(d) measure the slope of grain in structural lumber
(e) calculate areas of irregular plots of land

Solution

The D.M.D. method is used in the calculation of the land area within a closed traverse when the bearings and lengths of the sides are known ▶

Answer is (e)

11-6 Wt. 1

An alidade is used primarily in

(a) determining the B.O.D. (biochemical oxygen demand) of sewage
(b) a chemical laboratory
(c) plotting mass diagrams
(d) high-order triangulation
(e) plane table survey work

Solution

An alidade is a surveying instrument. It has a line of sight horizontally parallel to a straightedge which rests upon a drawing board (plane table) ▶

Answer is (e)

11-7 Wt. 1

Stadia surveying eliminates the need of

(a) an alidade
(b) a plumb bob
(c) a leveling rod
(d) a transit
(e) a tape

Solution

Reading a stadia rod is a means of obtaining distances; therefore it eliminates the need for a tape ▶

Answer is (e)

11-8 Wt. 5

A distance from A to B was measured to be 5368.25 ft, with a steel tape. The temperature during this time was 22°F. The tape was a standard 100.00 ft at 68°F. If the coefficient of expansion is 0.0000065, what is the true distance between A and B?

Solution

The change in length of the tape caused by thermal contraction is

$$\Delta L = \alpha L \Delta T$$

where α = coefficient of expansion

$L = 100$ ft

$\Delta T = (22 - 68) = -46°$

$$\Delta L = 6.5 \times 10^{-6} \times 100(22 - 68) = -0.0299 \text{ ft}$$

Thus at 22°F the tape was 0.0299 ft shorter than 100 ft. The tape was used $5368.25/99.97 = 53.70$ times to measure the distance. $53.70 \times 0.0299 = 1.61$ ft. Therefore

$$\text{Distance } A \text{ to } B = 5368.25 - 1.61$$
$$= 5366.64 \text{ ft} \blacktriangleright$$

11-9 Wt. 1

The scale of an aerial map was given as 1/25,000. This is called the

(a) error factor
(b) representative fraction
(c) metes and bounds fraction
(d) manuscript factor
(e) plotting fraction

Solution

The scale of an aerial map may be expressed as a representative fraction with the numerator equal to 1 and the denominator equal to the number of units on the ground represented by one unit on the map $\blacktriangleright$

Answer is (b)

11-10 Wt. 1

The angle measured from the horizon to Polaris is equal to the

(a) latitude
(b) longitude
(c) right ascension
(d) vernal equinox
(e) summer solstice

530 / Other Problems

Solution

The angle is equal to the latitude ▶

Answer is (a)

11-11 Wt. 1

In reading a compass, the angle between true North and the needle on the compass at a given location is called the

(a) strike
(b) adjustment
(c) index error
(d) declination
(e) dip

Solution

The angle between true North and magnetic North is called the magnetic declination. It varies with location and time ▶

Answer is (d)

11-12 Wt. 1

A vertical curve that is introduced into the profile of a highway will usually take the form of a

(a) parabola
(b) ellipse
(c) catenary
(d) hyperbola
(e) cycloid

Solution

The usual form of a vertical curve is the arc of a parabola ▶

Answer is (a)

11-13

The average end area and prismoidal formulas are used for measuring

(a) volumes of earthwork
(b) entropy
(c) fineness modulus
(d) differential levels
(e) B.O.D. concentrations

Solution

The average end area and prismoidal formulas are used for measuring volumes of earthwork ▶

Average end area: It is assumed that the volume of earthwork between successive cross sections is the average of their end areas multiplied by the distance between them:

$$V = L \frac{A_1 + A_2}{2}$$

Prismoidal Formula: The volume of the prismoid is

$$V = \frac{L}{6}(A_1 + 4A_2 + A_3)$$

where A_1 and A_3 are the end areas, A_2 is the area midway between the ends, and L is the distance separating the end sections.

Answer is (a)

11-14 Wt. 1

A tellurometer is a device for measuring

(*a*) latitude
(*b*) azimuth
(*c*) distance
(*d*) current velocity
(*e*) declination

Solution

A tellurometer is a device for measuring distances. It is used in surveying ▶

Answer is (c)

11-15 Wt. 1

An instrument used to measure directly the area of cross sections that are plotted to scale is called

(*a*) a transit
(*b*) a stereoplanigraph
(*c*) an alidade
(*d*) a Kelsh plotter
(*e*) a polar planimeter

Solution

The instrument is called a polar planimeter ▶

Answer is (e)

11-16 Wt. 1

Which of the following instruments is not used in drafting practice?

(a) proportional dividers
(b) Wrico guide
(c) semicircular protractor
(d) Ames guide
(e) Beaman arc

Solution

A Beaman arc is found on some transits and alidades. It is used in conjunction with stadia distances to obtain differences in elevations ▶

Answer is (e)

11-17 Wt. 1

An ephemeris is used in

(a) geologic investigation
(b) soil testing
(c) surveying practice
(d) concrete mix design
(e) ground water measurements

Solution

An ephemeris is used in surveying practice. It contains data and tables on the positions of Polaris, the Sun, the Moon, and so forth, at any date ▶

Answer is (c)

11-18 Wt. 1

The use of stereo pairs is highly developed for map making in the field of

(a) photography
(b) transit-stadia surveys
(c) monoscopic engineering
(d) photogrammetry
(e) stroboscopic engineering

Solution

The perception of depth can be obtained from stereo pairs of photographs (photograms). This phenomenon is particularly useful in photogrammetry ▶

Answer is (d)

11-19 Wt. 1

A Geodimeter is an instrument for measuring

(a) distance
(b) current velocity

(c) seismic amplitudes
(d) strength of concrete
(e) thickness of rock formations

Solution

The Geodimeter is an electronic distance-measuring instrument that projects a highly modulated light beam to a passive reflector. The reflector returns the light beam to the Geodimeter where a phase comparision is then made ▶

Answer is (a)

11-20 Wt. 4

Given the following data for a closed traverse for which the length DA and the azimuth of DA have not been observed in the field:

Course	Azimuth (from North)	Distance (ft)
AB	240	900
BC	330	1600
CD	45	1630
DA	unknown	unknown

Determine mathematically the length and azimuth of DA.

Note: No credit will be allowed for a graphical solution. Slide-rule accuracy is acceptable.

Solution

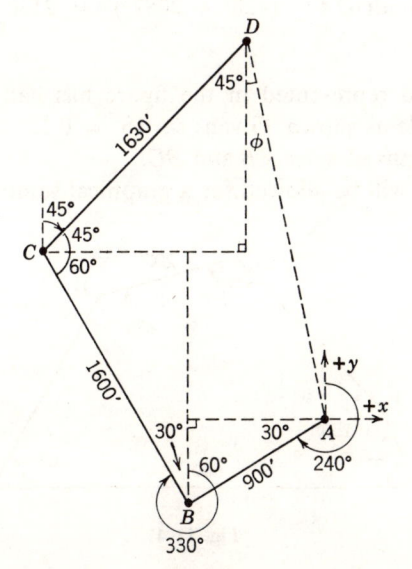

Fig. 11-2

534 / Other Problems

Determine the x and y coordinates of point D:

$$x = -900(\cos 30°) - 1600(\sin 30°) + 1630(\cos 45°)$$
$$= -780 - 800 + 1152 = -428 \text{ ft}$$
$$y = -900(\sin 30°) + 1600(\cos 30°) + 1630(\sin 45°)$$
$$= -450 + 1385 + 1152 = +2087 \text{ ft}$$

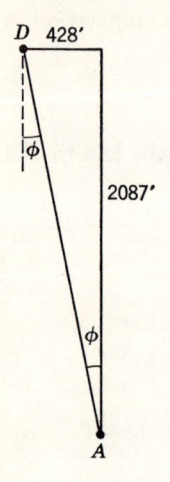

Fig. 11-3

$$\phi = \tan^{-1} \tfrac{428}{2087} = \tan^{-1} 0.205 = 11°35'$$
$$\text{Azimuth of } DA = 180° - 11°35' = 168°25' \blacktriangleright$$
$$\text{Length } DA = (428^2 + 2087^2)^{\frac{1}{2}} = 2130 \text{ ft} \blacktriangleright$$

11-21 Wt. 5

The plot of land represented in the figure has had linear and angular measurements made as shown. Given: $\sin 15° = 0.259$ and $\sin 75° = 0.966$. Determine the lengths of sides AB and BC.

Note: No credit will be allowed for a graphical solution.

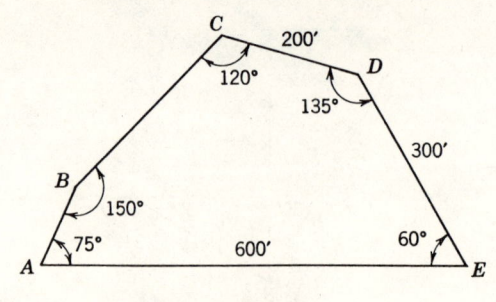

Fig. 11-4

Solution

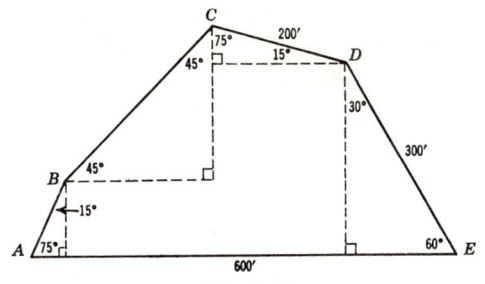

Fig. 11-5

$\sum_{\text{(vert displ)}}$: The vertical displacement of C from line AE is the same whether calculated from AB-BC or from ED-DC. Thus $AB_{\text{vert}} + BC_{\text{vert}} = ED_{\text{vert}} + DC_{\text{vert}}$.

$$AB(\sin 75°) + BC(\sin 45°) = 200(\sin 15°) + 300(\sin 60°)$$
$$0.966AB + 0.707BC = 200(0.259) + 300(0.866) = 311.6$$

Dividing by 0.966,

$$AB + 0.732BC = 322.6 \qquad (1)$$

$\sum_{\text{(horiz displ)}}$: The 600-ft length of line AE is equal to $AB_{\text{horiz}} + BC_{\text{horiz}} + CD_{\text{horiz}} + DE_{\text{horiz}}$.

$$AB(\sin 15°) + BC(\sin 45°) + 200(\sin 75°) + 300(\sin 30°) = 600$$
$$0.259AB + 0.707BC + 200(0.966) + 300(0.500) = 600$$

or

$$0.259AB + 0.707BC + 193.2 + 150 = 600$$

Dividing by 0.259,

$$AB + 2.73BC = 992 \qquad (2)$$

Substituting the value of AB from equation (2) into equation (1) gives

$$992 - 2.73BC + 0.732BC = 322.6$$
$$-2.00BC = -669.4 \qquad BC = 334.7 \text{ ft} \blacktriangleright$$

Substituting BC into equation (1) gives

$$AB = 322.6 - 0.732(334.7) = 77.6 \qquad AB = 77.6 \text{ ft} \blacktriangleright$$

11-22 Wt. 3

A television transmission tower is constructed on top of a 145-ft-high building, and it is desired to determine the tower height. A transit was set up

536 / Other Problems

near the building so that the instrument was 5 ft above the ground floor level of the building. The vertical angle to the tower base was measured and found to be 40°, and the vertical angle to the tower top was 60°. Find the tower height.

Solution

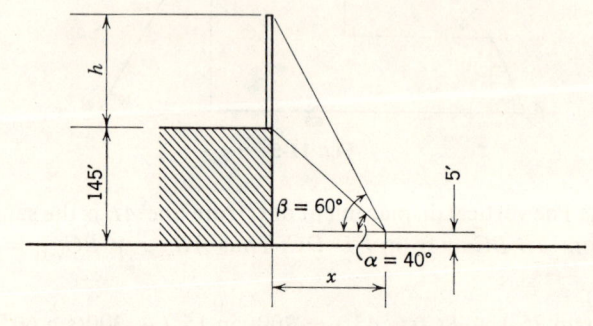

Fig. 11-6

$$\frac{140}{x} = \tan 40° = 0.84 \qquad x = 166.8 \text{ ft}$$

$$\frac{140 + h}{166.8} = \tan 60° = 1.73$$

$$h = 166.8(1.73) - 140 = 288.5 - 140 = 148.5 \text{ ft} \blacktriangleright$$

11-23 Wt. 6

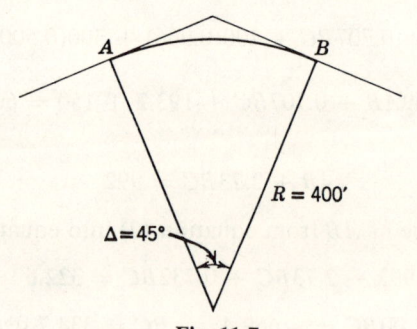

Fig. 11-7

Angle	Tangent	Sine	Cosine
11°15′	0.1989	0.1951	0.9808
22°30′	0.4142	0.3827	0.9239
45°00′	1.0000	0.7071	0.7071
67°30′	2.4142	0.9239	0.3827
90°00′	∞	1.0000	0.0000

A circular curve is required to connect two tangents on a highway location. If the central angle is 45° and the radius is 400 ft, determine each of the following:

(a) tangent distance
(b) long chord
(c) length of curve
(d) degree of curve
(e) middle of ordinate
(f) external distance

Solution

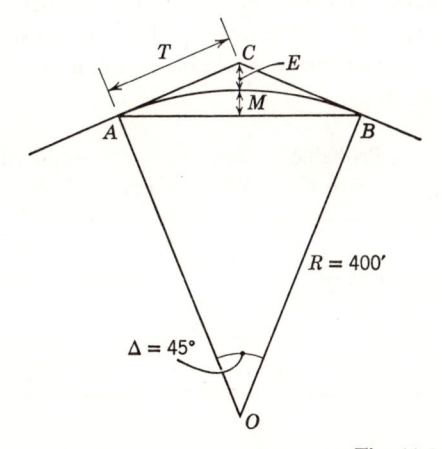

Fig. 11-8

(a) tangent distance

$$T = \text{dist } AC = R \tan \frac{\Delta}{2} = (400)(0.4142) = 165.68 \text{ ft} \blacktriangleright$$

(b) long chord

$$C = \text{dist } AB = 2R \sin \frac{\Delta}{2} = 2(400)(0.3827) = 306.16 \text{ ft} \blacktriangleright$$

(c) length of curve

$$L = 2\pi R(\tfrac{45}{360}) = 100\pi = 314.16 \text{ ft} \blacktriangleright$$

(d) degree of curve

D = central angle which subtends a 100-ft arc

$$100 = 2\pi R\left(\frac{D}{360}\right) \quad D = \frac{100(360)}{2\pi R} = \frac{5729.58}{R}$$

$$D = \frac{5729.58}{400} = 14.324° = 14°19' \blacktriangleright$$

(e) middle of ordinate

$$M = R - R \cos \frac{\Delta}{2} = 400(1 - 0.9239) = 30.44 \text{ ft} \blacktriangleright$$

(f) external distance

$$E = \frac{R}{\cos \Delta/2} - R = R\left(\frac{1}{\cos \Delta/2}\right) - R = 400\left(\frac{1}{0.9239} - 1\right) = 32.95 \text{ ft} \blacktriangleright$$

11-24 Wt. 6

In a given triangle ABC, each interior angle was measured by repetition, and in the table below the mean value of each angle and the number of repetitions used is given. Complete the table by adjusting the angles using relative weights. Assume every angle turned has a probable error of 30 sec. Give answers to the nearest whole second.

Angle	No. Observations	Mean Value	Probable Error	Rel. Wt.	Corr.	Adjusted Value
A	6	40°10′10″				
B	8	63°42′10″				
C	10	76°8′20″				
Σ						

Solution

Angle	No. Observations	Mean Value	Probable Error	Rel. Wt.	Corr.	Adjusted Value
A	6	40°10′10″	30″	6	−17″	40°9′53″ ▶
B	8	63°42′10″	30″	8	−13″	63°41′57″ ▶
C	10	76°8′20″	30″	10	−10″	76°8′10″ ▶
Σ		180°0′40″	40″	24	−40″	180°0′0″

Assuming that each observation was made with equal care, we would expect that angle B with 8 observations is more reliable than angle A with only 6 observations. The observations, then, are degrees of reliability, and the relative weights of the values are therefore based on the number of observations. The corrections to be applied are inversely proportional to the weights.

$$\text{Corr}_A + \text{Corr}_B + \text{Corr}_C = -40 \qquad \text{Corr}_A = \tfrac{8}{6}\text{Corr}_B = \tfrac{10}{6}\text{Corr}_C$$

Solving for Corr_A,

$$\text{Corr}_A + \tfrac{6}{8}\text{Corr}_A + \tfrac{6}{10}\text{Corr}_A = -40$$

$$(1 + 0.75 + 0.6)\text{Corr}_A = -40 \qquad \text{Corr}_A = \frac{-40}{2.35} = -17 \text{ sec}$$

$$\text{Corr}_B = \tfrac{6}{8}\text{Corr}_A = \tfrac{6}{8} \times (-17) = -13 \text{ sec}$$

$$\text{Corr}_C = \tfrac{6}{10}\text{Corr}_A = \tfrac{6}{10} \times (-17) = -10 \text{ sec}$$

11-25 Wt. 6

A surveyor wishes to find the horizontal distance across a cove of a still lake to a point A at the water edge from a point B on a cliff 75.80 ft above the water surface. He establishes a point C, 200.00 ft horizontally from B at an elevation of 10.50 ft above the water surface. His notes indicate that, with the transit set at B, the horizontal angle between A and C is 120°00′ and the angle from the horizontal to A is 14°59′. With the transit at C, the horizontal angle from A to B is 36°35′. With the instrument at B, the height was 4.50 ft; and at C, 4.24 ft.

(a) In what two ways can the horizontal distance from A to B be found from the information available?
(b) Which method gives the most accurate measure of the distance AB?
(c) What is the horizontal distance from A to B?

Solution

(a) Method 1. Horizontal measurements:

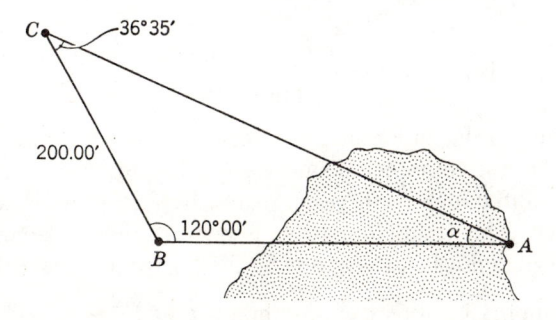

Fig. 11-9

540 / Other Problems

An inaccessible distance may be determined by establishing a base line BC and measuring the included angle from each end of the base line to the inaccessible point A. We can calculate angle α.

$$\alpha = 180° - 120°00' - 36°35' = 23°25'$$

Using the law of sines,

$$\frac{AB}{BC} = \frac{\sin 36°35'}{\sin 23°25'} = \frac{0.596}{0.397}$$

Since $BC = 200.00$ ft,

$$AB = \frac{0.596 \times 200.00}{0.397} = 300 \text{ ft} \blacktriangleright$$

Method 2. Vertical measurements:

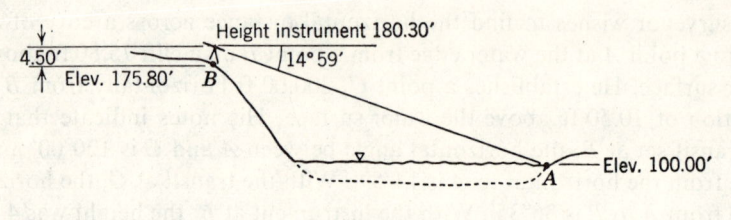

Fig. 11-10

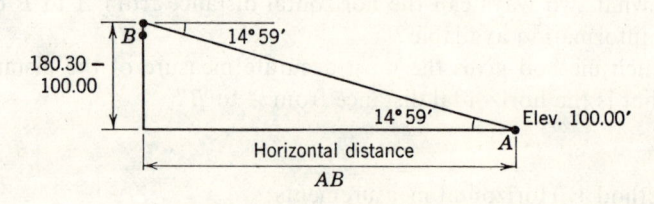

Fig. 11-11

$$\text{Horizontal distance } AB = \frac{180.30 - 100.00}{\tan 14°59'} = \frac{80.30}{0.268} = 300 \text{ ft} \blacktriangleright$$

(b) 1. Because of the larger included angle and base line, we would expect a smaller error in the computed distance AB using horizontal measurements.

2. The horizontal angles can be measured by repetition, that is, mechanically multiplied. This is not possible for vertical angles; hence the vertical angle readings may not be as accurate as the horizontal angle readings.

Therefore Method 1 is more accurate than Method 2 $\blacktriangleright$

(c) From the calculations above, we find the horizontal distance AB to be 300 ft $\blacktriangleright$

11-26 Wt. 4

A traverse was run around a tract of land. The courses and distances (in sequence) are as follows:

East	200.00 ft
N 30°00' E	149.60 ft
N 60°00' W	173.20 ft
S 30°00' W	249.60 ft

What is the area of the tract in acres?

Solution

The problem can be solved by the rather long process of D.M.D. In this case, however, the geometry of the problem leads to a quick and easy solution.

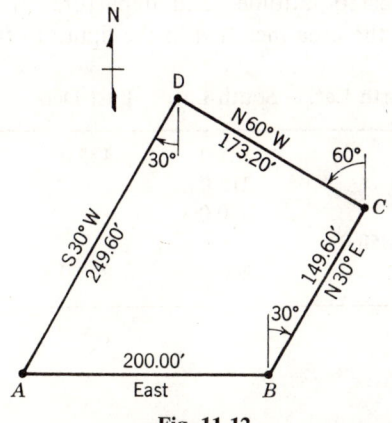

Fig. 11-12

If a careful sketch is made of the traverse, one can readily see that the area described is a trapezoid. Therefore,

$$A = \left(\frac{249.60 + 149.60}{2}\right)(173.20)$$

$$A = 34{,}571 \text{ ft}^2 = \frac{34{,}571}{43{,}560}$$

$$= 0.794 \text{ acre} \blacktriangleright$$

An alternative method of calculation is to assume the traverse to be a

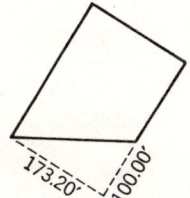

Fig. 11-13

portion of a rectangle. The area included in the traverse is the area of the rectangle less the area of the added triangle.

$$\text{Area of rectangle} = (249.60)(173.20) = 43,231 \text{ ft}^2$$

$$\text{Area of triangle (dashed)} = 0.5(173.20)(100.00)$$

$$= 8660 \text{ ft}^2$$

$$\text{Area of traverse} = 43,231 - 8660 = 34,571 \text{ ft}^2$$

$$= \frac{34,571}{43,560} = 0.794 \text{ acre} \blacktriangleright$$

11-27 Wt. 5

The following adjusted latitudes and departures were calculated for a closed traverse. Find the area included in the figure to the nearest foot.

Line	North Lat.	South Lat.	East Dep.	West Dep.
AB		125.0	425.0	
BC		215.0		300.0
CD		310.0		225.0
DE	850.0			100.0
EA		200.0	200.0	

Solution

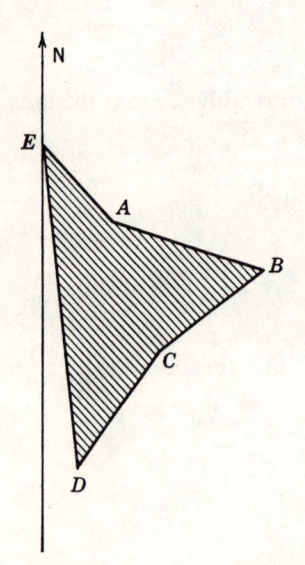

Fig. 11-14

The D.M.D. method will be used.

Course	Lat.	Dep.	D.M.D.	Double Area
EA	−200	200	200	200(−200) = −40,000
AB	−125	425	200 + 200 + 425 = 825	825(−125) = −103,125
BC	−215	−300	825 + 425 − 300 = 950	950(−215) = −204,250
CD	−310	−225	950 − 300 − 225 = 425	425(−310) = −131,750
DE	850	−100	425 − 225 − 100 = 100	100(+850) = 85,000
				Total = −394,125

$$\text{Area} = \tfrac{1}{2}(394{,}125) = 197{,}062 \text{ ft}^2 \blacktriangleright$$

11-28 Wt. 6

(a) Show a front view of this figure

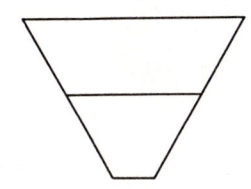

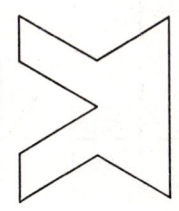

Fig. 11-15

(b) Show a top view of this figure

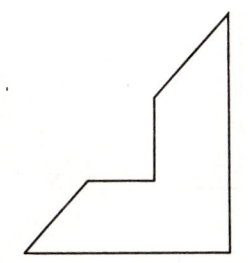

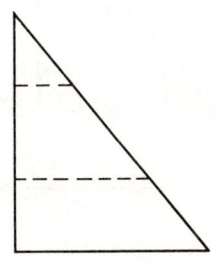

Fig. 11-16

544 / Other Problems

Solution

(a)

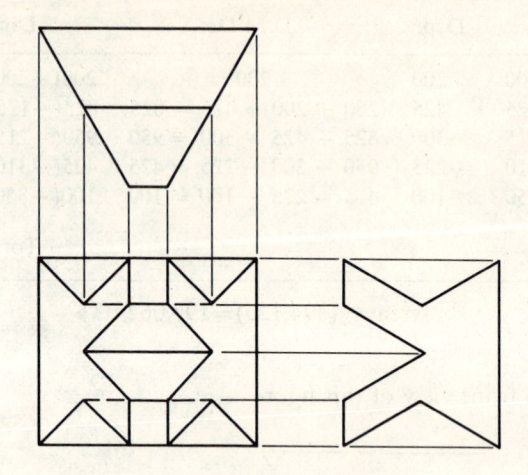

Fig. 11-17

(b)

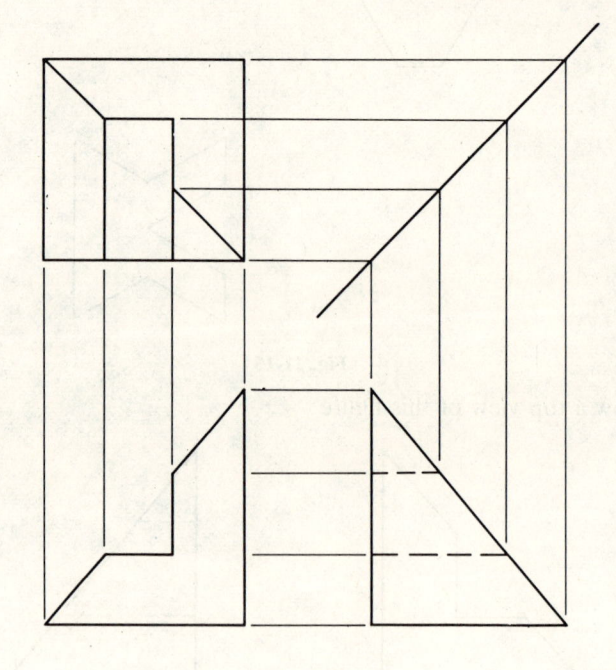

Fig. 11-18

11-29 Wt. 6

(*a*) Draw in the missing lines of the figure:

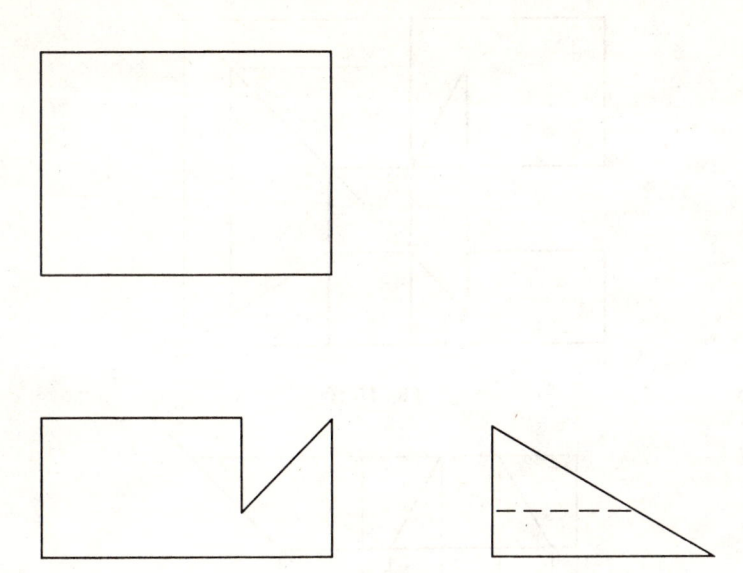

Fig. 11-19

(*b*) Draw a right side view of the figure:

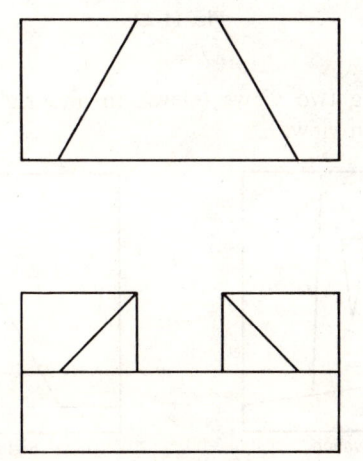

Fig. 11-20

Solution

(a)

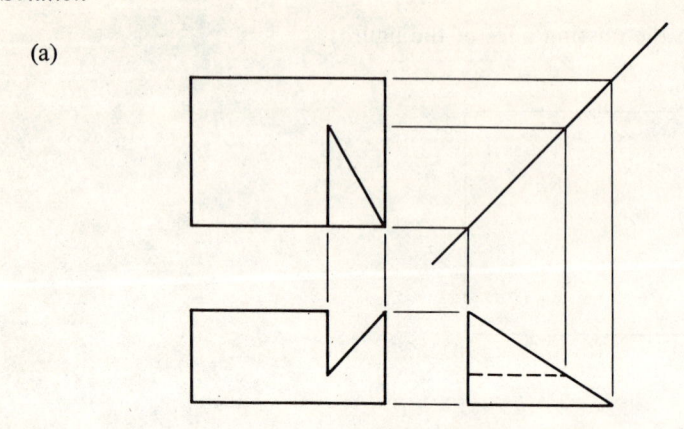

Fig. 11-21

(b)

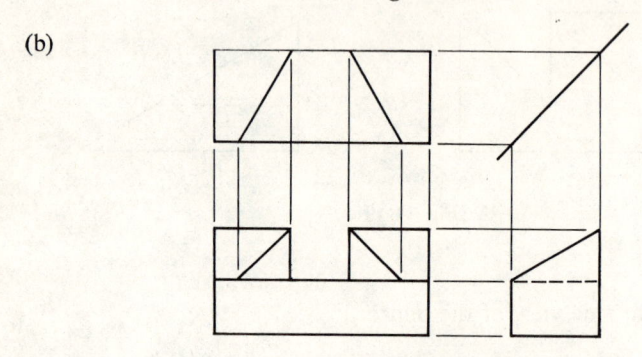

Fig. 11-22

11-30 Wt. 5

Given the following two views (elevation and right side) of an object. Draw an accurate plan view.

Front elevation

Right side

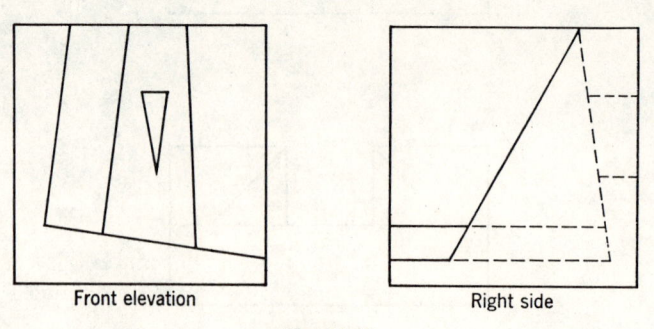

Fig. 11-23

Solution

The front elevation and right side views are not in agreement. If one assumes the front elevation is correct, the two horizontal hidden lines in the lower portion of the right side view appear to be improperly located. The relocation of these hidden lines depends on how one visualizes the object. One possible adjustment for the right side has been selected as the basis on which to construct the required plan view.

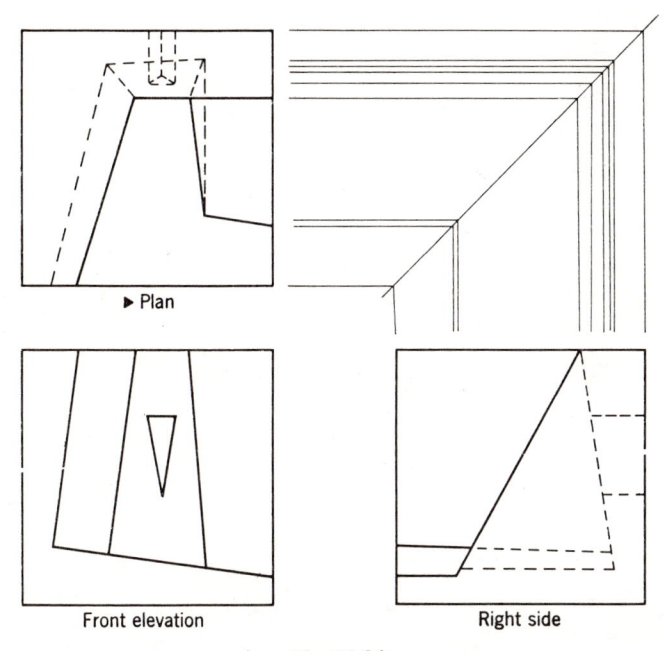

Fig. 11-24

The pictorial view (with hidden lines added) shows the object that is represented by the plan, front, and corrected right side views.

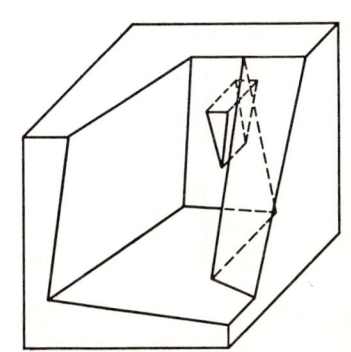

Fig. 11-25

548 / Other Problems

11-31 Wt. 6

Draw the right side view of the object shown.

Plan

Elevation

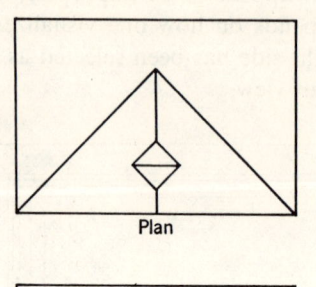

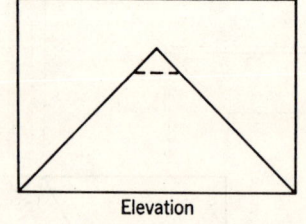

Fig. 11-26

Solution

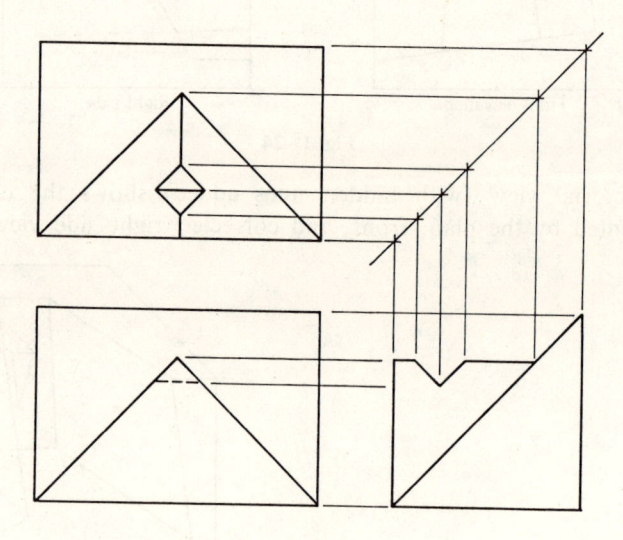

Fig. 11-27

PROBLEMS / 549

11-32 Wt. 6

The plan and right side views of an object are shown. Draw the front elevation.

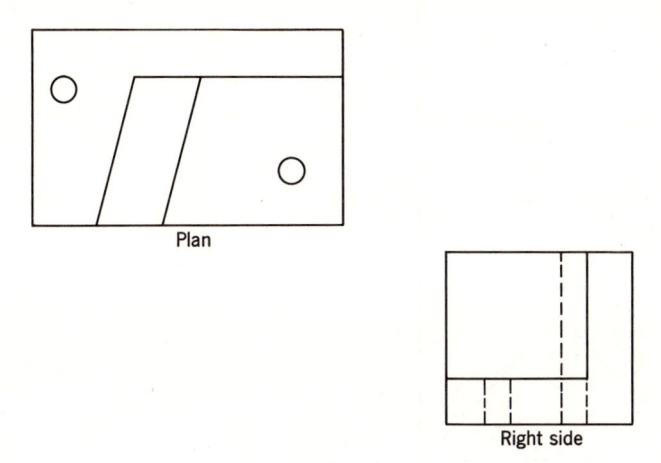

Fig. 11-28

Solution

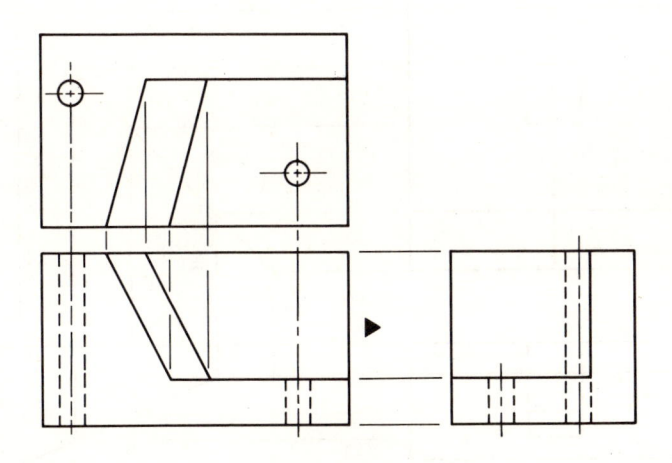

Fig. 11-29

11-33 Wt. 6

Draw the right side view of the object shown in Fig. 11-30.

550 / Other Problems

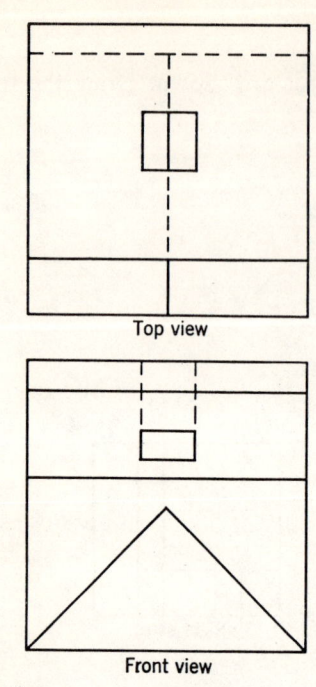

Top view

Front view

Fig. 11-30

Solution

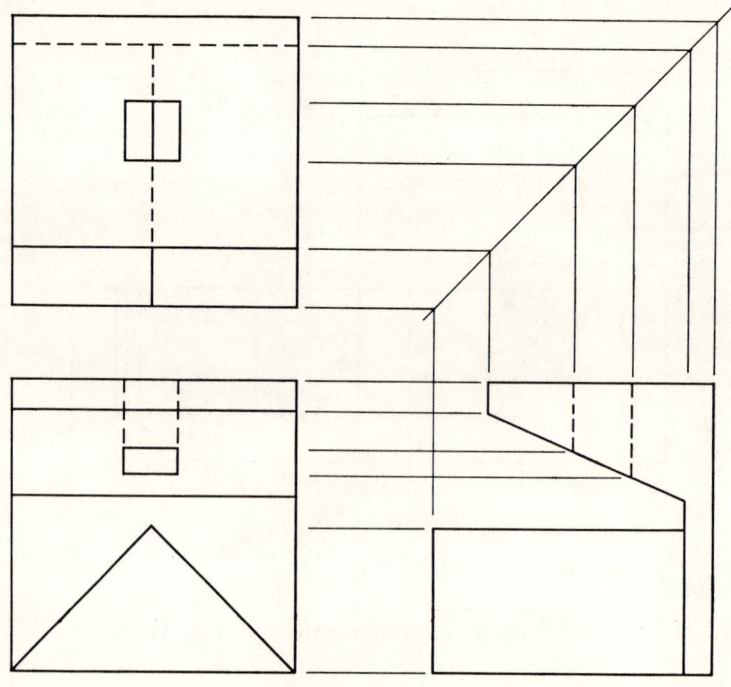

Fig. 11-31

11-34 Wt. 6

The elevation and 45° auxiliary views of an object are shown. Draw the plan and right side views.

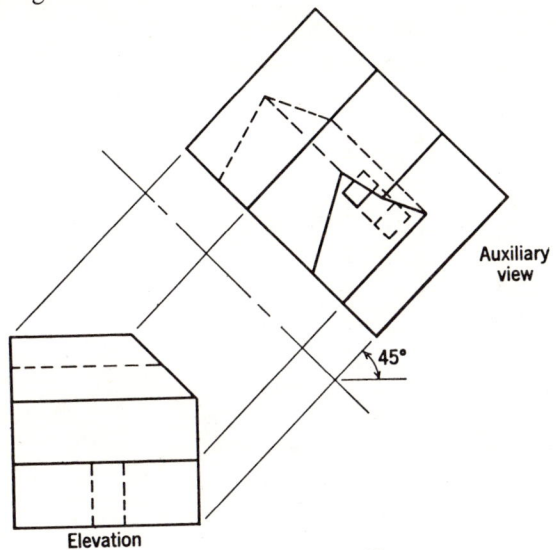

Fig. 11-32

Solution

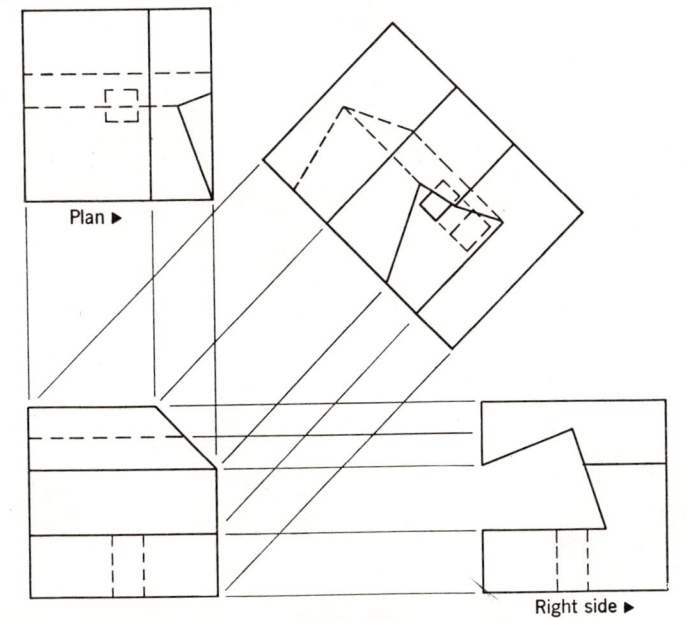

Fig. 11-33

552 / Other Problems

11-35 Wt. 6

Make an isometric drawing of the object shown in the plan and elevation views.

Plan

Elevation

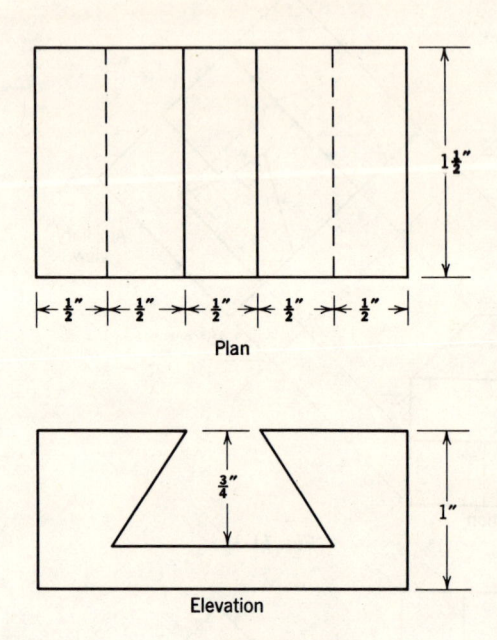

Fig. 11-34

Solution

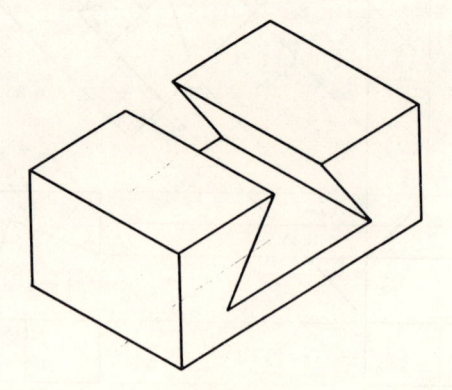

Fig. 11-35

11-36 Wt. 6

Make isometric drawings of the objects shown.

(a)

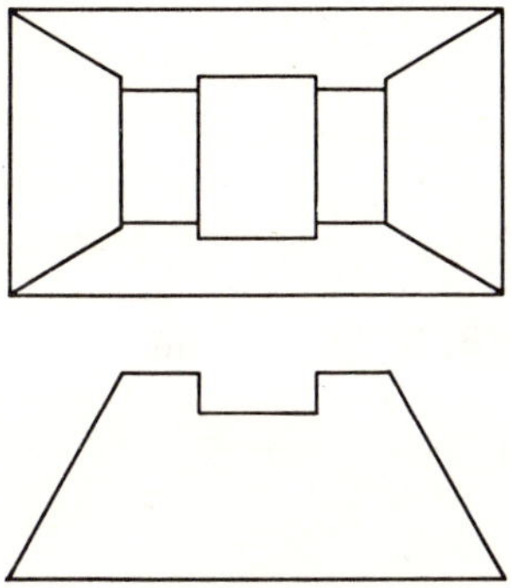

(b)

Fig. 11-36

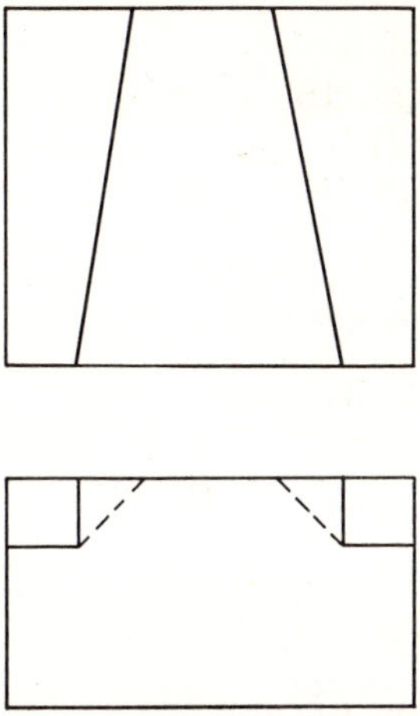

Fig. 11-37

Solution

(a)

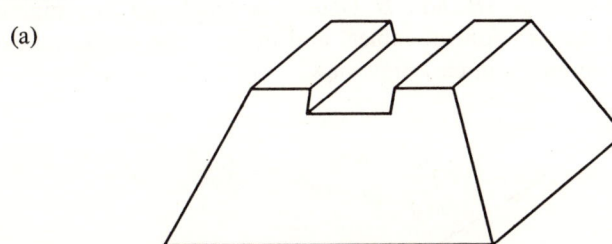

Fig. 11-38

(b)

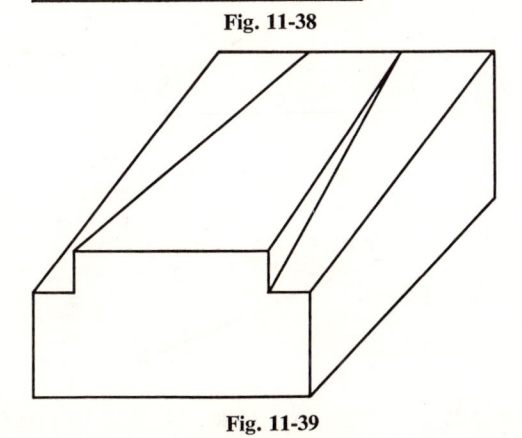

Fig. 11-39

554 / Other Problems

11-37 Wt. 1

Which grade of drafting pencil has the softest lead?

(a) H
(b) F
(c) HB
(d) B
(e) 3H

Solution

Hardest   Softest

$$9H, 8H, \ldots, 3H, 2H, H, F, HB, B, 2B, \ldots, 6B$$

Answer is (d)

11-38 Wt. 5

Two control cables for an aircraft are partially shown in the full-scale plan and elevation views as AB and CD. Determine graphically the minimum clearance between the two control cables, and indicate the minimum clearance as a line on the plan and elevation views.

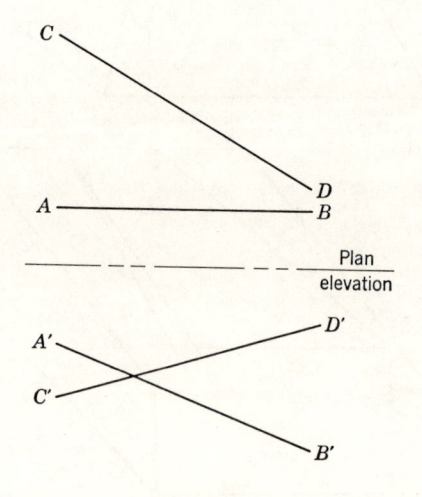

Fig. 11-40

Solution

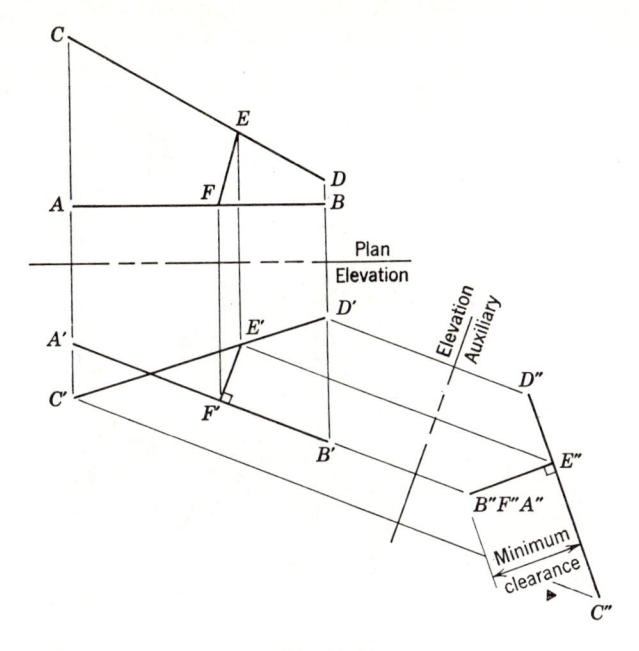

Fig. 11-41

11-39 Wt. 6

An oil company is preparing to drill a new well in an area favorable for the production of oil. There are three existing wells in the vicinity for which the depths to the top of the key bed of oil producing sands are known. The wells are located in relation to the proposed well as follows:

Well No.	Location	Depth of Key Bed (ft)
1	1 mile South	6000
2	2 miles South and 1 mile East	5000
3	2 miles East	7000

Determine the strike and dip of the key bed of oil producing sand. (Assume that the ground surface elevation at the existing wells is the same and the top of the key bed is a plane.)

Solution

To solve the problem we must define strike and dip.

556 / Other Problems

Strike: The direction of the intersection of the bed with a horizontal plane, that is, the direction of a level line along the bed.

Dip: The vertical angle between the plane of the bed and a horizontal plane, measured perpendicular to the strike.

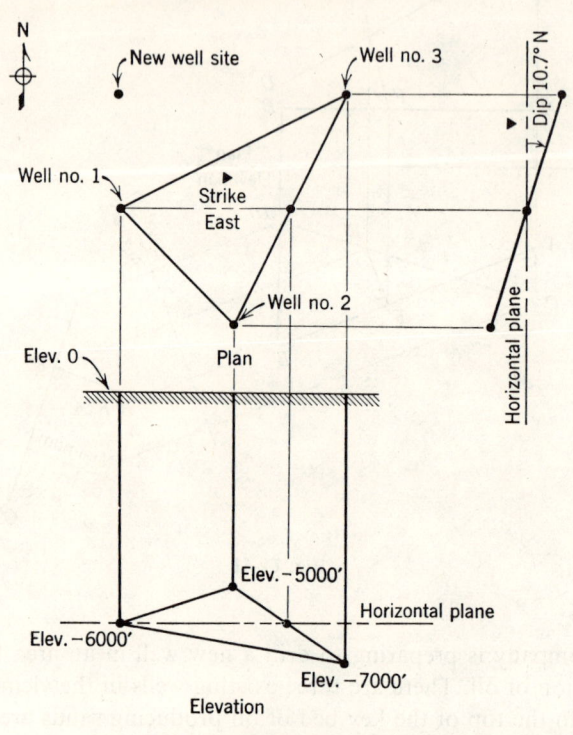

Fig. 11-42

11-40 Wt. 3

Using descriptive geometry, pass a plane through the given points aa', bb', and cc', which are shown below.

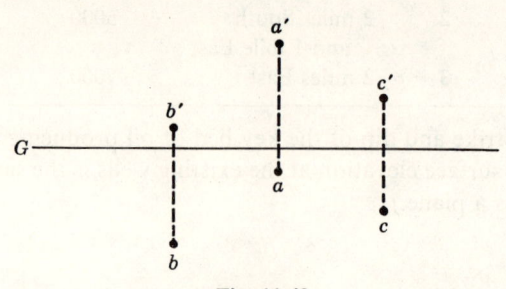

Fig. 11-43

Solution

A plane can be completely specified by identifying three points that lie in it. By connecting the given points with straight lines, the one plane that contains these three points is completely defined ▶

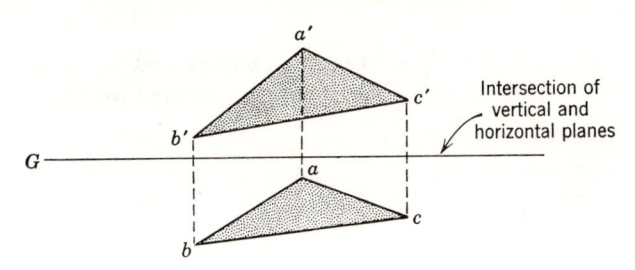

Fig. 11-44

11-41 Wt. 4

An athletic field 450 ft on each side is illuminated by six towers supporting banks of 1000-w incandescent lamps rated at 20 lumens/w. If the desired illuminance has been set at 20 lumens/ft², how many lamps are required in each bank if there is a 50% loss of luminous flux from the towers to the field?

Solution

$$\text{Lumens per lamp} = 20 \text{ lumens/w } (1000 \text{ w}) = 20{,}000$$

$$\text{Net lumens required for illuminating the field} = 20(450)(450)$$
$$= 4{,}050{,}000$$

$$\text{Total lumens required} = \frac{4{,}050{,}000}{0.50} = 8{,}100{,}000$$

$$\text{Total number of lamps} = \frac{8{,}100{,}000}{20{,}000} = 405$$

$$\text{Number of lamps per tower} = \frac{405}{6} = 68 \blacktriangleright$$

11-42 Wt. 1

An alpha particle consists of

(*a*) an electron
(*b*) a proton
(*c*) a neutron
(*d*) a proton and a neutron
(*e*) two protons and two neutrons

558 / Other Problems

Solution

Alpha particles are helium nuclei and therefore have two protons and two neutrons ▶

Answer is (*e*)

11-43 Wt. 1

If 5 g of mass were entirely converted to energy and the velocity of light equals 3×10^{10} cm/sec, the amount of energy liberated would be

(*a*) 15×10^{10} ergs
(*b*) 75×10^{10} ergs
(*c*) 22.5×10^{20} ergs
(*d*) 45×10^{20} ergs
(*e*) 45×10^{10} ergs

Solution

The energy liberated in ergs is found from the famous energy formula

$$E = mc^2$$

where m = converted mass in grams
c = speed of light in centimeters per second

$$E = 5(3 \times 10^{10})^2 = 45 \times 10^{20} \text{ ergs} \blacktriangleright$$

Answer is (*d*)

11-44 Wt. 1

The length of light waves is expressed in Angstrom units. One Angstrom unit is equivalent to

(*a*) 10^{-2} m
(*b*) 10^{-4} m
(*c*) 10^{-6} m
(*d*) 10^{-8} m
(*e*) 10^{-10} m

Solution

1 Angstrom unit $A = 10^{-8}$ cm $= 10^{-10}$ m ▶

Answer is (*e*)

11-45 Wt. 1

At sea level the air pressure is 14.7 psi or 29.92 in. of mercury. For every 1000 ft increase in elevation above sea level, the air pressure decreases approximately

(a) $\frac{1}{8}$ psi
(b) $\frac{1}{4}$ psi
(c) $\frac{1}{2}$ psi
(d) $\frac{3}{4}$ psi
(e) 1 psi

Solution

A table of standard atmosphere shows that the pressure decreases approximately 70 psf/1000 ft of altitude.

$$\frac{70}{144} = \tfrac{1}{2} \text{ psi, approximately} \blacktriangleright$$

Answer is (c)

11-46 Wt. 1

The wavelength of blue light is longer than the wavelength of

(a) green light
(b) orange light
(c) red light
(d) violet light
(e) yellow light

Solution

The visible spectrum is arbitrarily divided into six broad regions. In order of increasing wavelength the colors are

    Violet   Blue   Green   Yellow   Orange   Red

Hence the wavelength of blue light is longer than violet light $\blacktriangleright$

Answer is (d)

11-47 Wt. 1

If one observes, with a microscope, a single tiny clay particle which is suspended in water, he will see that it is zigzagging rapidly—never coming to rest. This phenomenon is known as the

(a) Doppler effect
(b) Brownian movement
(c) Lavoisier's movement
(d) Nelson effect
(e) Monomer movement

560 / Other Problems

Solution

The phenomenon described is the Brownian movement, first observed by Robert Brown in 1827 ▶

Answer is (*b*)

11-48 Wt. 1

The intensity of radiation for a black body or perfect radiator is a function of the absolute temperature of the body. If T is the absolute temperature, the intensity of radiation will be proportional to

(*a*) T
(*b*) T^2
(*c*) T^3
(*d*) T^4
(*e*) T^5

Solution

The Stefan-Boltzmann law states that the total emissivity of a black body is proportional to the fourth power of the absolute temperature ▶

Answer is (*d*)

11-49 Wt. 6

A 40-in. concave spherical mirror has a 24-in. radius. An object 8 in. high is centered 8 in. from the vertex of the mirror as shown in the figure.

(*a*) Using a scale of 1 in. = 10 in., make a drawing of the mirror and object, then sketch on the drawing the graphical construction of the image. If a virtual image results, use dotted lines; if real, use solid lines. Show if upright or inverted.

(*b*) Compute the distance S' of the image from the vertex V and the magnification of the image.

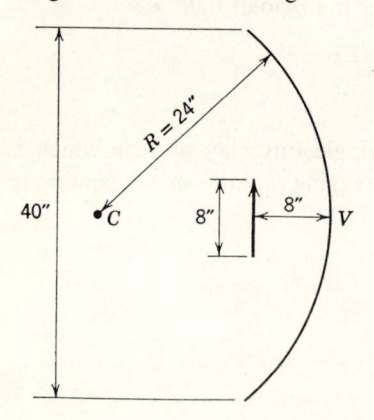

Fig. 11-45

Solution

On a spherical mirror the focal length f is approximately one-half the radius of curvature. To locate the image of point A, draw AB parallel to the principal axis, passing through the principal focus F after being reflected. Line AC drawn through the center of curvature of the mirror C is incident normally at D and retraces its path after being reflected. The point A' where the rays converge is the image of A. A similar construction at E gives us

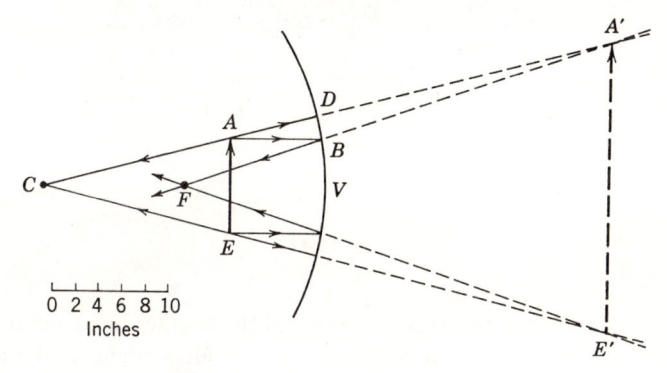

Fig. 11-46

point E'. The image equation is

$$\frac{1}{p} + \frac{1}{q} = \frac{1}{f}$$

where p = distance of the object from the mirror = 8 in.
q = image distance
f = focal length = 12 in.

$$\frac{1}{8} + \frac{1}{q} = \frac{1}{12} \qquad \frac{1}{q} = \frac{1}{12} - \frac{1}{8} = -\frac{1}{24}$$

$$q = -24 \blacktriangleright$$

This tells us that the distance from the mirror is 24 in. The negative sign indicates a virtual image, according to the sign convention used here.

$$\text{Magnification} = \left|\frac{q}{p}\right| = \frac{24}{8} = 3 \blacktriangleright$$

11-50 Wt. 4

A ray of light strikes the glass wall of an aquarium at an angle of 45° with the normal. After it penetrates the glass (index of refraction, 1.52) and enters the water (index of refraction, 1.33), what will be the angle made with the normal?

Solution

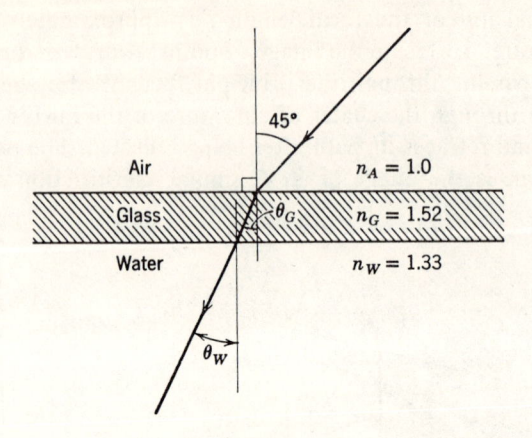

Fig. 11-47

Snell's law states that the ratio of the sine of the angle of incidence to the sine of the angle of refraction is a constant that is independent of the angle of incidence.

$$n_1 \sin \theta_1 = n_2 \sin \theta_2$$

For the air-glass interface,

$$\sin \theta_G = \left(\frac{n_A}{n_G}\right) \sin \theta_A$$

$$\sin \theta_G = \left(\frac{1.0}{1.52}\right) \sin 45°$$

$$= \left(\frac{1.0}{1.52}\right)(0.707) = 0.465$$

For the glass-water interface,

$$\sin \theta_W = \left(\frac{n_G}{n_W}\right) \sin \theta_G$$

$$= \left(\frac{1.52}{1.33}\right)(0.465) = 0.53$$

$$\theta_W = 32° \blacktriangleright$$

11-51 Wt. 3

A pistol is shot opposite a rock cliff, and the echo returns after 11 sec. How far away is the cliff?

Solution

Assuming standard temperature and pressure, the speed of sound is 1089 fps. In 11 sec the sound wave will travel a distance of

$$S = 1089 \text{ fps} \times 11 \text{ sec} = 11{,}980 \text{ ft}$$

Since the wave travels to the cliff and then back again, the distance to the cliff is half the distance traveled by the wave in 11 sec.

$$S_{1/2} = \tfrac{1}{2}(11{,}980) = 5990 \text{ ft} \blacktriangleright$$

11-52 Wt. 1

An observer standing at a railroad crossing notices that the pitch of the train's whistle lowers as the train rolls past him. This indicates that the frequency of the sound heard differs from the frequency of the moving source. The man who is generally credited with developing this principle is

(a) Archimedes
(b) Doppler
(c) Edison
(d) Einstein
(e) Newton

Solution

In 1842 Doppler noted that there are changes in frequency and in wavelength in any wave motion by movement either of the observer or of the source of the waves with respect to the medium by which they are carried ▶

$$\text{Answer is } (b)$$

11-53 Wt. 1

A hydraulic cement is defined as a cementing material that

(a) will harden under water
(b) requires water for mixing
(c) increases in strength if cured in moist air
(d) has a strength dependent upon the water-cement ratio
(e) can be used for structures to carry water, such as concrete pipes

Solution

"Hydraulic cement," or simply "cement" when the word is used in engineering, is a material that will harden under water ▶

$$\text{Answer is } (a)$$

564 / Other Problems

11-54 Wt. 1

A certain construction project requires that the concrete used will acquire a high strength as rapidly as possible. For this project, which type of cement should be used?

(a) Type I
(b) Type II
(c) Type III
(d) Type IV
(e) Type V

Solution

Types of Portland cement:

I   Normal
II   Modified (moderate low-heat and sulfate-resistant)
III   High-early-strength ▶
IV   Low-heat
V   Sulfate-resistant

Answer is (c)

11-55 Wt. 1

The principal difference between regular concrete and lightweight concrete is in the

(a) cement used
(b) water proportions used
(c) admixtures used
(d) relative mixing time
(e) aggregates used

Solution

Regular concrete weighs about 150 lb/ft^3, but by using special lightweight aggregates a concrete weighing about 90 lb/ft^3 can be produced. The difference is in the weight of the aggregates used ▶

Answer is (e)

11-56 Wt. 1

Air entrainment in concrete

(a) increases compressive strength
(b) decreases shrinkage
(c) reduces heat of hydration
(d) improves freeze and thaw resistance
(e) reduces permeability

Solution

Air entrainment in concrete improves workability and increases its resistance to freezing and thawing disintegration ▶

Answer is (*d*)

11-57 Wt. 1

The terms "pretensioned" and "post-tensioned" are applied primarily in the field of

(*a*) ceramic veneer
(*b*) prestressed concrete
(*c*) timber piles
(*d*) grouted brick masonry
(*e*) metal lath and plaster

Solution

In prestressed concrete, pretensioning means that the reinforcing steel has a high initial tensile stress when the concrete is cast. In post-tensioning, the reinforcing steel is protected to prevent bonding as the concrete is cast. After the concrete has hardened, tension is developed by end take-up of the reinforcing steel ▶

Answer is (*b*)

11-58 Wt. 1

In the design of a reinforced concrete T-beam, it is commonly assumed that

(*a*) the concrete has one-fourth of the tensile strength of the steel
(*b*) the modulus of elasticity for concrete is the same as the modulus of elasticity for steel
(*c*) the steel has a tensile stress of 1800 psi induced by shrinkage of the concrete
(*d*) the dead load of the beam may be neglected for long spans
(*e*) the tensile strength of the concrete is nearly zero

Solution

Conventional concrete design is done by a cracked section analysis; that is, it is assumed that the portion of beam in tension will crack with the result that the effective concrete tensile strength is zero ▶

Answer is (*e*)

566 / Other Problems

11-59 Wt. 1

Stirrups in a reinforced concrete beam are designed primarily to resist stresses caused by

(a) diagonal tension
(b) axial tension
(c) horizontal shear
(d) axial compression
(e) web crippling

Solution

At each point in a loaded concrete beam there are vertical or transverse shearing forces and, for equilibrium, longitudinal shearing forces of equal intensity. These forces tend to distort the beam so that tension exists along a diagonal plane. This diagonal tension acts along a line of action 45° from the beam's axis. Stirrups or other web reinforcement are designed to resist this diagonal tension ▶

$$\text{Answer is } (a)$$

11-60 Wt. 1

A concrete mix with an ultimate compressive strength f'_c of 3750 psi would have a n value of

(a) 15
(b) 12
(c) 10
(d) 8
(e) 6

Solution

f'_c = ultimate compressive strength of concrete, usually at 28 days
n = ratio of modulus of elasticity of steel to that of concrete
E_s = modulus of elasticity of steel = 30×10^6 psi
E_c = modulus of elasticity of concrete = $1000 f'_c$

For a concrete mix with $f'_c = 3750$ psi,

$$n = \frac{E_s}{E_c} = \frac{30 \times 10^6}{1000 \times 3750} = 8 \blacktriangleright$$

$$\text{Answer is } (d)$$

11-61 Wt. 4

The following batch weights have been used at a concrete batching plant for the production of 1 yd³ of concrete:

Material	Batch Weight (lb)
Cement	470
Water	175
Sand, with 4% free moisture	1404
Aggregate, with 1% free moisture	2121

If the free moisture content of the sand and aggregate increases to 6% and 2%, respectively, compute the batch weight for each material required to produce 1 yd³.

Solution

Initial conditions:

$$\text{Wt. of surface-dry sand} = 1404/1.04 = 1350 \text{ lb}$$
$$\text{Free moisture in sand} = 0.04 \times 1350 = 54 \text{ lb}$$
$$\text{Wt. of surface-dry aggregate} = 2121/1.01 = 2100 \text{ lb}$$
$$\text{Free moisture in aggregate} = 0.01 \times 2100 = 21 \text{ lb}$$
$$\text{Total water} = 175 + 54 + 21 = 250 \text{ lb}$$

Moisture in sand and aggregate increased to 6% and 2%:

$$\text{Desired wt. of surface-dry sand} = 1350 \text{ lb}$$
$$\text{Free moisture} = 0.06 \times 1350 = 81 \text{ lb}$$
$$\text{Wt. of sand in mix} = 1431 \text{ lb} \blacktriangleright$$

$$\text{Desired wt. of surface-dry aggregate} = 2100 \text{ lb}$$
$$\text{Free moisture} = 0.02 \times 2100 = 42 \text{ lb}$$
$$\text{Wt. of aggregate in mix} = 2142 \text{ lb} \blacktriangleright$$

Desired water		250 lb
less: water in sand	81	
water in aggregate	42	−123 lb
Wt. of water to add		= 127 lb ▶

568 / Other Problems

Summary:

Cement	470 lb ▶
Water	127 lb ▶
Sand	1431 lb ▶
Aggregate	2142 lb ▶

11-62 Wt. 1

Pile-driving formulas fall into what two general classifications?

(a) static and dynamic
(b) frictional and resistance
(c) shear and compressive
(d) bearing and resistance
(e) frictional and bearing

Solution

Pile-driving formulas are static and dynamic ▶

$$\text{Answer is } (a)$$

11-63 Wt. 1

A 6-in. × 6-in. wood post S4S has net actual dimensions of

(a) $5\frac{3}{8}$ in × $5\frac{3}{8}$ in.
(b) $5\frac{1}{2}$ in. × $5\frac{1}{2}$ in.
(c) $5\frac{5}{8}$ in. × $5\frac{5}{8}$ in.
(d) $5\frac{3}{4}$ in. × $5\frac{3}{4}$ in.
(e) 6 in. × 6 in.

Solution

The American Standard Dressed Size (S4S) of a 6 in. × 6 in. nominal size post is $5\frac{1}{2}$ in. × $5\frac{1}{2}$ in. ▶

$$\text{Answer is } (b)$$

11-64 Wt. 1

The standard measuring unit for lumber is

(a) ft³
(b) lb
(c) psi
(d) fbm
(e) fpm

Solution

Lumber is measured in feet board measure (fbm). The unit is generally called "board feet" ▶

$$\text{Answer is } (d)$$

11-65 Wt. 1

The chief difference between exterior grade plywood and interior grade plywood is

(a) in the number of laminations used
(b) in the weight per square foot
(c) in the size of knotholes permitted
(d) that only selected heartwoods are used for the exterior grade
(e) that different types of glues are used

Solution

The distinguishing feature between exterior and interior grade plywood is the difference in the durability of the adhesives employed ▶

Answer is (e)

11-66 Wt. 1

Warp, wane, and checks are all factors which help to determine the quality of

(a) brick masonry
(b) ceramic veneer
(c) structural glass
(d) concrete
(e) stress grade lumber

Solution

Warp (any variation from a true or plane surface), wane (bark or lack of wood on a corner or edge of a piece), and checks (a lengthwise separation of the wood) are irregularities or defects in wood ▶

Answer is (e)

11-67 Wt. 1

Which of the following is true in regard to untreated wood piles?

(a) They are always the cheapest type of foundation.
(b) Ordinarily they can support the same load as other types of piles.
(c) They cannot carry as much load as a properly treated wood pile.
(d) They are not practical in areas where the ground water level is close to the surface.
(e) They should be cut off and capped below permanent ground water.

570 / Other Problems

Solution

Untreated wood piles will rot if they are exposed to a changing water level. Therefore, they should be cut off and capped below permanent ground water ▶

Answer is (*e*)

11-68 Wt. 1

In geology or mining terminology, the term "strike" is used to designate

(*a*) the direction of a horizontal line in the plane of a stratum
(*b*) the depth below the earth's surface to the stratum
(*c*) the direction to drill or tunnel to hit the stratum in the shortest distance
(*d*) the point at which a valuable mineral is encountered
(*e*) the angle that the stratum dips from the horizontal

Solution

The strike is the direction of the intersection of the vein or stratum with a horizontal plane, that is, it is the direction of a level line along the stratum ▶

(The dip, on the other hand, is the vertical angle between the plane of the stratum and a horizontal plane, measured perpendicular to the strike.)

Answer is (*a*)

11-69 Wt. 1

The terms igneous, sedimentary, and metamorphic are commonly applied in

(*a*) geodesy
(*b*) matrix analysis
(*c*) physics
(*d*) chemistry
(*e*) geology

Solution

The terms igneous (congealed from a molten mass), sedimentary (originated by sedimentation or settling of particles), and metamorphic (preexisting rock masses in which new minerals or structures are formed at high temperatures and pressures) are geological terms ▶

Answer is (*e*)

11-70 Wt. 1

When earth is deposited as an embankment, it will assume a natural slope. The angle formed between this natural slope line and a horizontal plane is called the

(a) angle of incidence
(b) angle of refraction
(c) dip
(d) spectral angle
(e) angle of response

Solution

By definition this is the angle of repose. It is somewhat similar to the angle of internal friction ϕ, but the value may vary greatly from ϕ ▶

Answer is (e)

11-71 Wt. 1

A retaining wall which retains an earth embankment solely by its own weight is called a

(a) counterforted wall
(b) cantilevered wall
(c) gravity wall
(d) steel crib
(e) steel sheet pile cell

Solution

Gravity retaining walls depend primarily on their own weight for stability ▶

Answer is (c)

11-72 Wt. 1

The ratio between the amount of water vapor actually present in a given quantity of air and that which the air could hold if saturated at the same temperature is called the

(a) specific gravity
(b) density
(c) dew point
(d) relative humidity
(e) saturation point

Solution

The ratio is called the relative humidity ▶

Answer is (d)

11-73 Wt. 1

The gas that is released in greatest quantities from a sludge digestion tank is

(a) nitrogen
(b) hydrogen sulfide
(c) carbon monoxide
(d) carbon dioxide
(e) methane

Solution

All of the gases listed may be present. The gas is primarily methane, with a substantial amount of carbon dioxide ▶

Answer is (e)

11-74 Wt. 1

The white salt deposit which forms on the surface of brick masonry is called

(a) efflorescence
(b) effervescence
(c) fluorescence
(d) luminescence
(e) cavitation

Solution

The deposition of soluble salts on the surface of masonry is called efflorescence ▶

(Effervescence, on the other hand, refers to the emission of gas bubbles.)

Answer is (a)

11-75 Wt. 1

The Brinell number of a material is a measure of

(a) specific gravity
(b) weight
(c) specific heat
(d) density
(e) hardness

Solution

The Brinell test is a static indentation hardness test. The Brinell hardness number is found by dividing the applied load by the contact area indented by the hardened steel ball which applies the load ▶

Answer is (e)

11-76 Wt. 3

The following terms are used in the construction industry. Describe, briefly, any six.

1. pile	4. snatch block	7. driftpin
2. waler	5. slump	8. header
3. elephant trunk	6. Kelly ball	9. sleeper

Solution

1. Pile: A long, slender wood, steel, or reinforced-concrete member driven into the ground to support a vertical load ▶
2. Waler: A horizontal member transmitting the load from sheet piling to anchors or shoring ▶
3. Elephant trunk: A pipe or tube for placing concrete; also called a tremie ▶
4. Snatch block: A pulley block with a hinged side strap which is moved so that the rope can be placed in the pulley groove ▶
5. Slump: The distance in inches that concrete settles down after lifting the cone in a standard slump test ▶
6. Kelly ball: A 30-lb penetrator used as a test of the consistency of fresh concrete ▶
7. Driftpin: A smooth tapered steel tool for aligning rivet or bolt holes in steel construction ▶
8. Header: The upper member of a rough window or door frame opening. A header brick is laid with an end exposed on the face of a wall ▶
9. Sleeper: A heavy horizontal member, usually resting directly on the ground as a support for other members; also, a type of pipe support ▶

11-77 Wt. 2

In an office building containing 100 persons, the design allowed for 10 cfm per person of outside air for ventilation purposes. If the supply fan delivers 6000 cfm, what is the temperature of air entering the cooling coil if the outside air is 105°F and the recirculated air is 80°F?

(a) 75.3°F
(b) 80.6°F
(c) 84.2°F
(d) 93.7°F
(e) 95.0°F

Solution

Total flow: 6000 cfm
Outside air: 1000 cfm at 105°F
Recirculation: 5000 cfm at 80°F
Temperature of mixture: 1000 × 105 = 105,000
                        5000 × 80 = 400,000

                        6000   505,000

$$\frac{505,000}{6000} = 84.16°F \blacktriangleright$$

Alternative Solution

The outside air has a temperature difference over the recirculating air of 25°F. When diluted by a factor of 6, the resulting temperature of the mixture will be

$$80°F + \frac{25°F}{6} = 84.16°F \blacktriangleright$$

Answer is (c)

11-78 Wt. 1

A major classification group of stainless steels contains 6% to 22% nickel in addition to 16% to 26% chromium. This group differs from the other two major groupings in that its steels

(a) have a coefficient of linear expansion almost zero
(b) have a coefficient of linear expansion small enough to make it useful in the manufacture of precision measuring devices
(c) are nonmagnetic
(d) have their iron in the ferritic form
(e) have their iron in the martensitic form

Solution

The classification includes 18-8 stainless (18%Cr, 8%Ni). This group of stainless steels has sufficient chromium to make them austenitic and nonmagnetic ▶

Answer is (c)

11-79 Wt. 1

The high-strength bolts used in steel construction can be identified on the job by

(a) the button-shaped domed head
(b) the ten-sided flat head
(c) across flats head dimension
(d) three radial marks on the head
(e) the slope of the thread

Solution

ASTM A325 calls for three radial lines, 120° apart, on the bolt head plus a marking to identify the manufacturer ▶

Answer is (d)

11-80 Wt. 6

Two parallel lists are shown below. The first is a list of fundamental quantities. The second is a list of combinations of units, where L = length, F = force, and T = time. Show the letter from List 2 that corresponds with the proper number of List 1.

Example: Velocity = LT^{-1} Stress = FL^{-2}

List 1   List 2

List 1	List 2	
(1) absolute viscosity	(a) L^4	(k) $FL^{-4}T^2$
(2) mass density	(b) $FL^{-1}T^2$	(l) LT^{-2}
(3) mass	(c) FLT	(m) F^2TL^{-1}
(4) angular momentum	(d) FF^{-1}	(n) L^3
(5) work	(e) FT^3L^{-1}	(o) FL^2
(6) power	(f) $FL^{-2}T$	(p) FT^2
(7) acceleration	(g) FL	(q) L
(8) impulse	(h) FT	(r) FL^2T
(9) Reynolds number	(i) $T^2L^{1/2}$	(s) TL^2
(10) moment of inertia	(j) FLT^{-1}	(t) L^2
(11) radius of gyration		
(12) section modulus		

Solution

List 1	Units	Answer
(1) absolute viscosity	FTL^{-2}	▶ (f)
(2) mass density	FT^2L^{-4}	▶ (k)
(3) mass	FT^2L^{-1}	▶ (b)
(4) angular momentum	FLT	▶ (c)
(5) work	FL	▶ (g)
(6) power	FLT^{-1}	▶ (j)
(7) acceleration	LT^{-2}	▶ (l)
(8) impulse	FT	▶ (h)
(9) Reynolds number	dimensionless	▶ (d)
(10) moment of inertia	L^4	▶ (a)
(11) radius of gyration	L	▶ (q)
(12) section modulus	L^3	▶ (n)

11-81 Wt. 6

Using the fundamental units of

$$F = \text{force} \quad Q = \text{charge}$$
$$L = \text{length} \quad \theta = \text{temperature}$$
$$T = \text{time}$$

find the dimensional form of the following quantities.

Example: Pressure $= FL^{-2}$

(a) mass
(b) density
(c) energy
(d) momentum
(e) mass moment of inertia
(f) elastic strain
(g) specific heat
(h) absolute viscosity
(i) entropy
(j) voltage
(k) electric current
(l) electrical resistance

Solution

(a) $\dfrac{FT^2}{L} = FT^2L^{-1}$ ▶

(b) $\dfrac{FT^2}{L^4} = FT^2L^{-4}$ ▶

(c) FL ▶

(d) FT ▶

(e) FLT^2 ▶

(f) dimensionless (LL^{-1}) ▶

(g) $L^2T^{-2}\theta^{-1}$ ▶

(h) FTL^{-2} ▶

(i) $FL\theta^{-1}$ ▶

(j) FLQ^{-1} ▶

(k) QT^{-1} ▶

(l) $FLTQ^{-2}$ ▶

11-82 Wt. 1

Brass is basically an alloy of copper and

(a) zinc
(b) tin
(c) aluminum
(d) lead
(e) nickel

Solution

Brass is an alloy of copper and zinc ▶

Answer is (a)

11-83 Wt. 1

A critical path diagram is used for

(a) project planning and scheduling
(b) moment distribution
(c) solving indeterminate structures
(d) route surveys
(e) closing traverses

Solution

A critical path diagram is used in project planning and scheduling to determine the best order in which various tasks should be performed and their timing ▶

Answer is (a)

11-84 Wt. 5

Define the following terms:

(a) friction head
(b) present worth
(c) standard deviation (statistical)
(d) pH
(e) antilogarithm
(f) nomograph
(g) Bourdon tube
(h) circular mil
(i) dyne
(j) kinetic energy

Solution

(a) Friction head refers to the loss of head h_L due to frictional resistance in pipes. It is often calculated from the Darcy equation:

$$h_L = f \frac{L}{d} \frac{V^2}{2g} \quad \blacktriangleright$$

(b) Present worth is the discounted present value of a sum F of money n periods distant at some interest rate i.

$$\text{Present worth} = \frac{F}{(1+i)^n} \quad \blacktriangleright$$

(c) The standard deviation σ of a group of values is the square root of the arithmetic mean of the squares of the deviations of each value from the arithmetic mean ▶

$$\sigma = \sqrt{\frac{\Sigma(X - \bar{X})^2}{N}}$$

(d) The pH value of a solution is defined as the negative logarithm of the hydrogen ion concentration ▶

$$\text{pH} = -\log [\text{H}^+]$$

(e) The number N whose logarithm is k is called the antilogarithm of k. Thus if log 2 = 0.30103, then the antilogarithm of 0.30103 is 2 ▶

(f) A nomograph is the representation of two or more variables in graphical form ▶

(g) A Bourdon tube is a thin-walled tube flattened on two sides to produce a cross section which is approximately elliptical in shape. The tube is then bent into an arc of a circle. Pressure applied inside the tube tends to straighten out the tube. An application of this is the Bourdon tube pressure gage ▶

(h) A circular mil is an area equal to that of a circle with a diameter of 0.001 in. It is used for giving the cross-sectional area of wires. A wire 10 mils in diameter has an area of D^2 or 100 cir mils ▶

(i) The dyne is that unbalanced force which produces an acceleration of 1 cm/sec^2 when acting on a mass of 1 g ▶

(j) Kinetic energy is the energy possessed by a body by virtue of its motion ▶

Appendix A. Centroidal Coordinates and Moments of Inertia for Common Shapes

	Centroid	Moment of Inertia
Triangle	$\bar{x} = \dfrac{a+b}{3}$ $\bar{y} = \dfrac{h}{3}$	$I_X = \dfrac{bh^3}{12}$ $I_{X_0} = \dfrac{bh^3}{36}$
Rectangle	$\bar{x} = \dfrac{b}{2}$ $\bar{y} = \dfrac{h}{2}$	$I_X = \dfrac{bh^3}{3}$ $I_{X_0} = \dfrac{bh^3}{12}$
Quarter circle	$\bar{x} = \bar{y} = \dfrac{4R}{3\pi}$	$I_X = \dfrac{\pi R^4}{16}$ $I_{X_0} = R^4 \left(\dfrac{\pi}{16} + \dfrac{4}{9\pi} \right)$ $= 0.0549 R^4$

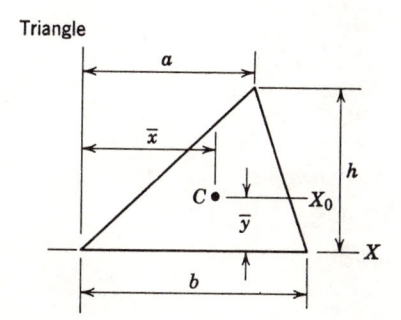

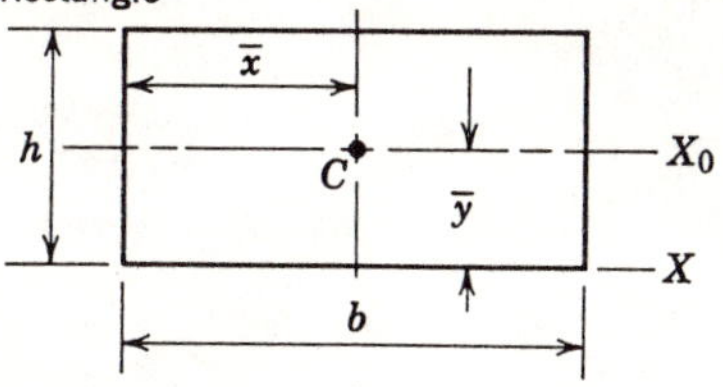

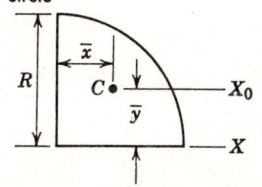

Semicircle

$$\bar{y} = \frac{4R}{3\pi}$$

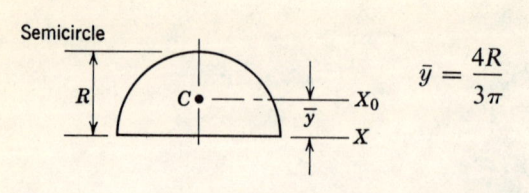

$$I_X = \frac{\pi R^4}{8}$$

$$I_{X_0} = R^4\left(\frac{\pi}{8} - \frac{8}{9\pi}\right)$$

$$= 0.1098 R^4$$

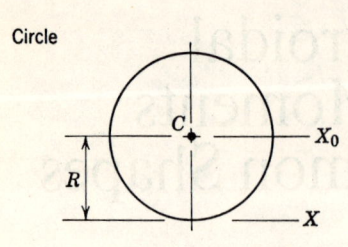

Circle

$$I_X = \frac{5\pi R^4}{4}$$

$$I_{X_0} = \frac{\pi R^4}{4}$$

Appendix B. Compound Interest Factors

Number of Interest Periods n	CA Single Payment Compound Amount Factor Given P To find F $(1+i)^n$	PW Single Payment Present Worth Factor Given F To find P $(1+i)^{-n}$	SF Sinking Fund Factor Given F To find A $\dfrac{i}{(1+i)^n - 1}$	CR Capital Recovery Factor Given P To find A $\dfrac{i(1+i)^n}{(1+i)^n - 1}$	SCA Series Compound Amount Factor Given A To find F $\dfrac{(1+i)^n - 1}{i}$	SPW Series Present Worth Factor Given A To find P $\dfrac{(1+i)^n - 1}{i(1+i)^n}$
			1% Compound Interest Factors			
1	1.010	0.9900	1.0000	1.0100	1.000	0.990
2	1.020	0.9802	0.4975	0.5075	2.010	1.970
3	1.030	0.9705	0.3300	0.3400	3.030	2.940
4	1.040	0.9609	0.2462	0.2562	4.060	3.901
5	1.051	0.9514	0.1960	0.2060	5.101	4.853
6	1.061	0.9420	0.1625	0.1725	6.152	5.795
7	1.072	0.9327	0.1386	0.1486	7.213	6.728
8	1.082	0.9234	0.1206	0.1306	8.285	7.651
9	1.093	0.9143	0.1067	0.1167	9.368	8.566
10	1.104	0.9052	0.0955	0.1055	10.462	9.471
12	1.126	0.8874	0.0788	0.0888	12.682	11.255
14	1.149	0.8699	0.0669	0.0769	14.947	13.003
15	1.160	0.8613	0.0621	0.0721	16.096	13.865
16	1.172	0.8528	0.0579	0.0679	17.257	14.717
18	1.196	0.8360	0.0509	0.0609	19.614	16.398
20	1.220	0.8195	0.0454	0.0554	22.018	18.045
22	1.244	0.8033	0.0408	0.0508	24.471	19.660
24	1.269	0.7875	0.0370	0.0470	26.973	21.243
26	1.295	0.7720	0.0338		30.048	29.525	22.795
28	1.321	0.7568	0.0311	0.0411	32.129	24.316
30	1.347	0.7419	0.0287	0.0387	34.784	25.807
35	1.416	0.7059	0.0240	0.0340	41.660	29.408
40	1.488	0.6716	0.0204	0.0304	48.886	32.834

Number of Interest Periods n	CA Single Payment Compound Amount Factor Given P To find F $(1+i)^n$	PW Single Payment Present Worth Factor Given F To find P $(1+i)^{-n}$	SF Sinking Fund Factor Given F To find A $\dfrac{i}{(1+i)^n - 1}$	CR Capital Recovery Factor Given P To find A $\dfrac{i(1+i)^n}{(1+i)^n - 1}$	SCA Series Compound Amount Factor Given A To find F $\dfrac{(1+i)^n - 1}{i}$	SPW Series Present Worth Factor Given A To find P $\dfrac{(1+i)^n - 1}{i(1+i)^n}$
\multicolumn{7}{c}{2% Compound Interest Factors}
1	1.020	0.9803	1.0000	1.0200	1.000	0.980
2	1.040	0.9611	0.4950	0.5150	2.020	1.941
3	1.061	0.9423	0.3267	0.3467	3.060	2.883
4	1.082	0.9238	0.2426	0.2626	4.121	3.807
5	1.104	0.9057	0.1921	0.2121	5.204	4.713
6	1.126	0.8879	0.1585	0.1785	6.308	5.601
7	1.148	0.8705	0.1345	0.1545	7.434	6.471
8	1.171	0.8534	0.1165	0.1365	8.582	7.325
9	1.195	0.8367	0.1025	0.1225	9.754	8.162
10	1.218	0.8203	0.0913	0.1113	10.949	8.982
12	1.268	0.7884	0.0745	0.0945	13.412	10.575
14	1.319	0.7578	0.0626	0.0826	15.973	12.106
15	1.345	0.7430	0.0578	0.0778	17.293	12.849
16	1.372	0.7284	0.0536	0.0736	18.639	13.577
18	1.428	0.7001	0.0467	0.0667	21.412	14.992
20	1.485	0.6729	0.0411	0.0611	24.297	16.351
22	1.545	0.6468	0.0366	0.0566	27.298	17.658
24	1.608	0.6217	0.0328	0.0528	30.421	18.913
26	1.673	0.5975	0.0296	0.0496	33.670	20.121
28	1.741	0.5743	0.0269	0.0469	37.051	21.281
30	1.811	0.5520	0.0246	0.0446	40.568	22.396
35	1.999	0.5000	0.0200	0.0400	49.994	24.998
40	2.208	0.4528	0.0165	0.0365	60.401	27.355
\multicolumn{7}{c}{4% Compound Interest Factors}
1	1.040	0.9615	1.0000	1.0400	1.000	0.961
2	1.081	0.9245	0.4901	0.5301	2.040	1.886
3	1.124	0.8889	0.3203	0.3603	3.121	2.775
4	1.169	0.8548	0.2354	0.2754	4.246	3.629
5	1.216	0.8219	0.1846	0.2246	5.416	4.451
6	1.265	0.7903	0.1507	0.1907	6.632	5.242
7	1.315	0.7599	0.1266	0.1666	7.898	6.002
8	1.368	0.7306	0.1085	0.1485	9.214	6.732
9	1.423	0.7025	0.0944	0.1344	10.582	7.435
10	1.480	0.6755	0.0832	0.1232	12.006	8.110
12	1.601	0.6245	0.0665	0.1065	15.025	9.385
14	1.731	0.5774	0.0546	0.0946	18.291	10.563
15	1.800	0.5552	0.0499	0.0899	20.023	11.118
16	1.872	0.5339	0.0458	0.0858	21.824	11.652
18	2.025	0.4936	0.0389	0.0789	25.645	12.659
20	2.191	0.4563	0.0335	0.0735	29.778	13.590
22	2.369	0.4219	0.0291	0.0691	34.247	14.451
24	2.563	0.3901	0.0255	0.0655	39.082	15.246
26	2.772	0.3606	0.0225	0.0625	44.311	15.982
28	2.998	0.3334	0.0200	0.0600	49.967	16.663
30	3.243	0.3083	0.0178	0.0578	56.084	17.292
35	3.946	0.2534	0.0135	0.0535	73.652	18.664
40	4.801	0.2082	0.0105	0.0505	95.025	19.792

COMPOUND INTEREST FACTORS / **583**

Number of Interest Periods n	CA Single Payment Compound Amount Factor Given P To find F $(1+i)^n$	PW Single Payment Present Worth Factor Given F To find P $(1+i)^{-n}$	SF Sinking Fund Factor Given F To find A $\dfrac{i}{(1+i)^n - 1}$	CR Capital Recovery Factor Given P To find A $\dfrac{i(1+i)^n}{(1+i)^n - 1}$	SCA Series Compound Amount Factor Given A To find F $\dfrac{(1+i)^n - 1}{i}$	SPW Series Present Worth Factor Given A To find P $\dfrac{(1+i)^n - 1}{i(1+i)^n}$
			5% Compound Interest Factors			
1	1.050	0.9523	1.0000	1.0500	1.000	0.952
2	1.102	0.9070	0.4878	0.5378	2.050	1.859
3	1.157	0.8638	0.3172	0.3672	3.152	2.723
4	1.215	0.8227	0.2320	0.2820	4.310	3.545
5	1.276	0.7835	0.1809	0.2309	5.525	4.329
6	1.340	0.7462	0.1470	0.1970	6.801	5.075
7	1.407	0.7106	0.1228	0.1728	8.142	5.786
8	1.477	0.6768	0.1047	0.1547	9.549	6.463
9	1.551	0.6446	0.0906	0.1406	11.026	7.107
10	1.628	0.6139	0.0795	0.1295	12.577	7.721
12	1.795	0.5568	0.0628	0.1128	15.917	8.863
14	1.979	0.5050	0.0510	0.1010	19.598	9.898
15	2.078	0.4810	0.0463	0.0963	21.578	10.379
16	2.182	0.4581	0.0422	0.0922	23.657	10.837
18	2.406	0.4155	0.0355	0.0855	28.132	11.689
20	2.653	0.3768	0.0302	0.0802	33.065	12.462
22	2.925	0.3418	0.0259	0.0759	38.505	13.163
24	3.225	0.3100	0.0224	0.0724	44.501	13.798
26	3.555	0.2812	0.0195	0.0695	51.113	14.375
28	3.920	0.2550	0.0171	0.0671	58.402	14.898
30	4.321	0.2313	0.0150	0.0650	66.438	15.372
35	5.516	0.1812	0.0110	0.0610	90.320	16.374
40	7.039	0.1420	0.0082	0.0582	120.799	17.159
			6% Compound Interest Factors			
1	1.060	0.9433	1.0000	1.0600	1.000	0.943
2	1.123	0.8899	0.4854	0.5454	2.060	1.833
3	1.191	0.8396	0.3141	0.3741	3.183	2.673
4	1.262	0.7920	0.2285	0.2885	4.374	3.465
5	1.338	0.7472	0.1773	0.2373	5.637	4.212
6	1.418	0.7049	0.1433	0.2033	6.975	4.917
7	1.503	0.6650	0.1191	0.1791	8.393	5.582
8	1.593	0.6274	0.1010	0.1610	9.897	6.209
9	1.689	0.5918	0.0870	0.1470	11.491	6.801
10	1.790	0.5583	0.0758	0.1358	13.180	7.360
12	2.012	0.4969	0.0592	0.1192	16.869	8.383
14	2.260	0.4423	0.0475	0.1075	21.015	9.294
15	2.396	0.4172	0.0429	0.1029	23.275	9.712
16	2.540	0.3936	0.0389	0.0989	25.672	10.105
18	2.854	0.3503	0.0323	0.0923	30.905	10.827
20	3.207	0.3118	0.0271	0.0871	36.785	11.469
22	3.603	0.2775	0.0230	0.0830	43.392	12.041
24	4.048	0.2469	0.0196	0.0796	50.815	12.550
26	4.549	0.2198	0.0169	0.0769	59.156	13.003
28	5.111	0.1956	0.0145	0.0745	68.528	13.406
30	5.743	0.1741	0.0126	0.0726	79.058	13.764
35	7.686	0.1301	0.0089	0.0689	111.434	14.498
40	10.285	0.0972	0.0064	0.0664	154.761	15.046

Appendix B

Number of Interest Periods n	CA Single Payment Compound Amount Factor Given P To find F $(1+i)^n$	PW Single Payment Present Worth Factor Given F To find P $(1+i)^{-n}$	SF Sinking Fund Factor Given F To find A $\dfrac{i}{(1+i)^n - 1}$	CR Capital Recovery Factor Given P To find A $\dfrac{i(1+i)^n}{(1+i)^n - 1}$	SCA Series Compound Amount Factor Given A To find F $\dfrac{(1+i)^n - 1}{i}$	SPW Series Present Worth Factor Given A To find P $\dfrac{(1+i)^n - 1}{i(1+i)^n}$
\multicolumn{7}{c}{8% Compound Interest Factors}
1	1.080	0.9259	1.0000	1.0800	1.000	0.925
2	1.166	0.8573	0.4807	0.5607	2.080	1.783
3	1.259	0.7938	0.3080	0.3880	3.246	2.577
4	1.360	0.7350	0.2219	0.3019	4.506	3.312
5	1.469	0.6805	0.1704	0.2504	5.866	3.992
6	1.586	0.6301	0.1363	0.2163	7.335	4.622
7	1.713	0.5834	0.1120	0.1920	8.922	5.206
8	1.850	0.5402	0.0940	0.1740	10.636	5.746
9	1.999	0.5002	0.0800	0.1600	12.487	6.246
10	2.158	0.4631	0.0690	0.1490	14.486	6.710
12	2.518	0.3971	0.0526	0.1326	18.977	7.536
14	2.937	0.3404	0.0412	0.1212	24.214	8.244
15	3.172	0.3152	0.0368	0.1168	27.152	8.559
16	3.425	0.2918	0.0329	0.1129	30.324	8.851
18	3.996	0.2502	0.0267	0.1067	37.450	9.371
20	4.660	0.2145	0.0218	0.1018	45.761	9.818
22	5.436	0.1839	0.0180	0.0980	55.456	10.200
24	6.341	0.1576	0.0149	0.0949	66.764	10.528
26	7.396	0.1352	0.0125	0.0925	79.954	10.809
28	8.627	0.1159	0.0104	0.0904	95.338	11.051
30	10.062	0.0993	0.0088	0.0888	113.283	11.257
35	14.785	0.0676	0.0058	0.0858	172.316	11.654
40	21.724	0.0460	0.0038	0.0838	259.056	11.924
\multicolumn{7}{c}{10% Compound Interest Factors}
1	1.100	0.9090	1.0000	1.1000	1.000	0.909
2	1.210	0.8264	0.4761	0.5761	2.100	1.735
3	1.331	0.7513	0.3021	0.4021	3.310	2.486
4	1.464	0.6830	0.2154	0.3154	4.641	3.169
5	1.610	0.6209	0.1637	0.2637	6.105	3.790
6	1.771	0.5644	0.1296	0.2296	7.715	4.355
7	1.948	0.5131	0.1054	0.2054	9.487	4.868
8	2.143	0.4665	0.0874	0.1874	11.435	5.334
9	2.357	0.4240	0.0736	0.1736	13.579	5.759
10	2.593	0.3855	0.0627	0.1627	15.937	6.144
12	3.138	0.3186	0.0467	0.1467	21.384	6.813
14	3.797	0.2633	0.0357	0.1357	27.974	7.366
15	4.177	0.2393	0.0314	0.1314	31.772	7.606
16	4.594	0.2176	0.0278	0.1278	35.949	7.823
18	5.559	0.1798	0.0219	0.1219	45.599	8.201
20	6.727	0.1486	0.0174	0.1174	57.274	8.513
22	8.140	0.1228	0.0140	0.1140	71.402	8.771
24	9.849	0.1015	0.0112	0.1112	88.497	8.984
26	11.918	0.0839	0.0091	0.1091	109.181	9.160
28	14.420	0.0693	0.0074	0.1074	134.209	9.306
30	17.449	0.0573	0.0060	0.1060	164.494	9.426
35	28.102	0.0355	0.0036	0.1036	271.024	9.644
40	45.259	0.0220	0.0022	0.1022	442.592	9.779

Index

Absolute pressure, 340–341, 362
Absolute temperature, 340–341, 362
Acceleration, 151, 161–174, 179–192, 205–210, 213–214
   angular, 158–160, 207–210
   curvilinear motion, 152
   normal component of, 152, 158–160, 176, 191–192
   rectilinear motion, 151–152
   tangential component of, 152, 158–160, 191–192
A–C circuits, 444–448, 483–493
Acoustics, 562–563
Adiabatic process, 343, 354, 365, 370
Air constituents, 396
Alcohol, 394
Algebra, 3–5, 39–45
   linear, 4, 39–41
   rules of, 3
Alidade, 528
Alkali metals, 392–393
Alpha particle, 557–558
Angle measurement, 5–6, 24, 538–539
Angle of repose, 570–571
Angstrom, 558
Angular momentum, 160, 211–212, 215
Angular motion, 158, 207–215
   acceleration, 158–160, 209–210
   velocity, 158
Annual cost, 497–499, 510–513
Arch, three-hinged, 111–112, 270–271
Archimedes' principle, 287–288, 302–305
Arithmetic progression, 77–78
ASTM material designations, 231–232
Atmospheric pressure variation, 558–559
Atomic number, 391
Average, arithmetic, 14, 70, 74
   geometric, 70–71
Average end area, 530–531
Average interest, 499–500, 513–520
Avogadro's number, 386–387
Azimuth, 533–534

Beam, concrete, 280–282
Beaman arc, 532
Beam deflection, 225–228, 272–274
Beam equilibrium, 221–224
Beam stress, 224–228; Stress, bending
Bearing, surveying, 527–528
Bending moment (diagram), 221–228, 253–267, 271
Bernoulli equation, 291–293, 295, 322–328
Binomial theorem, 5, 15
Black-body radiation, 351
Boyle's law, 347, 361, 386
Brass, 576
Brayton cycle, 349–350
Break-even point, 504, 506
Brinell test, 572
Brownian movement, 559–560
Bubble point, 406–409, 439–440
Buoyant force, 287–288, 302–308, 409

Calculus, 10–13; see also Differentiation, Integration, Limits
Cantilever beam, 255, 263–266, 272–274
Capacitance, 445–450, 455–457, 459
Capillarity, 299, 338–339
Capital recovery factor, 496–498, 501–502, 510–513
Carnot cycle, 349–350, 360
Catalyst, 393
Cathode ray, 452–453
Cavitation, 299

585

586 / Index

Cement, 563–564
  definition, 563
  types, 564
Center of gravity, 85, 100, 136–137
Center of mass, 136–137
Centigrade temperature scale, 340–341, 353
Centrifugal force, 154, 174–176
Centripetal acceleration, 176
Centroid, 63–64, 85–86, 135–139, 579–580
Charles' law, 347–348, 361–363, 386
Chemical balance calculations, 388–389, 412–424, 430, 435
Circle, 7–8, 32, 50–52, 58
Coefficient of friction, 84–85; see also Friction
Coefficient of restitution, 157
Column buckling, 228–229, 276–277
Combinations, 14, 75
Combustion, 389–390, 398–406, 414, 417
Complex numbers, 3, 17
Compound, 391
Compound amount factors, 495–496, 503, 523
Compound interest factor tables, 581–584
Compound interest formulas, 494–497
Concrete, 564–568
  batch design, 566–568
  prestressed, 565
Concrete beam, 280–282
Condenser, see Capacitance
Conic sections, 8, 47–53
Conservation of energy, fluid, 291–294, 322–328
Conservation of mass, 289–291, 307–308, 322–326, 329–331, 381
Conservation of momentum, 294–296, 333–334
Construction terms, 573
Continuity, 289–291, 307–308, 322–326, 329–331
Coordinate systems, 9–10, 34
Corrosion, 396–397
Coulomb's law, 453–454
Couple, 79
Cramer's rule, 4, 40–41
Critical depth, 334
Critical path diagram, 577
Critical velocity in pipe, 335

Current, electrical, 442–445, 455, 462–463, 469–470, 473–475, 478–479, 490–493
Cyanides, 411

Dalton's law, 365, 386–387, 423
Darcy-Weisbach equation, 291–293, 328–330
D–C circuits, 442–444, 459–463, 469–470, 472–473, 477–478
Declination, angle of, 527, 530
Deformation, 217
  stress, 217–218
Δ-Y circuits, 454–457, 492–493
Depreciation, sinking fund, 509–510
  straight-line, 499–500, 514–520
Descriptive geometry, 554–557
Determinant, 4–5, 39–42
Dew point, 365, 406–409, 439–440
Diesel cycle, 349–350
Differential equations, 13–14, 58, 67–70, 167, 475
Differentiation, 10–12, 54, 56–61, 65–67
  chain rule, 10–11, 66
  implicit, 54, 57
  partial, 66
Dimensions of engineering quantities, 575–576
Dip, 555–556, 570
Displacement, 151, 161–174, 179
D.M.D., 528, 541–543
Doppler effect, 563
Double meridian distance, 528, 541–543
Dyne, 160

Earth crust constituents, 395
Earthwork volumes, 530–531
Economic lot size, 504–506
Economy studies, 497–500, 504–524
Efflorescence, 572
Einstein energy formula, 558
Elastic curve (frames), 269–270
Electrical field, 449
Electrical machinery, 448–449, 478–482
  shunt motor, 448–449, 479–481
  synchronous motor, 478–480
Electrolysis, 397
Electromotive force, 442
Electromotive series, 431–432
Element, 391

Ellipse, 8–9, 53
emf, 442, 483
Emissivity, 351
Energy, conservation of, 155–156, 198
Energy equation, fluids, 291–296, 322–328
Engineering drawing, 543–553
 auxiliary view, 551
 isometric, 552–553
 orthogonal projections, 543–551
 pencil classification, 554
 pictorial, 547
Enthalpy, 343, 370–374, 379
Entropy, 343, 359, 370, 372–373, 376–377, 379
Ephemeris, 532
Equilibrium, 79–81, 88–100
 stable, 103
Equivalence, monetary, 494
Euler column formula, 228–229, 276–277
Exponents, 3–4, 17, 20–21

Factorial, 5, 38
Factor of safety, 230
Fahrenheit temperature scale, 340–341, 353
Faraday's law, 451
Flash vaporization, 406–409
Flow rate, 290–291, 322–326, 329–331, 433
 injection method, 433
Flow work, 343
Force, conservative, 155
 nonconservative, 155
Fourier law of heat conduction, 351
Frame, 82, 104–108, 111–114, 270–271
Free-body diagram, 82, 88, 91, 94–100
Frequency, 446–447, 475–476, 480
Friction, 84–85, 129–135, 164, 184, 188–190, 193, 206, 208
 belt, 133
 rolling, 134
Friction factor, fluid, 292–293, 329

Gage pressure, 341
Galvanizing process, 395–396
Gas constant, 347
Gas law relations, 346–349, 361–370, 386–390
Geodimeter, 532–533
Geology, 570

Geometric compatibility, 218–219, 236, 240–242
Geometry, 7–10, 45, 58, 554–557
 analytic, 8–10, 46–53
 descriptive, 554–557
 plane, 7–10
g-mole, 386, 390
Gram molecular weight, 386, 390
Graphics problems, *see* Engineering drawing
Gravitational force, 175–176

Halogen, 392
Hardness, water, 398–399
Head loss, 291–294, 327–331
Heat capacity, 341–342, 353–356
Heat exchanger, 383
Heat transfer, 341, 351–356, 358–359, 363–364, 368–371, 376–377, 379–380, 383–385
 conduction, 351–353, 383–385
 convection, 351–353
 radiation, 351
Highway curves, 530, 536–538
 horizontal, 536–538
 vertical, 530
Hooke's law, 218, 220, 235–242
Horsepower, 292, 331–333, 359, 361
Humidity, relative, 571
Hydraulic jump, 299
Hydraulic radius, 296–297, 335–336
Hydrogen ion concentration, 391, 410–411
Hydrostatics, 285–289, 302–321
Hydroxyl ion, 395, 410–411
Hyperbola, 8, 47–48
Hyperbolic paraboloid, 45–46

Illumination, 557
Impact, 157, 197–200
Impedance, 445–447, 458, 482, 484, 490–491
Impulse, 156–157, 160, 197, 215
 angular, 160, 215
Incremental economic analysis, 506–509
Inductance, 445–448, 451, 487
Influence line, 278–279
Instantaneous center of rotation, 158–159, 207
Integration, 12–13, 62–67
 area under curve, 12–13, 62–64
 limits of, 13
Interest factors, table, 581–584

Interest formulas, 494–497, 502
Internal energy, 342–343, 370, 374–375
Intersection of sets, 78
Isentropic efficiency, 377–378
Isotropy, 230

Joints, method of; *see* Method of joints
Joule's law, 450

Kelvin temperature scale, 340–341, 362
Kinematics, 151–152, 161–174, 203
Kinetic energy, 155–156, 160, 192–196, 198–203
 rotational, 160
 translational, 155–156
Kinetics, 151, 153–154, 160, 164, 167, 174–192, 206–214
Kirchhoff's laws, 442–444, 473–475, 478

Laser, 452
Latent heat, 342, 355
Latitude, 529–530
 and departure, 542–543
Law of cosines, 7
Law of sines, 7, 81
lb-mole, 386
Lease-purchase problem, 521–522
L'Hospital's rule, 11, 55–56
Light, wavelength, 558–559
Light refraction, 561–562
Limits, 11, 54–56
Line, 8, 46, 51
Logarithmic mean temperature difference, 383
Logarithms, 3–4, 18–23
 common, 4
 natural, 4
Lumber, 568–570

Magnetic field, 483
Manning equation, 296–297, 336–337
Manometer, 286–287, 292, 309–310
Mass number, 392
Maximum of function, 11, 59–61
Maxwell diagram, 127–128
Mean, arithmetic, 14, 70, 74
 geometric, 70–71
Measurement, 526–527, 538–539
 angles, 538–539
 precision, 527
Mechanical advantage, 101

Median, 14, 70
Method of joints, 82–83, 120–127
Method of sections, 82–83, 107, 119
Minimum of function, 11, 61–62
Mode, 14, 70
Modulus of elasticity, 218, 229
Mohr's circle, 275
Molar solution, 390, 432
Mollier diagram, 354, 371–372
Moment, 79, 89–94
Moment-area method, 226–228, 272–274
Moment of inertia, 85–87, 139–150, 160, 208–212, 215, 579–580
 polar, 140–141
Momentum conservation, 156–158, 197–204, 211–212, 215
 angular, 160, 211–212, 215
 fluids, 294–296, 333–334
 linear, 156–158
Money, time value, 494
Motion, equation of, 153, 164, 167, 174–192
 normal and tangential components, 154, 174–176

Newton, 161
Newton's law of heat convection, 351
Normal solution, 432–434

Offset method, 233
Ohm's law, 442–444, 459–460, 462–463, 482–483
Open channel flow, 296–297, 334–337
Optics, 560–562
Orsat analysis, 403–404
Otto cycle, 349–350
Oxidation, 393, 417–419, 431
Ozone, 396

Pan joist, 232
Parabola, 8–9, 48–50, 52–53, 56
Parallel axis theorem, 87, 142–150, 208, 212, 215
Parametric equation, 53
Partial pressures, law of, 386–387, 423
Pascal's principle, 298
Permittivity, 453–454
Permutation, 14
Phase, 361
Phasor, 445
Photogrammetry, 532

pH value, 391, 410–411
Pile-driving, 568–570
Pipe friction, 291–294, 328–330
Pipelines, 291–294, 300–301, 329–330
  parallel combination, 294
  series combination, 294
Plane, equation of, 47
Plane surveying, 527–543
Planimeter, 531
Poisson's ratio, 218, 233
Polygon, 7, 23–24
Post-tensioning, 565
Potential energy, 155–156, 194–196, 199
Pound molecular weight, 386
Power, electrical, 442, 446–447, 473–474, 477, 480–484, 488–489
  a-c, 446–447, 480–484, 488–489
  d-c, 442, 473–474, 477
  mechanical, 358
Power factor, 446, 480–481, 486–492
Present worth, 497–499, 521–522
Present worth factors, 495–496, 501–503, 521–522, 524
Pressure, fluid, 285–289, 300, 340–341
  absolute, 286, 340–341
  gage, 286, 341
Pretensioning, 565
Prismoidal formula, 531
Probability, 14–15, 71–73, 75
Process, adiabatic, 343–344, 354, 365, 370, 375, 378
  irreversible, 343
  isentropic, 344, 369–370, 373–375, 378, 380–381
  polytropic, 347, 366–369
  reversible, 343, 349–350, 370, 378
Product of inertia, 149–150
Pulleys, 179–182, 184–186
Pump efficiency, 293–294
Pump energy, 291, 331–333
Pump performance curve, 332–333

Quadratic equation, 5, 16, 23, 44
Quality (steam), 345–346, 370–374, 379

Radiation, heat, 560
Radius of gyration, 210, 228–229
Rankine cycle, 349–350, 378–379
Rankine temperature scale, 340–341, 360
Raoult's law, 440

Rate of return, 497–498, 523–524
Reactance, 445–447, 458, 478–479, 487–488
  capactive, 445, 487–488
  inductive, 445–447, 487–488
Reactions, 89–99; see also Equilibrium
Rectangle, 7
Reduction, 393
Refraction of light, 561–562
Refrigeration, 382
Relative motion, 170, 202
  velocity, 170
Representative fraction, 529
Resistance, 233, 442–447, 450–452, 454–455, 459–472
Resistivity, 442, 451–452
Resonant frequency, 446–447, 475–476, 486–488
Resultant force, 79, 94
Reynolds number, 292, 301–302, 329
Rigid body dynamics, 158–160, 207–216
Riveted connection, 243–245
Roughness coefficient, 296–297

Salvage value, 497, 512–513, 523
Section modulus, 283–284
Sections, method of, see Method of sections
Series, see Taylor's series
Set theory, 78
Shear force (diagram), 221–224, 253–267, 278–279
Shear modulus, 220
Significant figure, 15
Simultaneous equations, 41
Sinking fund depreciation, 509–510
Sinking fund factor, 496, 503, 509–510, 513, 523, 525
Slenderness ratio, 228–229, 276–277
Slope, 8, 10, 54, 56–57
Snell's law, 562
Sound, 562–563
Specific gravity, 286–288, 413
Specific heat, 200, 341–342, 355
  constant pressure, 342
  constant volume, 342, 363
Specific internal energy, 343
Specific volume, 343, 381
Sphere, 8
Spherical mirror, 560–561
Split ring, 232

Springs, 194–196, 216, 234, 272, 274, 279–280
Stadia rod, 528
Standard conditions, 386, 411
Standard deviation, 14, 74
Standard pressure, 386
Standard temperature, 386
State equation, gas, 347, 370, 376
Static indeterminacy, 226–228
Statistics, 14–15
Steam rate, 379
Steam tables, 345, 371–374, 379
Steel construction, 574–575
Stefan-Boltzmann law, 351, 560
Stiffness, 229, 282–283
Stirrups, 566
Straight-line depreciation, 499–500, 514–520
Strain, axial, 217
    lateral, 218
    shear, 220
    temperature, 218, 235–237
Stress, axial, 217, 235, 239, 249, 268, 275
    bending, 224, 252, 267–268, 283–284
    buckling, 276–277
    circumferential (hoop), 277–278
    combined, 268, 275
    flexure, *see* Stress, bending
    shear, 220, 224–225, 244–247, 268, 275
    torsional, *see* Stress, shear
Stress-strain curve, 234–235
Strike, 555–556, 570
Sublimation, 397–398
Surface tension, 338
Surveying, 527–543
    tape correction, 529

Taylor's series, 12, 16, 65
Tellurometer, 531
Temperature difference, logarithmic mean, 383
Temperature scales, 340–341
Thermal efficiency, 344, 359–360, 379
Thermal equilibrium, 341–342, 353–356
Thermal expansion, 218, 235–237
    coefficient of, 218, 231
Thermodynamics, first law of, 342–345, 356–359, 368, 373, 379–380
    second law of, 343–344, 359–361
Three-force body, 80–81
Ton of refrigeration, 382
Torsion, 220–221
Total head, 298
Transformer, 458–459
Trapezoid, 7
Traverse computations, 533–535, 538–543
Triangle, 7, 32, 36–37, 43
Trigonometry, 5–7, 24–37
    identities, 6, 30–32, 34, 53
    periodic functions, 6, 24
Truss, 82–83, 108–110, 115–128, 278
    graphical solution, 127–128
    plane, 119–128
    space, 108–110, 115–118
    see also Method of joints, sections
Turbine, 291, 372, 378–379
Two-force body, 80

Union of a set, 78
Universal gas law, 347, 361, 366–367, 386, 420–426

Valence, 393, 412
Variance, 14
Vector, 93
Velocity, 151, 158–159, 161–174, 179
    angular, 158–159
Velocity profile, fluid, 290
Venturi meter, 292–293, 324–325
Viscosity coefficient, 292, 301, 335
Voltage, 422, 444–448, 453, 457, 485, 492
Volt-amperes, 442, 446–447, 478–479, 483–484, 488–489

Water softening, 321–322
Watt, 442
Weir, 300, 336–337
Welded connections, 249–252
Wood, properties of, 229, 252, 568–570
Work, 156, 193, 203, 274, 342–343, 357–358, 368, 370, 372–378

Y-Δ circuits, 454–457, 492–493
Yield point, 233–234